The European Polysaccharide Network of Excellence (EPNOE)

Patrick Navard
Editor

The European Polysaccharide Network of Excellence (EPNOE)

Research Initiatives and Results

Springer

Editor
Patrick Navard
CEMEF-Centre de Mise en Forme des Matériaux
Mines ParisTech. UMR CNRS 7635
Sophia Antipolis cedex
France

ISBN 978-3-7091-0420-0   ISBN 978-3-7091-0421-7 (eBook)
DOI 10.1007/978-3-7091-0421-7
Springer Wien Heidelberg New York Dordrecht London

Library of Congress Control Number: 2012953922

© Springer-Verlag Wien 2012
This work is subject to copyright. All rights are reserved by the Publisher, whether the whole or part of the material is concerned, specifically the rights of translation, reprinting, reuse of illustrations, recitation, broadcasting, reproduction on microfilms or in any other physical way, and transmission or information storage and retrieval, electronic adaptation, computer software, or by similar or dissimilar methodology now known or hereafter developed. Exempted from this legal reservation are brief excerpts in connection with reviews or scholarly analysis or material supplied specifically for the purpose of being entered and executed on a computer system, for exclusive use by the purchaser of the work. Duplication of this publication or parts thereof is permitted only under the provisions of the Copyright Law of the Publisher's location, in its current version, and permission for use must always be obtained from Springer. Permissions for use may be obtained through RightsLink at the Copyright Clearance Center. Violations are liable to prosecution under the respective Copyright Law.
The use of general descriptive names, registered names, trademarks, service marks, etc. in this publication does not imply, even in the absence of a specific statement, that such names are exempt from the relevant protective laws and regulations and therefore free for general use.
While the advice and information in this book are believed to be true and accurate at the date of publication, neither the authors nor the editors nor the publisher can accept any legal responsibility for any errors or omissions that may be made. The publisher makes no warranty, express or implied, with respect to the material contained herein.

Printed on acid-free paper

Springer is part of Springer Science+Business Media (www.springer.com)

Foreword

This is a book on polysaccharides. It is not the first book on this topic. However, this is not any book. Rather, it is unique in its scope and approach not only in dealing in an integrated way with different aspects—chemistry, biochemistry, interfacial phenomena, material aspects, etc.—but also in comparing different polysaccharides and discussing their applications. In recent years, we have in all sciences seen an increasing tendency to fragmentation of research. Specialisation is of course a necessity in research, but a neglect of using information "around the corner" counteracts deep-going research and leads to duplication of research. We can certainly see this in research in the fields of macromolecules, with a specialisation not only on one type of macromolecules—synthetic, polypeptides, polysaccharides, polynucleotides, etc.—but even more so, for example, a group specialising on one protein or one polysaccharide, making little attention to the fact that mechanisms and interactions may have been clarified for a very similar system. In the scientific world of polysaccharides, this has been exemplified for polyglucoses, with groups working on cellulose and starch, for example, having little exchange of ideas or information. It is interesting to note that the European Polysaccharide Network of Excellence, with the acronym EPNOE, well known in a wide part of the scientific world, was created with a broad scientific scope.

Behind the ability to prepare a qualified treatise of a scientific topic, like this book, there must of course be a corresponding research effort, which is outside the capability of a normal research group. It requires instead a larger constellation of researchers, who are able to work together and interact constructively. EPNOE is such a research venture. It started from scratch on the initiative of Patrick Navard and a few insightful individuals with great visions. The opportunity of creating EPNOE was certainly dependent on the support from the European Commission. However, unlike many such analogous consortia which tend to dissociate into the individual starting groups as soon as the funding finishes, EPNOE built up a strength due to strong research to not only survive but also develop and expand with time. I would say that EPNOE in important respects is unique and should serve as an excellent example for future attempts to create strong collaborative research organisations. The important role played by EPNOE has been recognised by a large number of companies which have joined the organisation. This, in addition to its enormous impact on academic research, will make EPNOE a lasting venture as a centre of the international research on polysaccharides.

During more than 5 years, I have been able to follow EPNOE, first during 2 years as the European Commission appointed evaluator of the Network of Excellence and thereafter in several collaborative projects and as a participant in several conferences and workshops organised by the EPNOE community. Prior to the invitation to act as an evaluator, I had essentially no contacts with the EPNOE partners. As an evaluator, I was struck by one particular aspect in our evaluation meetings: this was the great interest in receiving critical comments to the various activities, even asking for additional critical comments and attempts to identify weak aspects of the consortium. Any such comment was seriously considered and commented on, frequently leading to follow-up questions.

Whereas an important aspect of the creation of EPNOE was the ability to identify an area of research of deep academic and fundamental interest as well as being highly industrially relevant, the critical point was certainly to bring leading research groups together and make them work truly together. This requires a strong and committed leadership. The demanding and sensitive task of bringing strong scientists together and working in the same direction is not easy to achieve, but Patrick Navard's and his colleagues' efforts in this respect have been spectacular. The introductory chapter of the book gives an interesting account of the "EPNOE saga", describing also the significant non-scientific obstacles in building a multinational network and the subsequent introduction of industrial companies as members.

For myself, the increasingly tighter contacts with EPNOE itself as well as with several partner groups individually have been most rewarding. Whereas I was involved rather extensively in polysaccharide research in the past, the contacts with EPNOE have taken my research into completely new directions. Thereby, I have benefited very much from the open atmosphere in EPNOE and the willingness to discuss controversial issues.

It was with enormous interest I received the manuscripts of this book. The breadth covered, from molecular and chemical characterisation, via manipulation and modification to various industrial aspects will make this book an important reading for students in several disciplines, as well as researchers in both academia and industry. The polysaccharide community is to be congratulated! This book marks a further step in the already extremely strong dissemination programme of EPNOE, including regular publications, courses, workshops and the biannual EPNOE polysaccharide conference. This conference was organised in Finland in 2009 and the Netherlands in 2011 and will be in France in 2013. EPNOE conferences are already an institution which is gathering all important scientists in the field.

Lund University and Coimbra University Björn Lindman
Kroksjön, Blistorp,
Sweden
July 2012

Contents

1 **Introduction: Challenges and Opportunities in Building a Multinational, Interdisciplinary Research and Education Network on Polysaccharides** 1
Julie Navard and Patrick Navard

2 **Etymology of Main Polysaccharide Names** 13
Pierre Avenas

3 **Polysaccharides: Molecular and Supramolecular Structures. Terminology.** 23
Thomas Heinze, Katrin Petzold-Welcke, and Jan E.G. van Dam

4 **Chemical Characterization of Polysaccharides** 65
Axel Rußler, Anna Bogolitsyna, Gerhard Zuckerstätter, Antje Potthast, and Thomas Rosenau

5 **Preparation and Properties of Cellulose Solutions** 91
Patrick Navard, Frank Wendler, Frank Meister, Maria Bercea, and Tatiana Budtova

6 **Cellulose Products from Solutions: Film, Fibres and Aerogels** 153
Frank Wendler, Thomas Schulze, Danuta Ciechanska, Ewa Wesolowska, Dariusz Wawro, Frank Meister, Tatiana Budtova, and Falk Liebner

7 **Polysaccharide Fibres in Textiles** 187
Lidija Fras Zemljic, Silvo Hribernik, Avinash P. Manian, Hale B. Öztürk, Zdenka Peršin, Majda Sfiligoj Smole, Karin Stana Kleinscheck, Thomas Bechtold, Barbora Široká, and Ján Široký

8 **Cellulose and Other Polysaccharides Surface Properties and Their Characterisation** 215
Karin Stana-Kleinschek, Heike M.A. Ehmann, Stefan Spirk, Aleš Doliška, Hubert Fasl, Lidija Fras-Zemljič, Rupert Kargl, Tamilselvan Mohan, Doris Breitwieser, and Volker Ribitsch

9 **Pulp Fibers for Papermaking and Cellulose Dissolution** ... 253
Pedro Fardim, Tim Liebert, and Thomas Heinze

10 **Cellulose: Chemistry of Cellulose Derivatization** 283
Thomas Heinze, Andreas Koschella, Tim Liebert, Valeria Harabagiu, and Sergio Coseri

| 11 | **Chitin and Chitosan as Functional Biopolymers for Industrial Applications** 329
Iwona Kardas, Marcin Henryk Struszczyk, Magdalena Kucharska, Lambertus A.M. van den Broek, Jan E.G. van Dam, and Danuta Ciechańska |
|---|---|
| 12 | **Polysaccharide-Acting Enzymes and Their Applications** . . . 375
Anu Koivula, Sanni Voutilainen, Jaakko Pere, Kristiina Kruus, Anna Suurnäkki, Lambertus A.M. van den Broek, Robert Bakker, and Steef Lips |

Index 393

Contributors

Pierre Avenas Representative of ARMINES as President of EPNOE Association, Paris, France

Öztürk Hale Bahar Lenzing AG, Textile Marketing, Business Unit Textile Fibers, Lenzing, Austria

Robert Bakker Wageningen UR Food and Biobased Research, Wageningen, The Netherlands

Široká Barbora Research Institute for Textile Chemistry and Textile Physics, University Innsbruck, Dornbirn, Austria

Maria Bercea Petru Poni Institute of Macromolecular Chemistry, Iasi, Romania

Anna Bogolitsyna Department of Chemistry, Wood, Pulp and Fiber Chemistry, University of Natural Resources and Life Sciences, Vienna, Austria

Doris Breitwieser Department for Rheology and Colloidal Chemistry, University of Graz, Institute of Chemistry, Graz, Austria

Tatiana Budtova Mines ParisTech, Centre de Mise en Forme des Matériaux – CEMEF, UMR CNRS 7635, Sophia-Antipolis, France

Danuta Ciechańska Institute of Biopolymers and Chemical Fibres, Łódź, Poland

Sergio Coseri Petru Poni Institute of Macromolecular Chemistry, Iasi, Romania

Aleš Doliška Laboratory of Characterization and Processing Polymers, University of Maribor, Maribor, Slovenia

Heike M.A. Ehmann Laboratory of Characterization and Processing Polymers, University of Maribor, Maribor, Slovenia

Department for Rheology and Colloidal Chemistry, University of Graz, Institute of Chemistry, Graz, Austria

Pedro Fardim Laboratory of Fibre and Cellulose Technology, Åbo Akademi University, Turku/Åbo, Finland

Hubert Fasl Department for Rheology and Colloidal Chemistry, University of Graz, Institute of Chemistry, Graz, Austria

Lidija Fras-Zemljič Laboratory of Characterization and Processing Polymers, University of Maribor, Maribor, Slovenia

Valeria Harabagiu Petru Poni Institute of Macromolecular Chemistry, Iasi, Romania

Thomas Heinze Centre of Excellence for Polysaccharide Research, Institute of Organic Chemistry and Macromolecular Chemistry, Friedrich Schiller University of Jena, Jena, Germany

Laboratory of Fibre and Cellulose Technology, Åbo Akademi University, Åbo, Finland

Široký Ján Research Institute for Textile Chemistry and Textile Physics, University Innsbruck, Dornbirn, Austria

Iwona Kardas Institute of Biopolymers and Chemical Fibres, Łódź, Poland

Rupert Kargl Department for Rheology and Colloidal Chemistry, University of Graz, Institute of Chemistry, Graz, Austria

Stana Kleinscheck Karin Laboratory of characterisation and processing of polymers, Institute for Textile Materials and Design, University of Maribor, Maribor, Slovenia

Anu Koivula VTT Technical research centre of Finland, VTT, Finland

Andreas Koschella Center of Excellence for Polysaccharide Research, Institute for Organic Chemistry and Macromolecular Chemistry, Friedrich Schiller University of Jena, Jena, Germany

Jaakko Pere VTT Technical research centre of Finland, VTT, Finland

Kristiina Kruus VTT Technical research centre of Finland, VTT, Finland

Magdalena Kucharska Institute of Biopolymers and Chemical Fibres, Łódź, Poland

Fras Zemljic Lidija Laboratory of characterisation and processing of polymers, Institute for Textile Materials and Design, University of Maribor, Maribor, Slovenia

Tim Liebert Center of Excellence for Polysaccharide Research, Institute of Organic Chemistry and Macromolecular Chemistry, Friedrich Schiller University of Jena, Jena, Germany

Falk Liebner Department of Chemistry, University of Natural Resources and Life Sciences, Tulln, Austria

Steef Lips Wageningen UR Food and Biobased Research, Wageningen, The Netherlands

Sfiligoj Smole Majda Laboratory of characterisation and processing of polymers, Institute for Textile Materials and Design, University of Maribor, Maribor, Slovenia

Avinash P. Manian Research Institute for Textile Chemistry and Textile Physics, University Innsbruck, Dornbirn, Austria

Frank Meister Thuringian Institute for Textile and Plastics Research, Rudolstadt, Germany

Tamilselvan Mohan Laboratory of Characterization and Processing Polymers, University of Maribor, Maribor, Slovenia

Patrick Navard Mines ParisTech, CEMEF – Centre de Mise en Forme des Matériaux, CNRS UMR 7635, Sophia Antipolis Cedex, France

Katrin Petzold-Welcke Centre of Excellence for Polysaccharide Research, Institute of Organic Chemistry and Macromolecular Chemistry, Friedrich Schiller University of Jena, Jena, Germany

Antje Potthast Department of Chemistry, Wood, Pulp and Fiber Chemistry, University of Natural Resources and Life Sciences, Vienna, Austria

Volker Ribitsch Department for Rheology and Colloidal Chemistry, University of Graz, Institute of Chemistry, Graz, Austria

Thomas Rosenau Department of Chemistry, Wood, Pulp and Fiber Chemistry, University of Natural Resources and Life Sciences, Vienna, Austria

Axel Rußler Department of Chemistry, Wood, Pulp and Fiber Chemistry, University of Natural Resources and Life Sciences, Vienna, Austria

Thomas Schulze Thuringian Institute for Textile and Plastics Research, Rudolstadt, Germany

Silvo Hribernik Laboratory for characterisation and processing of polymers, Institute for Textile Materials and Design, University of Maribor, Maribor, Slovenia

Stefan Spirk Laboratory of Characterization and Processing Polymers, University of Maribor, Maribor, Slovenia

Department for Rheology and Colloidal Chemistry, University of Graz, Institute of Chemistry, Graz, Austria

Karin Stana-Kleinschek Laboratory of Characterization and Processing Polymers, University of Maribor, Maribor, Slovenia

Marcin Henryk Struszczyk Institute of Security Technologies 'MORATEX', Lodz, Poland

Anna Suurnäkki VTT Technical research centre of Finland, VTT, Finland

Thomas Bechtold Research Institute for Textile Chemistry and Textile Physics, University Innsbruck, Dornbirn, Austria

Jan E.G. van Dam Wageningen UR Food & Biobased Research, Wageningen, The Netherlands

Lambertus A.M. van den Broek Wageningen UR Food & Biobased Research, Wageningen, The Netherlands

Sanni Voutilainen VTT Technical research centre of Finland, VTT, Finland

Dariusz Wawro Institute for Biopolymers and Chemical Fibres, Łódź, Poland

Frank Wendler Bozetto GmbH, Krefeld, Germany

Ewa Wesolowska Institute for Biopolymers and Chemical Fibres, Łódź, Poland

Peršin Zdenka Laboratory for characterisation and processing of polymers, Institute for Textile Materials and Design, University of Maribor, Maribor, Slovenia

Gerhard Zuckerstätter Kompetenzzentrum Holz GmbH, Linz, Austria

Introduction: Challenges and Opportunities in Building a Multinational, Interdisciplinary Research and Education Network on Polysaccharides

Julie Navard and Patrick Navard

Contents

1.1 Need for Organising Research and Education in the Polysaccharide Field 1
1.2 Building the EPNOE Network 3
1.3 Organisation of EPNOE 5
1.4 Facts and Figures: The Achievements of EPNOE 7
1.5 Challenges Around Networking and Opportunities Brought by EPNOE 10
References 11

1.1 Need for Organising Research and Education in the Polysaccharide Field

Polysaccharides represent by far the largest group of polymers produced in the world. Fully biodegradable, they are made by nature. They are the major source of carbon, on which our life and activities are based. Carbohydrates are the result of photosynthetic CO_2 fixation in plants and the central exchange and communication system between organisms. Polymeric carbohydrates (or polysaccharides) such as cellulose and chitin are natural polymers found abundantly in nature as structural building blocks. Other polysaccharides (starch, inulin) provide stored solar energy in the form of sugar for fuelling cells. Oil, gas and coal, made of (very) slowly modified biomass, have been cleverly used by humans to be a major energy source as well as a source of materials. Several factors are pushing for the use of the renewable biomass, i.e. the one that can be harvested in fields and forests. The first is the fact that 1 day or another, oil, gas and coal will be exhausted. Even before this time, the costs of exploitation will be higher and higher due to the fact that all easy-to-extract fossil biomass have been collected. The second is the push for preventing to send in the atmosphere the carbon present in fossil resources. A third driver for the use of renewable biomass is the more and more acute awareness of citizens about environmental issues that is influencing marketing departments of companies.

P. Navard (✉)
Mines ParisTech, CEMEF – Centre de Mise en Forme des Matériaux, CNRS UMR 7635, BP 207, 1 rue Claude Daunesse, 06904 Sophia Antipolis Cedex, France
e-mail: patrick.navard@mines-paristech.fr

P. Navard (ed.), *The European Polysaccharide Network of Excellence (EPNOE)*,
DOI 10.1007/978-3-7091-0421-7_1, © Springer-Verlag Wien 2012

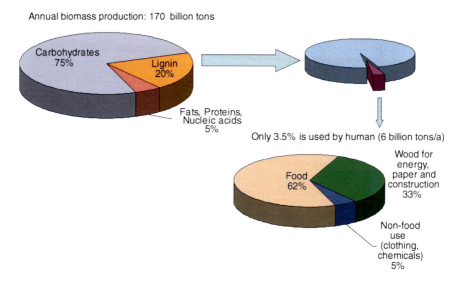

Fig. 1.1 Annual biomass production and use by humans (adapted by EPNOE partners Li Shen and Patel, Utrecht University from Thoen and Busch 2006)

Human beings are using a small portion of the whole biomass production (Fig. 1.1).

Biomass is mainly polysaccharides or molecules closely associated with them like lignin and proteins. All the new biorefinery concepts under development around the world and all the future trends in agriculture are linked, one way or another, to polysaccharides. From all sides, polysaccharides will be at the centre of a new emerging era in which sugar will be the value unit due to the emergence of a bioeconomy that will increase the contribution of bio-based products.

Polysaccharides are used in all sectors of human activities like materials science, nutrition, health care and energy. They are polymers with exceptional properties, far from being fully recognised, able to open routes for completely novel applications. In the global bioeconomy, carbohydrates, e.g. sugars and polysaccharides, are the central source of energy within which an economic value is intrinsically entrapped.

However, all the products present in nature are mixed with many other components in order to perform their biological role. Extraction, purification and treatment of these products are critical issues that have not been solved in a satisfactory way in most practical cases, leading to either polluting or energy-intensive treatments. But on the other hand, a widely unused source of polysaccharides is dormant in the waste of municipal water treatment, agricultural and food industries, leaving here huge potential sources of matter without conflicting with the food chain. The number of projects dedicated to and the amount of money poured in the development of polysaccharide-based products and fuel are the signs that a new industrial revolution might be emerging. Polysaccharides and polysaccharide-based polymers offer credible answers to the challenges faced by the world in terms of global sustainability.

For many reasons, including the fact that their structure is variable (depending on genetics, climate, location on Earth, soil, etc.), their use in highly engineered functional materials is in its infancy. Nevertheless, and withstanding all difficulties, polysaccharides are the sustainable source of polymeric materials for tomorrow. They offer numerous product development opportunities that are increasingly attractive in light of tightening oil supplies and rising concerns over environmental and biodegradability issues. The use of renewable raw materials such as polysaccharides is one of the targets of the European Union policies with objectives to increase the share of renewable energy and to promote biodegradation. Fixed targets of the

European Council of 8–9 March 2007 are the following: by 2020, at least 20 % reduction in greenhouse gas emissions compared to 1990 (30 % if international conditions are right, European Council, 10–11 December 2009); saving of 20 % of EU energy consumption compared to projections for 2020; and 20 % share of renewable energies in EU energy consumption, 10 % share in transport. On 15 December 2011, the European Commission adopted the Communication "Energy Roadmap 2050" (European Union 2011) that committed EU to reducing greenhouse gas emissions to 80–95 % below 1990 levels by 2050 in the context of necessary reductions by developed countries as a group.

In the same trend, one of the six EU Lead Market Initiatives targets bio-based products defined by the Ad-hoc Advisory Group for the Lead Market Initiative as products made from biological raw materials such as plants and trees that are renewable raw materials. It excludes food, traditional paper and wood products but also biomass as an energy source. Bio-based products can substitute fossil-based products. They are neutral in terms of greenhouse gas and leave a smaller ecological footprint, i.e. generate less waste and use less energy and water. Less consumption of natural resources lowers production cost and is better for the environment. This definition is totally fitting polysaccharide-based products.

These are the main reasons why the European Commission selected the polysaccharide topic for supporting the building of the European Polysaccharide Network of Excellence, better known as EPNOE.

1.2 Building the EPNOE Network

In 1999–2001, there was a dense informal network around cellulose with EU academia and industry. In 2002, the European Commission launched a manifestation of interest for creating networks in Europe. Several academic and industrial partners built a first consortium and submitted a network on cellulose called Cellnet. Considered too narrow by the European Commission, it was extended to polysaccharides. In 2003, the "*Polysaccharides*" proposal is submitted as a Network of Excellence. A Network of Excellence was a novel type of virtual research organisation at the level of the 25 countries of the European Union that was implemented by the European Commission. According to its definition, its purpose was to strengthen excellence on a particular research topic by networking together the critical mass of resources and expertise needed to be world force in that topic. A Network of Excellence was an instrument designed primarily to address the fragmentation of European research on a particular research topic, for which the main deliverable would be a restructuring and reshaping of the way research is carried out on that topic. It was thought to be the foundation stone for the construction of the European Research Area.

"Polysaccharides", very quickly called EPNOE for European Polysaccharide Network of Excellence, was very well ranked (first step: first over about 200 proposals at the first step and second over the remaining 36 proposals at the final stage) and was accepted. EPNOE started in May 2005 for 4 years and a half. EPNOE associated 16 European laboratories from 9 countries (Fig. 1.2).

The 16 institutions are composed of top-ranked universities and research centres, which have developed scientific expertise and state-of-the-art technologies in polysaccharide-related disciplines including chemistry, enzymology, biotechnology, chemical engineering, mechanics, materials science, microbiology, physics and life cycle assessment (Fig. 1.3). The following is the list of members (note that there is no number 14):

1. Centre de Mise en Forme des Matériaux CEMEF, ARMINES-Ecole des Mines de Paris/CNRS, France
2. Department of Chemistry, Universität für Bodenkultur, Austria
3. Centre of Excellence for Polysaccharide Research at the University of Jena, Germany
4. Fraunhofer-Institut für Angewandte Polymerforschung, Germany
5. VTT Technical Research Centre, Finland
6. Johann Heinrich von Thünen-Institute, Germany

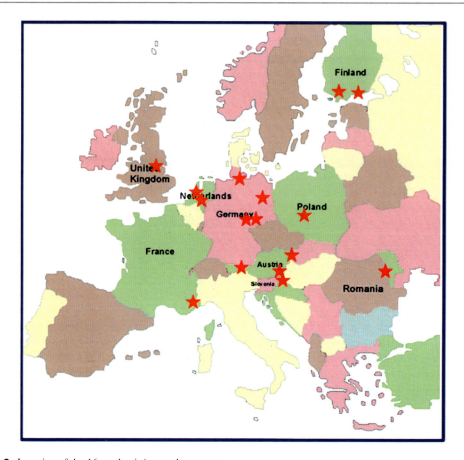

Fig. 1.2 Location of the 16 academic/research partners

Fig. 1.3 Expertise within EPNOE

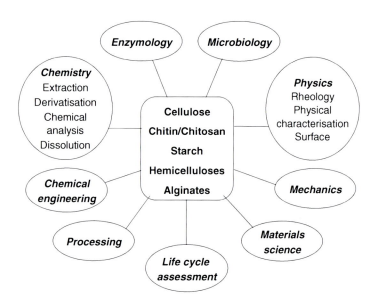

7. Process Chemistry Group, Laboratory of Forest Products Chemistry, Abo Akademi University, Finland
8. "Petru Poni" Institute of Macromolecular Chemistry, Romania
9. Laboratory for Characterization and Processing of Polymers, Faculty of Mechanical Engineering, University of Maribor-Univerza v Mariboru, Slovenia
10. DLO-FBR Stichting Dienst Landbouwkundig Onderzoek, Wageningen University and Research, The Netherlands
11. Thüringische Institut für Textil- und Kunststoff-Forschung (TITK), Germany
12. Institute of Biopolymers and Chemical Fibres – Instytut Biopolimerów i Włókien Chemicznych (IWCh), Poland
13. School of Biosciences, Division of Food Sciences, University of Nottingham, UK
14. Institute of Textile Chemistry and Textile Physics, Christian Doppler Laboratory for Textile and Fiber Chemistry, Universität Innsbruck, Austria
15. Department of Science, Technology and Society (STS), Universiteit Utrecht, The Netherlands
16. Institute of Chemistry, Colloid & Rheology Group, Universität Graz, Austria.

The main mission of EPNOE was to promote the use of polysaccharide renewable raw materials as industry feedstock for the development of advanced multifunctional materials.

The objectives were the following:
- To stimulate exchange and collaboration between the members through training and technology transfer activities
- To spread knowledge and excellence in the European Union scientific, industrial and public communities
- To develop a world-class research network.

From its initial structure, EPNOE evolved from an informal network composed of 16 academic/research institutions to a new formal structure. In 2007, the EPNOE network became a non-profit organisation called EPNOE Association. EPNOE Association is the current independent structure organising all the EPNOE activities. It is only funded by membership fees. Members are the initial 16 academic/research institutions plus companies. At the beginning of 2012, 25 SMEs and multinational companies working in various application fields such as food, paper, engineering and health are EPNOE members.

EPNOE is now a durable structure for organising Research and Education on polysaccharides at the European level. It is a complete, efficient and innovative research network on polysaccharide worldwide and a platform for bringing together companies and research centres.

1.3 Organisation of EPNOE

Although the main aspects of integration were clear at the beginning of the EC project, it took 2 years and a half to design and register the EPNOE Association. The EPNOE Association is a non-profit organisation under the French law "association loi 1901", registered in Paris on 14 December 2007. Its members are legal entities, physically represented in the various boards by persons they nominate. From the original concept to the design of the structure, many obstacles had to be overcome, the most complicated one being to find a structure in which the institutions would feel legally safe, considering that some of the partners not established in France would have positions implying responsibility like president or vice-president. At the present time, EPNOE is lead by the president, Dr Pierre Avenas (ARMINES, France), the vice-president for research, Prof. Karin Stana Kleinschek (University of Maribor, Slovenia) and the vice-president for education, Prof. Pedro Fardim (Abo Akademi, Finland). The absolute objective was to keep all partners on board, overcoming all administrative, legal or internal policy difficulties.

EPNOE Association has two types of members, regular members (academic/research institutions) and associate members (companies). It is organised with two legal documents (registered statute and association rules). Three boards are in charge of running EPNOE: a general assembly in which all regular and associate members vote for the plan of activity and approve budgets every year; a governing board, comprising one representative per regular member, that takes care

Fig. 1.4 Structure of EPNOE and EPNOE-related bodies

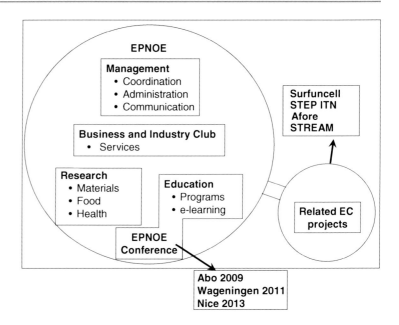

twice a year of the many decisions needed to fulfil the plan of activities; and the executive board, meeting about every 2 months (can be more often), that takes the operational decisions, implemented by the president and its representative.

The overall structure is given in Fig. 1.4. It shows the activities of EPNOE and of some EPNOE-related activities. Companies are mainly involved in the Business and Industry Club which was created to build a solid bridge between the 16 EPNOE academic/research members and industrial members. It is tailored to gain fast transfer and exploitation of knowledge, ideas and new processes and to offer its members a multidisciplinary and collaborative R&D platform. The Business and Industry Club offers four services:

1. Access to EPNOE Partner Databases: The objectives are to offer an easy access to and to give a complete general picture of all the 16 academic/research partners. The list and description of each partner, the list of all ongoing PhD and master's theses, a list of CVs of master and PhD students looking for employment and the name and details of all EPNOE researchers with their field of expertise are regularly updated.

2. EPNOE Research Information: In order to facilitate knowledge transfer, the maximum research information that is possible to be made available to other partners considering confidentiality issues and regulations is placed on the site like the full text of non-confidential EPNOE partners' PhD and master's theses, reports on the common basic and non-confidential research undertaken by EPNOE members.

3. Strategic and Technological Watch Data of EPNOE: The objectives are to offer information regarding the various stakeholders dealing with polysaccharides. An innovative and useful tool is the access to information on national-language conference papers. The title and details of communications dealing with polysaccharides of more than 150 conferences in 13 non-English languages are available. Several EPNOE market studies are also available.

4. Organisation of Dedicated Meetings.

EPNOE has been successful due to the strong links created during all these years of collaboration among the 16 partners. All partners are formally meeting at least three times per year all together and many times more in other meetings to discuss management, science and/or education. It is estimated that eight to ten meetings with the presence of all or part of EPNOE members are taking place every year since 2005.

EPNOE is now a well-known network, respected all over the world due to the highest quality of the research performed by its members, to its involvement in the international scientific life and to the development of its communication tools like the EPNOE Newsletter (20 issues published since November 2006 and distributed to now more than 400 subscribers). A look at Google shows that EPNOE is cited more than 8,000 times (65 times in Japan, 60 times in Brazil, 380 times in China, 1,800 in USA, etc.), the highest or one of the highest scores among all networks of excellence. Its contact address (contact@epnoe.eu) is attracting on average ten requests per week.

1.4 Facts and Figures: The Achievements of EPNOE

The main achievements of EPNOE members are the following:
- Establishment of a legal structure called "EPNOE Association" able to ensure a durable networking over the next 5–10 years.
- Creation of an active network involving 16 institutions, more than 20 companies, 100 scientists and more than 70 PhD students.
- Building of a research and education road map 2010–2020.
- Top-level scientific research (more than 40 on-going common research projects, about 20 PhD shared by two partners, round-robin testing, tool box with a set of 200 instruments available within the network).
- Education, with more than 50 exchanges of students, and creation of one EC-Intensive Programme on "Sustainable Utilization of Renewable Resources" (2009–2011).
- Active industrial membership with 25 members.
- Every year, more than 270 research projects are starting between EPNOE members and companies.
- High-level participants in many important stakeholder organisations (like organisation of meetings with scientific societies).

Some of these achievements are detailed below.

EPNOE Research Road Map: Partners prepared in 2009 a new joint EPNOE Road Map on polysaccharide research and development needs for the next decade (2010–2020), with a broad scope, encompassing materials, food and health. The research road map was prepared considering various social, political, industrial and scientific inputs (like market studies, EC documents and European Technology Platform strategic agendas), as well as other inputs from inside and outside EPNOE, mainly (1) results of four brainstorming sessions by EPNOE scientists and students, (2) individual contributions of EPNOE scientists and (3) individual contributions of scientists outside EPNOE through an internet review. The EPNOE Road Map has a research section structured around two main focus areas as shown in Fig. 1.5. The first, called "Fundamental basis of polysaccharide science", is where scientific challenges common to all application fields associated with major socio-economic and technological factors are reviewed. The second deals with the three selected "Application fields": materials, food and health care. For each application field, three levels of product cycle are considered: (1) extraction (disassembly), (2) conversion (reassembly) and (3) consumption (end of life cycle). In addition, the materials field has a section on economical and environmental assessments.

The EPNOE Road Map 2010–2020 has two versions: an extended one only available to EPNOE members on the internal website and a short public version only focusing on research (EPNOE Research Road Map 2010–2020). In order to disseminate this work, an article-like version of the short EPNOE Research Road Map was published in January 2011 in a major journal of the fields *Carbohydrate Polymers*: Z. Persin, K. Stana-Kleinschek, T. Foster, J. van Dam, C. Boeriu and P. Navard "Challenges and opportunities in polysaccharides research and technology: The EPNOE views for the next

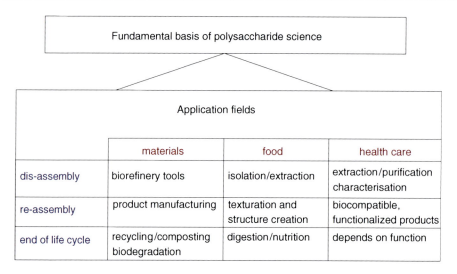

Fig. 1.5 Content of the EPNOE Research Road Map

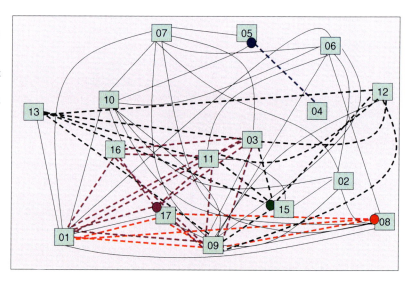

Fig. 1.6 Bilateral collaborations between the 16 academic/research laboratories (identified by numbers referring to the list given in the text) in 2011. The *four dots* are located on partner's institutions coordinating EC projects related to EPNOE

decade in the areas of materials, food and health", *Carbohydrate Polym*, vol 84 (2011) pp 22–32.

Applied R&D Research: R&D activities with industry have been increasing enormously within partner's institutions in large part, thanks to EPNOE. In 2007, 170 industrial projects were running, amounting to 13M€. In 2009, 270 industrial projects were running, amounting to 26M€.

Four running EC projects originate from the EPNOE activities: Surfuncell, STEP, Afore and STREAM.

Collaborative Research Activities and PhD/Postdoc Mobility: Collaborative research activities are very active among partners. At the beginning of EPNOE, only a few formal collaborations existed. In 2011, 62 such collaborations were active (Fig. 1.6) in various forms (EC projects, binational projects, visits of PhD students or postdoctoral scientists, basic science research, industrial projects).

Courses and e-Learning: Regular courses and meetings dedicated to industrial scientists or to postgraduate students have been or will be organised (last ones in May 2010 in Wageningen, The Netherlands, September 2010 in Sophia Antipolis, France, March 2011 in Lodz, Poland, August 2011 in Wageningen, September 2012 in Erfurt, Germany, September 2013 in Nice, France), and a set of tutorials on polysaccharides are posted in the EPNOE Web site (http://www.epnoe.eu).

Dissemination: scientists are very active in publishing their work and participating in conferences. More than 400 papers were published under the name of EPNOE, among which more than 50 papers were co-signed by at least two different EPNOE members. More than 400 communications at conferences were given with EPNOE name. EPNOE members organised or co-organised more than 40 conferences in Europe, China, India and USA (including a formal EPNOE—American Chemical Society collaboration), such as:

- *EPNOE—American Chemical Society session with cooperation of the US Department of Energy* (New Orleans, USA—April 6–10, 2008. ACS National Meeting)
- *EPNOE—Polymer Processing Society polysaccharide meeting*, Goa, India, 1 March 2009
- *1st international conference on Bamboo Fibre*, Quanzhou, China 23–25 March 2009
- *Polysaccharides as a sources of advanced materials*, EPNOE Conference, Abo, Finland, 21–24 September 2009
- *Narotech*, Erfurt (Germany), 9–10 September 2010
- *11th European Workshop on Lignocellulosics and Pulp*, Hamburg (Germany), 16–19 August 2010
- Series of *ACS meetings* every year

EPNOE Education Road Map: The targeted users of EPNOE education are students and postdocs, academic staff, industrial scientists, researchers and the general population. The aim of EPNOE education is to meet these demands with the help of education actions, which are divided into three action points: academic education, courses and e-learning and dissemination. EPNOE formally participated in education activities at the European level such as the EC-Intensive Programme on "Sustainable Utilization of Renewable Resources" (2009–2011).

EPNOE Tool Box: An infrastructure called "tool box" was developed by EPNOE in order to offer its members the possibility to figure out which member has instruments able to measure or estimate a given set of parameters. This database is associated to dedicated software enabling to search with different entries like the type of polysaccharides, the type of measures or the type of instruments. So far, more than 280 instruments are in the tool box database, with its major last update done in January 2011. This database is associated with a general agreement among members, part of the Association Rules of EPNOE Association that specifies the conditions of use of a piece of equipment from another member's institution. So far, there was in all cases a free access to equipment of other laboratories without any reported difficulty.

Joint Communication and Involvement in the Scientific, Policy and Industrial Communities: The EPNOE Newsletter is regularly published (20 issues so far) and has more and more subscribers (more than 450 now).

EPNOE scientists are (1) active participants in many important stakeholders' organisations like Technology Platforms (Suschem, Forestry, Food for Life), national organisations (pole de compétitivité in France, Polymer Institute in the Netherlands, Christian Doppler Laboratories in Austria, Zellcheming in Germany, etc.) and European organisations (European Renewable Raw Material Association, Advisory Group of the European Commission's Lead Market Initiative (LMI) on bio-based products); (2) board members of many organisations, that is, "Austrian Association of Textile Chemists and Colorists", "Electrokinetic Phenomena", "European Bioplastics", "Electrokinetic Society", "Forschungsvereinigung Werkstoffe aus nachwachsenden Rohstoffen", "Kunststoffnetzwerk

Brandenburg", etc.; and (3) members of editorial boards of journals "Cellulose", "Cellulose Chemistry and Technology", "Carbohydrate Polymers" and "Natural Fibers", "Holzforschung", "Arkivoc", "Current Organic synthesis", "Letters in Organic Synthesis", etc.

1.5 Challenges Around Networking and Opportunities Brought by EPNOE

To build a network, i.e. to construct links between the members of a group of institutions or persons, is most of the time thought to be the best way to increase efficiency. In most countries, networking different research institutions is compulsory for submitting projects to funding agencies. It is usually taken as granted that a network is much more efficient than the sum of its components. Although it can be true, and EPNOE is one example, we believe it may not be the case in many instances, especially when networking is compulsory. The advantages of a network are clear. Since no single researcher or no single research group can usually master the whole chain of expertise to treat a scientific problem, especially when several disciplines are at stake, there is no other choice than to bring together different research groups to solve a question. Often, when a research group is alone and obliged to use techniques, theories or modelling that are a bit far from its own scientific experience, errors and misinterpretations are numerous. All reviewers of submitted papers have experienced this very common effect. In this case, the search of a complementary competence outside the leading group is needed. This is exactly the driving force that hides behind the 62 collaborations shown in Fig. 1.6. In most of the cases (not all), these collaborations proceeded from the need to find a complementary expertise. In these cases, the existence of a network is very important: research groups know precisely what the others can do, under which time and under which resource conditions. But to reach such a state is a complicated task. Putting aside financial aspects, to build a research network with research groups that are competing for fame, are competing for money and are usually knocking at the same doors for funding, is complicated. Each member must feel that it will gain something from being a member of this network. In addition, it must also be sure that decisions are taken is a totally transparent and fair process. What we call an "*area of trustiness*" must be established. Contrary to what is usually thought and planned, this takes a lot of time. In the case of EPNOE, it took probably 5 years to start having research scientists willing to give some ideas without the risk of having them stolen by another member. Time, fairness and gain for members are the first ingredients for building a stable, long-lasting network. If these characteristics are not met, networks can exist, but they will be based on immediate opportunism, usually to solve one specific question or most often because networking is a request. *Opportunism networks* are short-life structures. In many European Commission consortia, some partners do not participate because they are the best to solve the scientific or technological challenge but because of requests like having SMEs as industrial members or researchers from certain countries.

Another difficulty with networks lies in the fact that the larger it is, the more time is lost to run it. It is like parallel computing where the speed of calculation is not the sum of the performances of the individual computers but lower due to the need for computers to exchange information. Exactly the same applies to research networks. It takes a lot of energy for individual members to perform activities dedicated to the network administration and to know the other groups. This loss of energy must be much lower than the gain of energy linked to the benefits of being a member. E-mails are a dramatic factor able to incredibly increase the loss of energy if too many are sent to too many people. Therefore, they must be limited to the minimum, despite all members must be aware of the running of the network. The same applies to meetings that must have a very rigorous format to be efficient in terms of energy lost. EPNOE meetings are structured in a very specific way. Each meeting has a "preparation document" that details all the points to be addressed and all the decisions that have to be taken point by point. The format of this "preparation document" is the precursor of the "minutes" of the meeting, with the same exact format. Below each point is the decision that is completed on line at the end of the discussion and vote. Except for phone meetings, the decision is shown on the screen. At the end of the meeting, minutes are nearly completed, and they can be sent for approval within days and posted, when approved, on the Web site. All "preparation documents" and thus all "minutes" have the same format, year after year. They contain all information pertaining to the meeting.

Network management must be optimised with very specific management tools far from the ones taught in management schools and with an effective *chain of command*, able to ensure minimum communication energy dissipation, maximum information flow and total transparency of decision and governance. EPNOE experienced many difficulties at its beginning and invented its own efficient management strategy. This participated to the fact that all the members that have built EPNOE over its first years are still present and active. This created a strong "*network*" that is recognised worldwide as an efficient and productive structure, able, for example, to build a long-lasting collaboration with the largest scientific society in the world. The fact that hundreds of research groups asked to join EPNOE is the sign of its success. As can be easily understood, above a certain size, the dissipation of energy to run a network is higher that its gain, and there is no more interest in networking. For EPNOE, the limit is close to the present number of 16 academic/research members, and so far, no other research institution was allowed to join for this reason, despite their clear scientific merits for most of them.

Conclusions

EPNOE is a success due to the commitment of its academic/research members, the initial support of the European Commission and the continuous support of its industrial members.

This network has managed to be very active and fruitful owing to its clear, easy-to-manage and transparent organisation. All partners are benefiting, one way or another, from EPNOE. Without such benefits for the institutions and more important for the individual members, EPNOE would not exist.

A new EC-funded project, called EPNOE CSA, started in March 2012 for a duration of 3 years. EPNOE CSA (CSA stands for Coordination and Support Action) is composed of the same 16 academic/research members of EPNOE. It aims at ensuring a durable financial viability to EPNOE Association while strengthening academia/research and industry relationship and promoting knowledge transfer with three objectives:

- Expanding EPNOE activities towards health-related materials and products
- Expanding EPNOE activities towards food-related materials and products
- Increasing financial viability via industrial participation and innovation by (1) installing the tools for increasing the financial viability of EPNOE during the 3 years of the EPNOE CSA project in order to ensure long-term sustainability of EPNOE Association activities after the project and (2) improving partnership with industry and boost innovation and knowledge transfer

References

European Union (2011) Energy Roadmap 2050, COM (2011) 885/2

Thoen J, Busch R (2006) Industrial chemicals from biomass-industrial concepts. In: Kamm B, Gruber PR, Kamm M (eds) Biorefineries-industrial process and products: status quo and future directions, vol 2. Wiley-VCH, Weinheim

Etymology of Main Polysaccharide Names

Pierre Avenas

Contents

2.1　Introduction: Etymology 13
2.2　*Saccharide* and *Sugar*: One Origin for Two Synonyms.. 14
2.2.1　*Sugar*, *Saccharide*, or *-ose*........................... 14
2.2.2　Mildness, Sweetness in Chemistry 15
2.3　A Large Variety of "-oses" 15
2.4　Cellulose 16
2.5　Indo-European Representation of Mildness or Sweetness... 16
2.5.1　From *Millstone* to *Mildness*?....................... 17
2.6　Starch.. 17
2.6.1　Starch in Greek, Latin, and Roman Languages.. 17
2.6.2　Starch in Germanic Languages and in Finnish.. 17
2.6.3　Starch in Slavic Languages......................... 18
2.7　Chitin 18
2.8　Other Polysaccharides 18
2.8.1　Carrageenan 18
2.8.2　Inulin 19
2.8.3　Pectin 19
2.8.4　Heparin 20
2.8.5　Pullulan 20
2.8.6　Hyaluronan, Murein, and Levan 20

References 21

P. Avenas (✉)
representative of ARMINES as President of EPNOE Association, 60 boulevard Saint-Michel, 75006 Paris, France
e-mail: pier.avenas@orange.fr

Abstract

This chapter deals with the etymology and history of names of the main polysaccharides and of some of their constitutive saccharides. The considered languages are mainly those which are used by the 16 academic EPNOE members, which are also the founders of EPNOE Association. Most of these nine languages belong to the Indo-European family (which includes also Greek and Latin), and they are distributed among the Germanic group (English, German, Dutch, Swedish), the Roman group (French, Romanian), and the Slavic group (Polish, Slovenian). Among the nine languages, the only non-Indo-European one is Finnish, which belongs to the Finno-Ugrian family.

2.1　Introduction: Etymology

Etymology studies the origin and history of words. The interest of this approach lies in the fact that the etymology (from Greek *etumos* "true") of a given word generally tells something about the reality which lies under this word. The present chapter deals with the etymology of the names given to the main saccharides and polysaccharides consumed and used by people.

For instance, the common name of a chemical substance is often related to the name of a plant from which the substance has been first isolated. That is true for saccharides like *sorbose* or *rhamnose*. The name of a substance can also be related to a specific part of plants, like cell wall for *cellulose* or fruit for *fructose*. In other cases, it can be related to an animal component, like liver for *heparin*, insect carapace for *chitin*, and milk for *lactose* and *galactose*. However, the link between the substance and its name can be quite different when it involves a particular property, like sweetening for *glucose*, stiffening for *starch*, and thickening for *pectin*. In any case, the common names of polysaccharides as chemicals are officially retained by IUPAC organization. Besides the names of some important mono or disaccharides, the complete list of which is huge, the following chapter will insist more on the well-known polysaccharides: **cellulose**, **starch**, **chitin**, **carrageenan**, **inulin**, **pectin**, **heparin**, and **pullulan**.

Let's begin this exploration with the name *polysaccharide* itself.

2.2 *Saccharide* and *Sugar*: One Origin for Two Synonyms

A polysaccharide is a polymer (from Greek *polu* "many" and *meros* "part"). A monomer is made of only one part, a dimer of two parts, and a polymer of many parts. Polysaccharides are sometimes called *polymerized sugars*. In other words, *saccharide* and *sugar* are roughly synonyms: the former being a scientific term, while the latter is also used in chemistry (for instance, in the phrase *sugar unit*) but mainly in common language, for ordinary *table sugar* in tea or coffee.

- The word (or element) *saccharide* is made of *sacchar-*, which means "sugar," and the suffix *-ide* (from Greek *eidos* "species"), which indicates the belonging to a family: a saccharide is a molecule of the family of sugars. It is the same name as English in German and French and nearly the same in Dutch (*sacharide*), Swedish (*sackarid*), Finnish (*sakkaridi*), Polish (*sacharyd*), Slovenian (*saharid*) and Romanian (*zaharid*).
- Both <u>saccharide</u> and <u>sugar</u> are derived from Sanskrit *çarkarā-* "gravel," and later "sugar" (originally, granulated sugar):

– <u>Saccharide</u> was recently derived, in scientific language, from the Latin word *saccharum*, continuing Greek *sakkharon*, which was the name of a product imported from India during antiquity. As a matter of fact, this product was something like cane sugar used only in small quantities and mainly for medicinal uses. At that time in Europe, the general way of sweetening food and beverages was with honey.

– *Sugar* was derived, through Old French, from Arabic *sukkar*, when the cultivation of sugar cane was introduced in southern Europe by the Arabian agronomists around the Middle Ages. Indeed, Arabic *sukkar* is, directly or indirectly, the origin of most European names for sugar, like Spanish *azucar* (from Arabic *al sukkar* "the sugar"), Italian *zucchero*, itself continued by German *Zucker*, Swedish *socker*, or French *sucre*, while English *sugar* and Dutch *suiker* are derived from Old French. We recognize the same Arabic origin in Polish *cukier* and Finnish *sokeri* (borrowed from Swedish), while in Romanian, *zahăr* is related to Modern Greek *zakharê* "sugar".

<u>N.B.</u>: the Slovenian name *sladkor* "sugar" as well as the verb *sladkati* "to sweeten" belongs to a Balto-Slavic family of words (including Polish *słodki* "sweet") related to the Old Norse *saltr* "seasoned, salted" (Buck 1988), itself related to the Indo-European root meaning "salt." This shows that seasoning can be done with salt or with sugar!

2.2.1 *Sugar*, *Saccharide*, or *-ose*

We know *monosaccharides*, like gluc<u>ose</u>, *disaccharides*, like lact<u>ose</u>, and *polysaccharides*, like cellul<u>ose</u> where we see the suffix *-ose*, which is still another way for designating a sugar or a sugar derivative. This suffix comes from the name gluc<u>ose</u> itself, attested in a French publication in 1838. The decision has been to name the other sugars after *glucose*, like *fructose* (from Latin *fructus* "fruit"), *lactose* (from Latin *lac*, *lactis* "milk"), *galactose* (from Greek *gala*, *galactos* "milk"), etc. Gluc<u>ose</u>, fruct<u>ose</u>, lact<u>ose</u>, galact<u>ose</u>, etc., were called *oses* in French, the word *ose* itself becoming another synonym for *sugar* and *saccharide*. Then *polyose* will be roughly a synonym of *polysaccharide*. But we still have to explain *glucose*.

2.2.2 Mildness, Sweetness in Chemistry

The name *glucose* is borrowed from the Greek name *gleukos* meaning, in Aristotle's works, "mild wine" or, in a figurative sense, "mildness." Then, *gleukos* itself is derived from the Greek adjective *glukus* "mild, sweet, delicious" in the literal as well as the figurative sense. From these names, we have many derivatives with the prefix *glyc(o)-* or *gluc(o)-*.

- The prefix *glyc(o)-* is used for a sweet substance, like *glycine* itself, or *glycerin* (made from the Greek adjective *glukeros*, nearly a synonym of *glukus*), and then derivatives like *glycol*.
- The prefix *glyc(o)-* or *gluc(o)-* represents glucose or any glucose-like molecules; for instance (cf. terminology in Chap. 3):
  - *Glycan*, as a synonym of *polysaccharide*, or oligosaccharide, is made only of sugar units.
  - *Glycogen* is a polymer of glucose, which can deliver (<u>generate</u>) <u>glucose</u>.
  - But *glucan* is a polysaccharide made only of glucose units, like cellulose and starch for instance.
  - In French food terminology, *glucide* is the word for *carbohydrate*.

To conclude this paragraph, we can say that *glucose* is a sort of pleonasm because *gluc-* means "sweet, mild as sugar" and, in chemistry, suffix *-ose* is a synonym of *sugar*! In the same way, the name *saccharose* (or *sucrose*) is a pleonasm as well!

2.3 A Large Variety of "-oses"

Starting from *gluc<u>ose</u>*, the suffix *-ose* is added, for designating different sugar units, to several elements related to:

Fig. 2.1 *Sorbus domestica*. Sorb tree, or service tree, or rowan. BotBln, Feb. 17, 2012 via Wikipedia, Creative Commons Attribution

- A chemical structure as for *hexose* (six carbons) and *pentose* (five carbons), or *aldose* (aldehyde function) and *ketose* (ketone function)
- An optical activity as for *dextrose* (Latin *dexter* "right") and *levulose* (Latin *loevus* "left")
- And, more frequently, the vegetal or animal origin of the molecule, as in *fructose*, *lactose*, and *galactose* already mentioned, or *xylose* (Greek *xulon* "wood"), *maltose* (from *malt*), and *fucose* (from Latin *fucus* "red alga").

The name *sorbose* is derived from the genus name *Sorbus* of several plants like sorb trees or rowans (Fig. 2.1).

The *rhamnose* was isolated from the buckthorn, a plant belonging to genus *Rhamnus*, created by Linnaeus in 1753 after the Greek name of this plant, *rhamnos* (Fig. 2.2).

The origin of *apiose* is not obvious. Could it be Latin *apis* "bee"? Indirectly yes, since the apiose has been extracted from parsley, and then *apiose* comes from Latin *apium* "parsley." But Latin *apium* originally is the name of celery, so named from *apis* "bee," because it was considered in antiquity as the *herb of bees*.

The origin of *mannose* is still more enigmatic: it comes from *manna*, the Hebraic name in the Bible for the miraculous food appearing in the desert but, in reality, a sweet secretion provided by some trees or bushes in favorable conditions.

Arabinose was extracted from *gum arabic*, an excretion of several species of Acacia, mainly in

Fig. 2.2 *Rhamnus frangula*. Alder buckthorn. David Perez, Feb. 17, 2012 via Wikimedia. Creative Commons Attribution

Arabic-speaking regions of northern Africa. Later on, the name *ribose* was used (in 1892) for a new isomer of arabinose, the change of letters from *arabinose* to *ribose* being a sort of literal representation of the chemical isomerization (likewise, in the same period, an isomer of xylose was named *lyxose*).

In the following decade, researchers of the Rockefeller Institute of Biochemistry (RIB in New York City) showed the crucial role of ribose in the chemistry of life (as part of ribonucleic acid or RNA). By chance, the initials RIB could then also be read in *ribose*.

We come now to a polymer of "-oses" of major importance.

2.4 Cellulose

Roughly speaking, cellulose is polymerized glucose, and this is the reason of the suffix *-ose* of *cellulose*. The first part of the name means that this natural polymer is an important constituent of vegetal cells, namely, the main constituent of cell walls. The French name *cellulose* is originally attested in a botanic course of Antoine de Jussieu in 1840, after French *cellule* "cell," derived earlier from Latin *cellula* "small room," diminutive of *cella* "room" (the etymology of which is perhaps related to the Latin verb *celare* "to hide"). In other Roman languages, the name of a cell is also linked to the Latin diminutive *cellula*, like Italian *cellula*, Spanish *célula*, Romanian *celulă*, but in other languages, the name comes directly from *cella*: English and Swedish *cell*, German *Zelle*, Dutch *cel*, as well as in Finnish *solu* or Polish *cela* (Slovenian is different with the diminutive *celica* "cell"). Nevertheless, in all languages, the name of cellulose is equivalent to the French word: Italian *cellulosa*, Spanish *celulosa*, Romanian *celuloső*, as well as English *cellulose*, German *Zellulose*, Swedish *cellulosa*, Dutch *cellulose* (besides *celstof*, which, in Dutch, means "constituent of cell"), Polish and Slovenian *celuloza*, and Finnish *selluloosa*.

Now, before coming to starch, the other important natural polysaccharide in terms of volume, let's examine different ways of expressing sweetness in European tongues.

2.5 Indo-European Representation of Mildness or Sweetness

We have seen the Greek adjective *glukus* "mild, sweet." Its Latin equivalent is *dulcis*, becoming *dulce* in Spanish, *dolce* in Italian, or *édulcorant* "sweetener" and *doux* "mild" in French. Can we relate *glukus* to *dulcis*? Yes, if we consider (Ernout and Meillet 1985) the probable existence of an Indo-European root, **dluku-*, and if we admit that its initial *d* becomes *g* in Greek (by attraction of *k*, since *g* is closer to *k* than *d*) while the element *-lu-* of **dluku-* becomes *-ul-* in Latin (in linguistics, such an exchange of letters is called *metathesis*, and it happens that the same word, *metathesis*, designates a chemical reaction exchanging one atom group of one molecule with one atom group of another molecule). However, we see that English *sweet* is quite different, as well as *mild*. First, we have another Indo-European root, **swad-* "mild, pleasant," which explains:

- English *sweet*, German *süss*, Dutch *zoet*, Swedish *söt*

- But also Latin *suavis* "mild, pleasant," then *suave* in French, borrowed as such by English, *soave* in Italian, more in the figurative meaning

Now, we have to deal with the adjective *mild*, which belongs to a completely different family of words.

2.5.1 From *Millstone* to *Mildness*?

As a matter of fact, there are connections (Onions 1992), even if they are not firmly established, between:
- Greek *mulê*, Latin *mola* "millstone," Latin verb *molere* "to grind, to mill," then in English *to mill*, and result of milling which is *meal* "flour," like in *wheatmeal*, for instance (nothing to do with *meal* "lunch or dinner", which is related to *measure*), and in Dutch *meel* "meal, flour" and in German *Mehl* "flour."
- Latin *mollis* "soft," because a milled product is no longer hard, then French *mou* "soft" and *mild* "not hard," in English as well as in German, Dutch, and Swedish.
- The Germanic root represented by English *to melt* "to become liquid" (and the variant *to smelt* "to fuse" in metallurgy, Dutch *smelten* "to melt," German *schmelzen* "to melt") since both melted and milled substances are fluid.
- Finally, English *malt* (then French *malt*, German *Malz*, Dutch *mout*) has something to do with *melt*, since the malt is produced by a digestion of barley grains in water, resulting in a sort of syrup.

In this process, barley starch is depolymerized by amylase: this leads us to the history of starch names in different languages.

2.6 Starch

2.6.1 Starch in Greek, Latin, and Roman Languages

This product was named *amylum* in classical Latin, itself borrowed from Greek *amulon*, coming from *amulos*, the first meaning of which is "not ground, not milled." Indeed, *amulos* is made (Chanteraine 1990) of the privative prefix *a-* "without" and the name *mulê* "millstone." The reason of this etymology is that starch was prepared with fresh grains, without any milling, as opposed to flour. Pliny the Elder (first century), in his *Natural History* (Liber XVIII, 76), explains it as follows: "*The invention of starch happened in Chios island, and still today the most estimated one is coming from there. It is so named because it is prepared without the use of any millstone.*" Of course, the production of starch has been much improved all along times: some crushing or grinding of the grains has been added in the process, and even milling, since the wheat starch can be produced today from flour. Nevertheless, the etymology of Latin *amylum* derivatives retains the footprint of the ancient process. In the medieval period, this name *amylum* was altered to *amidum*, whence the names of starch in Roman languages, like French and Romanian *amidon*, Italian *amido*, or Spanish *almidón*.

Let's make two remarks:
(1) The late Latin word *amidum* "starch" has nothing to do with the much more recent name *amide*. While *amidum* must be understood as *a + midum*, the name *amide* is made of *am + ide*, where *am-* is the beginning of *ammoniac*, since a molecule of amide is built around an atom of nitrogen. In the scientific nomenclature, no confusion can happen with *amide* since the starting point for expressing a relation with starch remains classical Latin in the prefix *amyl(o)-*, like in *amylose*, *amylase*, *amyloplast*, or, in French, *amylacé* "starchy."
(2) *Amyl-* is also synonym of *pentyl-*, like in *amyl alcohol*, probably because this alcohol can be produced from starch.

But then, what is the origin of *starch* in English, which is so different from the Latin form?

2.6.2 Starch in Germanic Languages and in Finnish

The English name of *starch*, as well as its equivalent in German, *Stärke*, or in Swedish, *Stärkelse*, is not related, as in Greek, to its manufacturing process but to its utilization. As a matter of fact, those names are related to the same Indo-European root as the adjective *stark*, in English "rigid, stiff" and in German and Swedish "solid, resistant," and this relation is due to the stiffness which is given to fabrics and clothes by the application of starch. Even more unexpected, the name of a famous bird, the *stork* in English and Swedish, *Storch* in German, is related to the same Indo-European root, just because this bird, so elegant while flying, looks stiff when it is landing and then walking on its nest. The name of the stork is totally different in Latin (*ciconia*, probably an onomatopoeic name, for this clattering bird), and then in the Roman languages (as *cigogne* in French). Coming back to starch, its name in Finnish, *tärkkelys*, is borrowed from Swedish *stärkelse*, with the fall of the initial s (cf. *Tukholma*, the Finnish name of Stockholm). Even if Finnish is not a Germanic language, and even not an Indo-European language, many borrowings happened between Finnish and Swedish all along the history of Finland.

N.B.: In this case, the Dutch names (van Veen 1989) are completely different from other Germanic languages:
- *Zetmeel* "starch" is made of *zet-*, meaning "making thick, setting," which is logical for starch in food use and of *-meel* "meal, flour," but the other Dutch name *stijfsel* "starch for nonfood use" is a derivative of *stijf* "stiff," which is logical for a stiffening product.
- *Ooievaar* "stork" is somewhat isolated and originally means "bringing luck."

2.6.3 Starch in Slavic Languages

The name of starch in Polish, *skrobia*, comes from the verb *skrobać* "to scrape," because starch has been produced from grated wheat grain and later on from grated potato or maize. In Slovenian, the name *škrob* "starch" is of the same origin. After cellulose and starch, a third important polysaccharide is chitin.

2.7 Chitin

The name *chitin* is attested (*chitine* in French in 1821) for designating the main constituent of the carapace of insects. This name, with the suffix *-in*, comes from Greek *khitôn*, which was used, in secondary meanings, for naming hard envelopes in anatomy and in botany.

This Greek name itself, probably of Semitic origin, designated originally a sort of tunic, generally short for men and long for women, still named *chiton* (in French too) in texts relating to antiquity. However, this Greek word took several derivates and other meanings, such as any item of clothing, an armored coat for a soldier, and, as we just said, the skin of animal organs, of snakes, or of fruit, or else the cork of some trees or the shell of mollusk shellfish.

Today, it is well known that chitin is the main constituent of carapace, not only of insects but of all arthropods, including particularly crustacean shellfish, and that it is also the main constituent of cell walls of fungi. Chitosan is obtained from chitin by partial deacetylation, resulting in the fact that chitosan is water soluble while chitin is not.

2.8 Other Polysaccharides

The last paragraph of this chapter will deal with the names of the following polysaccharides: *carrageenan*, *inulin*, *pectin*, *heparin*, *pullulan*, *hyaluronan*, *murein*, and *levan*.

2.8.1 Carrageenan

Carrageenan is extracted from red seaweed known under the common, and misleading, name of *Irish moss* or *carrageen moss* or *curly moss* (Fig. 2.3).

This English name *carrageen* (in French *carragheen*) probably comes from the old names (*Carrageen*, *Carragheen*) of the city now named *Carrigeen* (originally, in Irish *Carraigin*,

Fig. 2.3 Carrageen moss. *Chondrus crispus.* (from Greek *khondros* "cartilage" and Latin *crispus* "curly"). Franz Eugen Köhler, Feb. 17, 2012 via Wikimedia, Creative Commons Attribution

Fig. 2.4 *Inula helenium.* Elecampane. Eugene van der Pijll, Feb. 17, 2012 via Wikipedia, Creative Commons Attribution

which means "little rock," from *carraig* "rock"). This city is located near Waterford, on the southeast coast of Ireland, a region where this Irish moss was historically cropped and exploited for producing a sort of vegetal equivalent of the animal gelatin.

N.B.: There is another product similar to carrageenan: the *agar* or *agar-agar* (from its Indonesian or Malayan name), which is also extracted from a red seaweed. Purified agar-agar is a polysaccharide called *agarose*.

2.8.2 Inulin

Inulin is extracted from a large yellow-flowered plant called *elecampane* in English (Fig. 2.4). Since antiquity, this plant has been known for its medical properties. Pliny the Elder describes it under the name *inula*, itself derived from Greek *helenion*, which designates a plant supposed to be dedicated to the legendary, famous, and beautiful Helen of Troy. Namely, *helenion* was borrowed in primitive Latin as *elena*, or *enula*, due to the permutation of *l* and *n* (i.e., metathesis, as mentioned earlier in this chapter), and then *inula* in classical Latin. At the same time, *elena* (*alena*) evolved to *aunée* in French, *alant* in German and Dutch, *eolone* in Old English, and *elecampane* in Modern English, with Latin *campaneus* "living in the fields."

In 1753, Linnaeus established the genus *Inula* and named its main species *Inula helenium*, combining the Greek and Latin names. The name *inulin* is of course derived from *Inula*, with the suffix *-in*.

2.8.3 Pectin

This polysaccharide is known for its jelling properties, particularly in the composition of jams and jellies. Its name is made of the suffix *-in* after the element *pect-*, itself derived from the Greek adjective *pêktos* "coagulated," related to the irregular verb *pêgnumi* "to stick in, to fix, to solidify." In the same family, there is the adjective *pêgos* "compact, thick." It is interesting to note that the root *peg-* or *pek-* is also present in Latin words with the adjective *compactus* "compact," from the verb *pingere* "to stick in, to fix, to solidify" in the literal as well as the figurative sense, and even with the verb *pacere* "to pacify" in relation with *pax, pacis* "peace" because peace, thanks to a *pact*, brings stability and solidity to a human community. What a surprising semantic link between *pectin* and *peace*!

N.B.: In Dutch, pectin is called *pectine*, or *geleisuiker*, made of *gelei* "jelly" and *suiker* "sugar." The name *jelly* or *gel* (cf. *gelatin*) comes from the Latin verb *gelare* "to freeze," because of the analogy between solidification of water by freezing, resulting in translucent ice, and its solidification by gelling, with pectin, for example, resulting in some soft translucent gel. The link between the two concepts remains transparent in Roman languages, as in French where there are the following homonyms: *gel* "frost" and *gel* "gel, jelly." But this link is not obvious in a language where the verb meaning "to freeze" is not related to Latin *gelare*.

2.8.4 Heparin

This product has been discovered in extracts of liver showing antithrombosis effect. It is the reason why *heparin* is composed of the suffix *-in* after the element *hepar-*, from Greek *hêpar*, *hêpatos* "liver." However, the link between *heparin* and the name of liver is only historical since this polysaccharide is present in many organs and is produced today from other animal organs than liver.

About Liver: The designation of this organ is another case where we find very different names in all Europe. Greek *hêpar*, like Latin *jacur*, belong to the same Indo-European origin, linked to the role of this organ in religious prediction, while Polish *wątroba* and Slovenian *jetra* are related to Greek *entera* "entrails." But English *liver*, German *Leber*, and Dutch and Swedish *lever* are probably related to the Greek *lipos* "fat," perhaps because of the *foie gras* of goose, which was already popular in antiquity. This link between foie gras and the name of liver is still more certain in Roman languages: Italian *fegato*, Spanish *higado*, French *foie*, and Romanian *ficat* are all derivatives of Latin *ficatum* "foie gras," from *ficus* "fig, fig tree," because, already in ancient Greece, the geese were force-fed with figs, and also because liver was often cooked with figs. Even in Modern Greek, *sukôti* "liver" comes from *suko* "fig."

Let's end this list with the Finnish *maksa* "liver," which is linked to a very old Finno-Ugrian root (cf. Hungarian *máj* "liver") (Häkkinen 2007).

2.8.5 Pullulan

It is a polysaccharide which is produced from starch by a yeastlike filamentous fungus named *Aureobasidium pullulans*. The species name *pullulans* means "pullulating," from the Latin verb *pullulare* "to pullulate," because this fungus does pullulate in various environments. Then *pullulan* comes from a species name, unlike the different names seen until now, derived from a genus name, like *Sorbus*, *Rhamnus*, or *Inula*.

2.8.6 Hyaluronan, Murein, and Levan

These three last polysaccharides are also dealt with in Chap. 3:
- *Hyaluronan* is made of Greek *hualos* "glass," because this molecule is present in the vitreous humor of the eye and of *uronic* (*acid*) from Greek *ouron* "urine."
- *Murein* is a derivate of Latin *murus* "wall" (cf. French *mur* "wall"), because this molecule is present in cell walls of specific bacteria.
- *Levan* is made from Latin *laevus* "left," as opposed to *dextran*.

Conclusion

The huge biodiversity on the earth is partly visible in the large diversity of saccharides and polysaccharides which are provided by nature. The names of these products are linked to their natural origin, to their properties, or to some elements of their history, generally going back over antiquity. Apart from the case of starch, most of these names are very similar in all languages since they are generally derived from the same Greek or Latin words. For instance, the vernacular names of liver are quite different in different languages (*liver*, *foie*, *ficat*, *wątroba*, *maksa*), but the name of *heparin* remains the same, or nearly the same (*heparin, héparine, heparina, heparyna, heparini*), since it is derived from the Greek word *hêpar* "liver." Another example is the name of *inulin*, derived from the genus name *Inula* of a plant, regardless of the vernacular names of this plant, which are quite varied in Europe. Unfortunately, the Greek and Latin languages are less and less present in the education programs of European countries, but fortunately, they remain a sort of Esperanto, very useful for international communication in many scientific fields.

Acknowledgments Jan van Dam is thanked for the fruitful discussions that took place about the content of this chapter.

References

Buck CD (1988) A dictionary of selected synonyms in the principal Indo-European languages. The University of Chicago Press, Chicago, 1949

Chanteraine P (1990) Dictionnaire étymologique de la langue grecque. Éditions Klincksieck, Paris (1st edition 1968)

Ernout A, Meillet A (1985) Dictionnaire étymologique de la langue latine -histoire des mots-. Éditions Klincksieck, Paris (1st edition 1932)

Häkkinen K (2007) Nykysuomen etymologinen sanakirja, WS Bookwell Oy

Onions CT (1992) The Oxford dictionary of English etymology. Oxford University press, Oxford (1st edition 1966)

van Veen PAF (1989) Etymologisch woordenboek, Van Dale, Utrecht/Anvers

Polysaccharides: Molecular and Supramolecular Structures. Terminology.

Thomas Heinze, Katrin Petzold-Welcke, and Jan E.G. van Dam

Contents

3.1 **Structural Features** 24

3.2 **Glucans** 25
3.2.1 Cellulose 25
3.2.2 (1 → 3)-β-Glucans 28
3.2.3 Starch 31
3.2.4 Glycogen 34
3.2.5 Dextran 37
3.2.6 Pullulan 39

3.3 **Polyoses** 42
3.3.1 Xylans 42
3.3.2 Mannans 46
3.3.3 Xyloglucans 48
3.3.4 Mixed-Linkage β-Glucans 49

3.4 **Polysaccharides with Amino Functions** 49
3.4.1 Chitin and Chitosan 49
3.4.2 Hyaluronan or Hyaluronic Acid 50
3.4.3 Other Glycosaminoglycans 50
3.4.4 Murein 51

3.5 **Polysaccharides with Acid Functions** 52
3.5.1 Pectins 52
3.5.2 Alginate 55
3.5.3 Agar-Agar 56
3.5.4 Fucoidan or Fucogalactan 56
3.5.5 Carrageenan 56

3.6 **Miscellaneous** 57
3.6.1 Inulin 57
3.6.2 Levan 58
3.6.3 Xanthan gum 58

References 59

T. Heinze (✉)
Centre of Excellence for Polysaccharide Research, Institute of Organic Chemistry and Macromolecular Chemistry, Friedrich Schiller University of Jena, Humboldtstraße 10, 07743 Jena, Germany

Laboratory of Fibre and Cellulose Technology, Åbo Akademi University, Porthansgatan 3, 20500, Åbo, Finland
e-mail: thomas.heinze@uni-jena.de

Abstract

This chapter summarises important issues about the molecular and supramolecular structure of polysaccharides. It describes the terminology of polysaccharides systematically. The polysaccharides are divided regarding the molecular structures in glucans, polyoses, polysaccharides with amino functions, polysaccharides with acid functions and some miscellaneous. The most important glucans cellulose, (1 → 3)-β-D-glucans, starch, glycogen, dextran and pullulan are discussed. For polyoses, xylans, mannans, xyloglucans and mixed-linkage β-glucans are described. Polysaccharides with amino functions include the description of chitin and chitosan, hyaluronan or hyaluronic acid, glycosaminoglycans and murein. The polysaccharides with acid functions are described including pectins, alginates, agar-agar and carrageenan. Moreover, inulin, levan and xanthan gum are described.

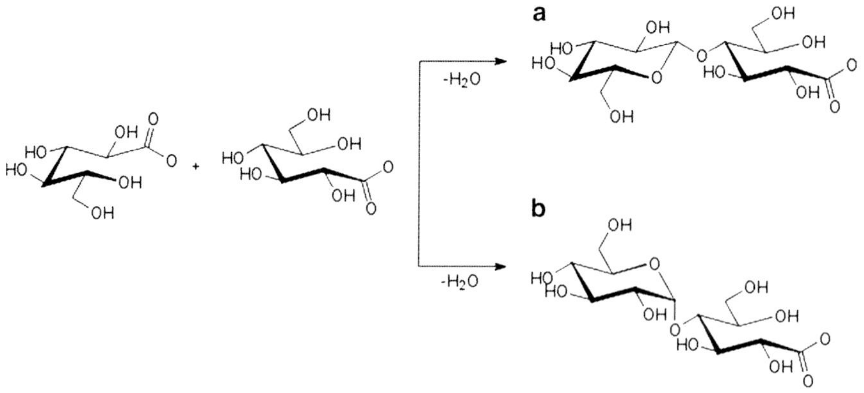

Scheme 3.1 Schematic formation of the acetal bond of D-glucose units in (**a**) (1 → 4)-β- and (**b**) in (1 → 4)-α bond

3.1 Structural Features

Polysaccharides (glycans) are the most widespread biopolymers in nature playing essential roles to sustain living organisms. Polysaccharides occur in all plants and animals and in many microorganisms and fungi, where they form structural building materials and energy stocks. The various polysaccharides possess many different properties that are related to their function. They serve as sources of energy and form the supporting and connecting tissue of plants and animals. The monosaccharide units (sugar units) that they are made up from are linked through acetal (or ketal) bonds, commonly called the glycosidic linkage (Scheme 3.1).

The sugar building blocks are commonly composed of five or six carbons. The monosaccharides are polyhydroxy aldehydes or ketones, with a high stereo-specific orientation. The sugar units in polysaccharides occur in the cyclic form (except the reducing end groups), where the aldehyde or keto group has formed an internal (hemiacetal) bond with one of the hydroxyls. The five- (furanose) or six-membered rings (pyranose) are the most common. Due to the high number of chiral carbon atoms in sugars, there are many possibilities of stereoisomeric monosaccharides. The aldohexoses, for example, have four chiral centres that result in a total of $2^4 = 16$ optical isomers. However, not all these sugars are found in nature.

The literature described a crude quantitative comparison of complexity and information content by considering an oligomer of four monosaccharides, in comparison with a tetrapeptide. By inclusion of only the potential amino acid variation in the oligopeptide (the only possible variation), and only the possible monosaccharide, linkage, location, ring size and absolute stereochemical variation (but not including the possibilities of branching, or of substituents) for the oligosaccharide, there are a couple of hundred thousand possible tetrapeptides and about 84 billion possible tetrasaccharides (Edgar 2009).

Because of the large diversity of polysaccharide structural and functional features found in nature, a simple classification system is not available.

The recommended nomenclature of carbohydrates and polysaccharides has been described in detail by the IUPAC commission (http://www.chem.qmul.ac.uk/iupac/2carb/39.html). For homoglycans (polysaccharides that are composed of one sugar residue), the general term is obtained by replacing the "-ose" ending in the sugar name by "-an", e.g. a polysaccharide build-up from glucose residues only is denoted "glucan" and a "fructan" is a polyfructose. The configuration of the residues is added (D-mannan or L-araban) as well as the type of glycosidic linkage between the sugar units, e.g. (1 → 4)-β-D-galactan. Common names for well-known polysaccharides are retained such as starch, cellulose and chitin, carrageenan and inulin

(Table 3.1). Heteroglycans may contain more than one sugar residue in the main chain (glucomannans) or may be branched by substitution with longer or shorter side chains (xyloglucans). Table 3.1 summarises the most important polysaccharides, which are described also in this chapter. Chapter 3 highlights the terminology and briefly the occurrence of the polysaccharides. It gives also an overview about the structure and superstructure.

3.2 Glucans

3.2.1 Cellulose

Cellulose (IUPAC name (1 → 4)-β-D-glucopyranan, CAS 9004-34-6), the most abundant renewable polymer, is a linear homopolymer composed of D-glucopyranose units (so-called anhydroglucose units, AGU) that are linked together by (1 → 4)-β glycosidic bonds (Fig. 3.1). The β-linkage induces a turning around of the cellulose chain axis by 180° of each second glucose unit. Consequently, the proper repeating unit is cellobiose with a length of 1.3 nm (Krässig 1993).

^{1}H-NMR spectroscopy confirmed the β-D-glucopyranose as the $^4C_1$ chair conformation, the lowest free energy conformation of the molecule (Krässig 1993). The hydroxyl groups are positioned in the ring plane (equatorial). The AGU within the cellulose chain exhibits three reactive hydroxyl groups, one primary (C6) and two secondary ones (C2, C3), which are able to undergo the typical reactions of hydroxyl groups including esterification, etherification and oxidation.

Cellulose forms the major structural polymer in higher plants. The primary occurrence of cellulose is in the abundant lignocellulosic materials of forests, with wood as the most important source. Wood and most other plant sources of cellulose like agricultural residues, water plants and grasses also contain hemicelluloses, lignin and a relatively small amount of extractives. Wood possesses about 40–50 wt% cellulose. Comparable amounts can be found in bagasse (35–45 wt%), bamboo (40–55 wt%) and straw (40–50 wt%) and even higher in bast tissues of flax (70–80 wt%), hemp (75–80 wt%), jute (60–65 wt%) and ramie (70–75 wt%) (Hon 1996; Klemm et al. 2002). Naturally highly pure cellulose occurs in seed hairs of cotton (95 %). In addition, several fungi exhibit cellulose in their cell walls as well as green algae (e.g. *Valonia ventricosa*, *Chaetamorpha melagonicum*, *Glaucocystis*) and some animals such as ascidians, which are marine animals that possess cellulose in the outer membrane. Several bacteria of the genera *Gluconacetobacter*, *Agrobacterium*, *Pseudomonas*, *Rhizobium* and *Sarcina* are able to synthesise bacterial cellulose from glucose (Vandamme et al. 1998; Jonas and Farah 1998).

The degree of polymerisation (DP) of native cellulose from various origins is in the range of 1,000–30,000, which is consistent with chain lengths from 500 to 15,000 nm. The cellulose that is gained after isolation possesses DP values ranging between 800 and 3,000 (Krässig 1993). However, the DP values must be regarded as average values due to the fact that cellulose samples are always polydisperse.

Both intra- and intermolecular hydrogen bonds occur in cellulose and they are known to have a significant influence on the properties (Kondo 1997). The limited solubility in most solvents, the low reactivity of the hydroxyl groups and the crystallinity of various cellulose samples are all governed mainly by the strong hydrogen bonding systems. In view of the fact that cellulose also contains hydrophobic areas (around the C-atoms), there is an ongoing discussion about the influence of hydrophobic interactions influencing the overall properties including insolubility (Lindman et al. 2010).

The three hydroxyl groups and the oxygen atoms of the ring and the glycosidic linkage can interact with each other internally or with a neighbouring cellulose chain by forming secondary valence bonds, namely, intramolecular and intermolecular hydrogen bonds. The versatile possibilities for the formation of the hydrogen bonds give rise to various three-dimensional structures. Solid-state ^{13}C-NMR spectroscopic measurements (Kamide et al. 1985) and IR

Table 3.1 Structure of polysaccharides of different origin

Polysaccharide type	Source	Structure	References
Cellulose	Plants	(1 → 4)-β-D-glucose	Klemm et al. (2002)
Curdlan	Bacteria	(1 → 3)-β-D-glucose	Nakata et al. (1998)
Scleroglucan	Fungi	(1 → 3)-β-D-glucose main chain, (1 → 6)-β-D-glucose branches	Giavasis et al. (2002)
Schizophyllan	Fungi	(1 → 3)-β-D-glucose main chain, D-glucose branches	Rau (2002)
Starch	Plants		Shogren (1998)
Amylose		(1 → 4)-α-D-glucose	
Amylopectin		(1 → 4)-α-D-glucose and (1 → 6)-α-D-glucose branches	
Glycogen	Animals	(1 → 4)-α-D-glucose and (1 → 6)-α-D-glucose branches	Melendez-Hevia et al. (1993)
Dextran	Bacteria	(1 → 6)-α-D-glucose main chain	Huynh et al. (1998)
Pullulan	Fungi	(1 → 6)-α linked maltotriosyl units	Shingel (2004)
Xylan	Plants	(1 → 4)-β-D-xylose main chain	Ebringerova et al. (2005) and Ebringerová (2006)
Mannan	Plants		Ebringerova et al. (2005)
Galactomannan		(1 → 4)-β-D-mannose main chain, D-galactose branches	
Glucomannan		(1 → 4)-β-D-mannose and (1 → 4)-β-D-glucose	
Galactoglucomannan		(1 → 4)-β-D-mannose and (1 → 4)-β-D-glucose, D-galactose branches	
Inulin	Plants	Mainly (1 → 2)-β-D-fructofuranose	Franck and De Leenheer (2002)
Levan	Plants, yeasts, fungi bacteria	(2 → 6)-β-D-fructofuranose, (2 → 1)-β-D-fructofuranose branches	Rhee et al. (2002)
Chitin	Animals	(1 → 4)-β-D-(N-acetyl)glucosamine	Roberts (1992)
Chitosan		(1 → 4)-β-D-glucosamine	
Hyaluronan	In all vertebrates	Alternating (1 → 4)-β and (1 → 3)-β linkages of disaccharide of D-glucuronic acid and D-glucosamine	Prehn (2002)
Murein	Bacteria	(1 → 4)-β-D-(N-acetyl)glucosamine and N-acetylmuramic acid	Heidrich and Vollmer (2002)
Alginate	Algae	(1 → 4)-α-L-guluronic acid	Sabra and Deckewer (2005)
		(1 → 4)-β-D-mannuronic acid	
Pectin	Plants	(1 → 4)-α-D-galacturonic acid main chain, with (1 → 2)-α or β-L-rhamnose	Ralet et al. (2002)
Xanthan	Bacteria	(1 → 4)-β-D-glucose, branches of D-glucuronic acid and D-mannose	Born et al. (2002)
Heparan sulphate	Animals	Sulphated (1 → 4)-α-L-iduronic acid and β-D-(N-acetyl)galactosamine	Gallagher et al. (1986)
Chondroitin sulphate	Animals	β-D-glucuronic acid and β-D-(N-acetyl) galactosamine, (1 → 3) linked	Rodén (1968)
Dermatan sulphate	Animals	L-iduronic acid and β-D-(N-acetyl)galactosamine, (1 → 3) linked	Malmström and Aberg (1982)
Carrageenan	Seaweed	Sulphated β- and α-D-galactose (1 → 4) and (1 → 3) linked	Van de Velde and De Ruiter (2002)

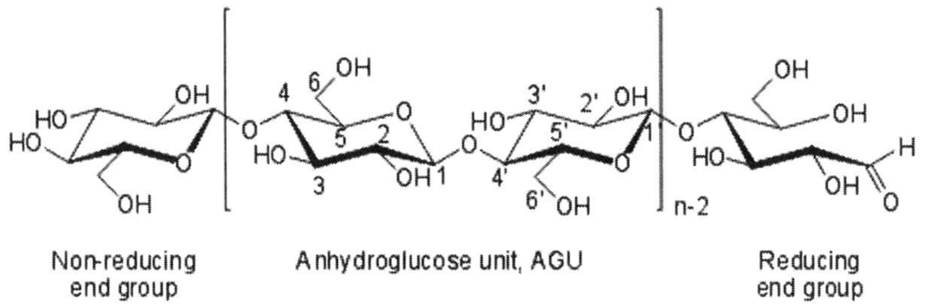

Fig. 3.1 Molecular structure of cellulose

spectroscopy (Liang and Marchessault 1959; Mitchell 1988) reveal intramolecular bonds between the OH of C3 and the adjacent ring oxygen of the AGU as well as a second one between the hydroxyl oxygen at position 6 and the adjacent OH at the position 2. The relative stiffness and rigidity of the cellulose molecules are a result of intramolecular hydrogen bonds in combination with the β-glycosidic linkage (Krässig 1993). The chain stiffness leads to highly viscous solutions (comparing to the α-glycosidically linked polysaccharides like starch or dextran), a high tendency to crystallise and the formation of fibrillar assemblies.

The order of the macromolecules in a cellulose fibre is not uniform throughout the whole structure. There exist low ordered regions (so-called amorphous regions) and crystalline regions. In the crystal lattice, the cellulose molecules are bound to one another by intermolecular hydrogen bonds, in particular between the OH of C6 and the oxygen at position 3 of a neighbouring chain along the (002) plane of native cellulose (cellulose I) (Gardner and Blackwell 1974). As a consequence, the cellulose molecules are linked together in a layer. The layers are kept together by hydrophobic interactions and weak C–H–O bonds, which could be verified by synchrotron X-ray and neutron diffraction data (Nishiyama et al. 2002). Native cellulose consists of two different crystal structures cellulose I_α and I_β as confirmed by using high-resolution, solid-state ^{13}C NMR spectroscopy (Atalla and VanderHart 1984).

Cellulose II obtained by regeneration of the dissolved polymer differs from native cellulose in the hydrogen bonding system. The intramolecular bonding of OH in position 2 is avoided and an intermolecular hydrogen bond of OH at C2 to OH at C2 of the next chain is formed (Kondo 2005). In comparison to cellulose I, the cellulose II molecules are more densely packed and strongly interbonded and, therefore, less reactive as commonly observed as a consequence of this extra intermolecular hydrogen bond (Krässig 1993). In Fig. 3.2, schematic view of the hydrogen bonding system in cellulose I and II is given.

Cellulose crystalline type III is obtained by treating native cellulose with liquid ammonia (below −30 °C) or an organic amine such as ethylenediamine followed by washing with alcohol. Small differences in lattice dimension exist between the two submodifications cellulose III_1 and III_2. By the treatment of cellulose in a suitable liquid at high temperature and under pressure, the fourth modification cellulose IV (cellulose IV_1 and IV_2, respectively) is formed (Gardiner and Sarko 1985; Klemm et al. 2002).

X-ray and NMR experiments confirmed the dimorphism in cellulose crystalline forms (Atalla and VanderHart 1984). The monoclinic unit cell of cellulose I_α with a space group $P2_1$ consists of two cellulose molecules containing each a cellobiose unit in the 002 corner plane and 002 centre plane in a parallel fashion (Gardner and Blackwell 1974). Cellulose I_β corresponds to a triclinic symmetry with space group P_1 containing one chain in the unit cell as schematically displayed.

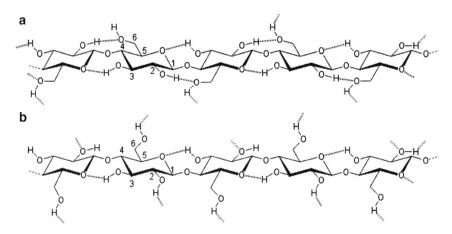

Fig. 3.2 Hydrogen bonding system of (**a**) cellulose I and (**b**) cellulose II (Reproduced with permission from Tashiro and Kobayashi (1991). © 1991, Elsevier)

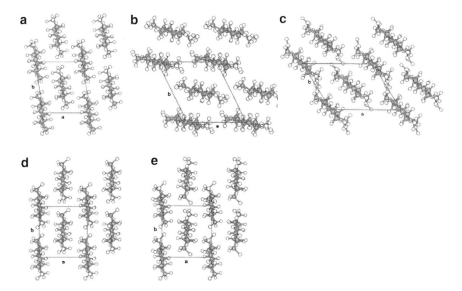

Fig. 3.3 Representation of the model of (**a**) cellulose I$_\beta$, (**b**) cellulose II, (**c**) cellulose III$_1$, (**d**) cellulose IV$_1$ and (**e**) cellulose IV$_2$ on the *a–b* plane (Reproduced with permission from Zugenmaier (2001). © 2001, Elsevier)

In cellulose II, two chains of cellulose are located antiparallel on the 2$_1$ axis of the monoclinic cell (Fig. 3.3), while the chains are displaced relative to each other by about one-fourth of the AGU (Langan et al. 1999). The crystalline structure of cellulose III$_1$ can be described with a one-chain unit cell and a P2$_1$ space group with the cellulose chain axis on one of the 2$_1$ screw axes of the cells (Wada et al. 2001). A single chain of cellulose III$_1$ is similar to one of the two chains existing in the crystal of cellulose II. The space group P$_1$ is assumed for the cellulose IV$_1$ and IV$_2$ (Gardiner and Sarko 1985).

3.2.2 (1 → 3)-β-Glucans

A number of structural variations exist within the class of polysaccharides classified as (1 → 3)-β-glucans that are found in several bacteria, yeasts and fungi and are well known to stimulate the human immune system. Also in higher plants,

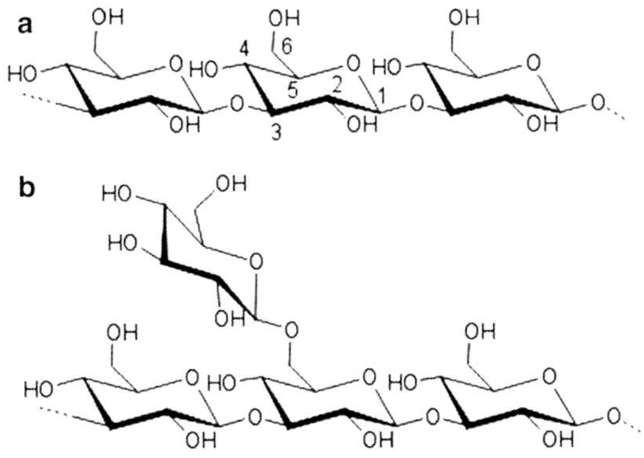

Fig. 3.4 Chemical structure of (1 → 3)-β-D-glucans: (**a**) curdlan, (**b**) repeating unit of scleroglucan

(1 → 3)-β-glucans are found, for example, callose and laricinan.

Curdlan (IUPAC name (1 → 3)-β-D-glucopyranan, CAS Nr. 54724-00-4) is a neutral, essentially linear homopolymer that is produced by *Alcaligenes faecalis* (var. myxogene 10C3) or *Agrobacterium radiobacter*. Curdlan consists of D-glucopyranose units that are linked together by (1 → 3)-β glycosidic bonds (Fig. 3.4). Curdlan has an average DP of ~450. The average molecular weight measured in 0.3 N NaOH is in the range of 5.3×10^4–2.0×10^6 g/mol (Nakata et al. 1998).

Scleroglucan is a linear (1 → 3)-β-D-glucan (Fig. 3.4) with at every third glucose unit regular branches of single (1 → 6)-β-linked D-glucopyranosyl unit produced by fungi of the genus *Sclerotium*. Depending on the cultures, the DP of scleroglucan is in the range of 110 (*S. glucanicum*) to 1,600. The commercially available polysaccharide has a DP of ≈800 (Wang and McNeil 1996; Powell 1979; Rodgers 1973).

The primary structure of schizophyllan is very similar to those of scleroglucan derived from *Schizophyllum commune* (Heinze et al. 2006a). The molecular weight is in the range from 6 to 12×10^6 g/mol (Rau 2002; Rau et al. 1990).

These (1 → 3)-β-D-glucans adopt a right-handed 6_1 triple-stranded helical structure in nature as one of their most characteristic features (Sakurai et al. 2005). Curdlan also adopts a six-fold, parallel, triple-stranded helical structure determined by X-ray diffraction (Bluhm and Sarko 1977; Deslandes et al. 1980). Regenerated curdlan exists in three forms (Fig. 3.5) (Kasai and Harada 1980; McIntosh et al. 2005). The anhydrous form obtained by vacuum heat has three intertwining chains forming a triple helix in which each chain has right-handed, sixfold conformation and a $P6_3$ space group. The hydrated form consists of triple helices with a P1 space group and a crystal hydrate with two molecules of water per glucosyl residue. Another form can be obtained by dialysing an alkaline solution of curdlan against water giving a 7_1 or 6_1 helical conformation. Whether this room-temperature form consists of loose intertwined triple helices with ~2 molecules water per glucosyl unit or a mixture of single helices with ~20 molecules water per glucosyl unit and triple helices is not clear (Koreeda et al. 1974; Marchessault and Deslandes 1979; Deslandes et al. 1980; Fulton and Atkins 1980; Kasai and Harada 1980; Chuah et al. 1983; Okuyama et al. 1991; McIntosh et al. 2005).

Whereas the coordinates of the carbon and oxygen atoms are provided by the X-ray diffraction, the hydrogen bond organisation has not been clarified. A possible, widely believed model is the helical structure stabilised though interstranded hydrogen bonding among the three O_2 atoms (Fig. 3.6) (Bluhm and Sarko 1977; Sakurai et al. 2005). Three secondary hydroxyl groups protrude towards the centre of the helix and form a closed triangle hydrogen bonding network. Since this hexagon uses three OH groups that are located in the same x–y plane along the helix, the hexagon is arranged

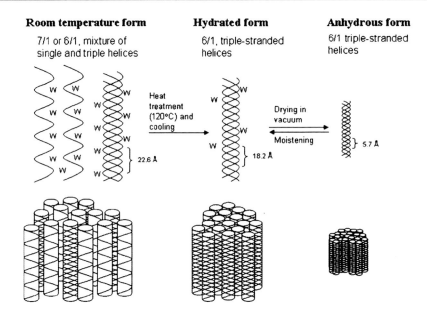

Fig. 3.5 Schematic representation of the structural changes between the three forms of curdlan (Reproduced with kind permission from Springer Science + Business Media (McIntosh et al. 2005, Fig. 3.2)

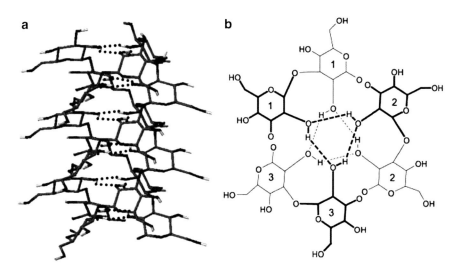

Fig. 3.6 The hexagonal H-bonds of curdlan (anhydrous form). (**a**) A *side view* for the full triple helix; *dotted lines* indicate the H-bonds. (**b**) A *cross-sectional view* of the helix (perpendicular to the helix). For convenience, only two glucose units in each chain are shown. The glucose rings marked with the same number belong to the same chain. The *bold lines* denote that the atoms are situated at the *upper position* than those of the *thin lines*. The O(2)–H(2)–O(2) angle is 132.6° and the O(2)–O(2) distance is 2.7 Å (Reproduced with permission of Miyoshi et al. (2004). Copyright Wiley-VCH Verlag GmbH & Co. KGaA)

perpendicularly to the helix axis (Miyoshi et al. 2004; Sakurai et al. 2005).

Semiempirical quantum mechanics calculation (MOPAC) showed a possibility that the triple-stranded helical structure could be stabilised through unique "continuous" hydrogen bonding networks along the helical structure. In this proposed model, three single strands are connected by interstranded hydrogen bonding networks among oxygen atoms (O_2 and O_2') on different *x–y* planes (Miyoshi et al. 2004; Sakurai et al. 2005).

X-ray diffraction shows that scleroglucan has a triple-helical backbone conformation similar to that of curdlan (Wang and McNeil 1996; Bluhm et al. 1982). Dissolved scleroglucan chains have a rod-like triple-helical structure in which the D-glucosidic side groups are on the outside and prevent the helices from coming close to each other and aggregating (Yanaki et al. 1981; Yanaki and Norisuye 1983).

Callose (CAS No 9041-22-9 and 9012-72-0) is a (1 → 3)-β-D-glucan that occurs widely in higher plants and is found to be deposited at the sieve plates in phloem tissue, in response to wound, thermal and mechanical stresses, regulating the flow rate at varying pressures (Jaffe et al. 1984). The occurrence in compression wood cell walls of callose (or laricinan, which is similar in structure), Altaner et al. 2007 suggest a role for callose biosynthesis in mechanoperception in plants (Timell 1986; Telewski 2006). Callose has been extracted from various sources such as the bark of Scots pine (*Pinus sylvestris*) or the pollen grain of rye (*Secale cereale*) and as minor component in cotton seed hair (*Gossypium hirsutum*). When this type of β-D-glucan is isolated from tamarack (*Larix laricina*) compression wood, it is known as laricinan. It occurs also by this name in other coniferous trees.

Paramylon is a (1 → 3)-β-D-glucan produced as storage polysaccharide in *Euglena* sp. and other related monads that are present as rigid rods.

Laminaran (CAS No 9008-22-4) is a relatively low-molecular-weight linear (1 → 3)-β-D-glucan with nonreducing mannitol end groups found in brown algae species of the genus *Laminaria* sp. (*L. digitata*) but also in *Chorda filum*, *Saccharina longicruris*, *Eisenia bicyclis* or *Ascophyllum nodosum*. In laminaran of other brown algae (e.g. *Cystoseira* sp.), the mannitol residue is not present. Laminaran may be occasionally branched at position 2 or 6 and has a typical molecular weight in the range of 3,000 kD (Rioux et al. 2010).

3.2.3 Starch

Starch (CAS No 9005-84-9), the primary plant storage polysaccharide, is widely present in grains, seeds and tubers. Starch is a polyglucan that consists of two major components containing D-glucose: amylose and amylopectin (CAS No 9005-82-7 and CAS No 9037-22-3). Amylose is a (1 → 4)-α-linked linear glucan. Amylopectin is composed of (1 → 4)-α-linked D-glucose and (1 → 6)-α branches (Fig. 3.7).

The molecular mass of amylose is in the range 10^5–10^6 g/mol, while amylopectin shows significantly higher values of 10^7–10^8 g/mol (Shogren 1998). Starch is the major energy reserve stored in tubers, roots, fruits, stems and seeds (grains) of plants and is one of the basic food ingredients because it is easy digestible. Starch is arranged in nature in granules with diameters ranging from around 0.1 to 200 μm (Srichuwong et al. 2005; Tester et al. 2004; Hoover 2001; Jane et al. 1994; Gallant et al. 1992). Starch granules are densely packed with semi-crystalline structures, the crystallinity varying from 15 to 45 %, the density being about 1.5 g/cm^3 (Pérez and Bertoft 2010). Starch exists in original in A, B or C X-ray patterns, which were described firstly by Katz (Katz 1928; Zobel 1988).

In freshly prepared aqueous solutions, amylose chain adopts a random coil structure, which is however not stable (Hayashi et al. 1981, Pérez and Bertoft 2010). Amylose forms single-helical inclusion complexes with suitable complexing agents (Takeo et al. 1973; Pérez and Bertoft 2010). Dough develops an X-ray diffraction pattern after cooking, which differs from the A, B or C form that was named V-pattern after the German word "verkleisterte Stärke" meaning gelatinisation (Zobel 1988, Pérez and Bertoft 2010).

A- and B-type starch form double helices, which are left handed and parallel stranded. They have a perfectly sixfold structure, with a crystallographically repeating unit of 10.5 Å. The symmetry of the double helices of the A-form differs slightly from the B-structure. In the A-structure, the repeating unit is a maltotriosyl unit, whereas the B-type has a maltosyl unit (Fig. 3.8). Chains of the A-type are crystallised in a monoclinic lattice. In such a unit cell, 12 glucopyranose units are located in 2 left-handed, parallel-stranded double helices, packed in a parallel fashion. For each unit cell, four water molecules (closed circles) are located between the helices. Chains of the B-type are crystallised in

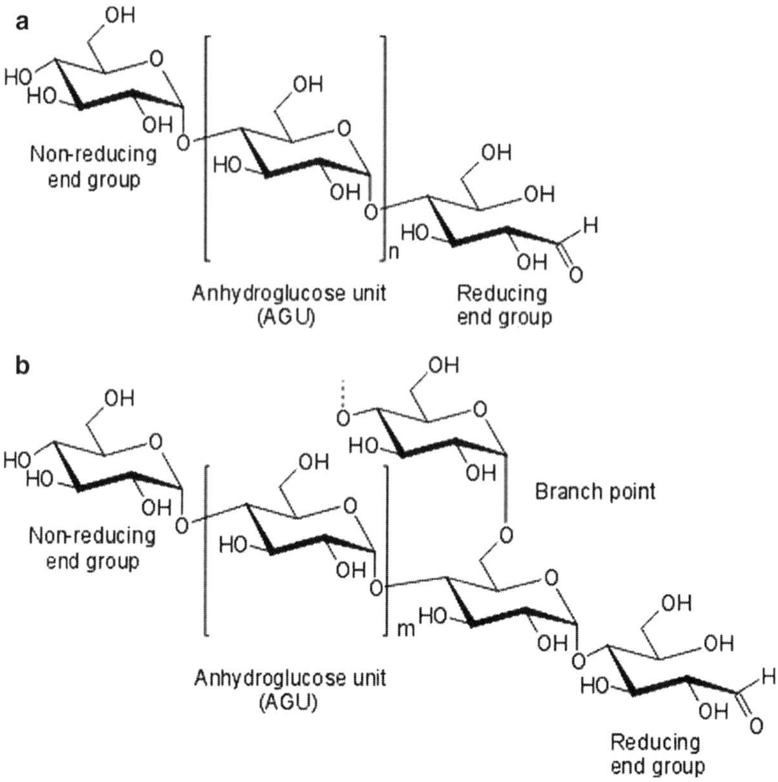

Fig. 3.7 Molecular structure of starch: (**a**) amylose, (**b**) amylopectin

a hexagonal lattice, where they pack as an array of left-handed parallel-stranded double helices in a parallel fashion. Thirty-six water molecules represent 27 % of hydration. Half of the water molecules are tightly bound to the double helices, and the remaining ones form a complex network centred around the sixfold screw axis of the unit cell (Pérez and Bertoft 2010).

For the description of the unit chain composition of amylopectin, the chains are grouped into certain categories (Peat et al. 1952). A-chains are unsubstituted by other chains and connected through a (1 → 6) linkage to the rest of the polysaccharide. B-chains are substituted by one or several other chains. In addition, each molecule contains a single C-chain, which exhibits the sole reducing end group (Fig. 3.9) (Pérez and Bertoft 2010).

The amylopectin chains have a DP range of 10–130. However, the average chain length of most amylopectins is 17–26 depending on the crystallinity type. A-crystalline starches show typically shorter chains than B-type (Pérez and Bertoft 2010; Hizukuri 1985).

Amylopectin chains possess a characteristic periodicity in length (approximately DP 25–30) and are subdivided the long chains (presumably representing B-chains) into B2- (DP ~ 45), B3-chains (DP ~ 70), etc. (Hizukuri 1986). The major group of short chains is a mixture of short B-chains (B1) and A-chains (DP 11–15). A shorter periodicity (~12) was found for A-chains (Hanashiro et al. 1996). The shortest group (named fa) has a DP of 6–12. The groups named fb_1 with DP 13–24 are B1-chains and the group fb_2 (DP 25–36) are presumably B2-chains. Chains with a DP > 37 (group fb_3) are not resolved into subgroups (Fig. 3.10) (Pérez and Bertoft 2010).

In addition to the groups of chains of DP up to ~100, "extra-long" or "super-long" chains with chain length in the same order like amylose have been reported (Pérez and Bertoft 2010).

The cluster model of amylopectin was introduced around 1970 (Pérez and Bertoft 2010; Nikuni 1978; French 1972). A great diversity in the suggested size of clusters is found in literature ranging from 4.22 chains up to 34 chains.

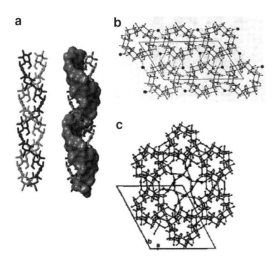

Fig. 3.8 3D structures of A-type and B-type crystalline starch polymorphs. (**a**) The left-handed and parallel-stranded double helices (**b**) structure of A starch. Projection of the structure onto the (a, b) plane. Hydrogen bonds are indicated as *broken lines* (**c**) structure of B starch. Projection of the structure onto the (a and b) plane. Hydrogen bonds are indicated as *broken lines* (Reproduced with permission of Pérez and Bertoft (2010). Copyright Wiley-VCH Verlag GmbH & Co. KGaA)

An amylopectin macromolecule with DP in the range of $10–16 \times 10^3$ is built from 60 to 120 units of clusters (Pérez and Bertoft 2010). The long B-chains are involved in the interconnection of the clusters of short chains. Bender et al. (1982) found three major size groups of clusters that as β-limit dextrins had DP from 40 to 140 by using cyclodextrin transferase. Generally, A-crystalline samples have larger clusters than B-crystalline samples.

The branches of amylopectin are apparently not evenly distributed within the clusters but found in small groups. These groups were called building blocks (Pérez and Bertoft 2010; Kong et al. 2009; Bertoft 2007a; Gérard et al. 2000; Bertoft et al. 1999). The DP of the building blocks varies from 5 to 40, and they are divided into types 2–7 with a corresponding estimated number of chains 2–7 (Bertoft et al. 1999; Gérard et al. 2000). The branching density is high within the building blocks; the internal chain length is therefore short (e.g. in amaranth and potato 2.0–2.8) (Kong et al. 2009; Bertoft 2007a).

In the native configuration, amylopectin participates in both crystalline and amorphous structures. The external chains form double helices that crystallise into thin lamella, behind which an amorphous lamella is found (Pérez and Bertoft 2010; Jenkins and Donald 1995). The major part of branches in amylopectin is found within the amorphous part. The chain length of the internal segments is 4–9 units. Three types of internal segments exist, namely, between clusters, between building blocks inside clusters and inside building blocks (Pérez and Bertoft 2010).

Figure 3.11 shows a model of the building block structure of the Φ,β-limit dextrin of a cluster from amaranth starch based on the structural data described by Kong et al. (2009 and Pérez and Bertoft (2010).

A cluster with a DP of 82 consists of six building blocks, one longer B1b- or B2-chain and the short rest are B1a- and A-chains (Pérez and Bertoft 2010). For the organisation of the clusters in amylopectin, alternative models exist (Fig. 3.12).

Hizukuri proposed that a cluster is built from A-chains and B-chains. The clusters are interconnected by long chains, so that B2-chains are involved in the interlinkage of two clusters and B3-chains of three clusters, etc. (Fig. 3.12a) (Hizukuri 1986; Pérez and Bertoft 2010). Figure 3.12b shows the two-dimensional backbone model, in which the direction of the clustered chains is perpendicular to the direction of the backbone (Bertoft 2004a; Pérez and Bertoft 2010). The entire long B-chains are amorphous. The "fingerprint A-chains" were also suggested to be amorphous. These chains might therefore preferentially be found associated with the long B-chains in the backbone (Bertoft 2004a, b; Pérez and Bertoft 2010). In a slightly modified backbone structure (for potato amylopectin), some long B2-chains participate in both the clusters and in the amorphous backbone, whereas others (together with B3-chains) are only found in the backbone (Fig. 3.12c) (Bertoft 2007b, Pérez and Bertoft 2010). Another modification of the backbone model is shown in Fig. 3.12d. The cluster itself constitutes a part of the backbone through its interblock segments (IB-S), rather attached to the backbone as a separate entirety (Laohaphatanaleart et al. 2010; Pérez and Bertoft 2010). The building blocks are directly attached to the backbone and, depending

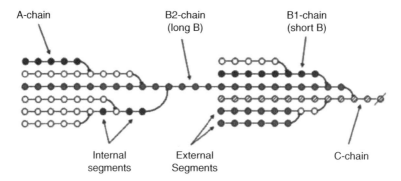

Fig. 3.9 Basic labelling of chains in amylopectin. *Circles* denote glucosyl residues, *horizontal lines* (1 → 4) and *bent arrows* (1 → 6) linkages. The reducing-end residue is to the right (Reproduced with permission of Pérez and Bertoft (2010). Copyright Wiley-VCH Verlag GmbH & Co. KGaA)

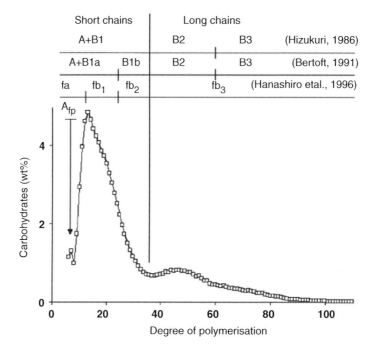

Fig. 3.10 Unit chain distribution by HPAEC of disbranched sweet potato amylopectin. Values at DP > 60 are approximate. Division of chains into categories presently mostly used as suggested by different authors are indicated. A_{fp} denotes "fingerprint A-chains", a subgroup of A-chains at DP 6–8 (Reproduced with permission of Pérez and Bertoft (2010). Copyright Wiley-VCH Verlag GmbH & Co. KGaA)

on the number of IB-S, the length of the backbone chains varies extensively.

3.2.4 Glycogen

Glycogen (CAS No 9005-79-2) is the major fuel storage polysaccharide molecules in animals. It is a highly branched homoglycan. It is built of (1 → 4)-α-linked glucose residues to form chains with 11–14 units, with (1 → 6)-α branches (Fig. 3.13). Glycogen has a relatively small protein part associated with it (Goldsmith et al. 1982).

The glycogen molecule is formed by three different kinds of chains: A-chains are unbranched,

3 Polysaccharides: Molecular and Supramolecular Structures. Terminology.

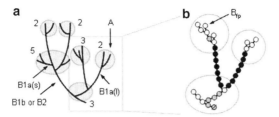

Fig. 3.11 A model of the building block structure of a cluster in amaranth amylopectin as a Φ,β-LD based on data given by Kong et al. (2009). (**a**) Building blocks are encircled and numbered according to their number of chains. Different categories of chains are indicated. (**b**) An enlargement of a part of the cluster. *Circles* symbolise glucosyl residues, of which *black* are residues in the IB-S and *grey* are internal residues of the blocks (Reproduced with permission of Pérez and Bertoft (2010). Copyright Wiley-VCH Verlag GmbH & Co. KGaA)

B-chains are branched, and the C-chain is the single chain in the molecule having a free reducing end group regarding the accepted model of Whelan (Gunja-Smith et al. 1970) mainly derived from data on enzymatic degradation of glycogen (Fig. 3.14) (Melendez-Hevia et al. 1993; Calder 1991; Goldsmith et al. 1982). The branching of the B-chains is uniformly distributed (degree of branching, DB 2), so every B-chain has two branches, which create further A- or B-chains. Between the branches, four anhydroglucose units exist and a tail after the second branch in the B-chains. A- and B-chains are of uniform length. The glycogen molecule is spherical and organised into concentric layers. Every layer has the same length (1.9 nm). The whole molecule is named as β-particle. In a β-particle, 12 layers are arranged with a total radius of 21 nm. Every A-chain is in the most external layer. As a consequence of the degree of branching (DB 2), the number of chains in any layer is twice that of the previous one (Melendez-Hevia et al. 1993).

Although the interchain linkages are randomly grouped in the glycogen molecule, the structure may be described in terms of three parameters: the average chain length (CL), the average exterior chain length (ECL) and the average interior chain length (ICL). The first represents the statistical average of the length of the individual chains in the molecule; an exterior chain is defined as that part of chain between the terminal group and the outermost interchain linkage, whereas an interior chain is that part between two branch points (Calder 1991). One parameter of interest is the β-amylolysis limit (β-limit); the β-limit is the degree of hydrolysis of the glycogen by β-amylase. This value along with CL, ECL and ICL has been used routinely to characterise and compare glycogens. β-Amylase is able to remove almost the entirety of molecule, which is exterior to the outermost branch points, while the inner part of the molecule resides intact. Due to the steric hindrance, the removing stops within a few glucose units of the branch point, leaving intact a high-molecular-weight β-limit dextrin. The β-limit is determined as proportion of the molecule released as maltose (Calder 1991). The CL is determined by a combination of glycogen phosphorylase (EC 2.4.1.1) and amylo-1,6-glucosidase (EC 3.2.1.33) (Illingworth et al. 1952), or pullulanase and β-amylase (Lee and Whelan 1966), or β-amylase (Manners and Wright 1962) or isoamylase (Gunja-Smith et al. 1971). The CL may then be calculated (Calder 1991):

$$CL = \frac{\text{total moles of glucose residues present}}{\text{moles of interchain linkages revealed after debranching}}$$

ECL and ICL can be calculated if β-limit and CL are known. Beta-amylolysis produces maltose and a higher-molecular-weight β-limit dextrin with the exterior chains reduced to "stubs". The size of stubs must be known of assessing the length of exterior chains. Manners (1962) suggested that "the uncertainty in length of the chain stubs does not cause serious error". The exterior chain length can be calculated by following equation (Calder 1991):

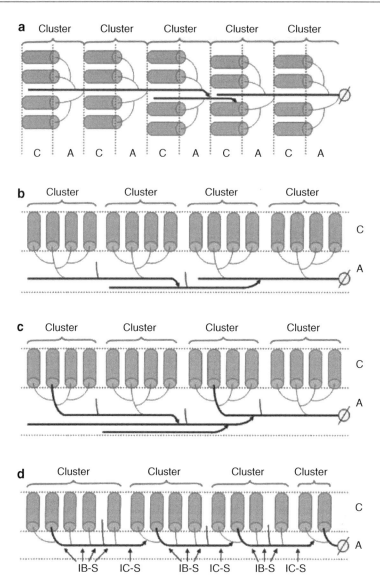

Fig. 3.12 Alternative models of the organisation of clusters in amylopectin. Long chains (*black, bold lines*), short chains (*grey, thin lines*), the reducing end (illustrated as *Stroked O*), and the amorphous (A) and crystalline (C) lamellae inside starch granules are indicated. External segments form double helices symbolised by *grey cylinders*. (**a**) The cluster model by Hizukuri (1986), in which the long chains form an integrated part of the clusters. (**b**) A two-directional backbone model, in which the clusters are anchored in perpendicular direction to a backbone of completely amorphous long chains. Some A_{fp}-chains (*black, thin lines*) are associated with the backbone with the backbone rather than the clusters (Bertoft 2004a). (**c**) A backbone model, in which some B2-chains participate in the crystalline lamella through their external segment. Additional long chains can be found in the amorphous backbone (Bertoft 2007b). (**d**) An alternative model in which the clusters are integrated parts of the backbone through their IB-S. Clusters are separated by somewhat longer IC-S. A_{fp}-chains can introduce defects in the crystalline lamella (Laohaphatanaleart et al. 2010) (Reproduced with permission of Pérez and Bertoft (2010). Copyright Wiley-VCH Verlag GmbH & Co. KGaA)

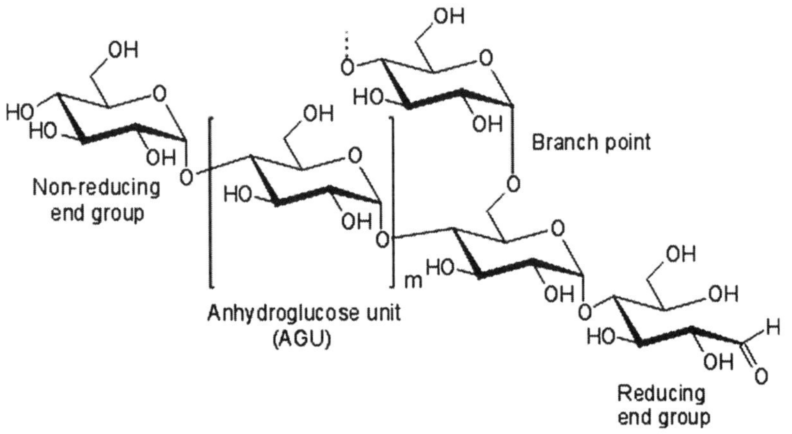

Fig. 3.13 Molecular structure of glycogen

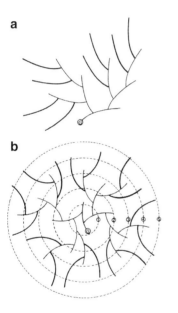

Fig. 3.14 Structure of the glycogen molecule as stated in Whelan's model (Goldsmith et al. 1982; Gunja-Smith et al. 1970, 1971), with two kinds of chains: A and B. (**a**) Extended structure to show the branching structure; (**b**) a more realistic drawing showing the disposition of the successive branches forming concentric layers (*numbered circles*). Both schemes show a simplified molecule with only four layers in (**a**) and five in (**b**): A-chains, in the outermost layer, are more thickly drawn (With permission of Portlandpress Melendez-Hevia et al. 1993, Fig. 1)

$$ECL = (\beta - \text{limit} \times CL) + 2.0,$$

where 2.0 is the average stub length.

Similarly, ICL = CL-ECL-1 (Manners 1962; Calder 1991).

Table 3.2 lists values of the fine structural parameters of a number of different glycogens indicating that glycogens have a β-limit of 45–55 %, CL of 10–14, ECL of 6–9 and ICL of 3–4 (Calder 1991).

3.2.5 Dextran

Dextran (CAS No 9004-54-0) is a homopolymer of glucose with predominantly $(1 \rightarrow 6)$-α linkages (50–97 %) (Heinze et al. 2006b; Naessens et al. 2005). It is the most important polysaccharide for medical and industrial applications produced by bacterial strains. Figure 3.15 shows a part of the dextran main chain with branching points in the 2, 3 and 4 positions.

The degree and nature of branching units depends on the dextran-producing bacterial strain (Jeanes et al. 1954). Different bacterial strains are able to synthesise dextran mainly from sucrose. Dextran can be formed by several strains mostly Gram-positive, facultative anaerobe cocci, e.g. *Leuconostoc* and *Streptococcus* strains (Table 3.3) (Jeanes et al. 1954). Dextran structures were evaluated by optical rotation, infrared spectroscopy and periodate oxidation reactions and by methylation analysis (Harris et al. 1984; Slodki et al. 1986; Seymour et al. 1977). Furthermore, structural investigations were carried out by the use of degradative enzymes of known specificity followed by means of thin-layer chromatography, HPLC and

Table 3.2 Fine structural parameters for glycogen from different sources [adopted from Calder (1991)]

Glycogen	β-limit	CL[a]	ICL[b]	ECL[c]	References
Rat liver	56 ± 3	13.2 ± 0.6	2.8 ± 0.3	9.4 ± 0.7	Calder and Geddes (1986)[d]
Rat muscle	54 ± 3	12.7 ± 1.8	2.8 ± 0.3	8.9 ± 0.6	Calder and Geddes (1985)[d]
Hen liver	48	15	3	11	Kjolberg et al. (1963)[e]
Fish liver	47	13	3	8	Kjolberg et al. (1963)[e]
Pig liver (n = 3)	48–56	15–16	3–4	9–11	Kjolberg et al. (1963)[e]
Pig muscle (n = 3)	45–51	11–16	2–5	7–10	Kjolberg et al. (1963)[e]
Cat liver	54	13	2–3	9–10	Manners (1957)[e]
Human placenta	41 ± 4	11.6 ± 0.7	3.8 ± 0.4	6.8 ± 0.4	Blows et al. (1988)[d]
Human muscle	40	11	3	7	Manners (1957)[e]
Human liver	46	15	5	9	Bathgate and Manners (1966)[f]

[a]Average chain length in nm.
[b]The average interior chain length in nm.
[c]The average exterior chain length; chain length determined using [d,f]pullanase or [e]β-amylase and assuming a stub length of [d]2.0 or [e,f]2.5 nm.
[d]Data are mean ± SD (n ≥ 4 separate samples).
[e,f]Data indicate the range of values obtained for the indicated number of samples or are for a single sample.

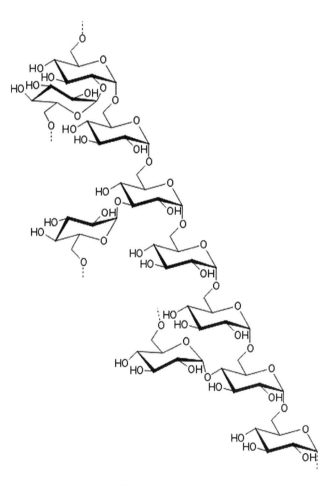

Fig. 3.15 Structure of dextran: part of the (1 → 6)-α-linked glucose main chain with branching points in position 2, 3 and 4

3 Polysaccharides: Molecular and Supramolecular Structures. Terminology.

Table 3.3 Occurrence, percentage of different glycosidic linkages in dextran of different bacterial strains obtained by methylation analysis [adopted from Heinze et al. (2006b)]

Strain number[a]	α-Linkages (%) 1 → 6	1 → 2	1 → 3	1 → 4	Solubility in water	References
Lm NRRL B-512F	95		5		+	Slodki et al. (1986) and VanCleve et al. (1956)
Lm NRRL B-1355 fraction 1	54		46		+	Seymour et al. (1977)
Lm NRRL B-1355 fraction 2	95		5		−	Seymour et al. (1979)
Lm NRRL B-1299 fraction 1	68	29	3		+	Dols et al. (1997)
Lm NRRL B-1299 fraction 2	63	27	8		−	Dols et al. (1997)
Lm NRRL B-742 fractions 1	50		50		+	Seymour et al. (1979)
Lm NRRL B-742	87			13	−	Seymour et al. (1979)
S. mutans 6715 fraction 1	64		36		+	Hare et al. (1978)
S. mutans 6715 fraction 2	4		96		−	Shimamura et al. (1982)
S. mutans GS5	70		30		+	Honda et al. (1990)
S. downei	90		10		+	Gilmore et al. (1990)

[a]*Lm Leuconostoc mesenteroides, S. Streptococcus.*

^{13}C NMR spectroscopy (Naessens et al. 2005; Hare et al. 1978). Examples of the linkage analysis of several dextran fractions produced by different bacterial strains are summarised in Table 3.3. The length of the side chains in dextran produced by *Lm* NRRL B-512F has been studied by sequential alkaline degradation (Larm et al. 1971). The procedure is based on the substitution of the terminal nonreducing glucopyranose (Glc*p*) at position 6 with *p*-toluenesulfonylmethyl groups. Analysis by GLC-MS reveals that about 40 % of the side chains contain one glucose residue, 45 % are two glucose units long and the remaining 15 % are longer than two. These results confirm HPLC studies of enzymatically hydrolysed dextran from *Lm* NRRL B-512F (Taylor et al. 1985).

Native dextran is generally of a high average molecular weight ranging from 9×10^6 to 5×10^8 g/mol with a high polydispersity (Alsop et al. 1977; Antonini et al. 1964; Bovey 1959; Senti et al. 1955). The polydispersity of dextran increases with the molecular weight as a result of increasing branch density (Ioan et al. 2000).

Anhydrous dextran structure crystallises in a monoclinic unit cell with the space group P2$_1$ with the b axis unique. The unit cell has two antiparallel-packed molecular chains, and the asymmetric unit contains two glucopyranose residues (Fig. 3.16) (Guizard et al. 1984). The conformation of the dextran molecule is relatively extended and ribbon-like. The chain conformation is stabilised by one intramolecular hydrogen bond per glucose residue. The chains of like polarity are packed into sheets with extensive intrasheet hydrogen bonding.

3.2.6 Pullulan

Pullulan (CAS No 9057-02-7) is a homopolysaccharide of glucose. Pullulan is the exopolysaccharide synthesised by a yeast-like fungus *Aureobasidium pullulans*. It is well established that the regularly repeating structural unit of pullulan is a maltotriose trimer (1 → 4)-α-Glc*p*-(1 → 4)-α-Glc*p*-(1 → 6)-α-Glc*p*-, which is excreted by *A. pullulans* (Singh et al. 2008;

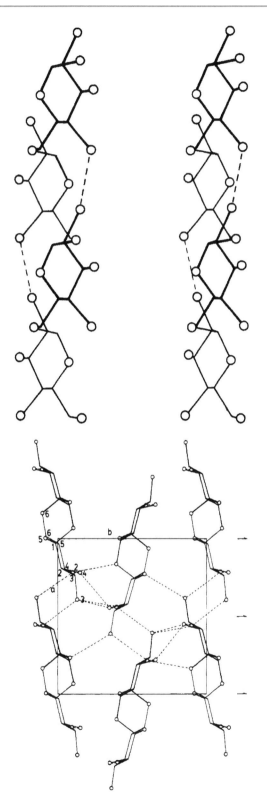

Fig. 3.16 *Top*: view of dextran chain seen from the *a* direction of the unit cell. *Bottom*: projection of unit cell down the *c* axis. Hydrogen bonds are indicated by *dashed lines*; hydrogen atoms are not shown. Corner chain has "up" polarity; centre chain is "down" (With permission of Guizard et al. (1984). Copyright 1984 American Chemical Society)

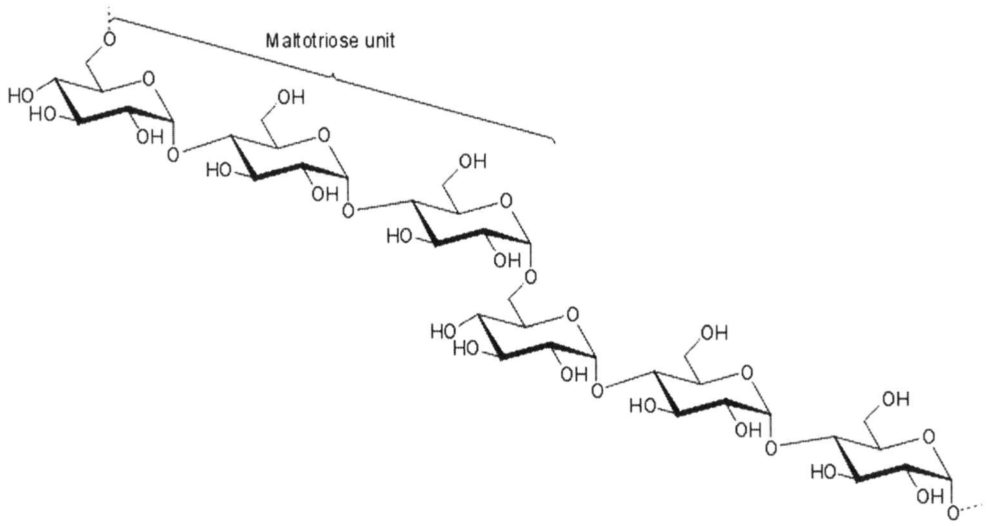

3 Polysaccharides: Molecular and Supramolecular Structures. Terminology. 41

Fig. 3.17 Schematic chemical structure of pullulan with maltotriose as repeating unit

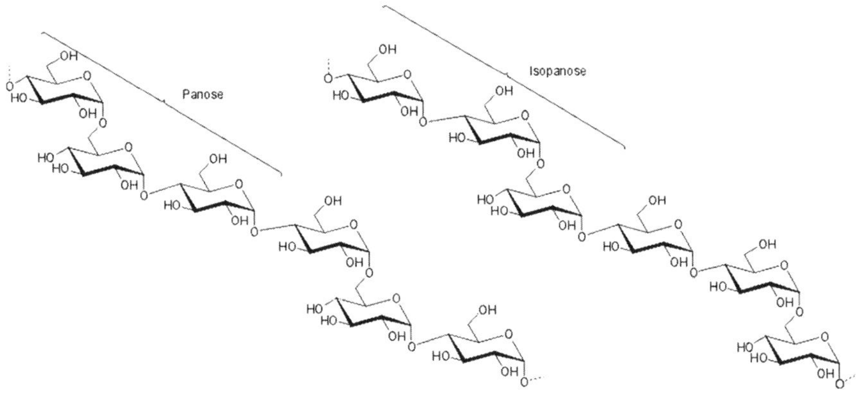

Fig. 3.18 Schematic chemical structure of pullulan with panose (*left*) and isopanose (*right*) as repeating unit

Leathers 2003; Gibbs and Seviour 1996). Other structures particularly the tetramer or maltotetraose (1 → 4)-α-Glc*p*-(1 → 4)-α-Glc*p*-(1 → 4)-α-Glc*p*(1 → 6)-α-Glc*p*- are described to be present in the polymer chain in a maximum of 7 % (Singh et al. 2008; Catley et al. 1986; Wallenfels et al. 1965). Moreover, (1 → 3)-α and even (1 → 3)-β as well as (1 → 6)-β linkages were determined in the polymer backbone of pullulan produced by some strains (Singh et al. 2008; Fujii et al. 1984).

An extracellular enzyme from *Aerobacter aerogenes*, i.e. pullulanase (EC 3.2.1.41), proved to be a critical tool for the analysis of pullulan structure (Singh et al. 2008; Bender and Wallenfels 1961). Pullulanase converted yeast (*A. pullulans*) α-glucan containing linkages in the polymer into maltotriose, thereby describing pullulan as a biopolymer of (1 → 6)-α-linked maltotriose subunits (Fig. 3.17).

Sometimes, partial acid hydrolysis of pullulan results in isomaltose, maltose, panose and isopanose (Singh et al. 2008; Bouveng et al. 1963; Sowa et al. 1963; Bender et al. 1959). So pullulan chain is often viewed with panose or isopanose as subunits (Fig. 3.18) (Singh et al. 2008).

Panose [*O*-α-D-glucopyranosyl-(1 → 6)-*O*-α-D-glucopyranosyl-(1 → 4)-*O*-α-D-glucopyranose]

(CAS No 25193-53-7/33401-87-5) and isopanose [*O*-α-D-glucopyranosyl-(1 → 4)-*O*-α-D-glucopyranosyl-(1 → 6)-*O*-α-D-glucopyranose] are the D-glucose containing trisaccharide, which has the (1 → 4)- and (1 → 6)-α-D-glycosidic linkage.

The number average molecular weight (M_n) of pullulan is about 100,000–200,000 g/mol and the weight average molecular weight (M_w) is about 362,000–480,000 g/mol (Okada et al. 1990; Singh et al. 2008). Values of M_w/M_n lie between 2.1 and 4.1 (Singh et al. 2008; Petrov et al. 2002; Roukas and Montzouridpu 2001; Wiley et al. 1993).

3.3 Polyoses

The structural plant tissues contain in their cell walls, besides cellulose and sometimes lignin, also noncrystalline polysaccharides that are commonly referred to as the hemicelluloses. These hemicelluloses comprise a group of different noncellulosic polysaccharides that occur as intermediate matrix between lignin and cellulose in the different plant cell wall layers. These components can be liberated in the cellulosic fibre production process. Depending on the botanical source, the hemicellulose fraction that is extracted under alkaline conditions contains almost always xylan of variable structure (number and type of attached sugar side chains, *O*-acetylation) and often glucomannan, but also other polysaccharides may be co-extracted. The group of hemicellulose polysaccharides was recently reviewed extensively (Ebringerová 2006; Ebringerova et al. 2005).

The polyoses that are known as hemicelluloses are usually divided into four general groups of structurally different polysaccharide types:
1. Xyloglycans (xylans)
2. Mannoglycans (mannans)
3. Xyloglucans
4. Mixed-linkage β-glucans.

All of them occur in many structural variations differing in polymerisation degree, side chain type, distribution, localisation and/or types and distribution of glycoside linkages in the main macromolecular chain (Ebringerová 2006; Ebringerova et al. 2005).

3.3.1 Xylans

Xylans or xyloglycans can be grouped into several structural subclasses:
1. Homoxylans
2. Glucuronoxylans
3. (Arabino)Glucuronoxylans
4. Arabinoxylans
5. (Glucurono)Arabinoxylans
6. Heteroxylans (Ebringerová 2006; Ebringerova et al. 2005; Ebringerová and Heinze 2000).

3.3.1.1 Homoxylans

Homoxylans are linear polymers consisting only of D-xylopyranosyl (Xyl*p*) residues linked by (1 → 3)-β-linkages (X3), (1 → 4)-β-linkages (X4) and/or mixed (1 → 3, 1 → 4)-β-linkages (Xm) (Fig. 3.19).

Types X3 and Xm are common in some seaweeds such as in red algae (*Nemiales* and *Palmariales* sp.) and green algae (*Caulerpa* and *Bryopsis* sp.), where they replace cellulose as the skeletal polymer of the cell walls. Homoxylans X4 can be isolated from guar seed husks (*Cyamopsis tetragonoloba*) (Ebringerová and Heinze 2000).

3.3.1.2 Glucuronoxylans

In dicots, plants of the highest evolutionary level, the main component of the secondary cell walls is the D-glucurono-D-xylans (CAS No 37317-38-7). Most of the glucuronoxylans are (1 → 4)-β-linked having single 4-*O*-methyl-α-D-glucopyranosyl uronic acid residues (MeGlcA) always at position 2 of the main chain Xyl*p* units (Fig. 3.20) (Ebringerová 2006; Ebringerova et al. 2005).

This structural type is usually named as 4-*O*-methyl-D-glucurono-D-xylan (MGX). However, the glucuronic acid side group may be present in both as 4-*O*-methylated and non-methylated unit (GlcA). MGX is the main hemicellulose component of hardwoods, showing Xyl:MeGlcA ratios from 4:1 to 16:1 depending on the source and isolation conditions used; on average, the

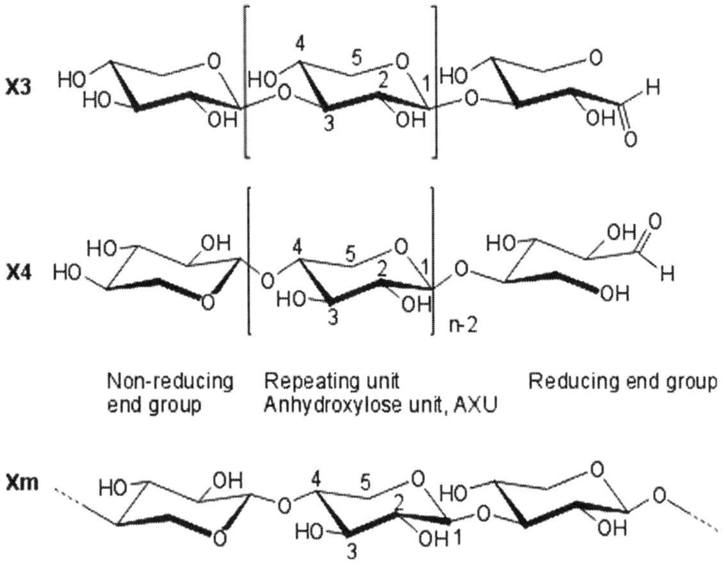

3 Polysaccharides: Molecular and Supramolecular Structures. Terminology.

Fig. 3.19 Structural features of homoxylan chains with (1 → 3)-β linkages (X3) and (1 → 4)-β linkages (X4) and mixed (1 → 3, 1 → 4)-β linkages (Xm)

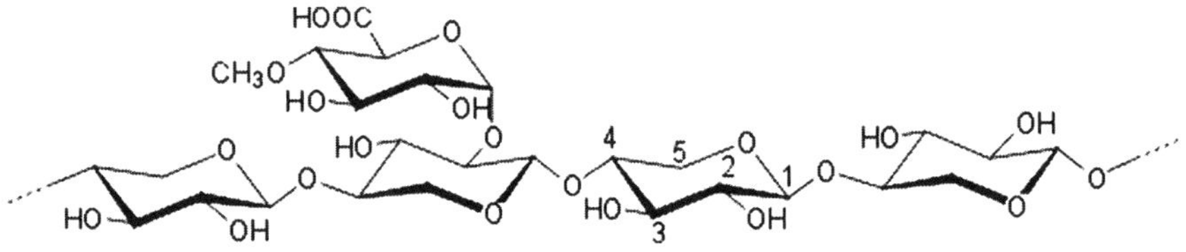

Fig. 3.20 Primary structure of 4-*O*-methyl-D-glucurono-D-xylan (MGX)

ratio is about 10:1. Fractional precipitation of the bulk MGX or fractional extraction of the xylan component from the plant source usually leads to xylan preparations with a broad range of Xyl:MeGlcA ratios (Ebringerová 2006; Ebringerova et al. 2005; Stephen 1983). An unusual MGX was found in the wood of *Eucalyptus globulus* as it contained, in addition to terminal MeGlcA units, some of these residues substituted at position 2 with α-D-galactose (Ebringerova et al. 2005; Evtuguin et al. 2003; Teleman et al. 2002). In native state, xylan is supposed to be *O*-acetylated. The content of acetyl groups of MGX isolated from hardwoods of temperate zones varies in the range 3–13 % (Ebringerová 2006; Ebringerová et al. 2005; Ebringerová and Heinze 2000). The acetyl groups are split during the alkaline extraction resulting in partial or full water insolubility of the xylan. The acetyl groups may be, at least in part, preserved by treating with hot water or steam. Water-soluble acetylated MGXs have been isolated from NaClO$_2$-delignified wood of birch and beech with DMSO (Ebringerova et al. 2005; Teleman et al. 2002). The degrees of acetylation ranged between 0.3 and 0.6 and the acetyl group appeared at various positions of the Xyl*p* residues. The composition of MGX is depending on the pulp origin and steeping conditions. The MGX of the β-fraction from press lye had a low uronic acid content (ratio of MeGlcA to Xyl is about 1:20). The molecular weight of the MGX fractions varied between 5,000 and 10,000 g/mol. In MGX from kenaf, additional terminal rhamnopyranosyl and arabinofuranosyl side chains and <10 % of acetyl groups were shown to branch the xylan

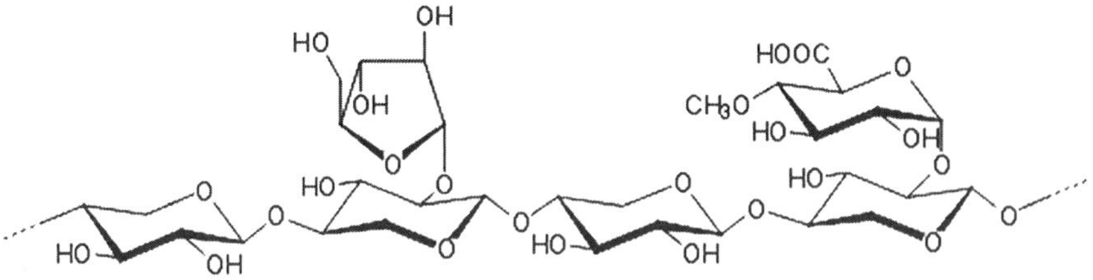

Fig. 3.21 Primary structure of (L-arabino)-4-O-methyl-D-glucurono-D-xylan

backbone. Acidic xylans with the glucuronic acid side chains both in the 4-O-methylated and non-methylated forms were isolated from olive pits, rape stem, red gram husk and jute bast fibre (Ebringerova et al. 2005; Ebringerová and Heinze 2000). The backbone of the xylan from the mucilage of quince seeds (*Cydonia cylindrica*) (Vignon and Gey 1998) is exceptionally heavily substituted with MeGlcA possessing a ratio of Xyl to MeGlcA of 2:1 and carries acetyl groups at various positions.

3.3.1.3 (Arabino)Glucuronoxylan and (Glucurono)Arabinoxylan

Both (arabino)glucuronoxylan (AGX) and (glucurono)arabinoxylan (GAX) (CAS No 9040-27-1) have single MeGlcA and α-L-arabinofuranosyl (Ara*f*) moieties linked at position 2 and 3, respectively, to the (1 → 4)-β-D-xylopyranose backbone (Fig. 3.21), which might also be slightly acetylated (Ebringerova et al. 2005).

The AGX type can be found in coniferous species (Ebringerova et al. 2005; Stephen 1983). Generally, the backbone of AGX is more heavily substituted by MeGlcA than that of the hardwood MGX, with 5–6 Xyl units per uronic acid in the former and 10 on average in the latter. AGX is the dominant polyose in the cell walls of lignified supporting tissues of grasses and cereals (sisal, corncobs and the straw from various wheat species) (Ebringerova et al. 2005; Ebringerová and Heinze 2000). Corncob xylans showed the presence of a linear, water-insoluble polymer (wis-AGX) with <95 % of the backbone unsubstituted and a water-soluble xylan (ws-AGX) having more than 15 % of the backbone substituted (Ebringerová et al. 1998, 2003, 2005). The uronic acid content was lower in the wis-AGX (~4 %) than in ws-AGX (about 9 %). A small proportion of the Xyl*p* residues of the backbone are disubstituted by α-L-Ara*f* residues. A peculiar structural feature of the ws-AGX is the presence of disaccharide side chains (Formula 3.1). This sugar moiety has been found usually esterified by ferulic acid (FA) at position 5 of the Ara*f* residue and appears as a widespread component of grass cell walls (Ebringerova et al. 2005; Ishii 1997). Similar to acetyl groups, ester-linked FA is cleaved during the alkaline extraction of AGX. However, FA-containing ws-AGX preparations can be obtained by ultrasonically assisted extraction of corncobs using hot water and very dilute alkali hydroxide solutions (Ebringerova et al. 2005; Hromádková et al. 1999). In contrast to AGX, the GAX has an arabinoxylan backbone, which contains about ten times fewer uronic acid side groups than α-L-Ara*f* ones, and has some Xyl*p* units doubly substituted with these sugars. The degree and pattern of substitution of GAX depend on the source from which they are extracted. These differences are the ratio of Ara to Xyl, the content of MeGlcA and the presence of disaccharide side chains (3.1) as well as the dimeric arabinosyl side chains (3.2) (Ebringerova et al. 2005; Ebringerová and Heinze 2000; Ishii 1997; Stephen 1983).

$$\beta\text{-D-Xyl}p\text{-}(1\rightarrow 2)\text{-}\alpha\text{-L-Ara}f\text{-}(1\rightarrow \quad (3.1)$$

$$\alpha\text{-L-Ara}f\text{-}(1\rightarrow 3)\text{-}\alpha\text{-L-Ara}f\text{-}(1\rightarrow \quad (3.2)$$

In wheat bran GAX, fractions of lowly and highly substituted GAX exist, which greatly differ in the amount and type of substitution by the Ara*f* units, but not in the content of the

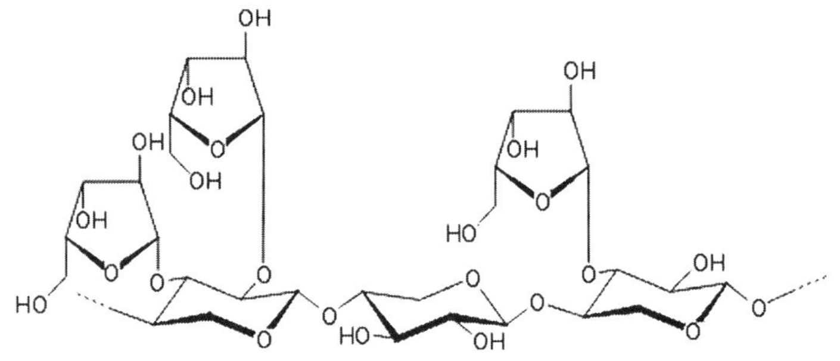

Fig. 3.22 Primary structure of L-arabino-D-xylan

glucuronic acid, half of which occurs as the 4-*O*-methyl ether (Schooneveld-Bergmans et al. 1999; Ebringerova et al. 2005). Feruloylated GAX fractions were isolated from wheat bran by a treatment with cold water, steam and dilute alkali (Ebringerova et al. 2005; Schooneveld-Bergmans et al. 1998).

3.3.1.4 Arabinoxylans

Arabinoxylans (AXs) appear as neutral or slightly acidic polymers, the latter usually being included in the GAX group. AX has a linear (1 → 4)-β-xylopyranan backbone that is, in part, substituted by L-α-Ara*f* units either in position 3 or 2 in case of monosubstitution or in both (disubstitution) of the Xyl*p* repeating units (Fig. 3.22) (Ebringerová 2006; Ebringerova et al. 2005).

Moreover, phenolic acids (ferulic and coumaric acids) have been found to be esterified to position 5 of some Ara*f* units in AX (Ebringerová 2006; Ebringerová et al. 2005; Rao and Muralikrishna 2001; Ebringerová and Heinze 2000). Depending on the location and degree of integration in the different tissues of the grain, a part of AX can be extracted with water, but the bulk can only be isolated by alkali solutions. Both AXs consist of a family of related polymers differing in DS by α-Ara*f* units and substitution patterns (Nilsson et al. 1996, 2000; Cleemput et al. 1995). According to the DS, AXs have been divided into three groups, depending on the isolation and fractionation procedures used as well as the distribution patterns (Ebringerová 2006; Ebringerova et al. 2005; Vinkx et al. 1995). The first group includes water-insoluble monosubstituted AXs (Ara:Xyl up to <0.2–0.3) with α-L-Ara*f* found mainly at position 3 of the xylan backbone. The second group includes water-soluble xylans with a ratio of Ara to Xyl between 0.3 and 1.2. AXs with an Ara to Xyl ratio of 0.5–0.9 have shorter sequences of disubstituted Xyl*p* units than that of the third group (Ebringerova et al. 2005; Nilsson et al. 1996).

AXs are located in a variety of the main commercial cereals: wheat, rye, barley, oat, rice and corn, sorghum as well as rye grass, bamboo shoots and pangola grass. They constitute the major polyose of cell walls of the starchy endosperm (flour) and outer layers (bran) of the cereal grain. AX contents differ from 0.15 % in rice endosperm to ~13 % in whole grain flour from barley, and rye, and up to 30 % in wheat bran (Ebringerová et al. 2005; Ebringerová and Hromádkova 1999).

3.3.1.5 Complex Heteroxylans

The complex heteroxylans (CHX) attendant in cereals, seeds, gum exudates and mucilages are structurally more complex (Ebringerova et al. 2005; Stephen 1983). They have a (1 → 4)-β-D-xylopyranan backbone branched with various mono- and oligoglycosyl side chains including single uronic acid and arabinosyl units. The xylan backbone of CHX from corn bran is heavily substituted (at both positions 2 and 3) with β-D-Xyl*p*, β-L-Ara*f*, α-D-Glc*p*A residues and oligosaccharide side chains (3.1), (3.3) and (3.4) (Ebringerova et al. 2005; Saulnier et al. 1995).

$$\beta\text{-D-Gal}p\text{-}(1 \to 5)\text{-}\alpha\text{-L-Ara}f\text{-}(1 \to \qquad (3.3)$$

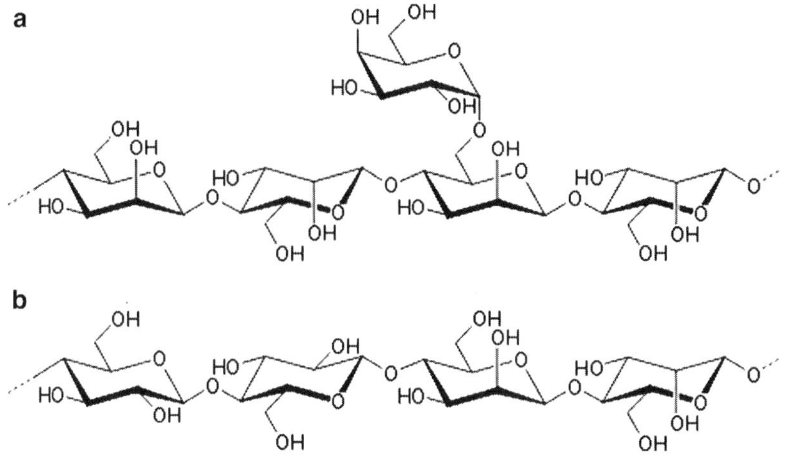

Fig. 3.23 Primary structures of (a) D-galacto-D-mannan and (b) D-gluco-D-mannan

$$\text{L-Gal}p\text{-}(1 \to 4)\text{-}\beta\text{-D-Xyl}p\text{-} \atop (1 \to 2)\text{-}\alpha\text{-L-Ara}f\text{-}(1 \to \quad (3.4)$$

The mucilage-forming seeds of *Plantago* sp. have very complex heteroxylans (Ebringerova et al. 2005; Fischer et al. 2004; Samuelsen et al. 1999a). For the CHX from *Plantago major* seeds, a $(1 \to 3, 1 \to 4)$-mixed-linkage xylopyranan backbone has been proposed possessing short side chains attached to position 2 or 3 of some $(1 \to 4)$-linked D-Xyl*p* units (Ebringerova et al. 2005; Samuelsen et al. 1999a). The side chains are built of β-D-Xyl*p* and α-L-Ara*f* residues and disaccharide moieties (3.5) and (3.6). $(1 \to 4)$-linked xylotrisaccharide and $(1 \to 3)$-linked xylooligo-saccharides with DP of 6–11 have been also found (Samuelsen et al. 1999b). These oligosaccharides were recommended as building blocks for the backbone of CHX. In addition, single β-D-Xyl*p* *O*-2 linked to the backbone as well as to the acidic disaccharide (3.6) was confirmed.

$$\alpha\text{-L-Ara}f\text{-}(1 \to 3)\text{-}\beta\text{-D-Xyl}p\text{-}(1 \to \quad (3.5)$$

$$\alpha\text{-D-Glc}p\text{A-}(1 \to 3)\text{-}\alpha\text{-L-Ara}f\text{-}(1 \to \quad (3.6)$$

In contrast, the gel-forming CHX from psyllium (*Plantago ovata*) husks (Fischer et al. 2004) is a neutral, highly branched arabinoxylan with the $(1 \to 4)$-β-D-xylopyranan backbone decorated at position 2 with single Xyl*p* units and at position 3 with the trisaccharide moiety (3.7).

$$\alpha\text{-L-Ara}f\text{-}(1 \to 3)\text{-}\beta\text{-D-Xyl}p\text{-} \atop (1 \to 3)\text{-}\alpha\text{-L-Ara}f\text{-}(1 \to \quad (3.7)$$

The molecular weight shows considerable variations and may vary depending on the determination method even for the same sample (Ebringerová and Heinze 2000; Ebringerová and Hromádková 1999; Stephen 1983). For cereal AX and CHX values of 64,000–380,000 g/mol and for MGX and AGX 5,000–130,000 g/mol and 30,000–370,000 g/mol were measured (Ebringerova et al. 2005).

3.3.2 Mannans

The mannan-type polysaccharides of higher plants are classified in view of the backbone into:
1. Galactomannans (CAS No 11078-30-1) (Fig. 3.23a)
2. Glucomannans (CAS No 11078-31-2) (Fig. 3.23b) (Table 3.4) (Ebringerova et al. 2005)

Whereas the polysaccharide backbone of the first type is formed exclusively of $(1 \to 4)$-

Table 3.4 Occurence of various structural types of mannan-type polysaccharides in plants [adopted from Ebringerová (2006)]

Plant source	Mannoglycan type	Gal:Glc:Man
Softwood	GaGM	0.3:1:4–6
	GM	0.1:1:4
Hardwood, herbal plants, grasses	GaGM	0.2:1:2–3
	GM	0.1:1:1–2
Konjac mannan	GM	5:8
Ivory nut, date, coffee bean	wis-GaM	1:0:23–25
Legume seeds (guar, carob, cassia)	ws-GaM	1:0:1.1–5.7

β-linked D-mannopyranose (D-Man$p$) residues in linear chains, the second type has both (1 → 4)-β-linked D-Man$p$ and (1 → 4)-β-linked D-glucopyranose (D-Glc$p$) units in the main chain. D-Galactopyranose (α-D-Gal$p$) units are linked to the mannan backbone at position 6 in both mannan-type polymers in different proportions as single side chains. The biopolymers observed are named galactomannans and galactoglucomannans. A tolerance limit for the proportion of galactosyl groups (15 %) has been suggested for such classification (Ebringerova et al. 2005; Stephen 1983).

3.3.2.1 Galactomannans

Mannans free of galactosyl side units are rather rare (CAS No 9036-88-8). Slightly galactosylated mannans (~4 % galactose), considered as linear (1 → 4)-β-D-mannans, have been isolated from the seed endosperm of ivory nut (*Phytelephas macrocarpa*) and date (*Phoenix dactylifera*) (Ebringerova et al. 2005; Reid and Edwards 1995). The ripening (and hardening) of the seeds of *Palmae* species such as coconut (*Cocos nucifera*), *Arenga saccharifera*, *Borassus flabellifer*, *Hyphaene thebaica* and *Phoenix dactylifera* is associated with the removal of the Gal side chains in the galactomannan. High proportions of low-branched galactomannans (GaM), which have a Gal$p$ unit every 23 Man$p$ residues, were found in green arabica coffee beans (*Coffea arabica*) (Ebringerova et al. 2005; Navarini et al. 1999). GaM samples highly substituted with Gal$p$ units (30–96 %) are abundant in the cell walls of storage tissues (endosperm, cotyledons, perisperm) of seeds (Ebringerova et al. 2005; Reid and Edwards 1995). The ratio of Man to Gal is one of the main characteristics of GaM that determines their properties, such as solubility in water, density and the viscosity of the solution. This ratio varies from 1:1 to 5.7:1 in seed gums of leguminous plants (Ebringerova et al. 2005; Srivastava and Kapoor 2005). Commercially the seed gums of guar (*Cyamopsis tetragonoloba*) (CAS No 9000-30-0), carob (or locust bean gum, *Ceratonia siliqua*) (CAS 9000-0-2), tara (*Caesalpinia spinosa*) (CAS No 39399-88-4) and cassia (*Cassia tora*) (CAS No 11078-30-1) are most important. GaMs are accumulated in many seeds, for example in the seeds of *Ipomoea turpethum*, a perennial plant (Singh et al. 2003), and various *Cassia* species (Kapoor et al. 1998; Chaubey and Kapoor 2001), date (Ishrud et al. 2001), *Mimosa scabrella* (Ganter et al. 1995) and some *Strychnos* species (Corsaro et al. 1995; Adinolfi et al. 1994). The highly branched GaM of *Mimosa* seeds showed a Man:Gal ratio of 1.1:1 (Ganter et al. 1995). The Gal$p$ units were found to be distributed regularly (*Cassia* GaM), block-wise (guar GaM) or randomly (carob, tara GaM) (Ebringerová 2006).

The slightly galactosylated mannans are linear polymers. They tend towards self-association, insolubility and crystallinity as a result of their cellulose-like (1 → 4)-β-D-mannan backbone. GaM of *Cassia spectabilis* seed with a Man:Gal ratio 2.65:1 shows an orthorhombic unit cell with lattice constants of $a = 9.12$, $b = 25.63$ and $c = 10.28$; the dimension b was shown to be sensitive to the degree of galactose substitution and the hydration conditions (Kapoor et al. 1995, 1998).

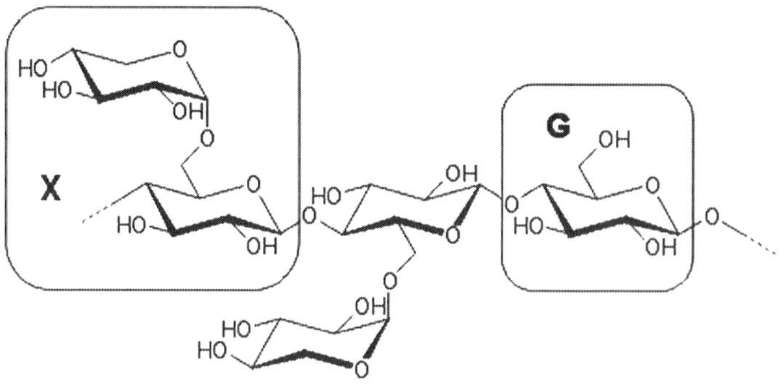

Fig. 3.24 Primary structure of D-xlyo-D-glucan

3.3.2.2 Glucomannans

D-Gluco-D-mannans (GM) and (D-galacto)-D-gluco-D-mannans (GGM) exhibit in contrast to GaM both (1 → 4)-β-linked Man$p$ and Glc$p$ units in the main chain (Fig. 3.23b). Gal$p$ residues are linked to the glucomannan chain at position 6 of Man$p$ residues. The galactose content of 15 % is the limit for typical GM and GGM types (Table 3.4). These polymers represent the main polyose component of the secondary cell wall of softwoods and the minor ones in that of hardwoods (3–5 %), herbal plants and grasses. They compose together with xylan and xyloglucan the primary cell wall of these plants (Ebringerová 2006; Ebringerova et al. 2005). The mannoglycans are easily extractable only by alkaline treatment from previously or simultaneously delignified materials. The most accessible is GGM, whereas GM according to its cellulose-like backbone is strongly associated to cellulose and extractable only using alkaline solvents of higher concentrations. Slightly O-acetylated GM of low molecular mass composed of galactose, glucose and mannose in the mole ratio of 1:8–15:30–60 was isolated by microwave treatment of softwoods (Lundqvist et al. 2002).

The storage GM in the tubers of *Amorphophallus konjac*, known as "konjac mannan" (CAS No 37220-17-0) in food industry, is composed of glucose and mannose in the ratio 1:1.6. Its backbone is slightly O-acetylated and branched (about 8 %) at position 6 of Glc$p$ residues (Ebringerová 2006; Zhang et al. 2001). The mucilaginous polysaccharide (acemannan) isolated from the parenchyma (fillet) of *Aloe vera* has more than 90 % of mannose and very low contents of glucose and galactose (Ebringerová 2006; Ebringerova et al. 2005). Its (1 → 4)-β-mannopyranan backbone is heavily O-acetylated at 2, 3 and/or 6 positions.

3.3.3 Xyloglucans

D-Xylo-D-glucan (XG) (CAS No 39379-72-1) consists of a regular repeating cellulosic β-(1 → 4)-β-glucopyranan backbone with α-D-Xyl$p$ units attached at position 6 of the Glc$p$ moieties (Fig. 3.24) (Ebringerová 2006).

XG is a polyose that is the major component of primary cell wall of all higher plants, where it is associated to cellulose surfaces. In the primary wall of dicotyledonous angiosperms, XG may occur up to 20–25 % (*Sycamore*, *Arabidopsis thaliana*), while in monocotyledonous cell walls, relative lower amounts are present like between 2 and 5 % in grasses and about 10 % in gymnosperm softwoods (Ebringerová 2006; Fry 1989).

In the XG of grasses, about 30–40 % of the Glc$p$ residues contain Xyl$p$ moieties, in those of dicotyls about 60–75 % (Ebringerová 2006; Ebringerova et al. 2005). Two main XG types have been characterised based on the distribution of the Xyl$p$ branches along the main chain:

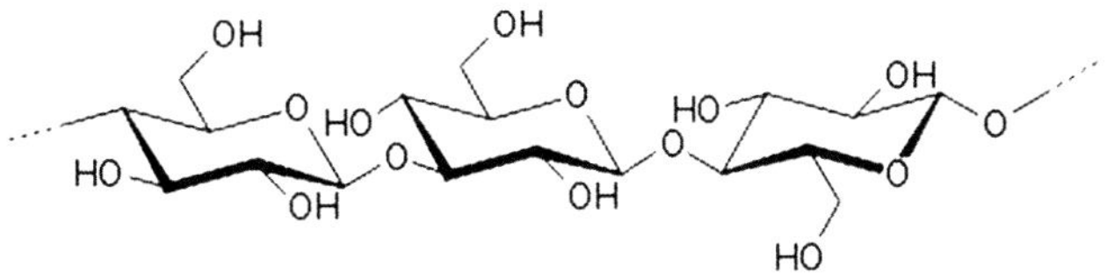

Fig. 3.25 Primary structure of mixed (1 → 3, 1 → 4)-β-linkage D-glucans

- Type I: –X–X–X–G– formed by blocks composed of three xylosylated (X) units and one non-substituted (G) Glc*p* unit
- Type II: –X–X–G–G– blocks built of two alternating X and G units (Ebringerová 2006).

Type I is present in the primary cell wall of herbal plants, some grasses and softwoods, while type II is typically found in the primary cell wall of *Solanaceae* sp. (such as tomato and tobacco). Both types include a variety of XG structures, which have additional oligomeric glycosyl side chains, composed of galactose, fucose, arabinose and/or xylose, which are attached to the backbone and/or branch the Xyl*p* units (Ebringerová 2006; Ebringerova et al. 2005; Hoffman et al. 2005). The XG side chain substitution pattern varies between species, and typical termination patterns by α-Ara*f* or α-Fuc*p* are commonly observed (Hoffman et al. 2005).

3.3.4 Mixed-Linkage β-Glucans

Mixed-linkage β-glucans (Mix-G) (CAS No 9041-22-9) are present in grasses (cereals) and some lichens. Mix-G consist of an unbranched backbone composed of Glc*p* units linked by (1 → 3)-β and (1 → 4)-β linkages (Fig. 3.25) (Ebringerová 2006). Single (1 → 3)-linked Glc*p* units separate blocks of (1 → 4)-linkage sequences (cellotriosyl, C3 as well as cellotetraosyl, C4, and longer cellulose-like segments, C9–14). The mole ratios of C3/C4 segments range between 4.2 and 4.5 for wheat, 2.8 and 3.3 for barley and 2.1 and 2.4 for oat, whereas in lichenan, a structurally similar Mix-G commonly found in lichen *Cetraria islandica*, the ratio is much higher (24–49) (Ebringerová 2006; Wood et al. 1994). A random distribution of the blocks along the chains is suggested. On commercial scale cereal β-D-glucans such as oat gum, barley gum and wheat bran gum are produced for food ingredients and stabiliser and moisturising agent in cosmetic industry.

The Mix-G are known as cereal β-glucans located in the subaleurone and endospermic cell wall and are unique to the Poales, the taxonomic order that includes cereal grasses (Ebringerová 2006; Ebringerova et al. 2005).

3.4 Polysaccharides with Amino Functions

3.4.1 Chitin and Chitosan

Chitin (CAS No 1398-61-4) is a major structural polysaccharide in the cell walls of fungi and the exoskeleton of insects and molluscs. Like cellulose, chitin is a homopolysaccharide composed of a linear (1 → 4)-β-linked chain of D-(*N*-acetyl)glucosamine. Chitin microfibrils provide strength to the cell walls and are embedded in a protein and polysaccharide matrix that is more or less associated with calcium phosphate and/or calcium carbonate. Three different crystalline forms of chitin exist in nature. The most abundant is α-chitin with antiparallel chain orientation, found commonly both in arthropods and fungi. Parallel chain orientation in β-chitin was observed, for example, in some protozoans, diatoms and squid (*Loligo* sp.). A hybrid form (γ-chitin) with two parallel and one antiparallel chain has been described (Jang et al. 2004). The antiparallel α-chitin sheets give strong hydrogen bonding within and between the sheets (inter- and intrasheet hydrogen bonds, Fig. 3.26a) (Pillai et al. 2009). In β-chitin intrasheet bonding is weak (Fig. 3.26b) and it has a lower chemical and thermal stability. The relatively rare γ-chitin polymorphic form has intermediate properties. Chitin has a typical degree of acetylation of 0.9

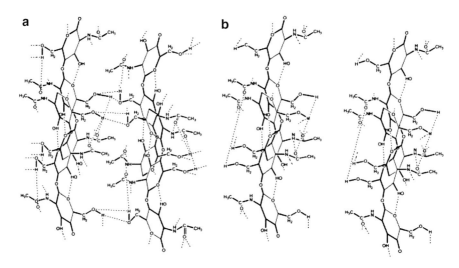

Fig. 3.26 Molecular structure and hydrogen bonding in (**a**) α-chitin and (**b**) β-chitin (Reproduced with permission of Elsevier, Pillai et al. 2009)

and is insoluble in water and difficult to extract without degradation. The degree of *N*-acetylation affects the ionic strength and physicochemical and biological properties. Chitin and its derivatives are used as wound dressing and antibacterial coating and medical implements.

→ 4)-β-D-GlcNAc*p*-(1 →

Chitin is commonly produced commercially from the shells of lobsters, crab or shrimp.

Chitosan (CAS No 9012-76-4) is the de-acetylated derivative of chitin and is soluble in dilute (acetic) acid and other solvents. Therefore, it finds wider industrial application than chitin as binder or film former. For example, in medical and cosmetic applications and in water filters, it binds metal ions, fats and proteins. In textile industries it is used to improve crease resistance.

3.4.2 Hyaluronan or Hyaluronic Acid

Hyaluronan or hyaluronic acid (CAS No 9067-32-7) is a linear polymer, which is composed of disaccharide units of *N*-acetyl-D-glucosamine (1 → 4) linked to D-glucuronic acid. It is found ubiquitous in animal connective tissues (cartilage) where it plays an important role in the control of mechanical wear and absorption of compression loads. In soft tissues, it controls hydration, plays a role in cell surface adhesion and lubricates synovial fluids and organs. Its applications are therefore mainly found in medical and pharmaceutical area or cosmetics. Hyaluronic acid is a low immunogenic substance and its derivatives can be used in orthopaedic surgery.

Hyaluronan can also be produced by bacterial fermentation, for example, with *Bacillus subtilis* (Widner et al. 2005). For commercial production, hyaluronic acid is extracted from rooster combs and umbilical cords. It is highly viscoelastic and commonly has a high polymerisation degree. Hyaluronic acid is obtained from synovial fluid and vitreous humour of the eye, umbilical tissue and cocks combs. The secondary structure of hyaluronan is a towfold helix structure with internal hydrogen bonds between the amide and carboxylate groups. The tertiary structure is composed of antiparallel stacked chains in a β-sheetlike structural arrangement with strong interchain hydrogen bonds (Scott and Heatley 1999).

3.4.3 Other Glycosaminoglycans

Other glycosaminoglycans (GAG) that are composed of linear disaccharide units similar to hyaluronan can be found in mammalian connective tissues. These GAGs or mucopolysaccharides are of lower polymerisation degree and commonly sulphated and often associated to protein moieties (proteoglycans). GAGs are classified into different groups: chondroitin, dermatan, keratin and heparin sulphates.

Chondroitin sulphate (CAS No 9007-28-7) has the similar basic disaccharide structure as hyaluronic acid, but instead of an *N*-acetylated glucosamine unit (D-GlcNAc) linked (1 → 4)-β to glucuronic acid, it contains a galactosamine (D-GalNAc) with sulphate substituents at *O*-4 or *O*-6.

Dermatan sulphate (CAS No 24967-94-0) also contains D-GalNAc and sulphate at *O*-4 but is alternating with L-iduronic acid units. Iduronic acid has a more flexible ring structure allowing for different structural conformations.

Keratan sulphate (CAS No 9056-36-4) is found in cartilage and differs from other GAGs, because it does not contain uronic acids. It is composed of (1 → 3)-β-linked disaccharide units with a linear repeating structure of galactose linked (1 → 4)-β to *N*-acetylglucosamine (poly-*N*-acetyl-lactosamine units), which are most often monosulphated at *O*-6 of GlcNAc.

In heparin and heparin sulphate are copolymers of the basic repeating structure D-GlcNAc*p*-α-(1 → 4)-D-GlcA*p*-β-(1 → 4) that show different modifications. During heparin and heparin sulphate biosynthesis, enzymatic sulphation and epimerisation reactions of D-GlcA into L-IdA occur. Typically D-GlcA is present with more or less (30–50 %) α-L-iduronic acid residues alternating with glucosamine sulphate units. Heparin (CAS No 9005-49-6) is produced by mast cells and is used for its anticoagulant and thrombolytic properties in clinical medicine (Esko et al. 2009).

3.4.4 Murein

The outer cell wall of many bacteria consists of a peptidoglycan murein. The carbohydrate part of murein consists of a linear polysaccharide of alternating *N*-acetylmuramic acid (MurNAc, CAS No 10597-89-4) linked (1 → 4)-β to *N*-acetylglucosamine. The polysaccharide consists of on average of 30 disaccharide units. The muramic acid residues are the 3-*O*-lactic acid ethers of *N*-acetylglucosamine which are interlinked by short peptide chains (4–5 residues). The terminal residues are 1,6-anhydromuramic acid residues. In this way a rigid structure is formed that protects bacteria from lysis (Vollmer and Bertsche 2008) (Fig. 3.27).

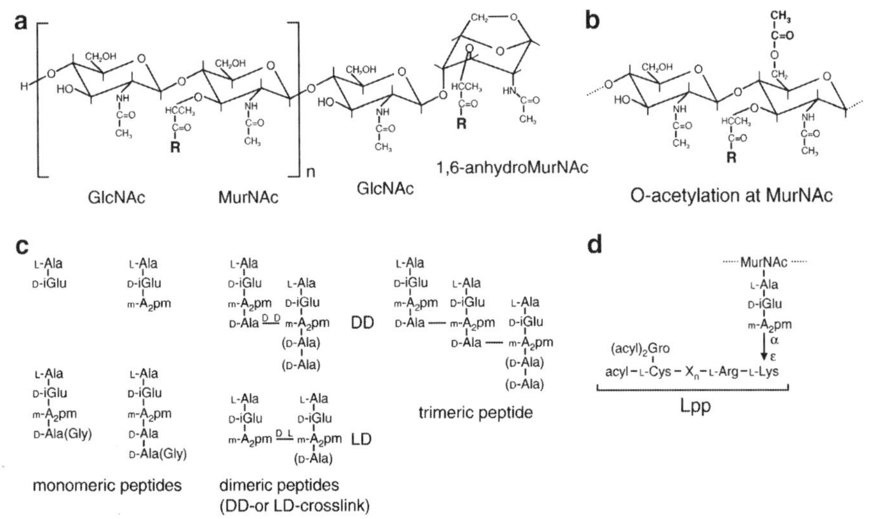

Fig. 3.27 Structures of the murein glycan strands with a terminal 1,6-anhydroMurNAc residue (**a**), the C6-*O*-acetyl modification (in *bold*) at MurNAc (**b**), the different peptide side chains and the DD- and LD-cross-links (**c**) and the attachment of Braun's lipoprotein (**d**) (Reproduced with permission of Elsevier, Vollmer and Bertsche 2008)

3.5 Polysaccharides with Acid Functions

3.5.1 Pectins

The pectic polysaccharides are a class of heteroglycans that are commonly found in the middle lamellae of plant cells but are also present in some plant gums or mucilages. Especially in the non-lignified tissues, pectin plays an important role in the cell adhesion and integrity of the plant structure. In fruits and vegetables, pectins are the prevailing polysaccharides. Pectins are involved in the plant cell development and undergo many chemical and physical changes during the process of maturation (Vincken et al. 2003).

Pectins are a complex group of high-molecular-weight polysaccharides that are widely used in food industries as gelling agent and thickener. The pectic polysaccharides all contain predominantly galacturonic acid residues (1 → 4)-α-D-Gal*p*A in the main chains with—depending on the plant source or method of extraction—different degree of methyl esterification, together with various amounts of incorporated L-rhamnose. Pectin may contain different regions that can be linear (homogalacturonan, smooth regions) or branched rhamnogalacturonans (RG) (ramified or hairy) regions, containing besides GalA the neutral sugars Rha, Gal and/or Ara) in the same molecule. There is a distinction between the two types of rhamnogalacturonan (RG-I and RG-II). In RG-I the D-GalA residues are alternating with L-Rha, while in RG-II the proportion of L-Rha in the backbone of the hairy regions is higher (see below).

Because of its physical properties, pectin is used as a gelling and thickening agent or stabiliser and emulsifier in food products. The solubility of pectins in water depends on the molecular weight and ionic strength. The degree of methoxylation of pectins influences its gelation properties. Highly methoxylated pectins will form gels at low pH (<3.5) in the presence of sucrose. Pectins with low methyl ester content form only gels in the presence of Ca^{2+}. The pectic acid chains were shown to form chelation complexes with an "egg box structure" including the calcium atoms (Fig. 3.28a, b).

Commercial sources of pectin are mainly found in by-products of industrial food production processes from fruits (apple pomace, citrus peels). These pectins have a high degree of methylation and low content of hairy regions. Other pectic residues such as potato fibre—a major residue of potato starch production—are containing too high amounts of branched RG-I structures and have a too low methyl ester content to make them a suitable gelling agent. Sugar beet pulp contains also substantial amounts (20–25 %) of pectin that can be extracted but has less suitable properties for use as gelling and thickening agent. The molecular weight of sugar beet pectin is much lower and the galacturonic acids are partly acetylated. The side chains contain relatively large amounts of neutral sugars, mostly arabinose, and are ester-linked to phenolic acids (ferulic acid). These properties prevent the complex formation under acidic conditions with Ca^{2+}.

In the plant cell wall, pectins play an important role in the control of cell wall growth and defence against microorganisms. The changing texture of fruits and vegetables due to ripening can be ascribed to the changing structure of pectins induced by enzymatic conversion and the formation of methyl esters in pectin.

In food industry, pectins are widely used as stabilisers in dairy products, canned foods, juices and jams. The conformation of pectins in relation to gelling properties has been reviewed (Rees 1970; Voragen et al. 2003). The role of pectins as dietary fibre has received much attention in nutritional science and other claimed pharmaceutical and medical roles (cholesterol, antitumour, etc.) (Willats et al. 2005).

Structural variations have been reported such as the insertion of longer or shorter chains of (1 → 2)-β or -α-linked L-Rha in the main chain (e.g. α in mucilage, β in seeds).

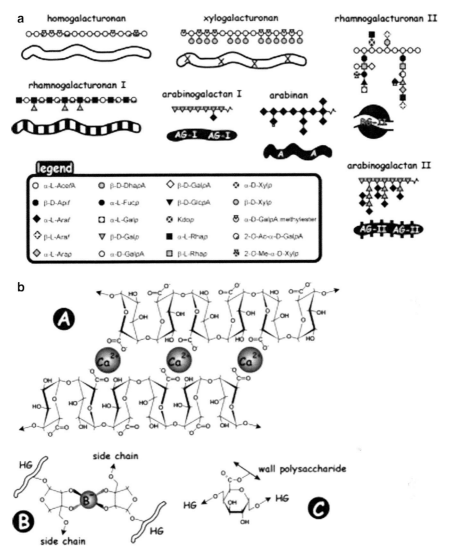

Fig. 3.28 (a) Schematic representative structures of the constituent polysaccharides of pectin. (b) Pectin molecules can be cross-linked in different ways [Reproduced from Vincken et al. (2003)]

3.5.1.1 Homogalacturonan (HG)

Polygalacturonan is a major constituent of pectins. Pure homogalacturonan is however not widely spread. In sun flower heads (*Helianthus annuus*), the presence of linear (1 → 4)-α-D-GalpA was reported with low methoxyl content (Sharma et al 2006). At least a small proportion of L-Rha residues is also present. By mild acid treatment of some pectins, practically pure galacturonan can be obtained due to the relative acid sensitivity of the Rha bond. Some of the uronic acids were found to be amidated.

→4)-α-D-GalpA-(1 → 4)-α-D-GalpA (1 → 4)-α-D-GalpA (1 → 4)-α-D-GalpA (1→

3.5.1.2 Rhamnogalacturonan Type I (RG-I) (CAS No 143795-00-0)

Most common pectins have the RG-I type structures, with alternating chains of galacturonic acid and rhamnose disaccharide units with large variation in the substitution patterns and length of side chains composed of most commonly arabinogalactan, arabinan or galactan. The substitution occurs at *O*-4 of the L-Rha residues and sometimes also at *O*-3 mainly with short (1 → 4)-β-D-Gal*p* and/or (1 → 5)-α-L-Ara*f* chains.

the pectins of some aqueous plant species (*Lemna*, *Potamogeton*), branched sugar residues (e.g. D-apiose) are found.

```
→4)-α-D-GalpA-(1 → 2)-α-L-Rhap(1 → 4)-α-D-GalpA (1 → 2)-α-L-Rhap(1→
                     4                                3
                     1                                1
          β-D-Galp(1 → 4)-β-D-Gal              β-D-Gal
```

RG-I is present in many vegetables and fruits such as cabbage, onion, potato, tomato and kiwi.

3.5.1.3 Rhamnogalacturonan Type II (RG-II)

The RG-II type of pectin resembles RG-I in its sugar composition but often contains other sugar residues. RG-II contains stretches of homogalacturonan together with highly branched (hairy) regions containing longer and shorter chains of arabinose and galactose inserted at L-Rha. The exact structure of RG-II is still under discussion and may be different for different sources, but the basic structural elements seem to be present and highly conserved in plant kingdom.

3.5.1.4 Xylogalacturonan and Other Pectic Structures

In some sources of pectins, also other substituents are found such as β-D-GlcA in soybean or β-D-Xyl in lemon peel. Xylogalacturonan is a homogalacturonan (HG) structure with attached xylose side chains that was found in watermelon, pea hulls and as minor fraction in apple pectin. Exotic residues such as apiose, aceric acid (AceA) or DHA (3-C-carboxy-5-deoxy-L-xylose, Dha*p*A) and KDO (3-deoxy-D-manno-octulosonic acid) may be present.

Pectins of commercial interest are derived from apple and citrus fruits.

Mucilages and exudate gums of different plant species can be of pectic nature (see karya gum and tragacanth). For example, the mucilages of *Abelmoschus manihot*, *Althaea officinalis*, paniculata (*Hydrangea paniculata*), mahogany (*Khaya senegalensis*) and mangle gum (*Rhizophora mangle*) are pectins with more or less unique structural features and branching (e.g. L-Fuc and L-Gal). In

Many other plants contain pectic polysaccharides in leaves, fruits and residues from food production, with more or less known structural features but with promising uses as gelling agents like okra (*Hibiscus esculentus*), cress (*Lepidium sativum*), sugar beet pulp, sisal leaves (*Agave sisalana*), linseed (*Linum usitatissimum*), tomato (*Solanum lycopersicum*), rapeseed (*Brassica* sp.), tobacco leaves (*Nicotiana tabacum*), papaya (*Carica papaya*), mango peels (*Mangifera indica*), coffee and cocoa and onion skins.

Gum tragacanth (CAS No 9000-65-1) is exudate gum of *Astragalus* sp. (Iran, Turkey). It is a highly branched polysaccharide and composed of two components, one acidic and one neutral: tragacanthic acid or bassorin (swelling in water) and tragacanthin (dissolving in water). The gum forms a stable gel at pH 3.0–8.5 and is relative resistant to acid hydrolysis. Tragacanthic acid is the major constituent of the gum (60–70 %) and contains a main chain of (1 → 4) D-GalA, with D-Xyl branches linked at *O*-3, which can be substituted at *O*-2 with L-Fuc or D-Gal. The neutral component tragacanthin (30–40 %) consists of highly branched structure of with a core of (1 → 6)-D-Gal and side chains of (1 → 2)/(1 → 3) and 1 → 5) L-Ara residues. Tragacanth is used as emulsifier and thickening agent in food and pharmaceutical industries.

Gum karya or karaya is an acidic polysaccharide gum with extremely high molecular weight (16×10^6) that is obtained commercially as exudate from *Sterculia* sp., especially *S. urens*, *S. caudate* and *S. striata*. The gum has emulsifying and thickening properties. The polysaccharide is of the rhamnogalacturonan type containing L-Rha (15–30 %), D-Gal (13–26 %) and D-GalA (40 %) residues that are partially acetylated (8 % *O*-acetyl). The rhamnogalacturonan backbone is

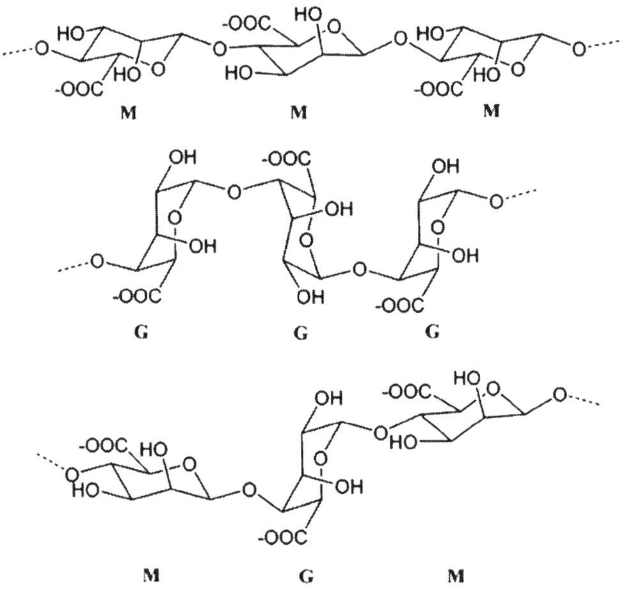

Fig. 3.29 Different alginate structures (Reproduced with permission of Bu et al. (2005). Copyright 2005 American Chemical Society)

branched with β-D-Gal at *O*-2 of L-Rha and with β-D-GlcA linked at the *O*-3 of GalA. The polymer contains both hydrophilic acidic regions and hydrophobic regions (acetylated neutral sugars). Karya gum swells in cold water but does not readily dissolve and is instable in alkali. After deacetylation in alkali, the gum can be solubilised in water. The gum is produced in India (Africa) (*S. striata* originates from Brazil).

Karaya gum is used as a laxative and denture adhesive because of its swelling capacity in water.

3.5.2 Alginate

Alginate (CAS No 9005-34-9) is extracted from brown seaweed species such as *Laminaria* sp., *Macrocystis* sp. (kelp), *Ecklonia* sp. and *Ascophyllum nodosum*. Alginate is a term used to describe a group of acidic polysaccharides and their salts and derived products. The basic structure of alginate (algin) consists of linear polyuronic acid chains containing β-D-mannuronic acid and α-L-guluronic acid in different proportions, depending on the species and the part of the plant. In some algae the mannuronic and guluronic acids are found alternating (poly MG), while in other sources homopolymeric sequences of both uronic acids occur (poly G and poly M). The ratio ManA:GulA in seaweeds is rather constant per species but may vary between 0.45 and 1.85 among different sources. The presence of mono- or divalent cations influences the three-dimensional helical structure and flexibility of the polymer. The presence of Ca^{2+} promotes the formation of hydrogels and the interchain binding of the G-blocks in the well-known egg box model of alginate gelation (Fig. 3.29) (Stokke et al. 1991). Alginates are used in films in food and pharmaceutical applications but also as adhesives and in (textile) printing, paper coating and different medical and dental uses. The gel is used as wound dressing but also in food as coating to prevent dehydration and as ion exchange medium.

The propylene glycol ester is produced from reaction of propylene oxide with alginate that finds large-scale use as thickening agent in ice cream fabrication.

Alginate is extracted from brown algae and is also produced by several bacteria. *Pseudomonas aeruginosa* and *Azotobacter vinelandii* produce exopolysaccharides that are of very similar structure as the seaweed alginates. The *Pseudomonas* alginate shows random distribution of

ManA and GulA residues (poly MG), while the alginate produced by *Azotobacter*—like many seaweeds—shows block copolymer features (poly M and poly G). The ratio ManA:GulA is higher than reported for seaweeds and may rise up to 4.0 in some *Pseudomonas* strains. An important difference between algal and bacterial alginate is that the latter are often partially *O*-acetylated at *O*-2 and/or *O*-3 of the ManA units.

Other algae and seaweeds may produce very different types of polysaccharides that are often sulphated.

3.5.3 Agar-Agar

Agar-agar (CAS No 9002-18-0) derived from seaweed species such as *Gelidium* and *Gracilaria*. Agar contains agarose (50–90 %) and agaropectin (10–50 %) fractions. Agarose is the commercially interesting fraction that is composed of a linear chain of $(1 \rightarrow 3)$-linked β-D-Gal$p$-sulphate, Ca^{2+} salt, alternating with $(1 \rightarrow 4)$-3,6-anhydro-α-L-Gal units. The agaropectin is also containing 4,6-*O*-(1-carboxyethylidene)-D-Gal$p$ units and is substituted with sulphates or methyl groups in variable amounts. Agar swells in cold water and dissolves at 95 °C, forming stiff gels at low concentrations after cooling to 38 °C. The molecular cross-linking involved in reversible sol-gel transitions and molecular conformation changes and gelling behaviour; spinodal demixing have been subject of detailed studies (Manno et al. 1999; Boral et al. 2008). The gelation mechanism of agar has been ascribed to the formation of double-helix networks in which each chain forms a left-handed threefold helix (Arnott et al. 1974). Agar is well known for its use as plating gel medium for microbial culture growth (Armisén and Galatas 2009).

3.5.4 Fucoidan or Fucogalactan

Fucoidan (CAS No 9072-19-9) or fucogalactan is isolated commercially from brown seaweeds of the genus *Undaria* but has been reported to occur in many different seaweed species. The fucan polysaccharide backbone most often contains mono- or disulphated $(1 \rightarrow 3)$-α-L-Fuc$p$ residues and $(1 \rightarrow 6)$-linked 3-*O*-sulphated D-Gal$p$ residues. It is currently investigated in pharmaceutical industry for its therapeutic value in treatment of a number of (viral) diseases. Fucoidan is obtained from other species of the *Laminaria* genus. Fucoidan is isolated from *L. saccharina*, *L. digitata* and *Phaeophyta* sp. that is high in its content of 4-*O*-sulphated $(1 \rightarrow 3)$-α-L-Fuc$p$ similar to what is obtained from *Fucus vesiculosus*. Other variations of the linkages have been reported as well. The sulphated fucan found in *Phaeophyta* sp. also may contain some Gal residues. These types of complex polysaccharides are widely investigated for their antimicrobial activity and positive effects on food digestion.

3.5.5 Carrageenan

Carrageenan (CAS No 9000-07-1) is a group of sulphated high-molecular-weight polysaccharides that are obtained from red seaweeds such as *Chondrus crispus* and *Gigartina stellata*. Carrageenans are containing sulphated units of alternating $(1 \rightarrow 4)$-α- and $(1 \rightarrow 3)$-β-D-Gal units (in μ, ν, λ and ξ carrageenan) and also varying amounts of 3,6-anhydro-D-Gal units (in κ, ι and θ-carrageenan). Carrageenans are polyelectrolytes and classified according to their degree of sulphation and presence of ring closure (anhydro form). The linear chains are sulphated at position *O*-2 and/or *O*-4 in $(1 \rightarrow 3)$-linked residues, while in $(1 \rightarrow 4)$ residues the *O*-2 and *O*-6 positions carry a sulphate group. Because of its self-assembly organisation, it is widely used in food industry as thickener, emulsifier and stabiliser. Its film-forming properties are used in pharmaceutical and cosmetic industries. The gelation process upon cooling of a carrageenan solution involves coil-helix transitions and aggregation of the double helices into rigid rods or network structures with flexible superstrands or microfibres. The gel formation properties are depending on the presence and concentration of cation type (e.g. Na^+, K^+ and/or Ca^{2+}) (Hermansson et al.

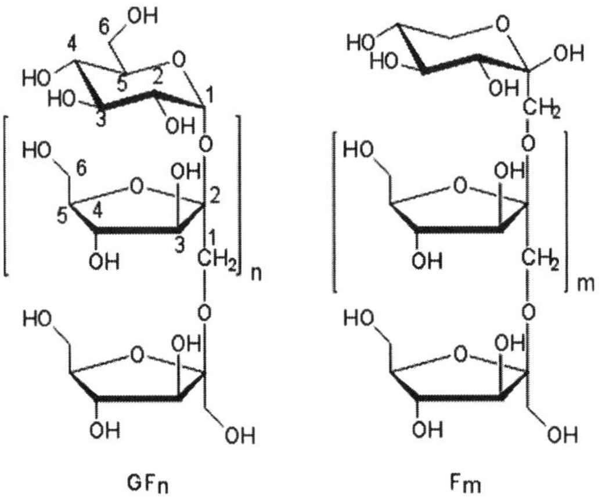

Fig. 3.30 Chemical structure of inulin, GF_n fructan molecule consisting of n fructofuranosyl units and containing terminal glucose, F_m fructanosyl-only fructan molecule with a DP of m

1991). The helix conformation is stabilised by inter- and intramolecular hydrogen bonds.

λ-Carrageenan is composed of 2,6-disulphated (1 → 4)-α linked D-Gal residues and 2-sulphated (1 → 3)-β-D-Gal units.

In κ-carrageenan, the 3,6-anhydro-form of (1 → 3)-α-D-Gal is present and the β-D-Gal is sulphated at O-3.

In ι-carrageenan also, the (1 → 3)-α-D-Gal 3,6-anhydro-unit is found but sulphated at O-2 and the β-D-Gal is sulphated at O-3.

Furcellaran (CAS No 9000-21-9) or Danish agar is also a sulphated galactan with some branching extracted from *Furcellaria fastigiata* and *F. lumbricalis*. Furcellaran structure is reported to be similar to κ-carrageenan, but with less sulphate groups (one per four sugar units). It is also composed of 1,4-linked 3,6 anhydro-D-Gal (1 → 3)-D-Gal units that are more or less sulphated at O-4. Minor amounts of methylation at position O-3 of D-Gal was reported (Yang et al. 2011).

3.6 Miscellaneous

3.6.1 Inulin

Inulin (CAS No 9005-80-5) is a polydisperse carbohydrate mainly composed of D-fructose (Fru) units, which are linked by (1 → 2)-β glycosidic bonds (Fig. 3.30) (Franck and De Leenheer 2002; Watherhouse and Chatterton 1993). A starting glucose unit can be present but is not necessary. Native inulin is a sucrose disaccharide (α-D-Glcp-(1 → 2)-β-D-Fruf) extended with a chain of additional (1 → 2)-β-linked Fru units. But when referring to the definition of inulin, both sucrose-containing polysaccharide (GF_n) and fructofuranosyl-only fructan molecule (F_m) are considered to be included under this nomenclature (Franck and De Leenheer 2002).

Inulin is widely found in plants as storage carbohydrate. Next to starch, it is the most widely spread nonstructural polysaccharide with relative low degree of polymerisation (DP 20-60). It occurs in crops such as Jerusalem artichoke (*Helianthus tuberosus*), chicory (*Cichorium intybus*), *Dahlia* sp., dandelion (*Taraxacum officinale*), garlic and onions (*Allium sativum* and *A. cepa*), leek, asparagus, banana, Jicama (*Pachyrhizus erosus*), yam (*Dioscorea* sp.) and *Agave* sp. (Franck and De Leenheer 2002). Inulin (extracted from chicory) finds application as fat replacing ingredient in processed foods. Inulin is one of the low-calorie dietary fibres that is considered to favour the development of microflora in the colon. It is used in clinical diagnosis of kidney infections.

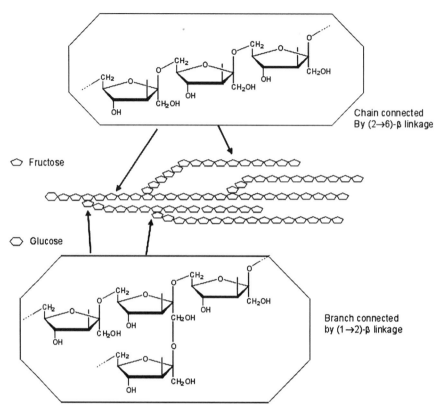

Fig. 3.31 Structure of levan. The main chain is linked by $(2 \rightarrow 6)$-β glycosidic bonds and the branch is linked by $(2 \rightarrow 1)$-β bonds, the branch then continues with $(2 \rightarrow 6)$-β linkages [Adopted from Rhee et al. (2002)]

3.6.2 Levan

Levan (CAS No 9013-95-0) is a polyfructan with a backbone of $(2 \rightarrow 6)$-linked β-D-Fru*p* as 6-kestose with extensive branches through $(2 \rightarrow 1)$-β linkages to the main chain (Fig. 3.31). It has a higher DP (100–200) than inulin. Levan is found in the expelled juice of grasses (*Dactylis glomerata*, *Poa secunda* and *Agropyron cristatum*), wheat, barley and fungi (e.g. *Aspergillus sydowii*) and is also produced by various microorganisms (Rhee et al. 2002).

The spheroidal molecular shape of levan indicates that the constituent chains are extended radially at the same growth rate (Newburn et al. 1971). Bacterial levans have typically a molecular weight in the range of $2 \times 10^6 - 10^8$ g/mol (Keith et al. 1991; Rhee et al. 2002).

3.6.3 Xanthan gum

Xanthan gum (CAS No 11138-66-2) is obtained by aerobic fermentation of *Xanthomonas campestri* cultures. The polysaccharide is composed of pentasaccharide units of $(1 \rightarrow 4)$-β-D-Glc*p* backbone with trisaccharide side chains on *O*-3 of alternate Glc units. These side chains consist of β-D-Man*p*-$(1 \rightarrow 4)$-β-D-Glc*p*A-$(1 \rightarrow 2)$-β-D-Glc*p*-$(1 \rightarrow 3)$ with pyruvate and acetyl groups attached. It is produced commercially on large scale because of its unique rheological properties. Xanthan gum is a common emulsifier, lubricant and thickener used in food industries. Because of its strong shear thinning behaviour, xanthan gum is also used in oil drilling.

The hydrocolloid is forming antiparallel double helical strands with an extended chain conformation.

References

Adinolfi M, Corsaro MM, Lanzetta R, Parrilli M, Folkard G, Grant W, Sutherland J (1994) Composition of the coagulant polysaccharide fraction from *Strychnos potatorum* seeds. Carbohydr Res 263:103

Alsop RM, Byrne GA, Done JN, Earl IE, Gibbs R (1977) Quality assurance in clinical dextran manufacture by molecular-weight characterization. Process Biochem 12:15–35

Altaner C, Knox JP, Jarvis MC (2007) In situ detection of cell wall polysaccharides in sitka spruce (*Picea sitchensis* (Bong. carrière) wood tissue. BioResources 2: 284–295

Antonini E, Bellelli L, Bruzzesi MR, Caputo A, Chiancone E, Rossi-Fanelli A (1964) Studies on dextran and dextran derivatives. I. Properties of native dextran in different solvents. Biopolymers 2:27–34

Armisén R, Galatas F (2009) Agar. In: Philips GO, Williman PA (eds) Handbook of hydrocolloids. Woodhead Publishing Ltd., Cambridge. ISBN 978-1-84569-414-2

Arnott S, Fulmer A, Scottl WE, Dea CM, Moorhouse R, Rees DA (1974) The agarose double helix and its function in agarose gel structure. J Mol Biol 90: 269–272

Atalla RH, VanderHart DL (1984) Native cellulose: a composite of two distinct crystalline forms. Science 223:283–285

Bathgate GN, Manners DJ (1966) Multiple branching in glycogens. Biochem J 101:3c–5c

Bender H, Wallenfels K (1961) Investigations on pullulan. II. Specific degradation by means of a bacterial enzyme. Biochem Z 334:79–95

Bender H, Lehmann J, Wallenfels K (1959) Pullulan, an extracellular glucan from Pullularia pullulans. Biochim Biophys Acta 36:309–316

Bender H, Siebert R, Stadler-Szöke A (1982) Can cyclodextrin glycosyltransferase be useful for the investigation of the fine structure of amylopectins? Characterisation of highly branched clusters isolated from digests with potato and maize starches. Carbohydr Res 110:245–259

Bertoft E (2004a) On the nature of categories of chains in amylopectin and their connection to the super helix model. Carbohydr Polym 57:211–224

Bertoft E (2004b) Lintnerisation of two amylose-free starches of A- and B-crystalline types, respectively. Starch-Starke 56:167–180

Bertoft E (2007a) Composition of building blocks in clusters from potato amylopectin. Carbohydr Polym 70:123–136

Bertoft E (2007b) Composition of clusters and their arrangement in potato amylopectin. Carbohydr Polym 70:433–446

Bertoft E, Zhu Q, Andtfolk H, Jungner M (1999) Structural heterogeneity in waxy-rice starch. Carbohydr Polym 38:349–359

Blows JMH, Calder PC, Geddes R, Wills PR (1988) The structure of placental glycogen. Placenta 9:493–500

Bluhm TL, Sarko A (1977) The triple helical structure of lentinan, a linear β-(1 $\to$ 3)-D-glucan. Can J Chem 55: 293–299

Bluhm TL, Deslands Y, Marchessault RH, Perz S, Rinaudo M (1982) Solid-state and solution conformations of scleroglucan. Carbohydr Res 100:117–130

Boral S, Saxena A, Bohidar HB (2008) Universal growth of microdomains and gelation transition in agar hydrogels. J Phys Chem 112:3625–3632

Born K, Langendorff V, Boulenguer P (2002) Xanthan. In: Vandamme E, De Baets S, Steinbüchel A (eds) Biopolymers: biology, chemistry, biotechnology, applications, vol 6, Polysaccharide I. Wiley-VCH, Weinheim, p 259

Bouveng HO, Kiessling H, Lindberg B, McKay J (1963) Polysaccharides elaborated by Pullularia pullulans. II. The partial acid hydrolysis of the neutral glucan synthesized from sucrose solutions. Acta Chem Scand 17:797–800

Bovey FA (1959) Enzymatic polymerization. I. Molecular weight and branching during the formation of dextran. J Polym Sci 35:167–182

Bu H, Kjøniksen A-L, Knudsen KD, Nyström B (2005) Effects of surfactant and temperature on rheological and structural properties of semidilute aqueous solutions of unmodified and hydrophobically modified alginate. Langmuir 21:10923–10930

Calder PC (1991) Glycogen structure and biogenesis. Int J Biochem 23(12):1335–1352

Calder PC, Geddes R (1985) The proteoglucan nature of mammalian muscle glycogen. Glycoconj J 2:365–373

Calder PC, Geddes R (1986) Digestion of the protein associated with muscle and liver glycogens. Carbohydr Res 148:173–177

Catley BJ, Ramsay A, Servis C (1986) Observations on the structure of the fungal extracellular polysaccharide, pullulan. Carbohydr Res 153:79–86

Chaubey M, Kapoor VP (2001) Structure of a galactomannan from the seeds of *Cassia angustifolia* Vahl. Carbohydr Res 332:439–444

Chuah CT, Sarko A, Deslandes Y, Marchessault RH (1983) Packing analysis of carbohydrates and polysaccharides. Part 14. Triple helical crystalline structure of curdlan and paramylon hydrates. Macromolecules 16: 1375–1382

Cleemput G, van Oort M, Hessing M, Bergmans MEF, Gruppen H, Grobet PJ, Delcour JA (1995) Variation in the degree of D-xylose substitution in arabinoxylans extracted from a European wheat flour. J Cereal Sci 22:73–84

Corsaro MM, Giudicianni I, Lanzetta R, Marciano CE, Monaco P, Parrilli M (1995) Polysaccharides from seeds of *Strychnos species*. Phytochemistry 39: 1377–1380

Deslandes Y, Marchessault RH, Sarko A (1980) Triple-helical structure of (1 $\to$ 3)-β-D-glucan. Macromolecules 13: 1466–1471

Dols M, Remaud-Simeon M, Willemot RM, Vignon M, Monsan PF (1997) Characterization of dextransucrases from *Leuconostoc mesenteroides* NRRL B-1299. Appl Biochem Biotechnol 62:47

Ebringerová A (2006) Structural diversity and application potential of hemicelluloses. Macromol Symp 232:1–12

Ebringerová A, Heinze T (2000) Xylan and xylan derivatives - biopolymers with valuable properties, 1. Naturally occurring xylans: structures, isolation, procedure and properties. Rapid Commun 21:542–556

Ebringerová A, Hromádková Z (1999) Xylans of industrial and biomedical importance. In: Harding SE (ed) Biotechnology and genetic engineering reviews, vol 16. Intercept, England, p 325

Ebringerová A, Hromádková Z, Alföldi J, Hříbalová V (1998) The immunologically active xylan from ultrasound-treated corn cobs: extractability, structure and properties. Carbohydr Polym 37:231–239

Ebringerová A, Kardšová A, Hromádková Z, Hříbalová V (2003) Mitogenic and comitogenic activities of polysaccharides from some European herbaceous plants. Fitoterapia 74:52–61

Ebringerova A, Hromadkova Z, Heinze T (2005) Hemicellulose, In: Polysaccharides I, Structure, characterization and use. Adv Polym Sci 186:1–67

Edgar KJ (2009) Polysaccharide chemistry: frontiers and challenges. In: Polysaccharide materials: performance by design. ACS Symp Ser 1017:3–12

Esko JD, Kimata K, Lindahl U (2009) Proteoglycans and sulfated glycosaminoglycans. In: Varki A et al (eds) Essentials of glycobiology. Cold Spring Harbor Lab Press, Cold Spring Harbor, NY, http://www.ncbi.nlm.nih.gov/books/NBK1908/?amp=&part=ch16

Evtuguin DV, Tomás JL, Silva AMS, Neto CP (2003) Characterization of an acetylated heteroxylan from Eucalyptus globulus Labill. Carbohydr Res 338:597–604

Fischer MH, Yu N, Gray GR, Ralph JR, Anderson L, Marlett JA (2004) The gel-forming polysaccharide of psyllium husk (Plantago ovata Forsk). Carbohydr Res 339:2009–2017

Franck A, De Leenheer L (2002) Polysaccharides. II. Polysaccharides from eukaryotes. In: Vandamme EJ, De Baets S, Steinbüchel A (eds) Biopolymers, vol 6. Wiley, Weinheim, pp 439–479

French D (1972) Fine structure of starch and its relationship to the organization of starch granules. J Jpn Soc Starch Sci 19:8–25

Fry SC (1989) The structure and functions of xyloglucan. J Exp Bot 40:1–11

Fujii N, Shinohara S, Ueno H, Imada K (1984) Polysaccharide produced by *Aureobasidium* sp. (black yeast). Kenkyu Hokuku-Miyazaki Daigaku Nogakubu, vol 31, pp 253–262

Fulton WS, Atkins EDT (1980) The gelling mechanism and relationship to molecular structure of microbial polysaccharide curdlan. In: French AD, Gardner KH (eds) Fibre diffraction methods. American Chemical Society, Washington, DC, pp 385–410

Gallagher JT, Lyon M, Steward WP (1986) Structure and function of heparan sulphate proteoglycans. Biochem J 236:313–325

Gallant DJ, Bouchet B, Buléon A, Pérez S (1992) Physical characteristics of starch granules and susceptibility enzymatic degradation. Eur J Clin Nutr 46:3–16

Ganter JLMS, Heyraud A, Petkowicz CLOM, Rinaudo M, Reicher F (1995) Galactomannans from Brazilian seeds: characterization of the oligosaccharides produced by mild acid hydrolysis. Int J Biol Macromol 17:13–19

Gardiner ES, Sarko A (1985) Packing analysis of carbohydrates and polysaccharides. 16. The crystal structures of celluloses IV_I and IV_{II}. Can J Chem 63:173–180

Gardner KH, Blackwell J (1974) The structure of native cellulose. Biopolymers 13:1975–2001

Gérard C, Planchot V, Colonna P, Bertoft E (2000) Relationship between branching density and crystalline structure of A- and B-type maize mutant starches. Carbohydr Res 326:130–144

Giavasis I, Harvey LM, McNeil B (2002) Scleroglucan. In: Vandamme E, De Baets S, Steinbüchel A (eds) Biopolymers: biology, chemistry, biotechnology, applications, vol 6, Polysaccharide II. Wiley-VCH, Weinheim, p 37

Gibbs PA, Seviour RJ (1996) Pullulan. In: Dimitiu S (ed) Polysaccharides in medicinal applications. Dekker, New York, pp 59–86

Gilmore KS, Russell RR, Ferretti JJ (1990) Analysis of the Streptococcus downei gtfS gene, which specifies a glucosyltransferase that synthesizes soluble glucans. Infect Immun 58:2452

Goldsmith E, Sprang S, Fletterick R (1982) Structure of maltoheptaose by difference Fourier methods and a model for glycogen. J Mol Biol 156:411–427

Guizard C, Chanzy H, Sarko A (1984) Molecular and crystal structure of dextrans: a combined electron and X-ray diffraction study. 1. The anhydrous high-temperature polymorph. Macromolecules 17:100–107

Gunja-Smith Z, Marshall JJ, Mercier C, Smith EE, Whelan WJ (1970) A revision of the Meyer-Bernfeld model of glycogen and amylopectin. FEBS Lett 12:101–104

Gunja-Smith Z, Marshall JJ, Smith EE (1971) Enzymatic determination of the unit chain length of glycogen and related polysaccharides. FEBS Lett 13:309–311

Hanashiro I, Abe J-I, Hizukuri S (1996) A periodic distribution of chain length of amylopectin as revealed by high-performance anion-exchange chromatography. Carbohydr Res 283:151–159

Hare MD, Svensson S, Walker GJ (1978) Characterization of the extracellular, water-insoluble α-D-glucans of oral streptococci by methylation analysis, and by enzymatic synthesis and degradation. Carbohydr Res 66:245–264

Harris PJ, Henry RJ, Blakeney AB, Stone BA (1984) An improved procedure for the methylation analysis of oligosaccharides and polysaccharides. Carbohydr Res 127:59–73

Hayashi A, Kinoshita K, Miyake Y (1981) The conformation of amylose in solution. Polym J 13:537–541

Heidrich C, Vollmer W (2002) Murein (peptidoglycan). In: Vandamme E, De Baets S, Steinbüchel A (eds) Biopolymers: biology, chemistry, biotechnology, applications, vol 6, Polysaccharide I. Wiley-VCH, Weinheim, p 431

Heinze T, Liebert T, Koschella A (2006a) Esterification of polysaccharides. Springer, Heidelberg, p 5

Heinze T, Liebert T, Heublein B, Hornig S (2006b) Functional polymers based on dextran. Adv Polym Sci 205:199–291

Hermansson A-M, Eriksson E, Jordansson E (1991) Effects of potassium, sodium and calcium on the microstructure and rheological behaviour of the kappa-carrageenan gels. Carbohydr Polym 16:297–320

Hizukuri S (1985) Relationship between the distribution of the chain length of amylopectin and the crystalline structure of starch granules. Carbohydr Res 141:295–306

Hizukuri S (1986) Polymodal distribution of the chain lengths of amylopectins, and its significance. Carbohydr Res 147:342–347

Hoffman M, Jia Z, Pena MJ, Cash M, Harper A, Blackburn AR II, Darvill A, York WS (2005) Structural analysis of xyloglucans in the primary cell walls of plants in the subclass Asteridae. Carbohydr Res 340:1826–1840

Hon DN-S (1996) Functional polymers: a new dimensional creativity in lignocellulosic chemistry. In: Hon DN-S (ed) Chemical modification of lignocellulosic materials. Dekker, New York, pp 1–10

Hoover R (2001) Composition, molecular structure, and physicochemical properties of tuber and root starches: a review. Carbohydr Polym 45:253–267

Hromádková Z, Kovačiková J, Ebringerová A (1999) Study of the classical and ultrasound-assisted extraction of the corn cob xylan. Ind Crop Prod 9:101–109

Huynh R, Chaubet F, Jozefonvicz J (1998) Carboxymethylation of dextran in aqueous alcohol as the first step of the preparation of derivatized dextrans. Angew Makromol Chem 254:61–65

Illingworth B, Lamer J, Cori GT (1952) Structure of glycogens and amylopectins. I. Enzymatic determination of chain length. J Biol Chem 199:631–640

Ioan CE, Aberle T, Burchard W (2000) Structure properties of dextran. 2. Dilute solution. Macromolecules 33:5730–5739

Ishii T (1997) Structure and functions of feruloylated polysaccharides. Plant Sci 127:111–127

Ishrud O, Zahid M, Zhou H, Pan Y (2001) A water-soluble galactomannan from the seeds of *Phoenix dactylifera L.*. Carbohydr Res 335:297–301

Jaffe MJ, Teleweski FW, Cooke PW (1984) Thigmomorphogenesis: on the mechanical properties of mechanically perturbed bean plants. Physiol Plant 62:73–78

Jane J-L, Kasemsuwan T, Leas S, Zobel H, Robyt JF (1994) Anthology of starch granule morphology by scanning electron microscopy. Starch-Starke 46:121–129

Jang M-K, Kong B-G, Jeong Y-I, Lee CH, Nah J-W (2004) Physicochemical characterization of α-chitin, β-chitin and γ-chitin separated from natural resources. J Polym Sci 42:3423–3432

Jeanes A, Haynes WC, Wilham CA, Rankin JC, Melvin EH, Austin MJ, Cluskey JE, Fisher BE, Tsuchiya HM, Rist CE (1954) Characterization and classification of dextrans from ninety-six strains of bacteria. J Am Chem Soc 76:5041–5052

Jenkins PJ, Donald AM (1995) The influence of amylose on starch granule structure. Int J Biol Macromol 17:315–321

Jonas R, Farah LF (1998) Production and application of microbial cellulose. Polym Degrad Stabil 59:101–106

Kamide K, Okajima K, Kowsaka K, Matsui T (1985) CP/MASS ^{13}C NMR spectra of cellulose solids: an explanation by the intramolecular hydrogen bond concept. Polym J 17:701–706

Kapoor VP, Chanzy H, Taravel FR (1995) X-ray diffraction studies on some seed galactomannans from India. Carbohydr Polym 27:229–233

Kapoor VP, Taravel FR, Joseleau J-P, Milas M, Chanzy H, Rinaudo M (1998) Cassia spectabilis DC seed galactomannan: structural, crystallographical and rheological studies. Carbohydr Res 306:231–241

Kasai N, Harada T (1980) Ultrastructure of curdlan. In: French AD, Gardner KH (eds) Fiber diffraction methods, vol 141. ACS Symposium, Washington, DC, pp 363–383

Katz JR (1928) In: Walton RP (ed) A comprehensive survey of starch chemistry. Reinhold, New York, p 68

Keith K, Wiley B, Ball D, Arcidiacono S, Zorfass D, Mayer J, Kaplan D (1991) Continuous culture system for production of biopolymer levan using *Erwinia herbicola*. Biotechnol Bioeng 38:557–560

Kjolberg O, Manners DJ, Wright A (1963) α-1,4-Glucosans. XVII. The molecular structure of some glycogens. Comp Biochem Physiol 8:353–365

Klemm D, Schmauder H-P, Heinze T (2002) Cellulose. In: Vandamme E, De Baets S, Steinbüchel A (eds) Biopolymers: biology, chemistry, biotechnology, applications, vol 6, Polysaccharide II. Wiley-VCH, Weinheim, p 275

Kondo T (1997) The relationship between intramolecular hydrogen bonds and certain physical properties of regioselectively substituted cellulose derivatives. J Polym Sci B Polym Phys 35:717–723

Kondo T (2005) Hydrogen bonds in cellulose and cellulose derivatives. In: Dumitriu S (ed) Polysaccharides: structural diversity and functional versatility. Dekker, New York, pp 69–98

Kong X, Corke H, Bertoft E (2009) Fine structure characterization of amylopectins from grain amaranth starch. Carbohydr Res 344:1701–1708

Koreeda A, Harada T, Ogawa K, Sato S, Kasai N (1974) Study of the ultrastructure of gel-forming $(1 \rightarrow 3)$-β-D-glucan (curdlan type polysaccharide) by electron microscopy. Carbohydr Res 33:396–399

Krässig HA (1993) Cellulose - structure, accessibility, and reactivity. Gordon & Breach, Amsterdam

Langan P, Nishiyama Y, Chanzy H (1999) A revised structure and hydrogen bonding scheme in cellulose II from a neutron fibre diffraction analysis. J Am Chem Soc 121:9940–9946

Laohaphatanaleart K, Piyachomkwan K, Sriroth K, Bertoft E (2010) The fine structure of cassava amylopectin. Part 1: Organization of clusters. Int J Biol Macromol 47:317–324

Larm O, Lindberg B, Svensson S (1971) Studies on the length of the side chains of the dextran elaborated by Leuconostoc mesenteroides NRRL B-512. Carbohydr Res 20:39–48

Leathers TD (2003) Biotechnological production and applications of pullulan. Appl Microbiol Biotechnol 62:468–473

Lee EYC, Whelan WJ (1966) Enzymic methods for the microdetermination of glycogen and amylopectin, and their unit-chain lengths. Arch Biochem Biophys 116:162–167

Liang CY, Marchessault RH (1959) Infrared spectra of crystalline polysaccharides. I. Hydrogen bonds in native celluloses. J Polym Sci 37:385–395

Lindman B, Karlström G, Stigsson L (2010) On the mechanism of dissolution of cellulose. J Mol Liq 156:76–81

Lundqvist J, Teleman A, Junel L, Zacchi G, Dahlman O, Tjerneld F, Stålbrand H (2002) Isolation and characterization of galactoglucomannan from spruce (*Picea abies*). Carbohydr Polym 48:29–39

Malmström A, Aberg L (1982) Biosynthesis of dermatan sulphate. Assay and properties of the uronosyl C-5 epimerase. Biochem J 201:489–493

Manners DJ (1957) The molecular structure of glycogens. Adv Carbohydr Chem Biochem 12:261–298

Manners DJ (1962) Enzymic synthesis and degradation of starch and glycogen. Adv Carbohydr Chem Biochem 17:371–430

Manners DJ, Wright A (1962) α-1,4-D-Glucosans Part XIII. Determination of the average chain length of glycogens by α-amylolysism. J Chem Soc, 1597–1602

Manno M, Emanuelle A, Martorana V, Bulone D, San Biagio PL, Palma-Vittorelli MB, Palma MU (1999) Multiple interactions between molecular and supramolecular ordering. Phys Rev E59:2222–2230

Marchessault RH, Deslandes Y (1979) Fine structure of (1 → 3)-β-D-glucans: curdlan and paramylon. Carbohydr Res 75:231–242

McIntosh M, Stone BA, Stanisich VA (2005) Curdlan and other bacterial (1 → 3)-β-D-glucans. Appl Microbiol Biotechnol 68:163–173

Melendez-Hevia E, Waddell TG, Shelton ED (1993) Optimization of molecular design in the evolution of metabolism: the glycogen molecule. Biochem J 295:477–483

Mitchell AJ (1988) Second derivative F.t.-i.r. spectra of celluloses I and II and related mono- and oligosaccharides. Carbohydr Res 173:185–195

Miyoshi K, Uezu K, Sakurai K, Shinkai S (2004) Proposal of a new hydrogen-bonding form to maintain curdlan triple. Chem Biodivers 1:916–924

Naessens M, Cerdobbel A, Soetaert W, Vandamme EJ (2005) Leuconostoc dextransucrase and dextran: production, properties and applications. J Chem Technol Biotechnol 80:845–860

Nakata M, Kawaguchi T, Kodama Y, Konno A (1998) Characterization of curdlan in aqueous sodium hydroxide. Polymer 39:1475–1481

Navarini L, Gilli R, Gombac V, Abatangelo A, Bosco M, Toffanin R (1999) Polysaccharides from hot water extracts of roasted Coffea arabica beans: isolation and characterization. Carbohydr Polym 40:71–81

Newburn E, Lacy R, Christie TM (1971) The morphology and size of extracellular polysaccharide from oral streptococci. Arch Oral Biol 16:863–872

Nikuni Z (1978) Studies on starch granules. Starch-Starke 30:105–111

Nilsson M, Saulnier L, Andersson R, Åman PM (1996) Water unextractable polysaccharides from three milling fractions of rye grain. Carbohydr Polym 30:229–237

Nilsson M, Andersson R, Andersson RE, Autio K, Åman PM (2000) Heterogeneity in a water-extractable rye arabinoxylan with a low degree of disubstitution. Carbohydr Polym 41:397–405

Nishiyama Y, Langan P, Chanzy H (2002) Crystal structure and hydrogen-bonding system in cellulose Iβ from synchrotron X-ray and neutron fiber diffraction. J Am Chem Soc 124:9074–9082

Okada K, Yoneyama M, Mandai T, Aga H, Sakai S, Ichikawa T (1990) Digestion and fermentation of pullulan. Nippon Eiyo Shokoryo Gakkaishi 43:23–29

Okuyama K, Otsubo A, Fukuzawa Y, Ozawa M, Harada T, Kasai N (1991) Single-helical structure of native curdlan and its aggregation state. J Carbohydr Chem 10:645–656

Peat S, Whelan WJ, Thomas GJ (1952) Evidence of multiple branching in waxy maize starch. J Chem Soc Chem Commun, 4546–4548

Pérez S, Bertoft E (2010) The molecular structures of starch components and their contribution to the architecture of starch granules: a comprehensive review. Starch-Stärke 62:389–420

Petrov PT, Shingel KI, Scripko AD, Tsarenkov VM (2002) Biosynthesis of pullulan by Aureobasidium pullulans strain BMP-97. Biotekhnologiya 1:36–48

Pillai CKS, Paul W, Sharma CP (2009) Chitin and chitosan polymers: chemistry, solubility and fibre formation. Prog Biopolym Sci 34:641–678

Powell DA (1979) Structure, solution properties and biological interactions of some extracellular polysaccharides. In: Berkeley RCW, Gooday GW, Ellwood DC (eds) Microbial polysaccharides and polysacchareses. Academic, London, pp 117–160

Prehn P (2002) Hyaluronan. In: Vandamme E, De Baets S, Steinbüchel A (eds) Biopolymers: biology, chemistry, biotechnology, applications, vol 6, Polysaccharide I. Wiley-VCH, Weinheim, p 379

Ralet M-C, Bonnin E, Thibault J-F (2002) Pectines. In: Vandamme E, De Baets S, Steinbüchel A (eds) Biopolymers: biology, chemistry, biotechnology, applications, vol 6, Polysaccharide II. Wiley-VCH, Weinheim, p 345

Rao MVSSTS, Muralikrishna G (2001) Non-starch polysaccharides and bound phenolic acids from native and malted finger millet (Ragi, Eleusine coracana, Indaf -15). Food Chem 72:187–192

Rau U (2002) Schizophyllan. In: Vandamme E, De Baets S, Steinbüchel A (eds) Biopolymers: biology, chemistry, biotechnology, applications, vol 6, Polysaccharide II. Wiley-VCH, Weinheim, p 61

Rau U, Müller R-J, Cordes K, Klein J (1990) Process and molecular data of branched 1,3-D-glucans in comparison with Xanthan. Bioprocess Eng 5:89–93

Rees DA (1970) Structure, conformation, and mechanism in the formation of polysaccharide gels and networks. Adv Carbohydr Chem Biochem 24:267–332

Reid JSG, Edwards ME (1995) Galactomannans and other cell wall storage polysaccharides in seeds. In: Stephen AM (ed) Food polysaccharides and their applications. Dekker, New York, pp 155–186

Rhee S-K, Song K-B, Kim C-H, Park B-S, Jang E-K, Jang K-H (2002) Levan. In: Vandamme E, De Baets S, Steinbüchel A (eds) Biopolymers: biology, chemistry, biotechnology, applications, vol 6, Polysaccharide I. Wiley-VCH, Weinheim, p 351

Rioux L-E, Turgeon S, Baeulieu M (2010) Structural characterization of laminaran and galactofucan extracted from the brown seaweed Saccharina longicruris. Phytochemistry 71:1586–1595

Roberts GAF (1992) Chitin chemistry. Macmillan, London, p 185

Rodén L (1968) The protein-carbohydrate linkages of acid mucopolysaccharides. In: Quintarelli G (ed) Chemical physiology of mucopolysaccharides, vol 1968. J & A Churchill Ltd., London, pp 17–32

Rodgers NE (1973) Scleroglucan. In: Whistler RL, BeMiller JN (eds) Industrial gums, 2nd edn. Academic, New York, pp 499–511

Roukas T, Montzouridpu F (2001) Effect of aeration rate on pullulan production and fermentation broth rheological properties in an airlift reactor. J Chem Technol Biot 76:371–376

Sabra W, Deckewer W-D (2005) Alginate – a polysaccharide of industrial interest and diverse biological functions. In: Dumitriu S (ed) Polysaccharides: structural diversity and functional versatility. Dekker, New York, p 515

Sakurai K, Uezu K, Numata M, Hasegawa T, Li C, Kaneko K, Shinkai S (2005) β-1,3-Glucan polysaccharides as novel one-dimensional hosts for DNA/RNA, conjugated polymers and nanoparticles. Chem Commun, 35:4383–4398

Samuelsen AB, Lund I, Djahromi JM, Paulsen BS, Wold JK, Knutsen SH (1999a) Structural features and anti-complementary activity of some heteroxylan polysaccharide fractions from the seeds of *Plantago major L.*. Carbohydr Polym 38:133–143

Samuelsen AB, Cohen EH, Paulsen BS, Brull LP, Thomas-Oates JE (1999b) Structural studies of a heteroxylan from *Plantago major L.* seeds by partial hydrolysis, HPAEC-PAD, methylation and GC–MS, ESMS and ESMS/MS. Carbohydr Res 315:312–318

Saulnier L, Marot C, Chanliaud E, Thibault J-F (1995) Cell wall polysaccharide interactions in maize bran. Carbohydr Polym 26:279–287

Schooneveld-Bergmans MEF, Hopman AMCP, Beldman G, Voragen AGJ (1998) Extraction and partial characterization of feruloylated glucuronoarabinoxylans from wheat bran. Carbohydr Polym 35:39–47

Schooneveld-Bergmans MEF, Beldman G, Voragen AGJ (1999) Structural features of (glucurono)arabinoxylans extracted from wheat bran by barium hydroxide. J Cereal Sci 29:63–75

Scott JE, Heatley F (1999) Hyaluronan forms specific stable tertiary structures in aqueous solution: a ^{13}C NMR study. PNAS 96:4850–4855

Senti FR, Hellmann NN, Ludwig NH, Babcock GE, Tobin R, Glass CA, Lamberts BL (1955) Viscosity, sedimentation, and light-scattering properties of fraction of an acid-hydrolyzed dextran. J Polym Sci 17:527–546

Seymour FR, Slodki ME, Plattner RD, Jeanes A (1977) Six unusual dextrans: methylation structural analysis by combined g.l.c.—m.s. of per-O-acetyl-aldononitriles. Carbohydr Res 53:153–166

Seymour FR, Chen ECM, Bishop SH (1979) Methylation structural analysis of unusual dextrans by combined gas-liquid chromatography-mass spectrometry. Carbohydr Res 68:113

Sharma BR, Naresh L, Dhuldhoya NC, Merchant SU, Merchant UC (2006) An overview on Pectins. Times Food Process J 4:44–51

Shimamura A, Tsumori H, Mukasa H (1982) Purification and properties of *Streptococcus mutans* extracellular glucosyltransferase. Biochim Biophys Acta 702:72

Shingel KI (2004) Current knowledge on biosynthesis, biological activity, and chemical modification of the exopolysaccharide pullulan. Carbohydr Res 339: 447–460

Shogren RL (1998) Starch: properties and materials applications. In: Kaplan DL (ed) Biopolymers from renewable resources. Springer, Berlin, pp 30–46

Singh V, Srivastava V, Pandey M, Esthi R, Sanghi R (2003) Ipomoea turpethum seeds: a potential source of commercial gum. Carbohydr Polym 51:357–359

Singh RS, Saini GK, Kennedy JF (2008) Pullulan: microbial sources, production and applications. Carbohydr Polym 73:515–531

Slodki ME, England RE, Plattner RD, Dick WE (1986) Methylation analyses of NRRL dextrans by capillary gas-liquid chromatography. Carbohydr Res 156: 199–206

Sowa W, Blackwood AC, Adams GA (1963) Neutral extracellular glucan of *Pullularia pullulans* (de Bary) Berkhout. Can J Chem 41:2314–2319

Srichuwong S, Sunarti TC, Mishima T, Isono N, Hisamatsu M (2005) Starches from different botanical sources I:

contribution of amylopectin fine structure to thermal properties and enzymes digestibility. Carbohydr Polym 60:529–538

Srivastava M, Kapoor VP (2005) Seed galactomannans: an overview. Chem Biodivers 2:295–317

Stephen AM (1983) Other plant polysaccharides. In: Aspinall GO (ed) The polysaccharides, vol 2. Academic, New York, pp 97–193

Stokke BT, Smidsrød O, Bruheim P, Sjåk-Bræk G (1991) Distribution of uronate residues in alginate chains in relation to alginate gelling properties. Macomolecules 24:4637–4640

Takeo K, Tokumura A, Kuge T (1973) Complexes of starch and its related materials with organic compounds. X. X-ray diffraction of amylose-fatty acid complexes. Starch/Stärke 35:357–362

Tashiro K, Kobayashi M (1991) Theoretical evaluation of three-dimensional elastic constants of native and regenerated celluloses: role of hydrogen bonds. Polymer 32:1516–1526

Taylor C, Cheetham NWH, Walker GJ (1985) Application of high-performance liquid chromatography to a study of branching in dextrans. Carbohydr Res 137:1–12

Teleman A, Tenkanen M, Jacobs A, Dahlman O (2002) Characterization of O-acetyl-(4-O-methylglucurono) xylan isolated from birch and beech. Carbohydr Res 337:373–377

Telewski FW (2006) A unified hypothesis of mechanoperception in plants. Am J Bot 93:1466–1476

Tester RF, Karkalas J, Qi X (2004) Starch-composition, fine structure and architecture. J Cereal Chem 39: 151–165

Timell TE (1986) Compression wood in gymnosperms, Springer series in wood science. Springer, Berlin, 2150 p

VanCleve JW, Schaefer WC, Rist CE (1956) The Structure of NRRL B-512 Dextran. Methylation Studies. J AmChem Soc 78:4435

Van de Velde F, De Ruiter GA (2002) Carrageenan. In: Vandamme E, De Baets S, Steinbüchel A (eds) Biopolymers: biology, chemistry, biotechnology, applications, vol 6, Polysaccharide II. Wiley-VCH, Weinheim, p 245

Vandamme EJ, De Baets S, Vanbaelen A, Joris K, De Wulf P (1998) Improved production of bacterial cellulose and its application potential. Polym Degrad Stabil 59:93–99

Vignon MR, Gey C (1998) Isolation, ^{1}H and ^{13}C NMR studies of (4-O-methyl-image-glucurono)-image-xylans from luffa fruit fibres, jute bast fibres and mucilage of quince tree seeds. Carbohydr Res 307: 107–111

Vincken J-P, Schols HA, Oomen RJFJ, McCann MC, Ulvsko P, Voragen AGJ, Visser RGF (2003) If homogalacturonan were a side chain of rhamnogalacturonan I. Implications for cell wall architecture. Plant Physiol 132:1781–1789

Vinkx CJA, Stevens I, Gruppen H, Grobet PJ, Delcour JA (1995) Physico-chemical and functional properties of rye nonstarch polysaccharides. VI. Variability in the structure of water-unextractable arabinoxylans. Cereal Chem 72:411–418

Vollmer W, Bertsche U (2008) Murein (peptidoglycan) structure, architecture and biosynthesis in Escherichia coli. Biochim Biophys Acta 1778:1714–1734

Voragen AGJ, Schols HA, Visser R (eds) (2003) Advances in pectin and pectinase research. Kluwer, Dordrecht

Wada M, Heux L, Isogai A, Nishiyama Y, Chanzy H, Sugiyama J (2001) Improved structural data of cellulose III prepared in supercritical ammonia. Macromolecules 34:1237–1243

Wallenfels K, Keilich G, Bechtler G, Freudenberger D (1965) Investigations on pullulan. IV. Resolution of structural problems using physical, chemical and enzymatic methods. Biochem Z 341:433–450

Wang Y, McNeil B (1996) Scleroglucan. Crit Rev Biotechnol 16:185–215

Watherhouse AL, Chatterton NJ (1993) Glossary of fructan terms. In: Suzuki M, Chatterton NJ (eds) Science and technology of fructans. CRC, Boca Raton, FL, pp 2–7

Widner B, Behr R, Von Dollen S, Tang M, Heu T, Sloma A, Sternberg D, DeAngelis PL, Weiggel PH, Brown S (2005) Hyaluronic acid production in Bacillus subtilis. Appl Environ Microbiol 71:3747–3752

Wiley BJ, Ball DH, Arcidiacono SM, Sousa S, Mayer JM, Kaplan DL (1993) Control of molecular weight distribution of the biopolymer pullulan produced by Aureobasidium pullulans. J Environ Polym Degrad 1:3–9

Willats WGT, Knox JP, Mikkelsen JD (2005) Pectin: new insights into an old polymer are starting to gel. Trends Food Sci Techol 17:97–104

Wood PJ, Weisz J, Blackwell BA (1994) Structural studies of $(1 \rightarrow 3)(1 \rightarrow 4)$-β-D-glucans by ^{13}C-NMR and by rapid analysis of cellulose-like regions using high-performance anion-exchange chromatography of oligosaccharides released by lichenase. Cereal Chem 71: 301–307

Yanaki T, Norisuye T (1983) Triple helix and random coil scleroglucan in dilute solution. Polym J 15:389–396

Yanaki T, Kojima T, Norisuye T (1981) Triple helix of scleroglucan in dilute aqueous sodium hydroxide. Polym J 13:1135–1143

Yang B, Yu G, Zhao X, Ren W, Jiao G, Fangg L, Wang Y, Du G, Tiller C, Girouard G, Barrow CJ, Ewart HS, Zhang J (2011) Structural characterization and bioactivities of hybrid carrageenan-like sulphated galactan from red alga Furcellaria lumbricalis. Food Chem 124: 50–57

Zhang H, Yoshimura M, Nishinari K, Williams MAK, Foster TJ, Norton IT (2001) Gelation behaviour of konjac glucomannan with different molecular weights. Biopolymers 59:38–50

Zobel HF (1988) Starch crystal transformation and their industrial importance. Starch-Starke 40:1–7

Zugenmaier P (2001) Conformation and packing of various crystalline cellulose fibers. Prog Polym Sci 26:1341–1417

Chemical Characterization of Polysaccharides

Axel Rußler, Anna Bogolitsyna, Gerhard Zuckerstätter, Antje Potthast, and Thomas Rosenau

Contents

4.1	**General Part**	65
4.1.1	Introduction to the Chemical Analysis of Polysaccharides	65
4.1.2	Extraction and Purification	67
4.1.3	Monomers and Building Blocks	67
4.1.4	Substituents	68
4.1.5	Side Chains, Linkage Pattern, and Anomeric Configuration	69
4.1.6	Alien Groups and End Groups	70
4.1.7	DP and DP Distribution	70
4.1.8	Crystallinity	71
4.2	**Work Performed at BOKU**	71
4.2.1	The CCOA and FDAM Method for Profiling of Oxidized Groups	71
4.2.2	Analysis of Trace Chromophores in Cellulosic Materials	75
4.2.3	Capillary Electrophoresis for the Analysis of Carbohydrates	78
4.2.4	Assessment of Cellulose Morphology by Solid-State ^{13}C CP–MAS NMR	80
References		82

Abstract

The chemical characterization of polysaccharides is an absolute request for a multitude of scientific and industrial applications that go beyond the simple use of polysaccharides where the physical characterization and the knowledge of usage-dependent behavior by specific tests are sufficient. Successful process optimization and development are today only possible by knowing and controlling the details of the molecular basis. Hence, chemical analysis of polysaccharides covers a broad range of chemical problems and structural hierarchies within the molecules. This leads to a diversity of methods necessary for a complete chemical characterization of a polysaccharide sample. This chapter reviews the main methods that can be used for performing a detailed chemical characterization of polysaccharides.

4.1 General Part

4.1.1 Introduction to the Chemical Analysis of Polysaccharides

The chemical characterization of polysaccharides is an absolute request for a multitude of scientific and industrial applications that go beyond the simple use of polysaccharides where the physical characterization and the

T. Rosenau (✉)
Department of Chemistry, Wood, Pulp and Fiber Chemistry, University of Natural Resources and Life Sciences, Muthgasse 18, 1190 Vienna, Austria
e-mail: thomas.rosenau@boku.ac.at

knowledge of usage-dependent behavior by specific tests are sufficient. Successful process optimization and development are today only possible by knowing and controlling the details of the molecular basis. Since 1922, when Hermann Staudinger discovered the true nature of polymers, a large number of analytical tools have been developed in order to reveal the chemical composition of polymers in all its aspects and structural levels.

In fact, many aspects of the chemical characterization of polysaccharides are much wider and complex than found for conventional synthetic polymers and even for other biopolymers like peptides and nucleic acids. Monosaccharides, the basic modules, exhibit a polyfunctionality that allows multiple branching and substitution possibilities that lead to innumerable structural possibilities. In addition, the specific chemical nature of carbohydrates allows different ring sizes, such as furanosidic or pyranosidic ring structures, and different anomeric configurations, such as α- and β-glycosidic linkages.

The analysis addresses several aspects. Some complex carbohydrates involved, for example, in immunological recognition between cells and in other biological functional polysaccharides need to have their detailed structure identified while other questions arise in the analysis of bulk polysaccharides and carbohydrates used, for instance, in plants mostly as structural ingredients or for energy storage or material and construction purposes, such as cellulose, hemicelluloses, and starch and their derivatives. For these substances, the primary structure is well known, but changes due to isolation and processing are increasing analytical difficulties. Modern polysaccharide derivatives can display a high degree of complexity.

The degree of polymerization (DP) is a very important structural feature for these polysaccharides, but the DP distribution (molecular weight distribution, MWD) gives a lot of additional useful information. Supermolecular characteristics, such as crystallinity and polydispersity, are also very important. When substituents are present, not only their distribution between different OH-groups is of importance,

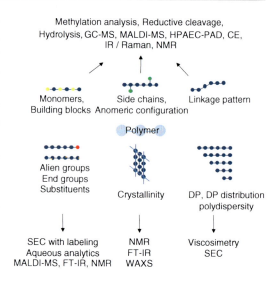

Fig. 4.1 Approaches to analyze different aspects of the chemical and physical nature of polysaccharides

but also their distribution pattern along and among polysaccharide chains plays an important role. These patterns can be controlled to certain extent by the reaction conditions. More and more, the analysis of alien groups, such as carbonyl and carboxyl groups present sometimes just in the ppb range, gains importance for the understanding of production processes and changes during storage as they have a huge influence on the structural integrity and the chemical behavior, out of proportion considering their very small amounts.

Especially for industrial applications, a good understanding of structure–property relationships is of increasing importance as the trend for more comprehensible, science-based design of material characteristics has largely replaced a more step-by-step optimization approach, and an improved reproducibility and product control is an economic as well as a regulatory necessity (Mischnick et al. 2010).

Hence, chemical analysis of polysaccharides covers a broad range of chemical problems and structural hierarchies within the molecules. This leads to a diversity of methods necessary for a complete chemical characterization of a polysaccharide sample. Figure 4.1 gives an overview on different approaches for the analysis of different structural features in polysaccharides.

This chapter will review the main methods that can be used for performing a detailed chemical characterization of polysaccharides.

2010; Sun et al. 2010; Yang et al. 2010; Nie and Xie 2011).

4.1.2 Extraction and Purification

Polysaccharides in nature are with some rare exceptions, such as cellulose in cotton, kapok, and some other seed hairs, always accompanied with many other substances. The purification of polysaccharides is therefore a crucial, complex, and laborious task, including - dependent on the substrate - grinding, extraction, treatment with enzymes, delignification, bleaching, and other processing steps. Many specific protocols can be found in literature. The determination of purity is thus the first step in the chemical analysis of an unknown polysaccharide after purification. To determine the content of impurities, it is necessary to analyze the total sugar content and if necessary the contents of the impurities, mostly uronic acids, proteins, lignin, ash, and moisture.

For the total sugar content, spectroscopic methods are generally used where the monomers, after a hydrolysis step, are reacted with another compound, a label, to form a light-absorbing molecule detectable in a UV–Vis spectrometer. Examples are the Sumner, the phenol–sulfuric acid, and the anthrone–sulfuric acid assays. For specific substrates, a number of enzyme-based assays have been developed.

Purification can be necessary on another level when several polysaccharides have to be separated from each other. This is always a very complex problem for which there is no general or universal procedure. Nevertheless, there exist, for example, protocols to fractionate the hemicellulosic and cellulosic part of wood after extraction of the holocellulose to more or less well-defined fractions. This is done by utilizing differences in solubility in different solvents or specific precipitation reactions, e.g., by use of bivalent ions, such as barium. However, polysaccharides isolated via such precipitation and dissolving steps can never be absolutely pure (Walter 1998; Chen et al. 2008; Willför et al. 2008; Allard and Derenne 2009; Ren and Sun

4.1.3 Monomers and Building Blocks

The analysis of the monomeric sugar units after a total hydrolysis can already offer a lot of information. It reveals the basic composition of a sample independent of its purity, either showing the nature of impurities (e.g., if pulps are analyzed according to their accompanying hemicelluloses) or identifying the monomeric composition (e.g., if a purified hemicelluloses sample is analyzed).

As already mentioned, the first step for an analysis of the monomers is to perform a total hydrolysis. This hydrolysis can be obtained with the use of strong acids: hydrochloric acid, trifluoroacetic acid, and sulfuric acid being most commonly used. Depending on different substrates, various hydrolysis protocols have been developed. Substrates that are easily degraded such as flour samples can be treated in one step with a lower acid concentration, whereas other materials such as wood pulp samples need a two-step protocol beginning with a more concentrated acid at ambient temperatures and a second step with a diluted acid at elevated temperatures in order to achieve a total hydrolysis (Albersheim et al. 1967; De Ruiter et al. 1992; Chandra et al. 2009; Katzenellenbogen et al. 2009; Rana et al. 2009; TAPPI 2009a, b; Willför et al. 2009; Melander and Tømmeraas 2010; Ruiz-Matute et al. 2010).

However, some new methods, for instance, acid hydrolysis with the help of microwaves, have been investigated and show promising results as they offer very mild hydrolysis conditions combined with short reaction times (Singh et al. 2006; Allard and Derenne 2009; Mazzarino et al. 2010).

Not only acids are used for the fragmentation of polysaccharides. Also enzymes are used either alone (Cohen et al. 2004; Mazumder et al. 2005; Willför et al. 2009; Yang et al. 2010) or in combination with a preliminary partial acidic hydrolysis (Yang et al. 2010). By this method, according to the conditions used, monomers as well as larger building units can be isolated.

Crucial points in hydrolyzing polysaccharides are the completeness of polymer degradation on the one hand and the prevention of losses by acidic degradation on the other hand. Degradationcan be monitored by time-depending kinetics where the sugar content should be continuously rising during hydrolysis and no degradation products should be present. For the verification of the absence of polymers and oligomers, a simple TLC check (thin layer chromatography) is usually sufficient. Enzymatic hydrolysis introduces protein residues into the substrate and requires normally longer reaction times but on the other hand avoids degradation of the carbohydrate monomers.

The hydrolysate obtained this way can then be further modified to suit a chromatographic analysis with GC, HPAEC, or CE.

For a GC separation, the nonvolatile monosaccharides have to be converted into volatile compounds. This is done either by acetylation or silylation or preliminary reduction of the monosaccharides to the corresponding alditols.

This procedure is not bound to very special or expensive equipment. The mixtures of peracetylated or persilylated alditols can be used in a simple GC separation. Only very small sample amounts are necessary, and the detection is very sensitive, even if no MS detection but just the frequently available FID detection is used.

A disadvantage of the GC method is the laborious sample preparation with the risk of sample losses and the difficulty of discriminations of molecules if hydrolyzation, reduction, or derivatization is incomplete.

Less preparative work needs to be done if HPAEC (high-performance anion exchange chromatography) is used for monosaccharide separation. In this case the neutralized hydrolysate can be used directly for injection into the chromatographic system. For a long time, this method was hampered by insufficiently sensitive detectors such as refractive index detectors. In order to use more sensitive detectors (UV or fluorescence), a pre- or post-column derivatization with all its disadvantages had to be established. With the upcoming of PAD (pulsed amperometric detection), a sufficiently sensitive analysis system is now available for unmodified monosaccharides (Black and Fox 1996; Dahlman et al. 2000; Schwikal et al. 2005; Sanz and Martínez-Castro 2007; Willför et al. 2009; Ruiz-Matute et al. 2010).

4.1.4 Substituents

The degree of substitution (DS) is a value that is relatively simple to obtain for stable side groups by the use of either elemental analysis or NMR measurements, while pyrolysis-GC can be applied in some cases. If the derivatives are labile, a previous chemical stabilization or secondary derivatization is necessary to apply these routine techniques (Alexandru and Rogovin 1953; Dunbrant 1965; Klemm et al. 1998; Vaca-Garcia et al. 2001; Sato et al. 2002; Rußler et al. 2005; Russler et al. 2005a, b).

NMR measurements allow in most cases distinguishing between derivatization of different OH-groups within the monomer building block, whereas the discrimination between un-, mono-, and multi-substituted monomers is mostly not possible. To overcome this restriction, an analysis of the monomeric units is necessary. Analysis techniques well established in this respect are the methylation analysis and the reductive cleavage (Lee and Gray 1995; Kiwitt-Haschemie et al. 1996; Mukerjea et al. 1996; Gohdes et al. 1997; Gohdes and Mischnick 1998; Klemm et al. 1998; Yu and Gray 1998a, b; Capitani et al. 2000; Mischnick et al. 2000; Senso et al. 2000; Dicke et al. 2001; Laine 2002; Bedouet et al. 2003; Thomas et al. 2003; Liebert et al. 2005; Rußler et al. 2005; Russler et al. 2005a; Petzold et al. 2006; Chandra et al. 2009; Rana et al. 2009; Yang et al. 2010).

However, with regard to substituents along a polysaccharide backbone, not only the absolute number and position within the monomer is of importance for understanding the properties of the polysaccharide but also the substituent distribution along the cellulose chain. This distribution can take many forms, a random pattern as well as regular or blocklike distributions or mixed forms.

To distinguish between these different substitution patterns, an enzymatic degradation of the samples is mostly used, since the hydrolysis of the polysaccharide backbone is hindered by substitutions. It therefore generates a mixture of different fragments that can be analyzed by a combination of chromatographic and spectroscopic methods, mostly GC or LC or MALDI hyphenated to MS, but also HPLC and CE methods are known (Erler et al. 1992; Steeneken and Woortman 1994; Arisz et al. 1995; Wilke and Mischnick 1995; Gohdes et al. 1998; Heinze et al. 1999, 2000; Horner et al. 1999; Saake et al. 2000; Senso et al. 2000; van der Burgt et al. 2000a, b, c; Levigne et al. 2002; Lee et al. 2003; Thomas et al. 2003; Oudhoff et al. 2004; Petzold et al. 2006; Adden et al. 2009; Mischnick et al. 2010). This allows to approximate the substitution pattern.

4.1.5 Side Chains, Linkage Pattern, and Anomeric Configuration

The elucidation of side chains, linkage pattern, and anomeric configurations is a difficult task that has not a critical relevance for most bulk polysaccharides and their derivatives but is most important for the structure identification as in the case of complex exopolysaccharides for intercellular recognition. The linkage pattern and structures of side chains are mostly very complex and can only be investigated after a previous fragmentation step. The techniques used for this fragmentation vary and so do the techniques for the subsequent analysis of the fragments. For the analysis of the monomer composition, chemical and enzymatic hydrolysis are used, but in order to get useful material from this step, the conditions have to be chosen in a way that the fragmentation leads not only to monomers but leaves the branching points intact and allows their identification (Heinrich 1999a, b; Mischnick 2001; Mischnick et al. 2001, 2010; Cui 2005).

The standard method to get first structural information is the methylation analysis where the polysaccharide is in a first step methylated to convert all free OH-groups into methoxyls. For this methylation reaction, several methods exists; most modern techniques however use solid sodium hydroxide as the base and methyl iodide as the methylation reagent. Subsequently, the methylated polysaccharide is hydrolyzed. The use of trifluoroacetic acid (TFA) for this purpose is most convenient as no neutralization step is necessary, but the acid can be evaporated from the hydrolysate. As the next step, the mixture of partly methylated sugars is reduced to alditols by the use of sodium borohydride or, to facilitate the distinction between C1 and C6, by the use of sodium borodeuteride in alkaline medium. An acetylation step with acetic anhydride is necessary to transfer the material into partially methylated alditol acetates (PMAA). The PMAAs are then subjected to GC-MS to get the spectroscopic information. However, this method cannot provide any information on the anomeric configuration, the sequence, and cannot distinguish between 4-O-linked aldopyranoses from 5-O-linked aldofuranoses (Ciucanu and Kerek 1984; Gohdes et al. 1997; van der Burgt et al. 1998, 1999, 2000a, b; Heinrich 1999a, b; Mischnick and Adden 2008).

The reductive cleavage analysis starting from the methylated polysaccharide provides more structural information as it does not convert the sugar units into PMAA but into partly methylated anhydroalditols with the anomeric position deoxygenated that can be subsequently acetylated to lead to different products depending on whether furanosides or pyranosides are reacted. For this reduction and simultaneous cleavage step, triethylsilane and TMS-O-mesylate/BF3-etherate are usually used (Mischnick 1991, 1997, 2002; Mischnick et al. 1994, 2010; Kiwitt-Haschemie et al. 1996; Osborn et al. 1999; Mischnick and Hennig 2001; Tüting et al. 2004b; Adden and Mischnick 2005; Nie and Xie 2011).

Nevertheless this method does not provide the information necessary for a full sequencing of a complex polysaccharide. Methods that go further are based on selective techniques that specifically degrade the polymer to oligomers by liberating repeating units or building blocks if present and leave branching points intact. The mixtures obtained can then be fractionated and subjected to further characterization.

The methods used in this context are based on three different principles with several modifications, so that they will only be mentioned shortly. The first principle is based on an acidic hydrolysis that is modified in a way to lead only to partial degradation due to different stability of glycosidic linkages within the molecule. Variants of this type are acetolysis and methanolysis as well as controlled acid hydrolysis.

The second principle is that of a selective degradation via an oxidation of the polysaccharide. The Smith degradation, a combination of periodate oxidation, reduction, and mild acidic hydrolysis, is such a method that is used in different variants. The chromium trioxide oxidation is another possibility for acetylated polysaccharides.

Specific enzymatic degradation protocols are the third way to obtain a meaningful fragmentation. In these methods, enzymes with a high purity and a high selectivity are used to cleave specific bonds while leaving others intact. The selection of appropriate enzymes strongly depends however on the polysaccharide subjected to hydrolysis (Heinrich 1999a, b; Tüting et al. 2004b; Adden et al. 2009).

For the spectroscopic analysis of the obtained fragments, different techniques are available, FAB-MS, ESI-MS, and MALDI-MS, sometimes with MS^2 (van der Burgt et al. 2000a, b, c; Kabel 2002; Kabel et al. 2003; Tüting et al. 2004a, b; Mazumder et al. 2005; Mischnick et al. 2005) and NMR methods (Gonera et al. 2002; Snyder et al. 2006; Nie and Xie 2011) being most frequently used.

4.1.6 Alien Groups and End Groups

The determination of alien groups, derived, e.g., from changes during polysaccharide processing via oxidative changes, introducing charges, possible cleavage points, and groups disturbing the initial molecular order of the substance, gets more and more into focus since these groups can have an overproportional influence on the properties of the polymer. However their detection is a very demanding task as they occur only in very small concentrations within the polymer structure. Specialized analysis procedures are used for their determination. Some methods from our group for the analysis of oxidized groups in pulp samples will be presented. As the formation of such groups is a phenomenon especially relevant for bulk carbohydrates, it is very useful if a combination with SEC is possible, as this helps to reveal differences in modification between different molecular weight fractions (Potthast et al. 2006; Kristiansen et al. 2009).

Some polysaccharides like some hemicelluloses contain specific end groups, containing sugar units and bond types that are not present in the main part of the polymer chain. They can be identified by use of the same spectroscopic methods after suitable hydrolysis, similar to the analysis of side chains and linkage patterns (Andersson et al. 1983).

4.1.7 DP and DP Distribution

One of the most important values for both fundamental research and many industrial applications is the degree of polymerization (DP). For example, the DP of a polysaccharide considerably influences its mechanical and rheological properties, as it is the case for all polymers.

For the determination of the DP, different analytical approaches are known, such as reducing end-group analysis, but the most common technique for a quick and convenient analysis is the viscosimetric approach, normally by utilization of an Ubbelohde viscosimeter. Contrary to these analytical methods, GPC of polysaccharides provides a separation of the molecules according to their hydrodynamic volume and therefore gives not only a single averaged parameter but a more complex picture of the degree of polymerization, showing the mass fractions according to their degree of polymerization. Only this technique can therefore ensure to track pattern and changes within the DP distribution like multimodal distributions or shifts in polydispersity that are not obvious from a simple average parameter (Vlasenko et al. 1998; Dupont and Mortha 2004).

The analysis of the sedimentation behavior of polysaccharide solutions via ultracentrifugation techniques can also provide a DP distribution and can give information on the conformation of the molecules analyzed. However the instrumental effort is rather high, and the method is more used for special scientific investigation than for routine measurements (Harding 2005).

The solvent system and column material used for GPC measurements depends on the polysaccharide investigated. For the calibration, two different approaches are possible. A calibration with adequate calibration substances and the use of a mass-sensitive detector, such as refractive index (RI) or ultraviolet light (UV) detector, are the more classical approaches. However, in some cases, it is hard to find calibration substances with molecular weight distributions narrow enough to be used in the calibration curve. The absolute or universal calibration uses, besides a mass-sensitive detector, also a detector that allows in combination with the IR detector a direct determination of the hydrodynamic radius of the molecules analyzed. This can be either a viscosimetric detector that makes an absolute calibration possible within the same polymer confirmation system by using the product of molecular weight and intrinsic viscosity (Mark–Houwink equation) or a light scattering detector that is independent of external calibration by direct determination of the molar weight via the Rayleigh equation. Examples for the use of GPC systems for different kinds of polysaccharides and its derivatives can be found in literature. (Gillespie and Hammons 1999; Eremeeva and Bykova 1998; Bikova and Treimanis 2002; Cellulose 2009; Saake et al. 2001; Schelosky et al. 1999; Wittgren and Porsch 2002; Eremeeva 2003; Rußler et al. 2006; Röder et al. 2009).

4.1.8 Crystallinity

Some polysaccharides like cellulose and starch are, due to a very regular molecular construction, able to form crystal structures. However, as these substances are polymers, they will never form perfect crystalline structures in their native state, but will also include more or less amorphous regions. The content of crystallinity is a crucial factor for the use of the polysaccharides, and some technological processes can change the portion of crystalline structures within the matrix. This is the case, for example, for regenerated cellulose or retrogradation of starch, forming the so-called resistant starch. Therefore it is very important to be able to measure the degree of crystallinity and to monitor possible changes (Schwikal et al. 2005; Röder et al. 2006a, b, 2009; Kumirska et al. 2010; Perera et al. 2010).

The classical method for this analysis is wide-angle X-ray scattering (WAXS), but this method requires some care with sample preparation and computation of the amorphous and crystalline peak areas (Manelius et al. 2000; Donald et al. 2001; Mihranyan et al. 2004; Calvini et al. 2006; Li et al. 2009).

Solid-state NMR (^{13}C CP–MAS NMR) is another method for the measurement of the degree of crystallinity, already quite established and with the advantage of not demanding any specific sample preparation. Deconvolution of the significant area of the spectrum is necessary. The accuracy of this method somewhat depends on the resolution and signal-to-noise ratio of the recorded spectrum (Liitiä et al. 2000; Hult et al. 2002).

FT-IR and FT-Raman spectroscopy for crystallinity determination are not easy to interpret and need some experience in interpretation as crystallinity is not only calculated from ratios of two bands, but several ones to be taken into account. On the other hand it demands only rather common analytical equipment (Oh et al. 2005; Calvini et al. 2006; Röder et al. 2009).

A comparison of different methods for the determination of the degree of crystallinity in different cellulosic materials can be found in Röder et al. (2006a, b).

4.2 Work Performed at BOKU

4.2.1 The CCOA and FDAM Method for Profiling of Oxidized Groups

The analysis of carbonyl (see Fig. 4.2) and carboxyl groups (see Fig. 4.3) is critical for the determination of the stability and quality of cellulose and

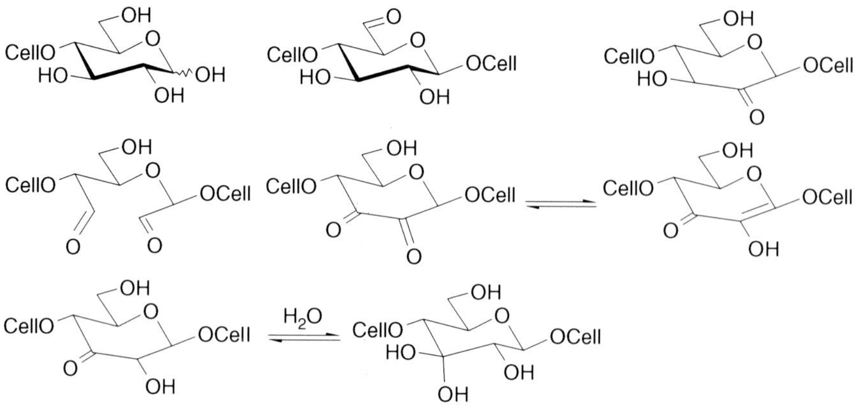

Fig. 4.2 Examples of different carbonyl structures in cellulose

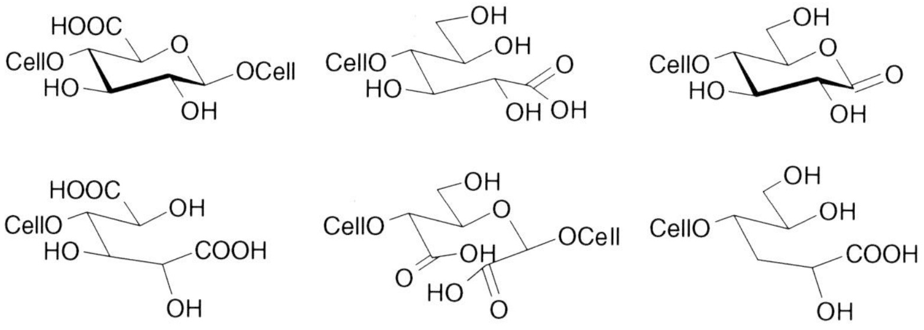

Fig. 4.3 Examples of different carboxyl structures in cellulose

paper samples. These two alien groups give a good indication of the oxidative status of the cellulose as they reflect the oxidative degradation of cellulose. Carbonyl and carboxyl groups can be introduced along the cellulose backbone on different occasions, such as pulping and bleaching processes, according to the respective conditions chosen (Schleicher and Lang 1994; Sixta 1995), high energy radiation, or natural or induced aging. Bleaching, both chlorine-based and oxygen-based, is known to affect the integrity of the cellulose backbone by generating oxidized positions and causing subsequent loss of DP. The creation of oxidized groups along the cellulose chain is typically a highly undesired process, as chain cleavage will preferentially occur at the destabilized positions of the oxidized groups. Therefore carbonyl and carboxyl contents have a disproportionately high influence even on macroscopic properties and quality characteristics of cellulose and its products.

Even the smallest oxidative damage in cellulose can decrease the quality of cellulose-containing materials drastically. Besides this, secondary performance characteristics are also affected. Oxidized cellulose gets more sensitive to β-alkoxy elimination that leads to degradation under alkaline conditions. Photoyellowing and photodegradation are favored due to the ability to undergo $n - \pi^*$ electron transitions upon UV irradiation (240–320 nm) (Gratzl 1985) as well as yellowing and hornification effects toward thermal stress (Haggkvist et al. 1998; Kato and Cameron 1999; García et al. 2002). The presence of air, air pollutants, and moisture leads to an accelerated aging of oxidatively pre-damaged celluloses (Lewin 1997; Blüher and Vogelsanger 2001).

Several conventional methods for the measurement of carbonyl and carboxyl contents in cellulose are known. However, all these methods suffer from general shortcomings, such as high limits of determination and detection—unacceptable for accurate methods—and rather large amounts of sample material required. As these methods provide only sum parameters, averaged over the whole molecular weight distribution, they are in addition not able to detect possible differences between different regions of the molecular weight distribution (MWD). But by far the major drawback is the low reproducibility: all these methods generate values of severely deficient comparability, both in interprocedural and in interlaboratory comparisons as was recently confirmed in a large round-robin test led by our groups within the European Polysaccharide Network of Excellence.

In industrial daily routine, the "copper number" is still the most common conventional method for easy and fast determination of the overall carbonyl group content (TAPPI 2009a, b). Nevertheless the reaction mechanism is still not understood in detail, and the responding alien groups are not known, but the "copper number" is able to provide reliable and reproducible values if performed under identical conditions. However, the sum parameter obtained cannot be directly linked to the quantity of a specific oxidized functionality, such as carbonyls. Also other semiquantitative methods using different reagents, such as hydrazine (Girard's reagent P, acetylhydrazide pyridinium chloride) (Wennerblom 1961), hydroxylamine (Cyrot 1957; Rehder et al. 1965), cyanide (Lewin 1972), or TTC (2,3,5-triphenyltetrazolium chloride) (Szabolcs 1961) have been developed, but neither of them has found wide application.

The basis of all conventional methods for carboxyl group determination is a special kind of acid–base titration, as all these methods have the conversion of free acids into salts in common. They depend on the cation exchange capacities of the material and differ mainly in the sample preparation, in the cation used, and in the direct or indirect measurement of the exchanged cations. The direct (Phillipp et al. 1965) and the reversible (Putnam 1964) methylene blue methods are the most common titrimetric methods used. They are based on a titration of methylene blue which is a colored organic base, measuring either its absorption to the polymer or its depletion in the titer, respectively. Accessibility, nonspecific adsorption, and reproducibility are well-known problems of this method that anyway was, due to the lack of alternatives, widely used. Presently, the reversible methylene blue method (Husemann and Weber 1942), the sodium bicarbonate procedure according to Wilson (1948), and the zinc acetate method (Doering 1956) are widely applied in pulp and paper laboratories.

The main obstacle for the accurate determination of oxidized functions in cellulose is the extremely low average contents of those groups within the cellulose matrix, which range in the order of μmol/g. This fact necessitates very sensitive means of detection since conventional, direct instrumental techniques, such as IR, Raman, UV, fluorescence, or NMR spectroscopy, fail to monitor such minor amounts. Another major problem lies within the insolubility of cellulose in conventional solvents. This requires either a heterogeneous reaction for derivatization with inherent problems of accessibility and reagent adsorption or unusual solvents like metal amine complexes, N,N-dimethylacetamide/LiCl, or N-methylmorpholine-N-oxide have to be used. However these solvents do not promote conventional carbonyl or carboxyl reactions. As there is no kinetic data available, the completeness of a derivatization reaction additionally is hard to predict.

Increasing analytical demands favored the development of two new methods in our laboratory, based on a different analytical strategy, the CCOA method for carbonyl profiling and the FDAM method for carboxyl profiling (see Fig. 4.4). Both methods fulfill all requirements to a reliable, comparable, and precise analysis, providing an unmatched level of information on the oxidation state of cellulosic samples. Both method names are derived from the abbreviations of the semi-systematic chemical terms of the respective selective labeling reagents: CCOA for carbazole–carbonyl–oxyamine and FDAM for fluorenyl–diazomethane. Both the CCOA method (Röhrling et al. 2001) for carbonyl profiling and the FDAM method (Bohrn et al. 2005) for carboxyl profiling were comprehensively discussed in

Fig. 4.4 Selective fluorescence labeling and carbonyl/carboxyl profiling with CCOA and FDAM

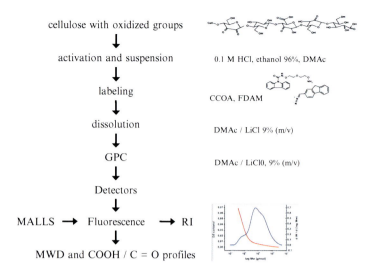

the literature (Röhrling et al. 2002a, b; Potthast et al. 2003, 2004; Bohrn et al. 2006), also with regard to comprehensive information on the gel permeation chromatography (GPC) system used; the eluent, calibration, validation of the method; and the detailed experimental procedures.

Both methods are based on a group-selective reaction of a fluorescence label reagent with the oxidized functionality which, in a subsequent GPC analysis, provides a concentration-equivalent signal. Only a method based on fluorescence markers provides enough sensitivity, which is not reached by other systems like UV label-based reactions. Extensive research on the improvement of the reactions selectivity and quantitativeness, as well as the optimization of the structure of the fluorescence label to match the spectroscopic requirements of the GPC–MALLS system, has been done. By the selectivity of the labeling reaction and the highest possible degree of conversion—comprehensively tested with model compounds (Röhrling et al. 2001; Bohrn et al. 2005)—an optimum accuracy and reproducibility were assessed, which are reflected by the validation parameters of the method (Röhrling et al. 2002b). Due to low sample amounts required, between 5 and 50 mg, the method can be applied to rare sources.

Both CCOA and FDAM labeling reactions are performed as derivatization steps prior to separation of cellulose in a modified gel permeation chromatography (GPC) system. By combining group-selective fluorescence labeling with size-exclusion chromatography, quantification of oxidized groups relative to the MWD of the cellulosic material became possible for the first time, displayed in carbonyl and carboxyl profile graphs. These DS plots have a typical exponentially decaying curve shape. While DS plots reflect the oxidation state of a given cellulosic material, dynamic oxidative changes are better visualized in ΔDS plots, which give the difference in DS_{CO} or DS_{COOH} between the starting material and the material after the respective treatment. Such a detailed monitoring of oxidative changes would have been simply impossible according to any conventional methodology based on averaged parameters. While conventional methods would report nearly no change in the overall carbonyl content, the new methodology shows that the decrease in low-molecular weight carbonyls is largely compensated by an increase in high-molecular weight carbonyls. It became evident that different regions of the MWD reacted quite differently and that there were significant differences according to the cellulosic substrates.

With the combination of CCOA and FDAM labeling, it is not only possible to follow changes in the oxidative status of cellulose samples within industrial processes or laboratory procedure, but new insights into structure–property relationships have been made possible. Degradation reactions can be monitored, and, for example, mechanical properties can be understood from the oxidative damage observed. The detailed view on the functional groups in relation to the molecular weight that is accessible by the two methods proved to be a very valuable tool especially in restoration sciences to validate the stability of historic papers and other cellulose-containing materials. In this context also, the possibility to use the labeling method for a macroscopic evaluation of different areas of paper, to observe oxidative damage in correlation with the migration behavior of metal ions, has to be mentioned (Henniges et al. 2006; Potthast et al. 2006).

4.2.2 Analysis of Trace Chromophores in Cellulosic Materials

An analytical problem, even more demanding than the analysis of oxidized groups in cellulosic materials is—due to their extremely low content in the ppm to ppb range—the isolation and identification of residual chromophores. These chromophores, comprising aromatic and quinoid structures, are generated during processing or degradation, e.g., aging. Due to their deep coloration, they have an important influence on the optical aspect of the materials, quantified as brightness, and on bleaching processes. Some of these chromophores are just absorbed to the polysaccharide surface, but others are covalently bond to the cellulose structure. In this case they can be considered as side groups with an extremely small degree of substitution.

The off-white discoloration of processed cellulosic materials is not due to only one specific chromophore but originates from the high extinction coefficient of a large number of individual compounds, either absorbed or covalently linked to the cellulose backbone in different ways as ester, ether, or secondary alkyls. Moreover, they might be present on the surface only or also in the interior parts of the sample structures. An approach toward chromophore analytics in cellulosics must regard all these difficulties, and the analysis must not affect the integrity of the cellulosic matrix in any negative way. In our laboratory, the so-called CRI (chromophore release and identification) method has been developed to face these challenges (Rosenau et al. 2004, 2005a).

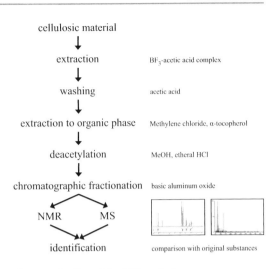

Fig. 4.5 Outline of the CRI method

As the amount of individual chromophore species is very small, the sample amount needed is quite high, between 50 g and several kg. The procedure itself is a multistep method as several individual steps are necessary to come to the pure substances.

The first part of the isolation is the extraction itself for which the sample is suspended in catalytic amount of BF_3-acetic acid complex in the presence of a reductant in an organic solvent (e.g., methylene chloride) and agitated for 48 h until it is separated and washed with acetic acid. This leads to the liberation of the chromophores and simultaneously to the acetylation of these compounds. The next step consists of the addition of water, methylene chloride, and α-tocopherol and subsequent extraction of the aqueous phase. This allows a transfer of the chromophores into the organic phase and at the same time inhibits oxidative changes. After deacetylation, the residue from this step is then chromatographed on basic aluminum oxide, using increasingly polar solvents and solvent mixtures as eluents. The fractions then can be analyzed by NMR and MS (Rosenau et al. 2004, 2005a) (Fig. 4.5).

For samples with even low chromophore contents like bleached pulps, the method has to be

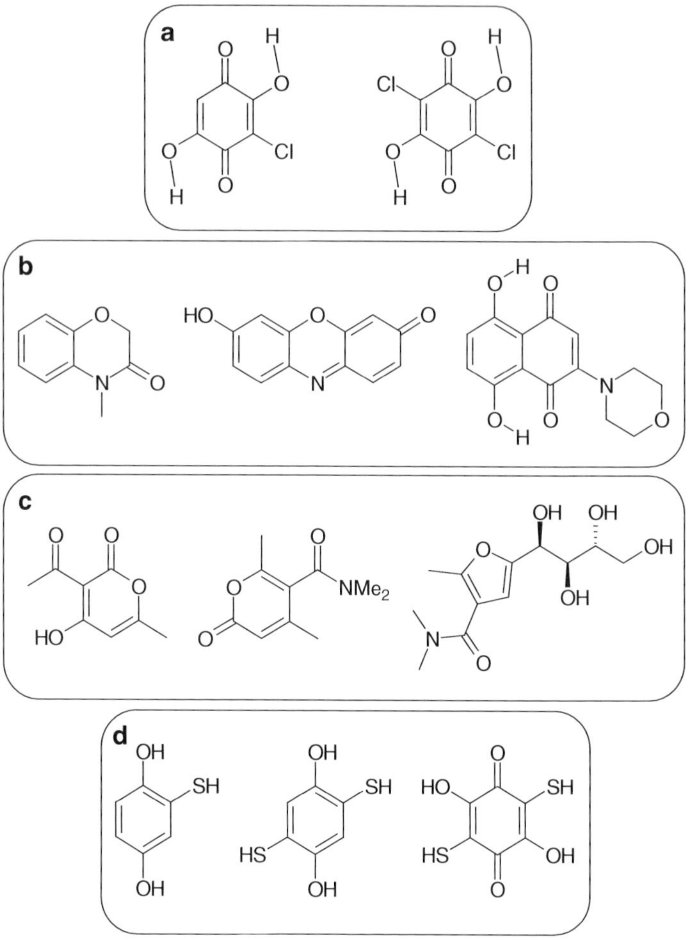

Fig. 4.6 Some selected chromophores from different cellulosic sources (**a**) chlorine dioxide bleached pulp, (**b**) lyocell fibers, (**c**) pulp after heating in DMAc, (**d**) viscose fibers

upscaled accordingly to obtain sufficient amounts of chromophores.

The method was validated and it proved not to promote the formation of chromophores itself, as no chromophores could be isolated using pure cotton linters as substrate. As well it was shown that a second, prolonged treatment of samples did not lead to isolable amounts of chromophores, which once more confirmed that the isolation procedure does not produce artifacts and that the chromophore isolation in the first step was rather complete.

A classical extraction approach just with common solvents does not lead to a separation of significant amounts of chromophores.

It was comprehensively demonstrated that the CRI method is suitable for all different kinds of cellulosic material, as well cellulose I and also cellulose II structures, pulps of different origin and purity, labeled pulps, regenerated fibers such as rayon and lyocell, and cellulose derivatives (Rosenau et al. 2004, 2005a, 2007).

According to the origin and history of the samples, different sets of chromophores were isolated and identified. These chromophores could be divided into so-called primary and secondary chromophores. While the first category is formed directly in reactions of carbohydrates upon thermal, acidic, or basic treatment, the formation of secondary chromophores involves the respective "process chemistry" by reactions of the primary chromophores with substances used in the individual processes, such as solvents, reagents, and bleaching agents. This can be seen very directly

4 Chemical Characterization of Polysaccharides

Fig. 4.7 Structure and stabilization of 2-hydroxy-1,4-benzoquinone

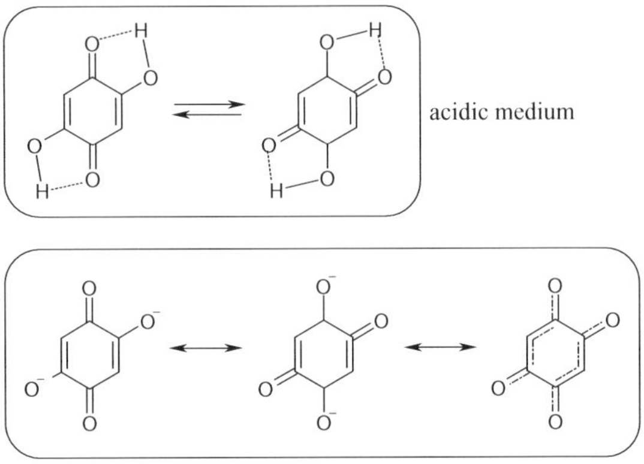

by the structure of some of the secondary chromophores. For example, some chromophores obtained from lyocell fibers contain nitrogen (from the amine N-oxide solvent), some chromophores in viscose contain sulfur (from CS_2), and some pulps bleached with chlorine dioxide contain chlorinated chromophores. As for these samples, the only source for these hetero-atoms lies in the related process chemicals; the origin of this modification is quite obvious.

Among the phenolic and chinoid structures, the primary chromophores with hydroxy-[1,4]-benzoquinone, 5-hydroxy-naphthoquinone, and 2-hydroxyacetophenone structures can be found in all analyzed samples (see Fig. 4.6) and must be considered key chromophores.

These substances and their derivatives form highly stabilized molecules due to the delocalization of the quinoid double bonds. By this stabilization effect, the molecule gets rather insensitive against reagents attacking the double bond system, such as most typical bleaching agents (Fig. 4.7).

Due to its omnipresence on processed or aged cellulosic materials and its structural features, 2-hydroxy-1,4-benzoquinone can be regarded both as key chromophore and as precursor structure from which other (primary and secondary) chromophores are derived due to its ability to fragment as well as to condensate under prolonged process conditions.

Like some other important chromophores found, 2-hydroxy-1,4-benzoquinone can be formed via aromatic condensation from hexoses as well as pentoses. So the cellulose bulk as well as hemicellulosic impurities can contribute to its formation.

Nevertheless the number of different chromophores, primary and secondary ones, does not correlate with the overall chromophore concentration or the brightness of the sample.

The CRI method has also been successfully used for the analysis of labeled celluloses. Highly pure pulp samples labeled by the CCOA procedure as it is described for the labeling of carbonyl groups in the previous chapter were subjected to the chromophore extraction method. In these cases, the used label, N-acetylcarbazol, was found as the only chromophore released and in amounts in satisfying agreement with conventional analysis method. This approach offers the possibility to get at least averaged values for the carbonyl content of labeled pulps that are insoluble in DMAc/LiCl as it is used in the conventional CCOA method.

Results from the CRI method helped to understand some of the complex processes that lead to the formation of chromophores. The revealed structure of the individual chromophore substances furthermore can lead to new, more effective, and sustained approaches for chromophore removal or for the inhibition of their formation during processing and re-formation during aging in different substrates (Rosenau et al. 2001, 2004, 2005a, b, 2007; Adorjan et al. 2005; Krainz et al. 2009).

4.2.3 Capillary Electrophoresis for the Analysis of Carbohydrates

Understanding the organization of polysaccharide structures is sometimes a tricky task when it comes to analytical issues. Celluloses and hemicelluloses are important abundant renewable resources currently finding more and more applications. In this context knowledge on cellulose and hemicelluloses composition is of great importance as their components differ significantly in content and structure depending on their origin. Hemicelluloses differ not only in monomer composition but also in side chain types, distribution, and types of glycoside linkages in the main macromolecular backbone (Ebringerova et al. 2005; Ebringerova 2006). The presence of uronic acid affects the molecular properties of polysaccharides as well. Therefore, several methods are required to describe their structure. To address the monosaccharide (neutral and acidic) composition or measure oligomers of polysaccharides up to a certain DP (normally below 10), a single analytical run capillary electrophoresis (CE) is enough.

In the recent decades, capillary electrophoresis (CE) was shown to be an effective analytic tool for carbohydrates analyses due to such advantages like robustness, very low amounts of sample and chemical reagents, short analysis time, and low running costs. Sensitivity of the method can be in many cases compared to the one achieved with ion chromatography which is known to be the most sensitive method of the carbohydrates separation.

Analysis of cello- and xylooligosaccharide mixtures faces difficult separation problems as the materials differ only in their hydroxyl group configuration showing rather similar separation properties. Deficiency of UV or fluorescence absorbance, as well as absence of easily ionizable groups, increases the difficulty. CE can be a successful tool for the analysis of such mixtures after certain manipulations resulting in an introducing of charge into the molecules. This can be achieved by chelation of the carbohydrate with a suitable ion. The most commonly used ion for such a purpose is borate (Sjöberg et al. 2004). The other option to achieve a migration of carbohydrates in the electric field could be deprotonation of hydroxyl groups (high pH values of the background electrolyte) or derivatization of the corresponding acidic sugars.

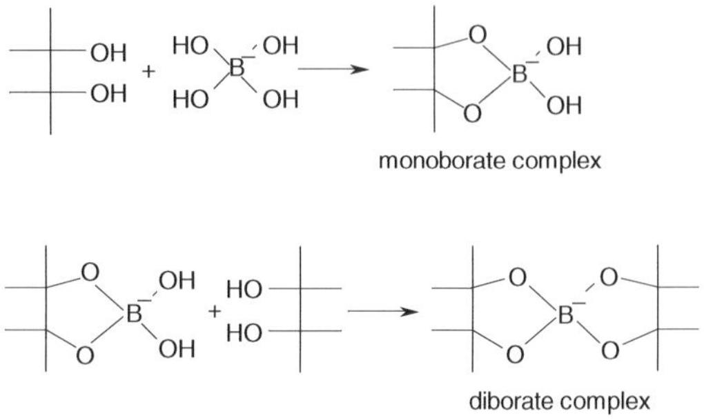

Complexation of carbohydrates with tetrahydroxyborate ions is well known (see Fig. 4.8) (Holfstetter-Kuhn et al. 1991). As carbohydrates exhibit numerous hydroxyl groups, they are able to form stable, negatively charged, five- or six-membered ring structures with tetrahydroxyborate. In case of underivatized carbohydrates, a low sensitivity is observed in direct detection mode. Another option for a direct detection of underivatized carbohydrates is an ionization of hydroxyl groups in highly alkaline buffers (Honda et al. 1991b; Rovio et al. 2007, 2008, 2011).

Mainly, analyses of underivatized carbohydrates by CE involve an indirect detection by applying background chromophores. Addition of the UV-active component to the background electrolyte leads to permanent absorption. When the analyte passes the detection window, it displaces the background chromophore, which leads to a decrease of absorption. Therefore, the observed signals exhibit negative absorption in the electropherogram. The use of various background chromophores has been reported in literature: 2,6-pyridinedicarboxylic acid (Soga and

Fig. 4.8 Borate complexation scheme

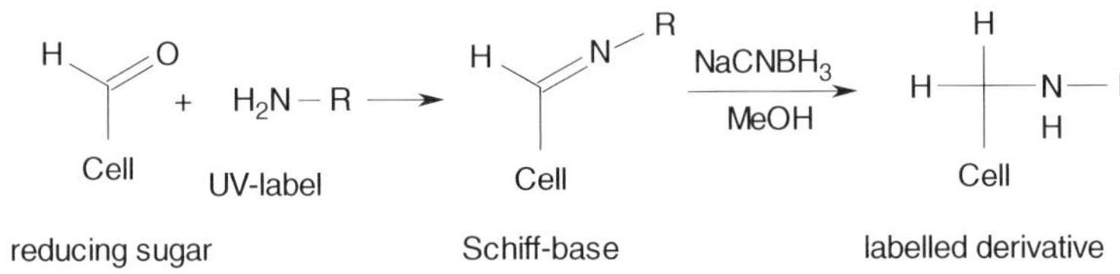

4 Chemical Characterization of Polysaccharides

Fig. 4.9 Derivatization of reducing sugars by reductive amination

Heiger 1998; Soga and Ross 1999; Soga and Serwe 2000), 2,6-naphthalenedicarboxylate (Dabek-Zlotorzynska and Dlouhy 1994), sorbic acid (Liu et al. 1997; Zemann et al. 1997; Jager et al. 2007), riboflavin (Liu et al. 1997), and *p*-nitrophenol (Plocek and Chmelik 1997). The most successful separation and detection of carbohydrates were achieved applying the *p*-nitrophenol which allowed the separation of monosaccharides, but the separation of oligosaccharides showed low efficiency and sensitivity.

The best way for the separation and detection of carbohydrates by CE is their labeling with UV-active tags with the following complexation with borate (for a scheme see Fig. 4.8). A large variety UV-active and fluorophore derivatization reagents have been suggested in the literature, e.g., 2-aminopyridine (Hasc et al. 1984; Kondo et al. 1990), 2-aminobenzamide (France et al. 2000), 8-aminonaphthalene-1,3,6-trisulfonate (Chiesa and Horvath 1993; Klockow et al. 1995, 1996; Mort and Chen 1996), 9-aminopyrene-1,4,6-trisulfonate (Chen and Evangelista 1995), 1-phenyl-3-methyl-2-pyrazolin-5-one (Honda et al. 1991a), 1-(2-naphthyl)-3-methyl-5-pyrazolone (You et al. 2008), 2-amino-3-phenylpyrazine (Yamamoto et al. 2003), 9-aminoacridone, 7-aminonaphthalene-1,3-disulfonic acid (Mechref et al. 1995, 1997), 8-aminopyrenesulfonic acid (Chiesa et al. 1996; Racaityte et al. 2005), 3-(4-carboxybenzoyl)-2-quinollnecarboxaldehyde (Liu et al. 1991), and *p*-hydrazinobenzenesulfonic acid (Wang and Chen 2001; Wang et al. 2002), etc.

Reductive amination (see Figs. 4.9 and 4.10) is the most widely used method of the carbohydrates UV labeling as it gives high yields even in the presence of water and leads to the formation of stable secondary amines. Satisfactory results of

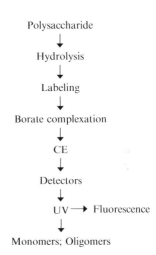

Fig. 4.10 Reaction scheme for the CE carbohydrate analysis

carbohydrates labeling have been achieved using 4-aminobenzoic acid (Grill et al. 1993; Cortacero-Ramirez et al. 2004), 4-aminobenzonitrile (Schwaiger et al. 1994; Nguyen et al. 1997; Ristolainen 1999; Sartori et al. 2003), 4-aminobenzoic acid ethyl ester (Rydlund and Dahlman 1996; Dahlman et al. 2000), and 6-aminoquinoline (Rydlund and Dahlman 1996, 1997).

Combination of reductive amination with further borate-ion complexation results in a successful baseline separation of mono- and oligosaccharide mixtures up to DP 6 (see Fig. 4.11).

In many cases, a better separation of carbohydrates can be achieved by the addition of organic solvents which are soluble in the aqueous phase (e.g., acetone, methanol, or propanol). These solvents have an effect on the intramolecular hydrogen bonding of the carbohydrates and thus influence their confirmation. This influence is dependent on the protophilic or protophobic character of the organic solvent and determines

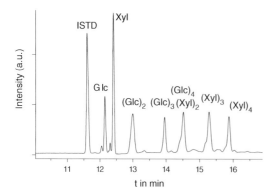

Fig. 4.11 Electropherogram of a mixture of derivatized cello- and xylooligosaccharides. *ISTD* internal standard (D-galactose). Reprinted from Carbohydrate Research, 2003 (Sartori et al. 2003), with permission from Elsevier

both how the solute becomes solvated and if intramolecular hydrogen bonds are established.

The most common detection method used in the analysis of carbohydrates by CE is a UV detection. Laser-induced fluorescence (Stefansson and Novotny 1994; Bui et al. 2008), electrochemical detection (Zhou and Baldwin 1996; Chen et al. 2005, 2006), conductivity detection (Zatkovskis Carvalho et al. 2003), amperometric detection (Chu et al. 2005; Dong et al. 2007; Cheng et al. 2008), refractometry detection (Ivanov et al. 2000), photometric detection (Momenbeik et al. 2006), and FTIR-spectroscopy (Koelhed and Karlberg 2005) are also used.

In the recent years, lots of attention has been devoted to developing a coupling (hyphenation) of CE with mass spectrometry (MS) detection systems for the carbohydrates analyses. In a CE–MS system, different buffers must be used as background electrolytes since the commonly used highly concentrated borate solutions (150–500 mM) are incompatible with MS detectors. There have been attempts to use a low concentrated borate buffer (10 mM) in the analysis of underivatized carbohydrates using negative ESI-MS detection (Li et al. 2000).

Highly alkaline salt-based background electrolytes cannot be introduced to the MS-detector. However, they can be substituted by volatile organic bases. For example, an application of diethylamine for the CE–ESI-MS analysis of carbohydrates in a negative ion mode showed a lower sensitivity compared to the analysis using pre-column derivatization but a higher one compared to the indirect detection of carbohydrates using background chromophores (Klampfl and Buchberger 2001).

As carbohydrates derivatization improves selectivity and sensitivity of the determination, new labeling reagents have been synthesized which are more suitable for the CE–MS analysis, for example, *N*-quaternized benzylamine (An et al. 2003) and 3-(acetylamino)-6-aminoacridine (Charlwood et al. 2000).

The most commonly employed interface for the on-line CE–MS coupling is electron spray ionization (ESI), mainly represented by ion trap (IT) (Campa et al. 2006). However, IT shows medium to high resolution and a mass accuracy in a range of 100 ppm (Huber and Hoelzl 2001; Klampfl 2006). Also this analytical technique has a limitation in mass range detection as highly complex oligosaccharides often exceed the mass limit of common IT spectrometers (Klampfl 2006). In this case, an ion fragmentation could provide comprehensive information on carbohydrates identification. A higher mass accuracy and better resolution compared to IT could be achieved with time-of-flight (TOF) mass analyzer.

Majority of the CE/MS methods apply an on-line sheath-flow interface (Campa et al. 2006) providing a stable electrical contact between the flow coming out of CE-capillary and electrode. The main disadvantage of this interface is a dilution of CE-flow with sheath liquid which decreases sensitivity significantly. This problem can be eliminated by using a sheathless-flow CE–ESI-MS interface (Kelli et al. 1997; Issaq et al. 2004; Zamfir and Peter-Katalinic 2004; Klampfl 2006). Compromise between a robustness of sheath-flow interface and high sensitivity of a sheathless mode was possible by micro- or nano-electrospray techniques (Chen et al. 2003).

4.2.4 Assessment of Cellulose Morphology by Solid-State ^{13}C CP–MAS NMR

Cellulose morphology influences strongly the chemical and physical properties of (ligno-) cellulosic materials. The supramolecular structure

4 Chemical Characterization of Polysaccharides

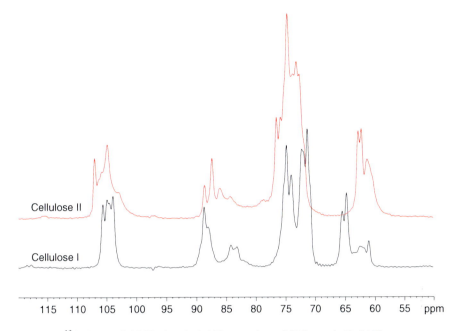

Fig. 4.12 Typical ^{13}C CP–MAS NMR chemical shift patterns of highly crystalline cellulose I (*black, bottom*) and II (*red, top*) samples. Note that cellulose I and II are usually analyzed in a wet, i.e., water-saturated state to enhance spectral resolution. It seems advisable to employ moderate MAS speeds (4–8 kHz on most spectrometers) to avoid local drying of the sample during the analysis. Relaxation delays in the order of 2–3 s are usually sufficient for a quantitative assessment of cellulose ^{13}C CP–MAS NMR spectra

of cellulose therefore plays an important role in industrial wood pulping, cellulose derivatization and dissolution, and the definition of product properties. Solid-state ^{13}C CP–MAS NMR spectroscopy is a well-established method for cellulose morphology assessment (Larsson et al. 1997; Wickholm et al. 1998; Newman and Davidson 2004; Zuckerstätter et al. 2009). Important structural parameters such as the degree of crystallinity, cellulose I vs. cellulose II ratio, cellulose I allomorphs, cellulose fibril, and fibril aggregate widths are accessible from solid-state NMR spectra.

For the pulp, paper, and regenerated fiber industry, the cellulose I and II structures are of prime importance. Both structures exhibit characteristic ^{13}C NMR chemical shift patterns (see Fig. 4.12), with separable contributions of crystalline modifications and distinct less-ordered forms of cellulose. Subcrystalline cellulose is mostly attributed to material located at cellulose fibril and crystallite surfaces (Wickholm et al. 1998; Newman and Davidson 2004). Common cellulose I models (Wickholm et al. 1998) also distinguish between solvent-accessible and solvent-inaccessible fibril surfaces, with the latter being introduced through fibril aggregation. Common cellulose II spectral assignments include no discrimination between crystallite surface signals according to solvent accessibility.

^{13}C CP–MAS NMR spectra of cellulose I exhibit up to seven separable resonance lines assigned to cellulose I allomorphs I$_\alpha$ and I$_\beta$, paracrystalline cellulose, as well as cellulose located at accessible and inaccessible fibril surfaces (Larsson et al. 1997; Wickholm et al. 1998) (see Fig. 4.13). Note here that hemicellulose signals could interfere with fibril surface signals. It is therefore suggested (Wickholm et al. 1998) to remove xylan and other hemicelluloses by mild acid hydrolysis prior to NMR analysis.

Experimental line shapes are typically analyzed through fitting procedures involving spectral deconvolution (Larsson et al. 1997). Deconvolution routines may in principle involve the optimization of all individual resonance

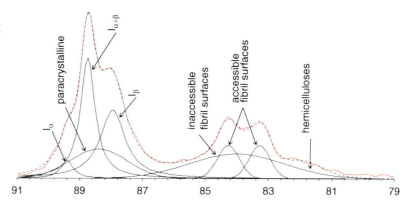

Fig. 4.13 Typical AGU-C4 resonance of cellulose I with signal assignments of deconvoluted peaks (Wickholm et al. 1998). The *red solid line* shows the experimental spectrum, and the *black dotted line* represents the superposition of individual fitted lines (*black solid lines*)

positions, line widths, and signal intensities. Note, however, that the chemical shifts and also line widths of certain signals vary negligibly for most samples on a given spectrometer and using constant experimental setup parameters (including MAS speed, ^{1}H decoupling scheme and field, acquisition time, spectral width, and pulse shapes/widths). A fixed parameter approach (Zuckerstätter et al. 2009) can thus help to reduce the computation effort and to avoid degenerate solutions for the fitting routine.

From the signal areas of all individual contributions, a number of important parameters can be derived. The fraction of well-ordered (core) structures including crystalline cellulose I$_\alpha$, crystalline cellulose I$_\beta$, and paracrystalline cellulose (PC) can be used to measure crystallinity, expressed as a crystallinity index $CrI_{NMR}[\%]$:

$$CrI_{NMR}[\%] = \frac{I_\alpha + I_\beta + PC}{\sum cellulose} \times 100\,\% \quad (4.1)$$

From the fraction q_{fibril} of cellulose at accessible fibril surfaces (AFS) and inaccessible fibril surfaces (IFS), the average lateral fibril dimension (LFD) can be computed. q_{fibril} is fed into a square cross-section model (Wickholm et al. 1998) and yields the number of glucan chains n along one side of the fibril. A conversion factor of 0.57 nm for the width of one glucan chain yields the lateral fibril dimension (LFD):

$$q_{fibril} = \frac{AFS + IFS}{\sum cellulose} = \frac{4n - 4}{n^2} \quad (4.2)$$

$$LFD[nm] = 0.57\,nm \times n$$
$$= 0.57\,nm$$
$$\times \left(\frac{4 + \sqrt{16 - 16 q_{fibril}}}{2 q_{fibril}}\right) \quad (4.3)$$

Analogously, the lateral fibril aggregate dimension (LFAD) can be computed from the fraction $q_{aggregate}$ of cellulose located only at solvent-accessible fibril surfaces (AFS).

Cellulose II ^{13}C CP–MAS NMR spectra can be analyzed and interpreted in a similar way. A superposition of five individual lines can be used to fit the experimental spectrum. Two inner-crystalline, two surface crystalline signals, and one resonance attributed to disordered cellulose have been commonly assigned in literature (Newman and Davidson 2004). From the fraction of material located at cellulose crystallite surfaces, average lateral fibril dimensions can be computed. The crystallinity of cellulose II is assessed in the same way as that of cellulose I, i.e., expressed as the fraction of well-ordered inner-crystalline (core) material.

References

Adden R, Mischnick P (2005) A novel method for the analysis of the substitution pattern of O-methyl-[alpha]- and [beta]-1,4-glucans by means of electrospray ionisation-mass spectrometry/collision induced dissociation. Int J Mass Spectrom 242(1):63–73

Adden R, Melander C et al (2009) The applicability of enzymes in cellulose ether analysis. Macromol Symp 280:36–44

Adorjan I, Potthast A et al (2005) Discoloration of cellulose solutions in N-methylmorpholine-N-oxide (Lyocell). Part 1: Studies on model compounds and pulps. Cellulose 12(1):51–57

Albersheim P, Nevins DJ et al (1967) A method for the analysis of sugars in plant cell-wall polysaccharides by gas–liquid chromatography. Carbohydr Res 5(3):340–345

Alexandru L, Rogovin ZA (1953) Verteilung der Thiocarbonatgru ppen zwischen den primären und sekundären Alkoholgru ppen im Cellulosexanthogenat. J Allg Chem (USSR) 23:1203–1205

Allard B, Derenne S (2009) Microwave assisted extraction and hydrolysis: an alternative to pyrolysis for the analysis of recalcitrant organic matter? Application to a forest soil (Landes de Gascogne, France). Org Geochem 40(9):1005–1017

An HJ, Franz AH et al (2003) Improved capillary electrophoretic separation and mass spectrometric detection of oligosaccharides. J Chromatogr A 1004:121–129

Andersson S-I, Samuelson O et al (1983) Structure of the reducing end-groups in spruce xylan. Carbohydr Res 111(2):283–288

Arisz PW, Kauw HJJ et al (1995) Substituent distribution along the cellulose backbone in O-methylcelluloses using GC and FAB-MS for monomer and oligomer analysis. Carbohydr Res 271(1):1–14

Bedouet L, Courtois B et al (2003) Rapid quantification of O-acetyl and O-methyl residues in pectin extracts. Carbohydr Res 338(4):379–383

Bikova T, Treimanis A (2002) Problems of the MMD analysis of cellulose by SEC using DMA/LiCl: a review. Carbohydr Polym 48(1):23–28

Black GE, Fox A (1996) Recent progress in the analysis of sugar monomers from complex matrices using chromatography in conjunction with mass spectrometry or stand-alone tandem mass spectrometry. J Chromatogr A 720(1–2):51–60

Blüher A, Vogelsanger B (2001) Mass deacidification of paper. Chimia 55:981

Bohrn R, Potthast A et al (2005) Synthesis and testing of a novel fluorescence label for carboxyls in carbohydrates and cellulosics. Synlett 20:3087–3090

Bohrn R, Potthast A et al (2006) The FDAM method: determination of carboxyl profiles in cellulosic materials by combining group-selective fluorescence labeling with GPC. Biomacromolecules 7(6):1743–1750

Bui A, Kocsis B et al (2008) Methodology to label mixed carbohydrate components by APTS. J Biochem Biophys Methods 70:1313–1316

Calvini P, Gorassini A et al (2006) FTIR and WAXS analysis of periodate oxycellulose: evidence for a cluster mechanism of oxidation. Vib Spectrosc 40(2):177–183

Campa C, Coslovi A et al (2006) Overview on advances in capillary electrophoresis-mass spectrometry of carbohydrates: a tabulated review. Electrophoresis 27:2027–2050

Capitani D, Porro F et al (2000) High field NMR analysis of the degree of substitution in carboxymethyl cellulose sodium salt. Carbohydr Polym 42(3):283–286

Chandra K, Ghosh K et al (2009) Chemical analysis of a polysaccharide of unripe (green) tomato (Lycopersicon esculentum). Carbohydr Res 344(16):2188–2194

Charlwood J, Birrell H et al (2000) A probe for the versatile analysis and characterization of N-linked oligosaccharides. Anal Chem 72:1453–1461

Chen F-TA, Evangelista RA (1995) Analysis of mono- and oligosaccharide isomers derivatized with 9-aminopyrene-1,4,6-trisulfonate by capillary electrophoresis with laser-induced fluorescence. Anal Biochem 230:273–280

Chen Y-R, Tseng M-C et al (2003) A low-flow ce/electrospray ionization MS interface for capillary zone electrophoresis, large-volume sample stacking, and micellar electrokinetic chromatography. Anal Chem 75:503–508

Chen G, Zhang L et al (2005) Determination of mannitol and three sugars in Ligustrum lucidum Ait. by capillary electrophoresis with electrochemical detection. Anal Chim Acta 530:15–21

Chen G, Zhang L et al (2006) Determination of glycosides and sugars in Moutan Cortex by capillary electrophoresis with electrochemical detection. J Pharm Biomed Anal 41:129–134

Chen Y, Xie M-Y et al (2008) Purification, composition analysis and antioxidant activity of a polysaccharide from the fruiting bodies of Ganoderma atrum. Food Chem 107(1):231–241

Cheng X, Zhang S et al (2008) Determination of carbohydrates by capillary zone electrophoresis with amperometric detection at a nano-nickel oxide modified carbon paste electrode. Food Chem 106:830–835

Chiesa C, Horvath C (1993) Capillary zone electrophoresis of malto-oligosaccharides derivatized with 8-aminonaphthalene-1,3,6-trisulfonic acid. J Chromatogr A 645(2):337–352

Chiesa C, O'Neil RA et al (eds) (1996) Capillary electrophoresis in analytical biotechnology. CRC Press, Boca Raton, FL

Chu Q, Fu L et al (2005) Fast determination of sugars in Coke and Diet Coke by miniaturized capillary electrophoresis with amperometric detection. J Sep Sci 28:234–238

Ciucanu I, Kerek F (1984) A simple and rapid method for the permethylation of carbohydrates. Carbohydr Res 131(2):209–217

Cohen A, Schagerlöf H et al (2004) Liquid chromatography-mass spectrometry analysis of enzyme-hydrolysed carboxymethylcellulose for investigation of enzyme selectivity and substituent pattern. J Chromatogr A 1029(1–2):87–95

Cortacero-Ramirez S, Segura-Carretero A et al (2004) Analysis of carbohydrates in beverages by capillary electrophoresis with precolumn derivatization and UV detection. Food Chem 87:471–476

Cui SW (2005) Structural analysis of polysaccharides. In: Cui SW (ed) Food carbohydrates - chemistry, physical properties and applications. CRC Press, Boca Raton, FL, p 56

Cyrot J (1957) Dosage des fonctions oximables de la cellulose dégradée. J Chim Anal 39:449

Dabek-Zlotorzynska E, Dlouhy JF (1994) Capillary zone electrophoresis with indirect UV detection of organic anions using 2,6-naphthalenedicarboxylic acid. J Chromatogr A 685(1):145–153

Dahlman O, Jacobs A et al (2000) Analysis of carbohydrates in wood and pulps employing enzymatic hydrolysis and subsequent capillary zone electrophoresis. J Chromatogr A 891(1):157–174

De Ruiter GA, Schols HA et al (1992) Carbohydrate analysis of water-soluble uronic acid-containing polysaccharides with high-performance anion-exchange chromatography using methanolysis combined with TFA hydrolysis is superior to four other methods. Anal Biochem 207(1):176–185

Dicke R, Rahn K et al (2001) Starch derivatives of high degree of functionalization. Part 2. Determination of the functionalization pattern of p-toluenesulfonyl starch by peracylation and NMR spectroscopy. Carbohydr Polym 45(1):43–51

Doering H (1956) Determination of carboxyl groups in cellulose by complexometry. Das Papier 10:140–141

Donald AM, Buschow KHJ, et al. (2001) Polysaccharide crystallization. In: Buschow et al. (eds) Encyclopedia of materials: science and technology. Elsevier, Oxford, pp 7714–7718

Dong S, Zhang S et al (2007) Simultaneous determination of sugars and ascorbic acid by capillary zone electrophoresis with amperometric detection at a carbon paste electrode modified with polyethylene glycol and Cu2O. J Chromatogr A 1161:327–333

Dunbrant SSO (1965) Determination of primary and secondary xanthate groups in cellulose-xanthate. J Appl Polym Sci 9:2489–2499

Dupont A-L, Mortha G (2004) Comparative evaluation of size-exclusion chromatography and viscometry for the characterisation of cellulose. J Chromatogr A 1026 (1–2):129–141

Ebringerova A (2006) Structural diversity and application potential of hemicelluloses. Macromol Symp 232:1–12

Ebringerova A, Hromadkova Z et al (2005) Hemicellulose. Adv Polym Sci 186:1–67

Eremeeva T (2003) Size-exclusion chromatography of enzymatically treated cellulose and related polysaccharides: a review. J Biochem Biophysl Methods 56 (1–3):253–264

Eremeeva TE, Bykova TO (1998) SEC of monocarboxymethyl cellulose (CMC) in a wide range of pH; Mark-Houwink constants. Carbohydr Polym 36 (4):319–326

Erler U, Mischnick P et al (1992) Determination of the substitution patterns of cellulose methyl ethers by HPLC and GLC - comparison of methods. Polym Bull 29:349–356

France RR, Cumpstey I et al (2000) Fluorescence labelling of carbohydrates with 2-aminobenzamide (2AB). Tetrahedron Asymmetry 11(24):4985–4994

García O, Torres AL et al (2002) Effect of cellulase-assisted refining on the properties of dried and never-dried eucalyptus pulp. Cellulose 9(2):115–125

Gillespie DT, Hammons HK (1999) Analysis of polysaccharides by SEC[3]. In: Provder T (ed) Chromatography of Polymers, vol 731. American Chemical Society, Washington, DC, pp 288–310

Gohdes M, Mischnick P (1998) Determination of the substitution pattern in the polymer chain of cellulose sulfates. Carbohydr Res 309(1):109–115

Gohdes M, Mischnick P et al (1997) Methylation analysis of cellulose sulphates. Carbohydr Polym 33 (2–3):163–168

Gonera A, Goclik V et al (2002) Preparation and structural characterisation of O-aminopropyl starch and amylose. Carbohydr Res 337(21–23):2263–2272

Gratzl JS (1985) Lichtinduzierte Vergilbund von Zellstoffen pp Ursachen und Verhütung. Das Papier 39(10A): V14–V23

Grill E, Huber C et al (1993) Capillary zone electrophoresis of p-aminobenzoic acid derivatives of aldoses, ketoses and uronic acids. Electrophoresis 14:1004–1010

Haggkvist M, Li T-Q et al (1998) Effects of drying and pressing on the pore structure in the cellulose fibre wall studied by 1 H and 2 H NMR relaxation. Cellulose 5(1):33–49

Harding SE (2005) Analysis of polysaccharides by ultracentrifugation. size, conformation and interactions in solution. In: Heinze T (ed) Polysaccharides I - structure, characterisation and use. Springer, Berlin, pp 211–254

Hase S, Ibuki T et al (1984) Reexamination of the pyridylamination used for fluorescence labeling of oligosaccharides and its application to glycoproteins. J Biochem 95:197–203

Heinrich J (1999a) Strukturaufklärung von Cellulosederivaten und Galactanen mittels chemischer, chromatographischer und massenspektrometrischer Methoden. Fachbereich Chemie, Universität Hamburg, Hamburg

Heinrich JPM (1999b) Determination of the substitution pattern in the polymer chain of cellulose acetates. J Polym Sci A Polym Chem 37(15):3011–3016

Heinze U, Heinze T et al (1999) Synthesis and structure characterization of 2,3-O-carboxymethylcellulose. Macromol Chem Phys 200(4):896–902

Heinze U, Schaller J et al (2000) Characterisation of regioselectively functionalized 2,3-O-carboxymethyl cellulose by enzymatic and chemical methods. Cellulose 7:161–175

Henniges U, Prohaska T et al (2006) A fluorescence labeling approach to assess the deterioration state of aged papers. Cellulose 13(4):421–428

Holfstetter-Kuhn S, Paulus A et al (1991) Influence of borate complexation on the electrophoretic behavior of carbohydrates in capillary electrophoresis. Anal Chem 63:1541–1547

Honda S, Suzuki S et al (1991a) Capillary zone electrophoresis of reducing mono- and oligo-saccharides as the borate complexes of their 3-methyl-1-phenyl-2-pyrazolin-5-one derivatives. Carbohydr Res 215(1):193–198

Honda S, Yamamoto K et al (1991b) High-performance capillary zone electrophoresis of carbohydrates in the presence of alkaline earth metal ions. J Chromatogr A 588(1–2):327–333

Horner S, Puls J et al (1999) Enzyme-aided characterisation of carboxymethylcellulose. Carbohydr Polym 40(1):1–7

Huber CG, Hoelzl G (eds) (2001) CapillaryElectrochromatography. Amsterdam, Elsevier

Hult E-L, Larsson PT et al (2002) A comparative CP/MAS 13C-NMR study of the supermolecular structure of polysaccharides in sulphite and kraft pulps. Holzforschung 56(2):179–184

Husemann E, Weber OH (1942) Der Carboxylgehalt von Faser- und Holzcellulosen. J Prakt Chem 159:334–342

Issaq HJ, Janini GM et al (2004) Sheathless electrospray ionization interfaces for capillary electrophoresis—mass spectrometric detection. Advantages and limitations. J Chromatogr A 1053:37–42

Ivanov AR, Nazimov IV et al (2000) Direct determination of amino acids and carbohydrates by high-performance capillary electrophoresis with refractometric detection. J Chromatogr A 894:253–257

Jager AV, Tonin FG et al (2007) Comparative evaluation of extraction procedures and method validation for determination of carbohydrates in cereals and dairy products by capillary electrophoresis. J Sep Sci 30:586–594

Kabel MA (2002) Characterisation of complex xylo-oligosaccharides from xylan rich by-products. PhD thesis. Department of Agrotechnology and Food Sciences, Wageningen University, Wageningen

Kabel MA, de Waard P et al (2003) Location of O-acetyl substituents in xylo-oligosaccharides obtained from hydrothermally treated Eucalyptus wood. Carbohydr Res 338(1):69–77

Kato KL, Cameron RE (1999) A review of the relationship between thermally-accelerated ageing of paper and hornification. Cellulose 6(1):23–40

Katzenellenbogen E, Kocharova NA et al (2009) Structure of an abequose-containing O-polysaccharide from Citrobacter freundii O22 strain PCM 1555. Carbohydr Res 344(13):1724–1728

Kelli JF, Ramaley L et al (1997) Capillary zone electrophoresis-electrospray mass spectrometry at submicroliter flow rates: practical considerations and analytical performance. Anal Chem 69:51–60

Kiwitt-Haschemie K, Renger A et al (1996) A comparison between reductive-cleavage and standard methylation analysis for determining structural features of galactomannans. Carbohydr Polym 30(1):31–35

Klampfl C (2006) Recent advances in the application of capillary electrophoresis with mass spectrometric detection. Electrophoresis 27:3–34

Klampfl CW, Buchberger W (2001) Determination of carbohydrates by capillary electrophoresis with electrospray-mass spectrometric detection. Electrophoresis 22(13):2737–2742

Klemm D, Philipp B et al (1998) Comprehensive cellulose chemistry, vol 1, Fundamentals and analytical methods. Wiley-VCH, Weinheim

Klockow A, Amado R et al (1995) Separation of 8-aminonaphthalene-1,3,6-trisulfonic acid-labelled neutral and sialylated N-linked complex oligosaccharides by capillary electrophoresis. J Chromatogr A 716:241–257

Klockow A, Amado R et al (1996) The influence of buffer composition on separation efficiency and resolution in capillary electrophoresis of 8-aminonaphthalene-1,3,6-trisulfonic acid labeled monosaccharides and complex carbohydrates. Electrophoresis 17:110–119

Koelhed M, Karlberg B (2005) Capillary electrophoretic separation of sugars in fruit juices using on-line mid infrared Fourier transform detection. Analyst 130:772–778

Kondo A, Suzuki J et al (1990) Improved method for fluorescence labeling of sugar chains with sialic acid residues. Agric Biol Chem 54(8):2169–2170

Krainz K, Potthast A et al (2009) Effects of selected key chromophores on cellulose integrity upon bleaching 10th EWLP, Stockholm, Sweden, August 25-28, 2008. Holzforschung 63(6):647–655

Kristiansen KA, Ballance S et al (2009) An evaluation of tritium and fluorescence labelling combined with multi-detector SEC for the detection of carbonyl groups in polysaccharides. Carbohydr Polym 76(2):196–205

Kumirska J, Czerwicka M et al (2010) Application of spectroscopic methods for structural analysis of chitin and chitosan. Mar Drugs 8(5):1567–1635

Laine CTTVAVT (2002) Methylation analysis as a tool for structural analysis of wood polysaccharides. Holzforschung 56(6):607–614

Larsson PT, Wickholm K et al (1997) A CP/MAS ^{13}C NMR investigation of molecular ordering in celluloses. Carbohydr Res 302:19–25

Lee CK, Gray GR (1995) Analysis of positions of substitution of O-acetyl groups in partially O-acetylated cellulose by the reductive-cleavage method. Carbohydr Res 269(1):167–174

Lee S-J, Altaner C et al (2003) Determination of the substituent distribution along cellulose acetate chains as revealed by enzymatic and chemical methods. Carbohydr Polym 54(3):353–362

Levigne S, Thomas M et al (2002) Determination of the degrees of methylation and acetylation of pectins using a C18 column and internal standards. Food Hydrocolloids 16(6):547–550

Lewin L (1972) In: Whistler RL, BeMiller JN (eds) Methods in carbohydrate chemistry, vol 6. Academic, New York, p 76

Lewin M (1997) Oxidation and aging of cellulose. Macromol Symp 118:715–724

Li DT, Sheen JF et al (2000) Structural analysis of chromophore-labeled disaccharides by capillary electrophoresis tandem mass spectrometry using ion trap mass spectrometry. J Am Soc Mass Spectrom 11:292–300

Li J, Wan Y et al (2009) Preparation and characterization of 2,3-dialdehyde bacterial cellulose for potential biodegradable tissue engineering scaffolds. Mater Sci Eng C 29(5):1635–1642

Liebert T, Pfeiffer K et al (2005) Carbamoylation applied for structure determination of cellulose derivatives. Macromol Symp 223(1):93–108

Liitiä T, Maunu SL et al (2000) Solid state NMR studies on cellulose crystallinity in fines and bulk fibres separated from refined kraft pulp. Holzforschung 54(6):618–624

Liu J, Shirota O et al (1991) Capillary electrophoresis of amino sugars with laser-induced fluorescence detection. Anal Chem 63:413–417

Liu Y, Shu C et al (1997) High-performance capillary electrophoretic separation of carbohydrates with indirect UV detection using diethylamine and borate as electrolyte additives. J Capillary Electrophor 4(3):97–103

Manelius R, Buléon A et al (2000) The substitution pattern in cationised and oxidised potato starch granules. Carbohydr Res 329(3):621–633

Mazumder S, Lerouge P et al (2005) Structural characterisation of hemicellulosic polysaccharides from Benincasa hispida using specific enzyme hydrolysis, ion exchange chromatography and MALDI-TOF mass spectroscopy. Carbohydr Polym 59(2):231–238

Mazzarino M, De Angelis F et al (2010) Microwave irradiation for a fast gas chromatography–mass spectrometric analysis of polysaccharide-based plasma volume expanders in human urine. J Chromatogr B 878(29):3024–3032

Mechref Y, Ostrander GK et al (1995) Capillary electrophoresis of carboxylated carbohydrates. Part 2. Selective precolumn derivatization of sialooligosaccharides derived from gangliosides with 7-aminonaphthalene-1,3-disulfonic acid fluorescing tag. Electrophoresis 16:1499–1504

Mechref Y, Ostrander GK et al (1997) Capillary electrophoresis of carboxylated carbohydrates. IV. Adjusting the separation selectivity of derivatized carboxylated carbohydrates by controlling the electrolyte ionic strength at subambient temperature and in the absence of electroosmotic flow. J Chromatogr A 792:75–82

Melander C, Tømmeraas K (2010) Heterogeneous hydrolysis of hyaluronic acid in ethanolic HCl slurry. Carbohydr Polym 82(3):874–879

Mihranyan A, Llagostera AP et al (2004) Moisture sorption by cellulose powders of varying crystallinity. Int J Pharm 269(2):433–442

Mischnick P (1991) Determination of the substitution pattern of cellulose acetates. J Carbohydr Chem 10(4):711–722

Mischnick P (1997) New developments in the analysis of the substitution pattern of polysaccharide derivatives. Macromol Symp 120:281–290

Mischnick P (2001) Challenges in structure analysis of polysaccharide derivatives. Cellulose 8(4):245–257

Mischnick P (2002) Challenges in structure analysis of polysaccharide derivatives. Cellulose 00:1–13

Mischnick P, Adden R (2008) Fractionation of polysaccharide derivatives and subsequent analysis to differentiate heterogeneities on various hierarchical levels. Macromol Symp 262:1–7

Mischnick P, Hennig C (2001) A new model for the substitution patterns in the polymer chain of polysaccharide derivatives. Biomacromolecules 2(1):180–184

Mischnick P, Evers B et al (1994) Analysis of oligosaccharides containing 2-deoxy-alpha-D-arabino-hexosyl residues by the reductive-cleavage method. Carbohydr Res 264(2):293–304

Mischnick P, Heinrich J et al (2000) Structure analysis of 1,4-glucan derivatives. Macromol Chem Phys 201:1985–1995

Mischnick P, Niedner W et al (2005) Possibilities of mass spectrometry and tandem-mass spectrometry in the analysis of cellulose ethers. Macromol Symp 223(1):67–78

Mischnick P, Momcilovic D et al (2010) Chemical structure analysis of starch and cellulose derivatives. Adv Carbohydr Chem Biochem 64:117–210

Momenbeik F, Johns C et al (2006) Sensitive determination of carbohydrates labelled with p-nitroaniline by capillary electrophoresis with photometric detection using a 406 nm light-emitting diode. Electrophoresis 27:4039–4046

Mort AJ, Chen EMW (1996) Separation of 8-aminonaphthalene-1,3,6-trisulfonate (ANTS)-labeled oligomers containing galacturonic acid by capillary electrophoresis: application to determining the substrate specificity of endopolygalacturonases. Electrophoresis 17:379–383

Mukerjea R, Kim D et al (1996) Simplified and improved methylation analysis of saccharides, using a modified procedure and thin-layer chromatography. Carbohydr Res 292:11–20

Newman RH, Davidson TC (2004) Molecular conformations at the cellulose-water interface. Cellulose 11:23–32

Nguyen DT, Lerch H et al (1997) Separation of derivatized carbohydrates by co-electroosmotic capillary electrophoresis. Chromatographia 46(3/4):113–121

Nie S-P, Xie M-Y (2011) A review on the isolation and structure of tea polysaccharides and their bioactivities. Food Hydrocolloids 25(2):144–149

Oh SY, Yoo DI et al (2005) FTIR analysis of cellulose treated with sodium hydroxide and carbon dioxide. Carbohydr Res 340(3):417–428

Osborn HMI, Lochey F et al (1999) Analysis of polysaccharides and monosaccharides in the root mucilage of maize (Zea mays L.) by gas chromatography. J Chromatogr A 831(2):267–276

Oudhoff KA, Buijtenhuijs FA et al (2004) Determination of the degree of substitution and its distribution of carboxymethylcelluloses by capillary zone electrophoresis. Carbohydr Res 339(11):1917–1924

Perera A, Meda V et al (2010) Resistant starch: a review of analytical protocols for determining resistant starch and of factors affecting the resistant starch content of foods. Food Res Int 43(8):1959–1974

Petzold K, Schwikal K et al (2006) Carboxymethyl xylan - synthesis and detailed structure characterization. Carbohydr Polym 64(2):292–298

Phillipp B, Rehder W et al (1965) Carboxylgruppenbestimmung in Chemiezellstoffen. Das Papier 19:1–9

Plocek J, Chmelik J (1997) Separation of disaccharides as their borate complexes by capillary electrophoresis with indirect detection in visible range. Electrophoresis 18:1148–1152

Potthast A, Röhrling J et al (2003) A novel method for the determination of carbonyl groups in cellulosics by fluorescence labeling. 3. Monitoring oxidative processes. Biomacromolecules 4(3):743–749

Potthast A, Schiehser S et al (2004) Effect of UV radiation on the carbonyl distribution in different pulps. Holzforschung 58(6):597–602

Potthast A, Rosenau T et al (2006) Analysis of oxidized functionalities in cellulose. In: Klemm D (ed) Polysaccharides II. Springer, Heidelber, pp 1–48

Putnam ES (1964) The exchange reaction between calcium and carboxyl groups in cellulose. TAPPI J 47:549–554

Racaityte K, Kiessig S et al (2005) Application of capillary zone electrophoresis and reversed-phase high-performance liquid chromatography in the biopharmaceutical industry for the quantitative analysis of the monosaccharides released from a highly glycosylated therapeutic protein. J Chromatogr A 1079:354–365

Rana V, Kumar V et al (2009) Structure of the oligosaccharides isolated from Dalbergia sissoo Roxb. leaf polysaccharide. Carbohydr Polym 78(3):520–525

Rehder W, Philipp B et al (1965) Ein Beitrag zur Analytik der Carbonylgru ppen in Oxycellulosen und technischen Zellstoffen. Das Papier 19(9):502

Ren J L, Sun R-C (2010) Hemicelluloses. In: Sun R-C (ed) Cereal straw as a resource for sustainable biomaterials and biofuels. Elsevier, Amsterdam, pp 73–130

Ristolainen M (1999) Characterization of totally chlorine-free effluents from Kraft pulp bleaching II. Analysis of carbohydrate-derived constituents after acid hydrolysis by capillary zone electrophoresis. J Chromatogr A 832:203–209

Röder T, Moosbauer J et al (2006a) Crystallinity determination of man-made cellulose fibers – comparison of analytical methods. Lenzinger Ber 86:132–136

Röder T, Moosbauer J et al (2006b) Crystallinity determination of native cellulose – comparison of analytical methods. Lenzinger Ber 86:85–89

Röder T, Moosbauer J et al (2009) Comparative characterisation of man-made regenerated cellulose fibres. Lenzinger Ber 87:98–105

Röhrling J, Potthast A et al (2001) Synthesis and testing of a novel fluorescence label for carbonyls in carbohydrates and cellulosics. Synlett 5:682–684

Röhrling J, Potthast A et al (2002a) A novel method for the determination of carbonyl groups in cellulosics by fluorescence labeling. 2. Validation and applications. Biomacromolecules 3(5):969–975

Röhrling J, Potthast A et al (2002b) A novel method for the determination of carbonyl groups in cellulosics by fluorescence labeling. 1. Method development. Biomacromolecules 3(5):959–968

Rosenau T, Potthast A et al (2001) Hydrolytic processes and condensation reactions in the cellulose solvent system N, N-dimethylacetamide/lithium chloride. Part 1. Holzforschung 55(6):661–666

Rosenau T, Potthast A et al (2004) Isolation and identification of residual chromophores in cellulosic materials. Polymer 45(19):6437–6443

Rosenau T, Potthast A et al (2005a) Isolation and identification of residual chromophores in cellulosic materials, vol 223. Wiley-VCH, Weinheim, pp 239–252

Rosenau T, Potthast A et al (2005b) Discoloration of cellulose solutions in N-methylmorpholine-N-oxide (Lyocell). Part 2: Isolation and identification of chromophores. Cellulose 12(2):197–208

Rosenau T, Potthast A et al (2007) Isolation and identification of residual chromophores from aged bleached pulp samples. Holzforschung 61:656–661

Rovio S, Yli-Kauhaluoma J et al (2007) Determination of neutral carbohydrates by CZE with direct UV detection. Electrophoresis 28:3129–3135

Rovio S, Simolin H et al (2008) Determination of monosaccharide composition in plant fiber materials by capillary zone electrophoresis. J Chromatogr A 1185:139–144

Rovio S, Siren K et al (2011) Application of capillary electrophoresis to determine metal cations, anions, organic acids, and carbohydrates in some Pinot Noir red wines. Food Chem 124:1194–1200

Ruiz-Matute AI, Hernández-Hernández O et al (2010) Derivatization of carbohydrates for GC and GC-MS analyses. J Chromatogr B Analyt Technol Biomed Life Sci 879(17–18):1226–1240

Rußler A, Lange T et al (2005a) A novel method for analysis of xanthate group distribution in viscoses. Macromol Symp 223(1):189–200

Russler A, Potthast A, et al (2005a) New methylation analysis of viscose. Institute of Chemistry, Slovak Academy of Sciences: 13th European carbohydrate symposium, 21–26 August, Bratislava; Book of Abstracts, Institute of Chemistry, Slovak Academy of Sciences

Russler A, Saake B, et al (2005b) A novel approach to assess xanthate group distribution in viscose. Japanese-European workshop on cellulose and functional polysaccharides. Department of Chemistry, University of Natural Resources and Applied Life Sciences, Vienna, p 74

Rußler A, Potthast A et al (2006) Determination of substituent distribution of viscoses by GPC. Holzforschung 60(5):467–473

Rydlund A, Dahlman O (1996) Efficient capillary zone electrophoretic separation of wood-derived neutral and acidic mono- and oligosaccharides. J Chromatogr A 738:129–140

Rydlund A, Dahlman O (1997) Oligosaccharides obtained by enzymatic hydrolysis of birch kraft pulp xylan: analysis by capillary zone electrophoresis and mass spectrometry. Carbohydr Res 30:95–102

Saake B, Horner S et al (2000) Detailed investigation on the molecular structure of carboxymethyl cellulose with unusual substitution pattern by means of an enzyme-supported analysis. Macromol Chem Phys 201(15):1996–2002

Saake B, Kruse T et al (2001) Investigation on molar mass, solubility and enzymatic fragmentation of xylans by multi-detected SEC chromatography. Bioresour Technol 80(3):195–204

Sanz ML, Martínez-Castro I (2007) Recent developments in sample preparation for chromatographic analysis of carbohydrates. J Chromatogr A 1153(1–2):74–89

Sartori J, Potthast A et al (2003) Alkaline degradation kinetics and CE-separation of cello- and xylooligomers. Part I. Carbohydr Res 338:1209–1216

Sato H, Mizutani S-I et al (2002) Determination of degree of substitution in N-carboxyethylated chitin derivatives by pyrolysis-gas chromatography in the presence of oxalic acid. J Anal Appl Pyrolysis 64(2):177–185

Schelosky N, Röder T et al (1999) Molmasseverteilug cellulosischer Produkte mittels Grössenausschlusschromatographie in DMAc/LiCl. Das Papier 53(12):728–738

Schleicher H, Lang H (1994) Carbonyl and carboxyl groups in pulps and cellulose products. Das Papier 12:765–768

Schwaiger H, Oefner PJ et al (1994) Capillary zone electrophoresis and micellar electrokinetic chromatography of 4-aminobenzonitrile carbohydrate derivatives. Electrophoresis 15:941–952

Schwikal K, Heinze T et al (2005) Cationic xylan derivatives with high degree of functionalization. Macromol Symp 232(1):49–56

Senso A, Franco P et al (2000) Characterization of doubly substituted polysaccharide derivatives. Carbohydr Res 329(2):367–376

Singh V, Tiwari A et al (2006) Microwave-promoted hydrolysis of plant seed gums on alumina support. Carbohydr Res 341(13):2270–2274

Sixta H (1995). Habilitation thesis, Zellstoffherstellung unter Berücksichtigung umweltfreundlicher, Aufschluß- und Bleichverfahren am Beispiel von Chemiezellstoffen. Technical University of Graz, Graz

Sjöberg J, Adorjan I et al (2004) An optimized CZE method for analysis of mono- and oligomeric aldose mixtures. Carbohydr Res 339:2037–2043

Snyder DS, Gibson D et al (2006) Structure of a capsular polysaccharide isolated from Salmonella enteritidis. Carbohydr Res 341(14):2388–2397

Soga T, Heiger DN (1998) Simultaneous determination of monosaccharides in glycoproteins by capillary electrophoresis. Anal Biochem 261:73–78

Soga T, Ross GA (1999) Simultaneous determination of inorganic anions, organic acids, amino acids and carbohydrates by capillary electrophoresis. J Chromatogr A 837:231–239

Soga T, Serwe M (2000) Determination of carbohydrates in food samples by capillary electrophoresis with indirect UV detection. Food Chem 69:339–344

Steeneken PAM, Woortman AJJ (1994) Substitution patterns in methylated starch as studied by enzymic degradation. Carbohydr Res 258:207–221

Stefansson M, Novotny M (1994) Separation of complex oligosaccharide mixtures by capillary electrophoresis in the open-tubular format. Anal Chem 66:1134–1140

Sun Y-X, Liu J-C et al (2010) Purification, composition analysis and antioxidant activity of different polysaccharide conjugates (A PPs) from the fruiting bodies of Auricularia polytricha. Carbohydr Polym 82(2):299–304

Szabolcs O (1961) A colorimetric method for the determination of reducing carbonyl groups in cellulose. Das Papier 15:41

TAPPI (2009) T 249 - Carbohydrate composition of extractive-free wood and wood pulp by gas–liquid chromatography, p 8

TAPPI (2009) TAPPI method T-430 om-99 Copper number of pulp, paper, and paperboard (Braidy)

Thomas M, Chauvelon G et al (2003) Location of sulfate groups on sulfoacetate derivatives of cellulose. Carbohydr Res 338(8):761–770

Tüting W, Adden R et al (2004a) Fragmentation pattern of regioselectively O-methylated maltooligosaccharides in electrospray ionisation-mass spectrometry/collision induced dissociation. Int J Mass Spectrom 232(2):107–115

Tüting W, Wegemann K et al (2004b) Enzymatic degradation and electrospray tandem mass spectrometry as tools for determining the structure of cationic starches prepared by wet and dry methods. Carbohydr Res 339(3):637–648

Vaca-Garcia C, Borredon ME et al (2001) Determination of the degree of substitution (DS) of mixed cellulose esters by elemental analysis. Cellulose 8(3):225–231

van der Burgt YEM, Bergsma J et al (1998) Distribution of methyl substituents over branched and linear regions in methylated starches. Carbohydr Res 312(4):201–208

van der Burgt YEM, Bergsma J et al (1999) Distribution of methyl substituents over crystalline and amorphous domains in methylated starches. Carbohydr Res 320(1–2):100–107

van der Burgt YE, Bergsma J et al (2000a) Substituent distribution in highly branched dextrins from methylated starches. Carbohydr Res 327(4):423–429

van der Burgt YEM, Bergsma J et al (2000b) Distribution of methyl substituents in amylose and amylopectin from methylated potato starches. Carbohydr Res 325(3):183–191

van der Burgt YEM, Bergsma J et al (2000c) FAB CIDMS/MS analysis of partially methylated maltotrioses derived from methylated amylose: a study of the substituent distribution. Carbohydr Res 329:341–349

Vlasenko EY, Ryan AI et al (1998) The use of capillary viscometry, reducing end-group analysis, and size exclusion chromatography combined with multi-angle laser light scattering to characterize endo-1,4-[beta]-glucanases on carboxymethylcellulose: a comparative evaluation of the three methods. Enzym Microb Technol 23(6):350–359

Walter RH (1998) Isolation, purification, and characterization. In: Steve T (ed) Polysaccharide dispersions. Academic, San Diego, CA, Chapter 7, pp 123–155

Wang X, Chen Y (2001) Determination of carbohydrates as their p-sulfophenylhydrazones by capillary zone electrophoresis. Carbohydr Res 332:191–196

Wang X, Chen Y et al (2002) Analysis of carbohydrates by capillary zone electrophoresis with on-line capillary derivatization. J Liq Chrom Rel Technol 25(4):589–600

Wennerblom A (1961) Determination of carbonyl groups in hydrocellulose. Sven Pap 14:519

Wickholm K, Larsson PT et al (1998) Assignment of non-crystalline forms in cellulose I by CP/MAS ^{13}C NMR spectroscopy. Carbohydr Res 312:123–129

Wilke O, Mischnick P (1995) Analysis of cationic starches: determination of the substitution pattern of O-(2-hydroxy-3-trimethylammonium)propyl ethers. Carbohydr Res 275(2):309–318

Willför S, Sundberg K et al (2008) Spruce-derived mannans - a potential raw material for hydrocolloids and novel advanced natural materials. Carbohydr Polym 72(2):197–210

Willför S, Pranovich A et al (2009) Carbohydrate analysis of plant materials with uronic acid-containing polysaccharides-A comparison between different hydrolysis and subsequent chromatographic analytical techniques. Ind Crop Prod 29(2–3):571–580

Wilson K (1948) Determination of carboxylic groups in pulp. Sven Pap 51:45–49

Wittgren B, Porsch B (2002) Molar mass distribution of hydroxypropyl cellulose by size exclusion chromatography with dual light scattering and refractometric detection. Carbohydr Polym 49(4):457–469

Yamamoto K, Hamase K et al (2003) 2-Amino-3-phenylpyrazine, a sensitive fluorescence prelabeling reagent for the chromatographic or electrophoretic determination of saccharides. J Chromatogr A 1004(1–2):99–106

Yang L, Wang Z et al (2010) Isolation and structural characterization of a polysaccharide FCAP1 from the fruit of Cornus officinalis. Carbohydr Res 345(13):1909–1913

You J, Sheng X et al (2008) Detection of carbohydrates using new labeling reagent 1-(2-naphthyl)-3-methyl-5-pyrazolone by capillary zone electrophoresis with absorbance (UV). Anal Chim Acta 609:66–75

Yu N, Gray GR (1998a) Analysis of the positions of substitution of acetate and butyrate groups in cellulose acetate-butyrate by the reductive-cleavage method. Carbohydr Res 312(4):225–231

Yu N, Gray GR (1998b) Analysis of the positions of substitution of acetate and propionate groups in cellulose acetate-propionate by the reductive-cleavage method. Carbohydr Res 313(1):29–36

Zamfir A, Peter-Katalinic J (2004) Capillary electrophoresis-mass spectrometry for glycoscreening in biomedical research. Electrophoresis 25:1949–1963

Zatkovskis Carvalho A, da Silva JAF et al (2003) Determination of mono- and disaccharides by capillary electrophoresis with contactless conductivity detection. Electrophoresis 24:2138–2143

Zemann A, Nguyen DT et al (1997) Fast separation of underivatized carbohydrates by coelectroosmotic capillary electrophoresis. Electrophoresis 18:1142–1147

Zhou W, Baldwin RP (1996) Capillary electrophoresis and electrochemical detection of underivatized oligo- and polysaccharides with surfactant-controlled electroosmotic flow. Electrophoresis 17:319–324

Zuckerstätter G, Schild G et al (2009) The elucidation of cellulose supramolecular structure by ^{13}C CP-MAS NMR. Lenzinger Ber 87:41–49

Preparation and Properties of Cellulose Solutions

5

Patrick Navard, Frank Wendler, Frank Meister, Maria Bercea, and Tatiana Budtova

Contents

5.1	Introduction	91
5.1.1	Derivatising Pathways	93
5.1.2	Non-derivatising Compounds	94
5.2	**Macroscopic Mechanisms of the Dissolution of Native Cellulose Fibres**	98
5.2.1	Gradient of Solubility in the Various Locations of the Cell Wall	99
5.2.2	Overall Dissolution Mechanism	100
5.3	**Cellulose–Sodium Hydroxide–Water Solutions**	101
5.3.1	Mercerisation	101
5.3.2	Dissolution in NaOH–Water	102
5.4	**Cellulose–Ionic Liquid Solutions**	120
5.4.1	Materials and Methods	122
5.4.2	Steady-State and Intrinsic Viscosity of Cellulose–EMIMAc and Cellulose–BMIMCl Solutions	122
5.4.3	Influence of Water on Cellulose–EMIMAc Viscosity	123
5.4.4	Influence of Cellulose on the Diffusion of $[EMIM]^+$ and $[Ac]^-$ Ions in Cellulose–EMIMAc Solutions	126
5.4.5	Phase Diagram and Solubility Limit of Cellulose in EMIMAc in the Presence of DMSO	127
5.4.6	Conclusions	128
5.5	**Stabilisation of Cellulose: NMMO Solutions (Lyocell Process)**	128
5.5.1	Solution Quality and Solution State	128
5.5.2	Thermostability of Cellulose Solutions	130
5.6	**Cellulose–Other Polysaccharide Blends**	133
5.7	**Cellulose–Lignin Mixtures in Solution (Sescousse et al. 2010b)**	139
5.7.1	Cellulose–Lignin Mixtures in Aqueous 8 % NaOH	140
5.7.2	Coagulation of Cellulose from Cellulose–Lignin–8 % NaOH–Water Gels and Aero-Lignocellulose Morphology	143
References		144

Abstract

Cellulose cannot melt and is not soluble in common organic solvents. The first part of this chapter is a review of the main aspects of cellulose dissolution. Research results obtained in several EPNOE laboratories are then described. This includes the mechanisms of the dissolution of native cellulose fibres, solution properties in sodium hydroxide water and ionic liquids, stabilisation of cellulose in *N*-methylmorpholine-*N*-oxide–water and mixtures of cellulose with other polysaccharide or lignin in solution.

5.1 Introduction

Cellulose is a linear, semi-flexible polymer (see Chap. 2) which arranges itself when condensed into crystalline and noncrystalline phases. This is the case when cellulose is biosynthesised in plants or other organisms or when it is

---

P. Navard (✉)
Mines ParisTech, Centre de Mise en Forme des Matériaux – CEMEF, UMR CNRS 7635, Rue Claude Daunesse, BP 207, 06904 Sophia-Antipolis, France
e-mail: patrick.navard@mines-paristech.fr

Fig. 5.1 Basic mechanisms for a solid polymer part situated at the interface between polymer and solvent

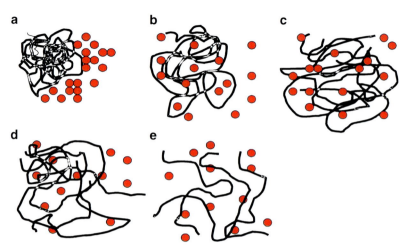

regenerated from a solution. As such, it is following the general rules that are applicable to long-chain molecules. Among these rules are the facts that noncrystalline phase (so-called amorphous phase) has different degrees of order and organisation, that long chains are more difficult to be dissolved than short ones for thermodynamic reasons and that they can entangle at high enough molar mass and concentration.

Since cellulose cannot melt, dissolution is a major issue. Many reviews have been devoted to cellulose dissolution (e.g. Warwicker et al. 1966; Liebert 2010). Cellulose solutions are used for processing directly cellulose in the form of fibres, films, membranes or other not too bulky objects like sponges or for performing chemical derivatisation (see Chap. 9). Since cellulose chains have no specific features, dissolving cellulose should then happen as it is occurring for any other flexible or semi-flexible polymer, and solutions should behave as normal polymer solutions. This is indeed almost the case with one major difference from most synthetic polymers: Cellulose is "synthesised" by nature in a complex environment where many other compounds (lignin, hemicellulose, fats, proteins, pectins, etc.) are present and interacting more or less strongly with cellulose chains. In addition and due to the biosynthesis mechanisms, the organisation of chains is usually complex, like in the secondary plant cell walls where cellulose chains are forming a sort of self-composite with many differently oriented layers. Due to the importance of the field of polymer dissolution in materials engineering where dissolution is used in many industrial areas (drug delivery, pulp and paper, membranes, recycling, etc.), there is a good knowledge of the mechanisms at stake when a solid polymer is placed in contact with a solvent (Miller-Chou and Koenig 2003). As a general basis, polymer chains will go into solution through the interface between the solid polymer and the solvent and will pass several phases as shown on Fig. 5.1.

When the solid phase is placed in contact with the solvent (Fig. 5.1a), the solvent is swelling the solid phase at the interface which goes above Tg (Fig. 5.1b), and this swelling is increasing up to the point of disentanglement (Fig. 5.1c). Chains can then move out of the swollen phase to the solvent phase (Fig. 5.1d) and the solubilisation front can advance inside the solid material (Fig. 5.1e). Such a scheme is indeed what is occurring when a regenerated cellulose fibre is placed in a solvent (Chaudemanche and Navard 2011). In this case, the swelling and dissolution mechanisms of dry, never-dried and rewetted lyocell fibres prepared from solutions in N-methylmorpholine-N-oxide were investigated by redissolving these cellulose fibres in the same solvent varying the contents of water (from monohydrate to 24 % w/w). As for any synthetic polymer fibre, a radial dissolution starting from the outer layers was observed.

It is usually said that cellulose is difficult to be dissolved and that a rather limited number of solvents are available. However, this is the case for many polymers. Dissolution is favoured since the entropy will increase in a solution state through the contributions of several entropic factors like the entropy of mixing, the entropy of conformation mobility and, if applicable, the entropy gain due to counter ions. The major difficulty for dissolving a polymer is well known and is due to the very large decrease of entropy that is happening when dissolving of the polymer is compared to its parent monomer, because of its long-chain character. If considering, for example, very common polymers like polyethylene or polypropylene, the number of the possible solvents is not larger than the one for cellulose. Since these polymers are melting, it is also true that efforts placed in developing their solvents were much smaller than the efforts for finding solvents for cellulose. There are solvents for cellulose, but they are not the simple, common organic solvents we are used of for the vast majority of polymers. A recent series of papers (Glasser et al. 2012a, b) discussed these issues with questions around the reason why cellulose is not soluble in water, Lindman et al. 2010 arguing that the main reason is not the existence of hydrogen bonds but due to hydrophobic interactions

5.1.1 Derivatising Pathways

Dissolving cellulose has always been a search since cellulose was isolated in the nineteenth century. Many compounds have been tested and several commercial pathways have been implemented to produce regenerated cellulose. Most of them are in fact going through cellulose derivatives or complexes (Liebert 2010). Many compounds and methods are able to produce derivatives that are then chemically reverted to cellulose after processing. After 7 years of research, one of the first to be industrially implemented in 1890 was the use of nitrocellulose by Count Hilaire de Chardonnet. Another example is the Fortisan process, not any more in production, which used cellulose acetate–acetone solutions for producing cellulose acetate that were saponified in caustic soda to obtain cellulose fibres (Segal and Eggerton 1961). Cellulose carbamate, obtained by the reaction of cellulose with urea, is another example. The first to report this reaction were Hill and Jacobsen (1938). But it was only at the beginning of the 1960s that Sprague and Noether (1961) found that this reaction leads to a cellulose derivative they named cellulose carbamate. The potential for using this derivative for producing fibres was explored by the Finnish company Neste Oy. Cellulose is treated with urea to produce cellulose carbamate that is dissolved in NaOH–water. After processing this solution, cellulose is recovered after an acidic treatment. Despite its interest and further developments (Kunze and Finck 2005), there is no industrial use at the present time.

The only large scale production of cellulose products through a cellulose derivative is the viscose process made from cellulose xanthate, dating from first patent on the use of cellulose xanthate preparations and subsequent regeneration by Cross et al. (1892). Since that time, after some first economical and technical difficulties, the viscose process proved to be a very efficient and powerful method for producing fibres, films and sponges, still in use today despite its difficulty for controlling air and water pollution. The viscose process is based on the treatment of cellulose fibres from wood or cotton with sodium hydroxide and carbon disulphite, forming a cellulose derivative called cellulose xanthate which has the interesting property to be soluble in sodium hydroxide–water mixtures. The solution can then be shaped, followed by an acid or a thermal treatment that reverts the cellulose derivative back to cellulose, with a noticeable change of cellulose crystal structure (cellulose I to cellulose II transformation). Among the many key chemical reactions and physical processes that are characterising this process, the dissolution step of the cellulose xanthate fibres into the sodium hydroxide–water mixture is of great importance since it is controlling the quality of the subsequent processing. As can be imagined, there has been numerous scientific works studying, for example, the influence of cellulose origin, purity, molecular weight, xanthate group distribution (Russler et al. 2005, 2006) or

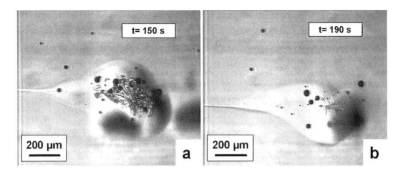

Fig. 5.2 Dissolution of cellulose xanthate in NaOH 8 %–water under flow (shear rate about 20 s^{-1}, flow from *right* to *left*) observed with the contra-rotating rheometer at two successive dissolution times: (**a**) 150 s and (**b**) 190 s. Cellulose xanthate fibres are producing a highly viscous phase that has difficulties to be dispersed into the solvent. Dispersion and final dissolution happen through the production of a thin filament of viscous phase that is extracted by the solvent friction

dissolution conditions, temperature (Musatova et al. 1972), soda concentration, etc., on the efficiency of the process. The way cellulose xanthate dissolves in NaOH–water (LeMoigne and Navard 2010a) is a good illustration of the difficulty to go through a gel phase before entering into the solution phase. Cellulose xanthate with 61 % CS$_2$ placed in NaOH 8 % water is producing a very viscous, gelly phase that will disperse only during shear through the extraction of cellulose xanthate viscous solution phase filaments as can be seen on Fig. 5.2. The dispersion and complete dissolution of this viscous phase can only be achieved by a strong flow.

5.1.1.1 Complexing Agents

Many aqueous-based compounds are complexing cellulose. They are not leading to a real solution, i.e. a dispersion of cellulose chains in a solvent. The first agent was discovered by Schweitzer in the middle of 1800s. He found that it is possible to prepare clear solutions when cellulose is mixed in a solution of copper salts and ammonia. This process called now cuprammonium is leading to good regenerated cellulose objects (fibres and membranes) but the cost of solvent and of depollution is very high. Today, the production is limited to a few tens of thousands tons per year. However, this class of compounds is very interesting in the sense that the cellulose complex can be easily studied. Two compounds of this class, cuoxen (Cu(NH$_2$(CH$_2$) 2NH$_2$)$_2$[HO]$_2$) and cadoxen (Cd(NH$_2$(CH$_2$) 2NH$_2$)$_2$[HO]$_2$), are still widely used to give an indication of the cellulose molar mass through the measurement of the intrinsic viscosity (these methods being normalised like in ASTM D1795 and ASTM D4243, for example). These metal complexes dissolve cellulose by deprotonating and coordinatively binding C2 and C3 hydroxyl groups (Liebert 2010). Many other inorganic salt hydrates combined or not with water have also been found to dissolve cellulose whilst complexing it. A list of aqueous salts is given in Liebert (2010). As an example (Hattori et al. 2004), ethylenediamine/thiocyanate salts have been shown to solubilise cellulose DP210 up to 16 %, leading to the formation of a mesophase for the highest concentrations.

5.1.2 Non-derivatising Compounds

There are not so many classes of compounds that are leading to more or less well-dispersed but non-derivatised cellulose solutions. Aside the fact that the increase of entropy for long chains going into solution is low, two other reasons are also at stake for cellulose. The first reason is chain rigidity. Degrees of freedom for undergoing chain conformational changes are very limited for a cellulose chain that can only turn around its 1,4 links. A lot of work was performed from 1950s till 1970s for estimating the

Table 5.1 Persistence length q and unperturbed dimension parameter A for cellulose and derivates [adapted from Kamide and Saito (1983)]

Polymer (DS)	Solvent	q (nm)	A (nm)
Cellulose	Cadoxen	4	0.16
Cellulose	FeTNa	9	0.19
Cellulose acetate (2.46)	Acetone	8	0.17
Cellulose nitrate (2.55)	Acetone	12	0.21
Hydroxyethyl cellulose	Water	9	0.19
Polyethylene	Ethylhexyl adipate	0.96	
Polystyrene	Cyclohexane	1	

Values of polyethylene and polystyrene are adapted from Kamide and Saito (1987)

mean conformation parameters. Benoît 1948 calculated the mean-square end-to-end distance of cellulose chains in the unperturbed state, showing that it is much larger than chains with cis or trans free rotations. A large number of papers were then devoted to the theoretical approaches needed to measure various conformation parameters of cellulose and cellulose derivatives by light and neutron scattering (Gupta et al. 1976), flow birefringence (Noordermeer et al. 1975) and viscometry (Holt et al. 1976). Aside debates on chain statistics theories, all data show that the cellulose backbone is semi-rigid. This can be appreciated in Table 5.1 where the persistence length of cellulose and cellulose derivatives is markedly larger than the one of polyolefins. Polymers having a rigid backbone have a further difficulty in going into solution since there is a further loss of conformational entropy increase compared to flexible polymer, when dissolved. A second reason is the large amount of intra- and inter-hydrogen bond connections present in cellulose, needed solvents able to break them and leading to a high tendency to self-aggregation in solution.

The last point of interest before detailing the properties of cellulose solutions is whether these mixtures of cellulose and a solvent give a true solution, i.e. if the state is a homogeneous dispersion of individual cellulose chains in the solvent. This is a critical issue with all cellulose solutions where nanoscale cellulose agglomerates (often called prehump, or micro- or nanogels) are often present in the solution. The method for ascertaining the presence of cellulose agglomerates in solution is light scattering (Seger et al. 1996). Many papers are showing that indeed, aggregates are present in what is usually considered as good, non-derivatising solutions. Drechsler et al. (2000) showed that cellulose is molecularly dispersed only in special mixtures of N-methylmorpholine-N-oxide (NMMO)/water with diethylenetriamine. The M_w dependence of the radii of gyration follows a power law behaviour with an exponent equal to the theoretical renormalisation group value for flexible, linear chains in a good solvent (i.e. 0.6). But solutions in pure N-methylmorpholine-N-oxide/water are not showing this. Röder and Morgenstern (1999) plotted Guiner–Zimm plots that showed that aggregates of the order of several million g/mol were observed. Aggregates comprising several 100 chains were present together with small aggregates, giving a bimodal structure. Activation of cellulose, i.e. pretreatments like NaOH or ammonia, was effective for decreasing the size of aggregates, but did not change the fact that aggregates were present. The same applies to solutions in N,N-dimethylacetamide/lithium chloride where aggregates with sizes above 100 nm are present (Röder et al. 2000), even if these results must be taken with care regarding the very important effect of water on chain aggregation (Potthast et al. 2002). Cellulose aggregates with a radius of gyration of 230 nm were found in solutions of NaOH–urea (Chen et al. 2007). All these results show that attaining molecularly dispersed cellulose solutions is very difficult, probably owing to the ability of cellulose to build intermolecular hydrogen bonds in water-based solutions.

5.1.2.1 Phosphoric Acid-Based Solvents

The possibility to dissolve cellulose in phosphoric acid has been known from a long time, starting probably with a patent filled by British Celanese (1925). It is only in the 1980s that interest for this solvent emerged again (Turbak et al. 1980). More recently, Boerstel et al. (2001) and Northolt et al. (2001) showed that concentrated solutions are anisotropic at rest and that their spinning is giving high modulus (44 GPa) and high strength (1.7 GPa) cellulose fibres. Despite that this solvent is not used, it may have an industrial potential to give specialty products.

5.1.2.2 LiCl-Based Solvents

The possibility to dissolve cellulose in a non-aqueous mixture of N,N-dimethylacetamide (DMAc) and LiCl was first published in 1979 by Charles McCormick (1979). It is a widely used tool to analyse cellulose chains. For example, McCormick et al. (1985) measured cellulose chain dimensions and conformation parameters, showing that this solvent is not degrading cellulose chains and is able to dissolve up to 15 % of cellulose. These authors proposed that the dissolution mechanism involves hydrogen bonding of the hydroxyl protons of cellulose with the chloride ions. They showed that concentrated solutions have a lyotropic character, but this must be taken with care since this liquid crystalline order is obtained by shearing, and not at equilibrium. In another paper, Matsumoto et al. (2001) compared the solution behaviour of plant and bacterial cellulose solution rheology in this solvent. Strangely, they found that plant cellulose behaves as a flexible polymer, while bacterial cellulose is like a rod-like chain. In a similar work, Tamai et al. (2003) reported that X-ray scattering data showing that the solutions have the characteristic of a two-phase system could explain the rod-like character of bacterial cellulose if it is not well dissolved and stays in bundles. Ramos et al. (2005a) found that not all cellulose can be easily dissolved down to molecular scale in this class of solvent and that activation is sometimes needed, as for cotton linters that can only dissolve if previously mercerised. This solvent is also used for preparing cellulose derivatives, sometimes in various ammonium variants like dimethylsulfoxide/tetrabutylammonium fluoride (Ramos et al. 2005b). None of the solvents of this class are commercially exploited to produce cellulose products.

5.1.2.3 N-Methylmorpholine-N-Oxide/Water

In 1939, Graenacher and Salman applied for a patent (Graenacher and Sallman 1939) where they described the possibility to dissolve cellulose in amine oxides, with aliphatic and cycloaliphatic amine oxides giving 7–10 % solutions of cellulose at 50–90 °C. This discovery was not exploited until the end of the 1960s when a series of patents (Johnson 1969; Franks and Varga 1979; McCormick and Lichatowich 1979) disclosed that one member of this series of compounds, N-methylmorpholine-N-oxide (NMMO) mixed with water, is able to dissolve cellulose (see Chap. 5 for more details). The most recent phase diagram of the mixture of NMMO and water is given in Fig. 5.3.

It is only within a rather limited range of composition and temperature that NMMO–water can dissolve cellulose. At too low water concentration, below 10 %, the required temperature for melting the solvent is very high and a very strong degradation of cellulose and solvent occurs, leading to very dangerous exothermic transitions. The instability of the solutions with cellulose has been studied in depth by the group of Rosenau (Rosenau et al. 2001). Above about 25 %, the mixture is not a solvent anymore (Cuissinat and Navard 2006a). Spinning cellulose in NMMO monohydrate is now a commercial process called lyocell by the Lenzing AG company in Austria or Alceru by Smart Fiber Company in Germany. Fibres are mainly used in the textile industry. Their main drawback is their strong tendency for fibrillation in the wet state (Fig. 5.4), thought to be due to their very high chain orientation (Ducos et al. 2006). More details can be found in Chap. 5 "Cellulose Products from Solutions: Film, Fibres and Aerogels".

Two other classes of solvent have been widely studied, ionic liquids and NaOH–water. They are described in details in next paragraphs.

5 Preparation and Properties of Cellulose Solutions

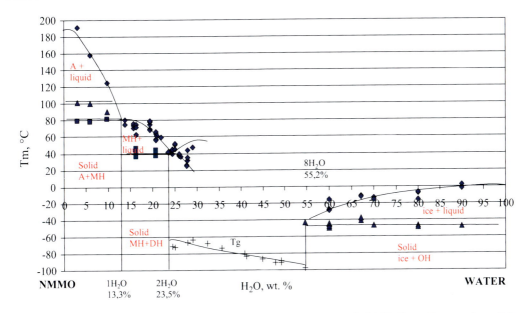

Fig. 5.3 NMMO–water phase diagram (adapted from Biganska O, Navard P (2003) Phase diagram of a cellulose solvent: N-methylmorpholine-N-oxide – water mixtures, Polymer 44: 1035-1039)

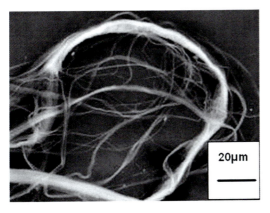

Fig. 5.4 Fibrillated cellulose fibres prepared from NMMO–water solution [adapted from Ducos et al. (2006)]

5.1.2.4 Rheology of Solutions

Rheology of cellulose solutions has been investigated from a very long time due to its use for measuring molar masses and to assess the behaviour of the cellulose chain in solution. To this end, the measurement of only the viscosity of dilute solutions has been performed extensively. Generally, in the dilute regime, noncharged polymers present a Newtonian flow behaviour and absence of viscoelasticity. The viscosity of polymer solutions depends on the concentration and size of the dissolved macromolecules, as well as on the solvent quality and temperature, supplying information about chain dimensions in solution, molecular shape, polymer–polymer interactions, excluded volume effects governed by polymer–solvent interactions and chain stiffness.

Generally, the relationship between the intrinsic viscosity $[\eta]$ and dilute solution viscosity η takes the form of a power series in concentration c:

$$\frac{\eta_{sp}}{c} = [\eta] + k_1 \cdot [\eta]^2 \cdot c + k_2 \cdot [\eta]^3 \cdot c^2 + k_3 \cdot [\eta]^4 \cdot c^3 + \ldots \quad (5.1)$$

where k_1, k_2, k_3, etc., are dimensionless constants. In order to evaluate the intrinsic viscosity, the theoretical analysis of the hydrodynamics of macromolecules with different flexibilities led to different relationships (Lovell 1989). The most frequently used one is the Huggins equation:

$$\eta_{sp}/c = [\eta]_H + k_H \cdot [\eta]_H^2 \cdot c \quad (5.2)$$

k_H is referred as the Huggins dimensionless constant and is correlated to the size and shape of polymer segments, as well as to hydrodynamic interactions between different segments of the same polymer chain. k_H equals to 0.5 means that solvent is theta solvent (no excluded volume expansion, polymer coils like ideal chains and excess chemical potential of mixing between a polymer and a theta solvent being zero); below 0.5 the solvent is thermodynamically good and above the solvent is bad.

Equation (5.2) is an approximation of (5.1), and it is applicable only for $[\eta] \cdot c < 1$. At higher concentrations, the experimental data show an upward curvature when plotted according to (5.2).

The Kraemer equation is an approximation of the Huggins equation, from which it may be derived assuming $\eta_{sp} \ll 1$:

$$\frac{\ln(\eta_{rel})}{c} = [\eta]_K - k_K \cdot [\eta]_K^2 \cdot c \quad (5.3)$$

Usually, the experimental data are plotted according to both Huggins and Kraemer equations, and the correct evaluation of $[\eta]$ is made by double extrapolation to infinite dilution. Theory predicts that $k_H + k_K \cong 0.5$ when the approximation is satisfactory valid. From a practical point of view, the determination of molar masses is performed in cadoxen, cuoxam or cupriethylenediamine according to norms (Kasaai 2002). The analysis of the viscosity is also giving indications on the rigidity and hydrodynamic behaviour of cellulose (Danilov et al. 1970; Kasaai 2002; Zhou et al. 2004).

Rheology of more concentrated solutions requires good solvents like N-methylmorpholine-N-oxide or ionic liquids (Blachot et al. 1998; Gericke et al. 2009b). It shows that cellulose is behaving as a normal semi-flexible polymer with a Newtonian region at low shear rate followed by a shear-thinning, non-linear region at high shear rates.

5.2 Macroscopic Mechanisms of the Dissolution of Native Cellulose Fibres

When placed in a swelling agent or a solvent, natural cellulose fibres show a nonhomogeneous swelling. The most spectacular effect of this

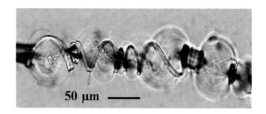

Fig. 5.5 *Gossypium hirsutum* cotton fibre swollen by ballooning in N-methylmorpholine-N-oxide with 20 % of water w/w

nonhomogeneous swelling is the ballooning phenomenon where swelling takes place in some selected zones along the fibres (Fig. 5.5). This heterogeneous swelling has been observed and discussed long ago by Nägeli (1864), Pennetier (1883), Flemming and Thaysen (1919), Marsh (1941), Hock (1950) or Tripp and Rollins (1952). One explanation for this phenomenon is that the swelling of the cellulose present in the secondary wall is causing the primary wall to extend and burst. According to this view, the expanding swollen cellulose pushes its way through the tears in the primary wall, the latter rolls up in such a way as to form collars, rings or spirals which restrict the uniform expansion of the fibre and forms balloons as described by Ott et al. (1954). This explanation assumes that cellulose is in a swollen state in each of the balloons.

Further studies of Chanzy et al. (1983), Cuissinat and Navard (2006a, b, 2008a) and Cuissinat et al. (2008b, c) have shown that the dissolution mechanism is strongly dependent on the solvent quality. Cuissinat and Navard performed observations by optical microscopy of free-floating fibres between two glass plates for a wide range of solvent quality (as an example, N-methylmorpholine-N-oxide (NMMO) with various amounts of water, w/w). They identified four main dissolution modes for wood and cotton fibres as a function of the quality of the solvent (the quality of the solvent decreases from mode 1 to mode 4):

Mode 1: Fast dissolution by fragmentation, occurring in good solvent (e.g. in NMMO with <17 % of water, 90 °C)

Mode 2: Swelling by ballooning and full dissolution, occurring in moderately good solvent (e.g. in NMMO with 19–24 % water, 90 °C)

Mode 3: Swelling by ballooning and no complete dissolution, occurring in bad solvent (e.g. in NMMO with 25–35 % water, 90 °C)

Mode 4: Low homogeneous swelling and no dissolution, occurring in very bad solvent (e.g. in NMMO with more than 35 % water, 90 °C)

These mechanisms also have been observed with NaOH–water with or without additives (Cuissinat and Navard 2006a), ionic liquids (Cuissinat and Navard 2006b) and for a wide range of plant fibres (Cuissinat and Navard 2008a) and some cellulose derivatives that had been prepared without dissolution (Cuissinat and Navard 2008, 2008b). From all these studies, it is shown that the key parameter in the dissolution mechanism is the morphology of the fibre. Indeed, as long as the original wall structure of the native fibre is preserved, the dissolution mechanisms are similar for wood, cotton, other plant fibres and some cellulose derivatives. Ballooning, often observed when cellulose native fibres are dissolving, originated from the specific cut followed by the rolling of the primary wall (LeMoigne et al. 2010a).

5.2.1 Gradient of Solubility in the Various Locations of the Cell Wall

In order to investigate in more depth the influence of the primary wall on the dissolution mechanism, cotton fibres with different maturity were studied (LeMoigne 2008). The cotton fruit is a capsule (commonly called cotton boll) composed of 4 or 5 carpels, each of them bearing about 10 seeds. Each cotton fibre is produced by the outgrowth of a single epidermal cell of the seed coat. A cotton fibre is a single cell, mainly made of cellulose microfibrils arranged in concentric walls. Fibre development can be divided in five main growth stages: initiation, elongation, transition, development and maturation (Figs. 5.6 and 5.7):

- The initiation stage corresponds to the differentiation of epidermal cells into fibre cells and takes place around 2 or 3 days preanthesis.
- The elongation stage corresponds to the synthesis of the primary wall and takes place between 1 and 15 days postanthesis (DPA)

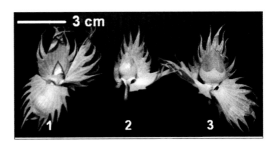

Fig. 5.6 Cotton bolls at elongation stage in days postanthesis (DPA). (1) 7 DPA, (2) 11 DPA and (3) 14 DPA. Reproduced with permission by Wiley from LeMoigne et al. (2008), Fig. 5.5.2

- The transition stage corresponds to the end of the primary wall synthesis and to the beginning of the secondary wall deposition (S1 wall) and takes place between 15 and 25 DPA. At this stage, secondary wall deposition is initiated while the cell is still elongating.
- The development stage corresponds to the massive deposition (without elongation) of cellulose forming the main body of the secondary wall (S2 wall) and takes place between 25 and 50 DPA.
- After about 50 days, the cotton bolls are mature, and it opens. After opening, the cotton fibres dry out. This stage is called maturation.

The dissolution of fibres taken at the elongation stage turned out to be impossible in both good and moderately good solvents. In NMMO with 16 and 20 % water, the reaction is very slow and leads to a uniform gel-like material with no measurable swelling. The elongation stage fibres dissolve only in the very good solvent like NMMO monohydrate (13.3 % of water). The reaction leads to a uniform gel-like material. A subsequent regeneration in water shows that this material was partially dissolved. These results show that the primary wall of cotton fibres is very resistant to dissolution in solvents that, as will be seen below, are dissolving the secondary wall. Another important observation is that the ballooning phenomenon is not present with these fibres. This is fully in agreement with the common explanation which attributes the balloons to the swelling of the secondary wall causing the extension and the bursting of the primary wall. Without secondary wall, there are no balloons

Fig. 5.7 Cotton bolls at transition stage in days postanthesis (DPA). (1) 20 DPA, (2) 23 DPA ; development stage, (3) 26 DPA, (4) 29 DPA, (5) 30 DPA, (6) just before opening and (7) mature fibres. Reproduced with permission by Wiley from LeMoigne et al. (2008), Fig. 5.3

(note that mature fibres without primary walls are also not showing balloons). At the transition stage, ballooning starts due to the presence of the secondary cellulose-based walls and a small number of balloons can be observed. These balloons are much smaller (swelling ratio around 200–300 %) than those observed for mature cotton fibres (swelling ratio around 450–600 %). At the mature stages, swelling and dissolution proceeds like described above.

These dissolution experiments on cotton fibres at different growth stages show that there is a gradient of dissolution capacity from the inside to the outside of the fibre. For fibres with enough cellulose inside the secondary wall, the inside of the fibre is the easiest part to dissolve. The dissolution mechanism is fragmentation where weak parts of the wall are quickly dissolved, leaving rod-like fragments floating, which completely dissolve later. Ballooning appears in fibres having a secondary wall, at least being at the transition stage. Balloons are formed by the expansion of the secondary wall due to the dissolution. When the secondary wall is swelling by ballooning, the primary wall breaks in localised places and rolls up to form helices and surrounds fibre sections that cannot be swollen. The primary wall does not dissolve easily and even sometimes does not dissolve at all as it is occurring in bad solvents like NaOH–water. The study of well-characterised cotton fibres in terms of growth stage has shown that most recent deposited cell wall layers (S2 wall) are most easily dissolved. The absence of non-cellulosic polysaccharide networks in younger wall layers may explain their higher dissolution capacity.

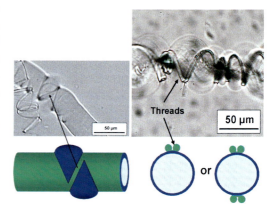

Fig. 5.8 *Left side*: initiation of the breakage of the primary wall leading to its rolling. *Right side*: rolling of the two sides of the primary wall leading to a double helical thread surrounding the balloon

5.2.2 Overall Dissolution Mechanism

As was shown (LeMoigne 2008), the secondary S2 wall is the easiest to dissolve as compared to the external walls which contain larger amount of non-cellulosic components. The solvent goes inside the fibre through the primary wall which is permeable to the solvent but not easy to dissolve and not extensible. Optical observations show that the primary wall breaks (Fig. 5.8) in one or more places under the pressure of the S1 wall swelling and then rolls up. Finally, balloons are formed due to the large swelling of the S1 wall. The primary wall that is cut and rolled up forms threads (seen as thin lines along the balloon surface) and collars (regroupment of the rolled primary wall around the fibre diameter, preventing fibres to swell and form collars).

The selection between threads and collars is depending on the way the primary wall is broken, i.e. depends on the shape of the initial cut. When cut occurs on the whole circumference of the fibre, the primary wall rolls up along the fibre direction in the two opposite directions and forms only collars. If cut is more local and directed along the fibre axis, the primary wall rolls up perpendicularly to the fibre axis and forms one or more threads attached to two collars. The collars block the swelling of the fibre, forming what has been called the unswollen sections (actually a region between two balloons) by Cuissinat and Navard (2006a). When the S2 wall is fully dissolved, the swelling of the balloons reaches its maximum size. The balloon is thus formed of the dissolved S2 wall cellulose inside an undissolved membrane composed of the swollen S1 wall, surrounded by one or more threads of the primary wall and delimited by two collars.

5.3 Cellulose–Sodium Hydroxide–Water Solutions

To prepare solutions of cellulose in a NaOH–water mixture is very attractive. It is simple, with reagents that are easy to recycle and cheap. It is thus not surprising that it attracted attention. Already in the 1930s, it was found that cellulose is soluble in NaOH–water mixtures in a certain range of rather low NaOH concentrations and low temperatures, but that this dissolution was difficult, with only partial dissolution of most untreated cellulose samples. Due to these difficulties, effects of other chemicals in NaOH–water were tested. It was found at that time that addition of compounds like ZnO or urea was helping dissolution (Davidson 1937).

The history of the relations between cellulose and sodium hydroxide dates back in the nineteenth century. It is the discoveries of the viscose process where a cellulose derivative is dissolved in NaOH–water and of mercerisation that were the real starts of the use of NaOH in the cellulose industry. In three instances, groups of researchers tried later to use NaOH–water for processing cellulose fibres. The first group in Japan worked in the 1980s and found that steam-exploded cellulose was more readily soluble. More recently, two processes based on enzymatically treated pulps for one and additives for the second were developed, but with no industrial development up to now. Difficulties of mixing (use of low temperatures), low stability of the mixtures, low maximum concentration of cellulose and moderate mechanical properties of the regenerated fibres are the main factors that explain why this dissolution method is not used. Other methods where cellulose is molecularly dispersed like N-methylmorpholine-N-oxide or ionic liquids are favoured over NaOH–water at the present time.

5.3.1 Mercerisation

Mercerisation is a process named after its inventor, John Mercer (The Editors 1903). Patented between 1844 and 1850, mercerisation is the treatment of native cellulose by concentrated (18–20 wt%) caustic alkaline solution. After immersion in the NaOH solution and then washing, the initial cotton fabric has improved properties like better lustre and smoothness, improved dye intake, improved mechanical properties and improved dimension stability. Mercerisation has been an industrial process from the beginning of the twentieth century up to now. Mercerisation is strongly acting on the cell wall cotton morphology, changing, for example, the crystalline structure from the native cellulose I to cellulose II (Okano and Sarko 1985a, b). Mercerisation is not a dissolution but a complex change of morphology and crystalline structure that occurs in a derivatised (production of various alkali–cellulose complexes) and highly swollen state. The accessibility of –OH groups depends on the crystallinity of cellulose (Tasker et al. 1994), the highly crystalline cellulose being more difficult to mercerise (Chanzy and Roche 1976).

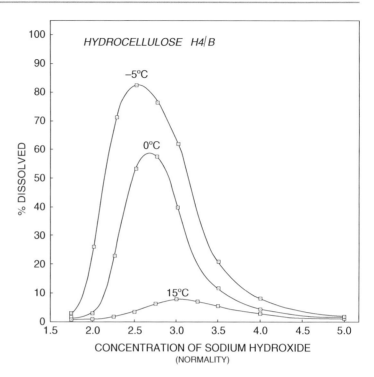

Fig. 5.9 Hydrocellulose solubility versus NaOH concentration and solution temperature. Adapted from Davidson (1934)

5.3.2 Dissolution in NaOH–Water

In the 1930s, Davidson (1934, 1936) studied cellulose dissolution, and he is to our knowledge the first to report dissolution, not swelling as in the mercerisation process. He looked for optimal conditions to dissolve modified cotton, termed hydrocellulose, which was cellulose hydrolysed in strong acid conditions, which is decreasing its molar mass. Such a product would be called microcrystalline cellulose now. Davidson showed that a decrease of temperature improves cellulose dissolution, as illustrated in Fig. 5.9.

Davidson found a maximum solubility at a very large value, 80 %, probably due to the low molar mass of his hydrolysed cellulose, that solubility occurred in a narrow range of NaOH–water concentration (he reported 10 % of NaOH) and that dissolution was possible at the low temperature of −5 °C. He noticed that solubility increases when the chain length decreases, leading Davidson to deduce that it will not be possible to dissolve unmodified (i.e. high molar mass) cellulose. With these results, we must consider that Davidson is the real inventor of cellulose dissolution in NaOH–water mixtures. However, it is often Sobue (Sobue et al. 1939) who is cited when cellulose dissolution in NaOH–water is concerned. The reason is that Sobue explored the whole ramie cellulose–NaOH–water ternary phase diagram, based on the work of his group and previously published data on cellulose–alkali mixtures by Saito (1939). He noticed that in a narrow range of NaOH, water and temperature, cellulose can be dissolved. The dissolution range was NaOH concentrations of 7–10 % and in low temperature range (−5 to +1 °C). They refer to this region and cellulose state by the term "Q-state". This phase diagram, reported in all reviews, is plotted on Fig. 5.10.

Apparently, the discovery of cellulose dissolution was not a major event and not many scientific papers reported its study. A revival of an

5 Preparation and Properties of Cellulose Solutions

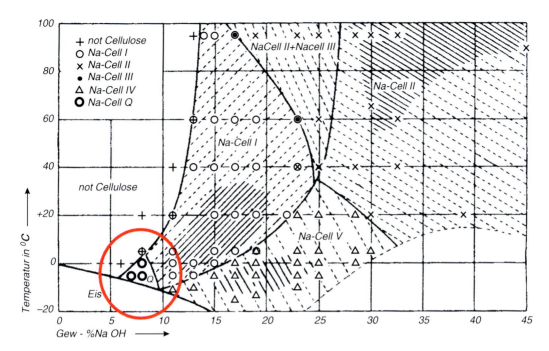

Fig. 5.10 NaOH–water–cellulose phase diagram, adapted from Sobue et al. (1939). The *circle* locates the dissolution range

interest in cellulose dissolution in NaOH–water came from Japan in the middle of 1980s. A team of Japanese researchers from Asahi Chemical Industry Co. made a breakthrough in the dissolution of cellulose in dilute aqueous solutions of sodium hydroxide. In a series of papers, Kamide et al. (1984, 1987, 1990, 1992), Yamashiki et al. (1988, 1990a, b, c, 1992), Yamada et al. (1992), Matsui et al. (1995), Yamane et al. (1996a, b, c, d) extensively studied cellulose dissolution mechanisms using different approaches, focusing on methods able to lead to produce materials, mainly fibres and films. The first paper by Kamide et al. (1984) reported that regenerated cellulose from a cuprammonium solution and ball-milled amorphous cellulose where intramolecular hydrogen bonds are completely broken or weakened dissolves in aqueous alkali and that the solution is stable over a long period of time. They identified the hydrogen bond intramolecular (O_3–H...O_5') as the one needing to be weakened. The authors found that if crystallinity has a role, the more crystalline cellulose being the more difficult to dissolve, as found already by Davidson in 1934, it is not by far the only factor governing dissolution. The authors conclude that "in other words, the solubility behaviour cannot be explained by only the concepts of 'crystal-amorphous' or 'accessible-inaccessible'". The lack of a strong correlation between the amount of amorphous phase and solubility was confirmed later (Kamide et al. 1992). The authors deduced that these solutions were molecularly dispersed. The solution was found to be a theta solvent at 40 °C. Cellulose has the behaviour of a semi-rigid chain (flexibility in NaOH–water lying between those in cadoxen and iron–sodium tartrate) having a partially free draining behaviour. The unperturbed dimensions decreased with temperature increase. A detailed study of cellobiose–NaOH interactions was conducted by Yamashiki et al. (1988). Based on ^{1}H and ^{23}Na NMR results, they proposed a model for explaining NaOH–cellulose interactions. The number of water molecules solvated to a NaOH molecule is maximum at 4 °C in the range 0–15 % of NaOH, decreasing at this temperature from 11 water molecules at very low concentration to 8 water molecules at

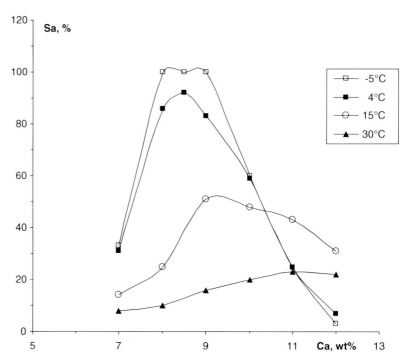

Fig. 5.11 Dependence of solubility Sa of alkali-soluble cellulose as a function of alkali concentration, at four temperatures (−5, 4, 15 and 30 °C). Plot adapted from Yamashiki et al (1990d)

15 % of NaOH. The authors concluded that provided that the proper intermolecular hydrogen bonds have been broken, the factor that controls dissolution is the structure of the alkali. If, as hypothesised, weakening intermolecular hydrogen bonds is a prerequisite for good solubility in NaOH–water, the same group turned towards steam explosion in order to avoid having to use regenerated cellulose. The solubility of steam-exploded pulp was found to be very high, always with a maximum solubility in the temperature–NaOH concentration window found by the first investigators (Davidson and Sobue), as shown in Fig. 5.11.

X-ray studies showed that dissolution occurs in this 8–9 % of NaOH in water at low temperature without any conversion of cellulose into Na-cell I, and that dissolution may occur first in the amorphous parts, resulting in a transparent, molecularly dispersed solution (Kamide et al. 1990). The processing and properties of fibres and films produced by this method have been studied and reported in several papers (Yamane et al. 1996a, b, c, d). Starting from previous knowledge, authors developed an industrial method for dissolving cellulose starting from steam-exploded cellulose pulp, wet pulverisation to increase the surface of cellulose particles, pretreatment in NaOH–water with 2–6 % NaOH at −2 °C and high-speed mixing followed by a dissolution at 6–9 % NaOH–water at −2 °C at 12,000 rpm during 1 mn (Yamane et al. 1996a). Tensile strength of maximum 2 g/den, elongation about 20 % (Yamane et al. 1994) and Young's modulus of about 110 g/den in the dry state were reported in Yamane et al. (1996c, d). These values are comparable to viscose fibres and inferior to lyocell or rayon fibres. Isogai and Atalla (1998) found difficult to dissolve microcrystalline cellulose using the procedure developed by the Asahi group. They found that dissolution of a large variety of cellulose origins (microcrystalline cellulose, cotton linter, softwood bleached and unbleached kraft pulp, groundwood pulp) and treatments (mercerised, regenerated) can be better performed if a cellulose solution in 8.5 % NaOH is frozen at −20 °C before thawing at room temperature while adding water to reach

5 % NaOH in the solvent. With this procedure, solutions with microcrystalline cellulose were stable at room temperature. Other cellulose samples were partially soluble. The authors found that the presence of hemicellulose did not seem to be a factor influencing dissolution since most hemicellulose fractions were soluble in NaOH–water mixtures. Molar mass was supposed to be the key point for explaining solubility, the higher masses being more difficult to dissolve, a fact explained by the authors though the concept of cellulose chain coherent domains.

However, despite huge efforts for understanding cellulose dissolution in NaOH–water and the processing of these solutions, this process did not reach an industrial stage. The reasons are multiple, but the stability of the spinning dope was a difficulty that prevented the use of this route for preparing regenerated cellulose objects. Increasing stability (i.e. preventing gelation) and improving the quality of the solution were then looked at by the addition of various compounds.

Additives like urea, thiourea or zinc oxide have been known to influence mercerisation from a long time. Davidson (1937) found that the addition of small amounts of zinc oxide to a solution of sodium hydroxide in water increases its swelling and dissolution capability towards cellulose. Additives like urea and thiourea were also studied and used from a long time for improving the viscose process or understanding Na-cell formation (e.g. Harrison 1928).

It was rather straightforward to investigate if the addition of compounds like urea, thiourea or ZnO would improve the state of dissolution of cellulose in NaOH. The first to report such attempts with thiourea is Laszkiewicz (Laszkiewicz and Cuculo 1993) following a previous work on mercerisation with additives like thiourea, ZnO, urea and aluminium hydroxide (Laszkiewicz and Wcislo 1990). He found that addition of thiourea improves the solubility of a cellulose III sample in NaOH–water. Later, the same author reported that the addition of 1 % urea in a solution of bacterial cellulose in NaOH–water allows dissolution of this cellulose if DP is lower than 560 (Laszkiewicz 1998). The Institute of Biopolymers and Chemical Fibres in Poland together with other research groups mainly in Finland developed methods to activate cellulose either by hydrothermal or enzymatic treatments (Ciechańska et al. 1996, 2007; Mikolajczyk et al. 2002; Struszczyk et al. 2000; Wawro et al. 2009). They developed a process called Biocelsol (see a detailed description in Chap. 5) where cellulose is treated by enzymatic activations and dissolved in NaOH–water with the addition of ZnO (Vehviläinen et al. 2008). Typical conditions are cellulose concentration of 6 %, 7.8 % NaOH and 0.84 % of ZnO. The best fibres have tenacity of 1.8 cNdtex-1 with 15 % elongation at break.

Starting in 2000, the group of L. Zhang revisited the dissolution of cellulose in NaOH–water using derivatives (Zhou and Zhang 2000; Zhang et al. 2002, 2004; Zhou et al. 2002a, b, 2004; Weng et al. 2004; Cai and Zang 2005). The preparation of solutions was inspired by Isogai and Atalla (1998) adding first urea and later thiourea, inspired by the work of Laszkiewicz (Laszkiewicz and Wcislo 1990; Laszkiewicz and Cuculo 1993; Laszkiewicz 1998). Solutions with 4–8 % were prepared and were stable enough to allow spinning fibres and preparing membranes. Cellulose cannot be dissolved at temperatures above 10 °C in NaOH–urea–water, and the lower the molar mass was, the higher was the dissolution yield (Qi et al. 2008a). They found that treating mechanically the cellulose fibres (Valley beating machine) was only slightly decreasing molar mass but decreasing more substantially crystallinity, which resulted in an increase of dissolution yield. The authors deduced that crystallinity is an important factor for increasing solubility, not in agreement with previous results (Kamide et al. 1984, 1992). One explanation for this discrepancy is that mechanical beating is also changing the structure of the fibres, mainly removing its primary wall and thus increasing accessibility to the solvent (LeMoigne 2008; LeMoigne and Navard 2010b).

The addition of urea (Zhou and Zhang 2000) has two advantages. It increases the dissolution yield, i.e. a larger fraction of a given pulp is dissolved and the solution is more stable (gelation delayed in time). Whether there is or not complexation between urea, NaOH and cellulose was a matter of debate. Kunze and Fink (2005) found that there exists a specific urea–NaOH–cellulose

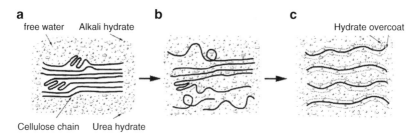

Fig. 5.12 Schematic dissolution process of the cellulose in LiOH/urea and NaOH/urea aqueous solutions pre-cooled to −10 °C: (**a**) cellulose bundle in the solvent, (**b**) swollen cellulose in the solution and (**c**) transparent cellulose solution. With kind permission from Springer Science + Business Media: Cai and Zang (2005), Fig. 5.10

complex different from the usual Na–cellulose complex by NMR, in the range of 0–8 % NaOH and 15–40 %, both at −25 °C and room temperature. Such complex is also formed when treating cellulose with only urea. As found by the Asahi group in the 1980s for cellulose dissolved in only NaOH, intramolecular hydrogen bonds are destroyed in the NaOH–urea–water mixture (Zhang et al. 2001). Further NMR work (Zhou et al. 2004) suggests that an interaction between urea and NaOH exists in the solution and plays a role in the solvation of cellulose, improving dissolution. Interactions between urea and cellulose block the self-association of cellulose, preventing gelation. But cellulose chains form aggregates of radius of gyration 200–300 nm (Chen et al. 2007), comparable to what is classically found for cellulose solutions (Röder and Morgenstern 1999; Röder et al. 2000). Chen et al. suggests that the polar amine and carbonyl groups of urea can be considered as hydrogen-bonding donors and acceptors, respectively, for the cellulose molecules. The presence of NaOH allows hydrogen bonds to be formed between urea and cellulose. Helped by DSC measurements, Cai and Shang (2005) suggest that when cellulose is placed in NaOH–urea/water at low temperatures, alkali hydrates, urea hydrates and free water surround the cellulose molecules as shown in Fig. 5.12 destroying intra- and intermolecular hydrogen bonds, solvating the cellulose molecules and protecting it with a sort of "overcoat", preventing reaggregation.

Further studies by Zhang et al. (Cai et al. 2008; Lu et al. 2011; Lue et al. 2011) show that in NaOH (as well as LiOH)–urea–water solutions, cellulose is present as isolated molecules and as aggregates. The fraction of aggregates increases with increasing temperature above −12 °C. Based on NMR, FT-IR, SANS, TEM and WAXS, Cai et al. (2008) found that low temperatures promote the formation of hydrogen-bonded networks of NaOH, urea and water, with the formation of these complexes being favoured by low temperatures. Using SANS, they suggest that NaOH is weakening the association of urea with water, a result found at the same time by Egal et al. (2008) using calorimetry. The chemical shifts of carbon atoms are similar in NaOH–urea–water and NaOH–water environments, showing that urea does not interact with cellulose, a result in agreement with Egal et al. findings (Egal et al. 2008). Cai et al. suggest a model where NaOH hydrates are hydrogen-bonded to cellulose molecules and that urea hydrates are bonded to NaOH hydrates at the surface of cellulose–NaOH complex. This is making a sort of envelope protecting cellulose chains to aggregate. These authors claim that this arrangement called inclusion complex is a non-stable arrangement that is slowly displaced with time and when increasing temperature, leading to the formation of large aggregates of radius of gyration larger than 200 nm (Lu et al. 2011; Lue et al. 2011).

Some authors claimed that using both urea and thiourea is improving further dissolution (Jin et al. 2007). ^{13}C NMR shows that this mixture is a direct solvent and that NaOH, urea and thiourea are all bound to cellulose which "brings cellulose molecules into aqueous solution to a certain extent" and prevents gelation. Solutions of up to 6 % of DP 510 can be prepared and are stable below 0 °C. NMR suggests similar interactions among NaOH, urea, thiourea and water in the solvent and the cellulose solution. The structure of the solvent does not change with the introduction of cellulose. The authors suggest that NaOH hydrates bound to cellulose to form a protective layer preventing cellulose chain aggregation. This solvent was studied also by Zhang et al. (2011) and Kihlman et al. (2011)

The properties of fibres prepared from solutions in NaOH–water–urea (Cai et al. 2004) and NaOH–water–thiourea (Ruan et al. 2004) give mechanical properties of about tensile strength of 1 cN/dtex and elongation at break of about 15 % for thiourea. A comparison between fibres prepared in the 1990s in Japan, in China by the group of Zhang and in Poland by the Biocelsol process is given in Table 5.2. It shows that processing in NaOH–water system brings properties lower than the ones of lyocell and viscose. One of the reasons is that solubility is low, with maximum cellulose concentration being in the order of 7–8 % (see below the results of Centre de Mise en Forme des Matériaux for an explanation of this concentration). Only low molar masses can be dissolved at these maximum concentrations. In addition, the stability of the solution is posing problems. These solutions are gelling and gelation is faster with temperature increase and when concentration and/or molar mass is large. Weng et al. (2004) and Ruan et al. (2008) studied gelation kinetics and found gelation to be an irreversible physical phenomenon. All these features are not helping in preparing a good spinning dope with high enough cellulose molar mass and concentration able to produce fibres with good mechanical properties. The fact that a cooling much below room temperature is needed is hampering the industrial use of this process.

We present below part of the work that has been conducted in Centre de Mise en Forme des Matériaux, Mines Paristech/CNRS/Armines, on cellulose–NaOH-solution properties, with and without additives. The use of NaOH–water to prepare cellulose materials is presented in Chap. 5.

5.3.2.1 Structure of Cellulose–NaOH–Water Solutions (Roy et al. 2001; Egal et al. 2007)

In order to understand better what is controlling the solubility of cellulose in NaOH–water, binary NaOH/water and the ternary cellulose/NaOH/water phase diagrams at NaOH concentrations in the region of cellulose dissolution (7–10 % NaOH below 0 °C) were studied by DSC. To avoid the complexity usually associated with natural fibres directly extracted from the cell wall (Cuissinat and Navard 2006b), microcrystalline cellulose Avicel®PH-101 (FMC Corporation, mean degree of polymerisation of 170) was used. NaOH was dissolved in water at concentrations around 12 % and cooled down to −6 °C. At the same time but in another vessel, water was added to cellulose pulp for cellulose swelling and the cellulose/water system was left at +5 °C for about 2 h. Cold NaOH/water solution was added to this swollen-in-water cellulose in such a proportion that in 100 g of final solution there were 0.5–7.6 g of cellulose and 7.6 or 8 g of NaOH. The weight proportions between the components in 100 g solution will be noted as Xcellulose–YNaOH–water which means X g of cellulose, Y g of NaOH and (100 − X − Y) grams of water. Cellulose–NaOH–water mixtures were placed into a thermobath at −6 °C and stirred at ~1,000 RPM for 2 h. Then the solutions were removed from the bath and stored at +5 °C.

DSC melting thermograms of NaOH + H$_2$O solutions at $T < 0$ °C and $C_{NaOH} = 0$–20 wt% are showing two peaks, characteristic of a eutectic phase diagram, as been shown by Cohen-Addad et al. (1960) and Rollet and Cohen-Addad (1964). The melting peak at low temperature, around −33/−34 °C, is independent on the NaOH concentration and is the trace of the melting of the crystalline eutectic mixture, composed of one metastable sodium pentahydrate and four water molecules (NaOH·5H$_2$O; 4H$_2$O). Its melting enthalpy is $\Delta H_{eut,\ pure} = 187$ J/g, measured at $C_{NaOH} = 20$ %. The high temperature peak corresponds to the melting of ice. The higher the sodium hydroxide concentration, the lower the ice melting temperature because of the decrease of ice fraction in solution. Applying the level rule

Table 5.2 Tensile properties of cellulose fibres spun from a NaOH–water solution and comparison with other fibres

Fibre	Fineness dtex	Modulus (GPa)	Tenacity (MPa)	Elongation at break (%)	DP	References
NaOH alone	–	0.45 (wet)	310	21	1,060	Yamane et al. (1994)
Asahi (Japan)		23 (dry)	310	15	>320	Okajima and Yamane (1997)
NaOH/ZnO Biocelsol (Poland)	1.4	–	270	15	268	Vehviläinen et al. (2008)
NaOH/urea L. Zhang	6	–	150	15	310	Ruan et al. (2004)
	–	–	255–315	2.2–1.9	590	Qi et al. (2008b)
	–	–	195–285	18–2	480	Cai et al. (2007)
NaOH/urea/thiourea	8.6×10^4		267	12		Zhang et al. (2009)
Viscose			~340	~15	~300	Adusumali et al. (2006) and Fink et al. (2001a, b)
		9.3	260	23		Northolt et al. (2001)
Modal	–		~440	~10		Adusumali et al. (2006)
Rayon tire cord			~780	11		Adusumali et al. (2006)
Lyocell		22–30	450–600	7–13	~600	Gindl et al. (2008)
Cuprammonium			300–440	7–23		Yamane et al. 1994 and Qi et al. (2008b)
Ionic liquid BMIMCl	1.46	10.2	800	13	514	Kosan et al. (2008a, b)
Carbamate	2.8	0.45–3 (wet)	195–390	8–27		Fink et al. (2001a, b)
Cellulose acetate saponified in caustic soda						
Fortisan	–	32	1,000	6.8		Northolt et al. (2001)
Kim et al.			220	36		Kim et al. (2006)
Phosphoric acid	–	45	1,300	5.1	800	Northolt et al. (2001)
LiCl/DMAc	–	0.23 (wet)	609	6.2		Turbak et al. (1981)

on NaOH/water phase diagram, it is possible to calculate the fractions of the eutectic mixture $f_{eut, calc}$ and of ice $f_{ice, calc}$ at any NaOH concentration. It was straightforward to determine the corresponding fractions from the experimental data, $f_{eut, exp}$ and $f_{ice, exp}$, i.e. from experimentally measured melting enthalpies ΔH_{eut} and ΔH_{ice} at a given NaOH concentration, and to know the melting enthalpies of pure compounds (ice and eutectic), $\Delta H_{eut, pure}$ and $\Delta H_{ice, pure}$, respectively. Fractions calculated and determined from experimental data coincide within experimental errors. This match is good for the enthalpy of the eutectic compound which has a well-defined melting peak. The match is less good for the melting of ice which occurs over a large range of temperatures with a peak having a "tail" at low temperatures difficult to extract from the baseline.

Structure of Cellulose–NaOH–Water Solutions at $T < 0\ °C$: An example of DSC melting thermograms for solutions of different cellulose concentrations is given in Fig. 5.13.

Whatever the concentrations of cellulose and sodium hydroxide are, the melting temperature of the peak at lower temperatures is constant. It coincides with the melting temperature of the eutectic mixture in pure aqueous sodium hydroxide solution of 7.6 % but with a systematic shift of about 1 °C towards lower temperatures. This means two things. First, the same eutectic mixture (NaOH·5H$_2$O; 4H$_2$O) is present in cellulose–sodium hydroxide aqueous

5 Preparation and Properties of Cellulose Solutions

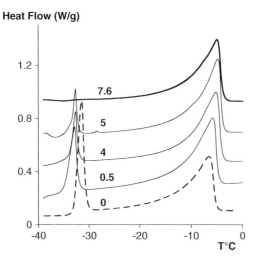

Fig. 5.13 DSC melting thermograms of XAvicel/7.6NaOH/water solutions with X = 0, 0.5, 4, 5 and 7.6 g in 100-g solution. *Dashed line* corresponds to X = 0 (solution without cellulose). *Curves* are shifted vertically for clarity. Reprinted with permission from Egal et al. (2007). Copyright 2007 American Chemical Society

solutions in this region of temperatures and NaOH concentrations, owing to the fact that the temperature shift is very small. The presence of cellulose is not changing its composition. Second, because there is a change in molecular environment due to the presence of cellulose, there is a slight shift in the melting temperature. However, the presence of cellulose is drastically decreasing the amount of the water–NaOH eutectic mixture, as can be seen in Fig. 5.13.

The higher is the cellulose concentration, the smaller is the amount of eutectic compound that can crystallise and then melt at -34 °C. Since NaOH is present only in the eutectic compound, the decrease of its melting enthalpy allows the calculation of the number of NaOH molecules linked to cellulose and thus not able to participate to the NaOH·5H$_2$O crystal fraction of the eutectic mixture. The fact that ΔH_{eut} is reaching zero at a certain cellulose concentration means that all sodium hydroxide molecules have been trapped by the cellulose chains. This corresponds to the dissolution limit since there is no more NaOH molecules able to solvate any additional cellulose chain that could be brought in the mixture. At this dissolution limit, we can calculate the proportion between cellulose anhydroglucose unit (AGU) and NaOH molecules. In our case, the eutectic peak disappears when weight ratio $M_{cell}/M_{NaOH} = 1$. In moles, this proportion is four NaOH per AGU ($m_{AGU} = 162$ g/mol and $m_{NaOH} = 40$ g/mol). It is also possible to calculate the proportion between linked AGU and NaOH at any weight ratio of the components. The amount of NaOH molecules linked to one anhydroglucose unit as a function of the weight ratio M_{cell}/M_{NaOH} for the systems Avicel–7.6-NaOH–water (dark squares) and Avicel–8-NaOH–water is shown in Fig. 5.14 (open squares). At $M_{cell}/M_{NaOH} > 0.4$–0.5, the proportion between AGU and NaOH molecules is constant and equal to the limit of cellulose dissolution which is four NaOH per AGU. At low cellulose concentrations, $M_{cell}/M_{NaOH} < 0.25$, up to 20 NaOH molecules seem to be linked to one anhydroglucose unit.

DSC experiments and careful analysis of experimental data allowed understanding the thermodynamic behaviour and structure of microcrystalline cellulose–sodium hydroxide aqueous solutions at temperatures below 0 °C, in the region of cellulose dissolution. It was possible to determine the limit of cellulose dissolution in NaOH–water as being at least four NaOH molecules per one anhydroglucose unit or the weight ratio of cellulose/NaOH being

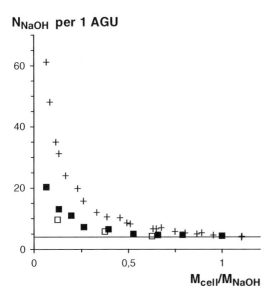

Fig. 5.14 Number of NaOH molecules linked to one anhydroglucose unit, calculated from melting enthalpy data, as a function of cellulose/NaOH weight ratio in solution. *Dark squares*: cellulose/7.6 NaOH/water, *open squares*: cellulose/8 NaOH/water, *crosses*: data from Kuo and Hong (2005) for NaOH concentrations of 4.5, 6.2 and 7.9 %. Reprinted with permission from Egal et al. (2007). Copyright 2007 American Chemical Society

one. If the concentration of cellulose is higher, it will not be dissolved. Because cellulose can be dissolved only in a narrow range of sodium hydroxide concentrations, from 7 to 8–10 %, this means that maximal amount of cellulose that can be dissolved in NaOH–water solutions is 8–10 %. This result is compatible with what is found experimentally when preparing cellulose/NaOH/water solutions for processing.

When the amount of water is too large (below 6–7 % of NaOH), we can speculate that the size of the hydrate is too large to penetrate the cellulose fibres. When the amount of water is low (NaOH concentrations higher than 18–20 %), NaOH is forming Na–cellulose crystals (mercerisation process). When the amount of water is in between these two cases (i.e. NaOH is 7–10 %), NaOH hydrates penetrate into the cellulose fibres and bind to each chain, but without forming Na–cellulose crystal. This pushes the detachment of individual chains out of the cellulose fibre and makes a solution. At these concentrations, there is an unstable equilibrium between NaOH hydrates bound to each other and to cellulose, which makes all the entities to be in solution.

5.3.2.2 Viscosity of Solutions of Cellulose in NaOH–Water (Roy et al. 2003)

The shear rheology of a microcrystalline cellulose dissolved in a 9 % NaOH aqueous solution was studied in the steady and oscillatory modes. Dilute solutions, below 0.8–1.2 % (the value of the critical overlap concentration depending on temperature, as it will be shown in the following), flow like Newtonian fluids. Semi-dilute solutions below 5 % show a Newtonian plateau and a shear-thinning regime. Above 5 %, in the studied shear rate range, the Newtonian plateau disappears. The flow index n of viscosity–shear rate curves ($\eta \sim \dot{\gamma}^n$) in the shear-thinning region monotonously increases with temperature from 0.09 to 0.12 for the 3 % cellulose solution and from 0.13 to 0.17 for the 5 % cellulose solution. These values are 4–5 times smaller than the ones of either typical "normal" polymer solutions (experimental data or theoretical predictions both for flexible and rigid chain polymers) or, in particular, polysaccharides like guar gum, λ-carrageenan, hyaluronate (see, e.g. Morris et al. 1981) or entangled physical networks (Clark and Ross-Murphy 1987). The viscosity–shear rate dependencies of cellulose–9 % NaOH solutions do not also obey the empirical law found by Morris (1990) and Haque and Morris (1993) for several polysaccharides. The flow of cellulose–9 % NaOH solutions is close to the one of suspensions with a flow index lower than 0.5. Indeed, as shown by several authors (Röder and Morgenstern 1999, 2000; Drechsler et al. 2000; Schulz et al. 2000), the dynamic and static light scattering studies of cellulose–NMMO, viscose and several cellulose derivative solutions demonstrated that cellulose chains can aggregate in solutions. Thus, the very low flow index value of cellulose–9 % NaOH solutions is an indirect proof that cellulose in 9 % NaOH aqueous solution is also aggregated, as was proved later by light scattering (Chen et al. 2007).

It was also shown that cellulose intrinsic viscosity $[\eta]$ decreases with temperature increase.

As a consequence, the overlap concentration, C^*, which is a critical parameter for gelation, is also temperature dependent: $C^* = 0.83\%$ for cellulose–9 % NaOH below 20 °C and $C^* = 1.25\%$ for cellulose–9 % NaOH at 40 °C.

<u>Gelation of Cellulose–NaOH Solutions Seen by Dynamic Rheology</u> (Roy et al. 2003; Egal 2006, Gavillon and Budtova 2008; Budtova et al. 2010): Gelation is the transition between a fluid liquid state and an elastic solid state. According to the gelation theory of Winter and Chambon (1986), in the fluid-state viscous modulus $G' \sim \omega^a$ and elastic modulus $G'' \sim \omega^b$ (with ω being angular frequency), ideally the gel point is reached when $a = b = 0.5$. The development of the slopes of $G' = f(\omega)$ and $G'' = f(\omega)$ curves in time is thus a privileged method for looking at gelation kinetics. This works well for chemically cross-linked networks but may be difficult to apply for more complicated systems like physical gels. In the latter case a simplified approach is used: G' and G'' evolution as a function of time t is monitored at a fixed frequency and gel point is considered to be reached when $G'(t) = G''(t)$.

The evolution of $G'(t)$ and $G''(t)$ at various cellulose concentrations and temperatures T was recorded and the simplified approach, as described above, was used to deduce gelation time as a function of solution temperature and cellulose concentration. At $t = 0$, G'' is larger than elastic modulus G': The system is in solution state and behaves like a viscous liquid. With time, G' increases more rapidly than G'': It crosses G'' at a certain gelation time t_{gel} and becomes larger than G''. The system gradually transforms from a viscous liquid to an elastic network. It should be noted that while the exponent a decreases with time as expected (it should ideally reach $a = 0$ or $G' = $ const indicating the formation of a stable network), G' of cellulose–NaOH–water solutions never reached the ideal constant value. Together with G'', G' was slowly increasing in time after the gel point. The formed gel is opaque; the reason could be a microphase separation into polymer-rich and polymer-poor phases, as it occurs for some gelling polysaccharide solutions such as methylcellulose. In a few hours after gel point, the experimental points become scattered because of syneresis: Solvent is released from the gel and the sample slides leading to bad data reproducibility.

Whatever is the temperature, the gelation of semi-dilute cellulose–NaOH solutions takes place, being faster at higher temperatures. Gelation time t_{gel} exponentially decreases with a temperature increase $t_{gel} \sim \exp(-aT)$, varying from 0.35 to 0.4 for 8–9 % NaOH–water solvent.

Gel strength at the gel point $G_{gel} = G' = G''$ was studied as a function of cellulose concentration (Gavillon and Budtova 2008). It was found that G_{gel} is power law dependent on cellulose concentration with exponent being from 3 to 4. This is due to a progressive increase of the number of contacts between cellulose chains leading to a stronger network structure. Such a strong power law concentration dependence is not typical for gelling polysaccharides, which is reported to be square dependent on polymer concentration.

The thermally induced gelation of cellulose–NaOH solutions can be interpreted as follows. As shown for dilute solutions, solvent thermodynamic quality decreases with temperature increase. This leads to the preferential cellulose–cellulose and not cellulose–solvent interactions. In dilute solutions below polymer overlap concentration, the coils contract. Above the overlap concentration, gelation occurs via intra-chain interactions. Both time and temperature are acting on cellulose–NaOH–water solutions in the same "destabilising" way.

5.3.2.3 Structure of Cellulose/ NaOH–Urea–Water Solutions (Egal 2006; Egal et al. 2008)

Cellulose is better dissolving in NaOH–water when a certain amount of urea is added. In order to understand the mechanisms of this dissolution and the interactions between the components, the binary phase diagram of urea/water, the ternary urea–NaOH–water phase diagram and the influence of the addition of microcrystalline cellulose in urea–NaOH–water solutions were studied by DSC.

Binary Urea/Water Phase Diagram: The full urea–water phase diagram was plotted using DSC experiments and X-ray scattering. Apart for 30 % urea where only one peak is present, all DSC thermograms reveal two melting peaks more or less well separated. The first one, at low temperature, has the same position whatever the urea concentration from 6 to 30 % is. At the onset of the peak, the temperature is about -12.5 °C. This peak corresponds to the melting of urea–water eutectic mixture. The second one, at higher temperatures, is shifted towards lower temperatures when the urea content increases. This peak corresponds to the gradual melting of free ice. The peak of free ice disappears at about 30 % of urea. This means that the urea–water eutectic mixture corresponds to 30 % of urea. As the molar masses of urea—$CO(NH_2)_2$—and water are 60 g/mol and 18 g/mol, respectively, 30 % of urea in weight corresponds to about a molar ratio $n_{H_2O}/n_{urea} = 7.8$. Thus, the total urea–water eutectic mixture consists of 1 urea and 8H_2O molecules. This is confirmed by X-ray scattering that shows that at higher urea concentrations, above 30 %, only two types of crystals are present, ice and crystalline urea.

Ternary NaOH–Water–Urea Phase Diagram: The composition of cellulose solvents studied was in the range from 7.6 NaOH–6 urea–water to 7.6 NaOH–25 urea–water. DSC melting thermograms were plotted for several urea concentrations. Four peaks can be seen on Fig. 5.15.

Peak no. 1 corresponds to the melting of the metastable eutectic mixture NaOH·5H_2O + 4 H_2O. Peak no. 2 appears at about -30 °C, but it is not always present. It corresponds to the melting of the stable eutectic mixture NaOH·7H_2O +2H_2O that was not present in NaOH/water solution when heated/cooled with a low temperature rate (Pickering 1893; Cohen-Adad et al. 1960). It seems that the presence of urea in a 7.6 NaOH/water is favouring the crystallisation of this stable eutectic, while this is not occurring in the same conditions of cooling and heating rates without urea. Since we have no access to conditions where only the stable eutectic compound is present without urea, we do not know its pure melting enthalpy. In the following, we will use the same specific enthalpy

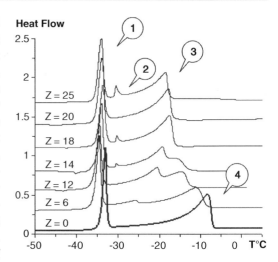

Fig. 5.15 DSC melting thermograms for 7.6 NaOH/Zurea/water solutions ($Z = 0$, 6, 12, 14, 18, 20 and 25). DSC traces are vertically shifted for clarity. With kind permission from Springer Science + Business Media: Egal et al. (2008), Fig. 5.3

for the stable eutectic compound as the known one of the metastable eutectic compound (187 J/g). Peak no. 3 appears as soon as urea is added. This peak should correspond to the melting of a urea compound. As will be seen below, peak no. 3 represents the melting of urea eutectic mixture and peak no. 4, the melting of free ice. The temperature of the melting peak no. 4 decreases when urea concentration increases from 0 to 14 g of urea in 100 g of solution. This peak corresponds to the gradual melting of free ice.

We have shown that urea does not react with NaOH molecules and does not modify the NaOH eutectic. Thus we assume in the presence of NaOH, urea and water behave as if they were alone: They should form the eutectic mixture containing one urea and eight water molecules, as determined for the binary urea/water solution. The fact that peak no. 3 corresponds to the melting of the urea eutectic mixture can be checked by plotting the melting temperature of both peaks as a function of urea concentration. It shows that peaks no. 3 and no. 4 are joining at a composition corresponding at 7.6 NaOH–18 urea–water. If the assumption of noninteraction between NaOH and urea is correct, this means that at this composition there should be two independent, coexisting eutectic compounds, NaOH + water and urea + water, with no more free water in solution. In a 100-g solution containing 7.6-g NaOH + 18-g urea + 74.4-g water, there are 4.13 moles of water, 0.19 moles of NaOH which corresponds to 0.19 moles of NaOH eutectic mixture (there is one NaOH mole per mole of the NaOH eutectic compound) and 0.3 moles of urea. The NaOH eutectic mixture is not modified by the addition of urea, and it contains 1 NaOH and 9 H_2O molecules. 1.71 moles of water is thus trapped in the NaOH eutectic mixture, and it remains $4.13 - 1.71 = 2.42$ moles of water. The urea eutectic mixture is composed of 1 urea and 8 H_2O molecules, and thus, $0.3 \times 8 = 2.4$ moles of water is trapped in the urea eutectic. There is no more water in the system (the remaining 0.02 moles of free water is within the experimental error of determination of the crossing composition). As a conclusion, 100 g of 7.6 NaOH/18 urea/water solution contains 18 g of urea and 43.56 g of water, i.e. the local concentration of urea in water is $\frac{18}{18+43.56} \times 100\% = 29.2\%$ which corresponds to the composition of the urea eutectic mixture, as shown above for urea/water binary solution. At this composition, the solution is composed of only the two eutectic mixtures: NaOH + water and urea + water. These results show that both NaOH and urea eutectic mixtures are formed independently and thus are competing for water. NaOH, urea and water behave as in their binary systems.

Solutions of Cellulose in NaOH–Urea–Water: The addition of cellulose does not modify or change the position of the peaks as compared to the case without cellulose. The DSC melting curves have three peaks. The peak at the lowest temperature, at about $-35\ °C$, corresponds to the melting of the NaOH + water eutectic mixture. The next one, at about $-25\ °C$, corresponds to the melting of the urea + water eutectic mixture. The peak at the highest temperature, at about $7\ °C$, corresponds to the melting of free ice. Whatever the concentration of cellulose is, the melting temperature of the NaOH + water eutectic mixture is equal to the one obtained for a pure 7.6 NaOH–6 urea–water (without cellulose). During the dissolution process, cellulose is trapping some NaOH molecules that cannot participate anymore to the formation of the eutectic mixture. It is remarkable to see that the addition of cellulose is trapping the same amount of NaOH with or without urea. The addition of urea does not change the interactions between cellulose and NaOH. Whatever the concentration of cellulose is, the peak of the urea + water eutectic mixture is seen. The enthalpy of the urea eutectic does not depend on cellulose concentration. The addition of cellulose does not change the urea–water interactions. Since the melting enthalpy of free ice is approximately constant for all cellulose concentrations (the same was observed for cellulose–7.6 NaOH–water without urea, see Egal 2006; Egal et al. 2007), we conclude that the amount of free ice remains the same with and without cellulose.

Three main facts are obtained from these series of experiments and their interpretations:
1. NaOH and urea do not interact when mixed together with water. Their eutectic mixtures are formed independently.
2. Urea and cellulose do not interact when mixed with NaOH and water since the urea + water eutectic is not changed.
3. The interactions between NaOH hydrates and cellulose are not changed by the presence of urea (the decrease of the eutectic peak enthalpy of NaOH + water with cellulose content is the same with or without urea).

This means that urea is interacting neither with NaOH nor with cellulose and is not changing the NaOH–cellulose interactions. A possible origin of the role of urea as a dissolution promoter can be that it strongly interacts with water. Cellulose chains have the tendency to aggregate as soon as the proportion 4 NaOH/AGU does not hold any more for any reason (temperature increase, cellulose concentration increase). The addition of urea is decreasing the amount of free water, thus helping cellulose chains to stay in solution.

5.3.2.4 Influence of ZnO on Cellulose–NaOH–Water Solution Properties (Liu et al. 2011)

Zinc oxide is an additive improving cellulose dissolution known since the pioneering work of Davidson (1937). Recently, it was revisited for processing fibres (Vehviläinen et al. 2008), and

Table 5.3 Gelation times of cellulose–8 % NaOH–water solutions at different temperatures and concentrations of ZnO

	Gelation times (min)				
	4 % cellulose–8 % NaOH		6 % cellulose–8 % NaOH		
T (°C)	0 % ZnO	0.7 % ZnO	0 % ZnO	0.7 % ZnO	1.5 % ZnO
−5	>5 days	>3 days			
0	>2 days	>2 days			
10			1,106	>2 days	
15	4,320		87.75	791	>2 days
20	1,384		15		2,190
25	32			74	1,032
30	6.7	2,980		7.7	156

Yang et al. reported that the maximal cellulose dissolution is reached at 0.5 % ZnO in the mixture of 7 % NaOH–12 % urea. The reason of the improved solubility was supposed to be the stronger hydrogen bonds between cellulose and Zn(OH)$_4^{2-}$ as compared with cellulose–NaOH and cellulose–urea. The goal of our work was to make a comprehensive investigation of the influence of ZnO on used Avicel® PH-101 microcrystalline cellulose–NaOH–water solution properties and especially on its gelation delaying effect using viscometry and rheology.

Cellulose solutions were prepared following the procedure described in Egal et al. (2007, 2008). pH strongly influences ZnO solubility (Liu and Piron 1998). It was shown that ZnO is practically insoluble in water (solubility below 10^{-6} g L^{-1}) and becomes more soluble in strong acidic or basic media (Liu and Piron 1998). In our work the preparation of solutions is started by making 18–20 % NaOH–water (pH ≈ 14.7) and adding a certain amount of ZnO. At this pH, ZnO can be dissolved up to 27 g L^{-1} (Liu and Piron 1998). The pH of the final cellulose solutions is 14.3 in which ZnO solubility decreases to about 4 g L^{-1}. Thus, the saturation of ZnO is reached in the final solutions containing cellulose and part of ZnO is in suspended state. Seen by optical microscopy, dispersions of undissolved ZnO were homogeneous, with two-size populations, i.e. ZnO particles of about 1 μm and a few aggregates of about 10–20 μm. Calculations of sedimentation times imposed to limit ZnO concentrations to 1.5 %.

Different concentrations of components in cellulose solutions will be noted as X % cellulose–Y % NaOH–Z % ZnO–water, in which X % is the weight concentration of cellulose, calculated as $X\% = 100 \times M_{cell}/(M_{cell} + M_{water} + M_{NaOH} + M_{ZnO})$ where M_i is the weight of each component. Y and Z % are the weight concentrations of NaOH and ZnO, respectively, in solvent only, calculated as $Y\% = 100 \times M_{NaOH}/(M_{water} + M_{NaOH} + M_{ZnO})$ and $Z\% = 100 \times M_{ZnO}/(M_{water} + M_{NaOH} + M_{ZnO})$. The concentration of NaOH in the final solution was fixed to 8 %, and the concentration of ZnO was varied from 0 to 1.5.

A steady decrease of the intrinsic viscosity [η] with increasing temperature was observed for solutions with and without ZnO (here 0.7 wt%) and is consistent with the results reported above for cellulose of another molecular weight dissolved in 8 % NaOH–water without any additive (Egal 2006) and in 9 % NaOH–water (Roy et al. 2003). The drop of [η] signifies that the thermodynamic quality of solvent decreases with temperature. The presence of ZnO did not bring any noticeable change to the intrinsic viscosity of cellulose. When 0.7 % ZnO is mixed with 8 % NaOH–water, only part of ZnO is dissolved. The unaffected value of [η] indicates that neither dissolved nor suspended ZnO influences the conformation and behaviour of cellulose chains at molecular level in dilute region. As shown above, gelation of cellulose–NaOH–water solution is thermally induced and irreversible. The results on gelation time of cellulose–8 % NaOH–water solutions at temperatures from −5° to 50 °C at various cellulose and ZnO concentrations are summarised in Table 5.3.

Gelation is significantly delayed in the presence of ZnO: For example, a 6 % cellulose solution at 20 °C is gelling in 15 min in 8 % NaOH–water and in 36 h in the presence of 1.5 % ZnO. The results presented in Table 5.1 also show that gelation time depends on solution temperature and cellulose and ZnO concentrations

Influence of Temperature on Cellulose–NaOH–Water Gelation in the Presence of ZnO: Gelation time of cellulose solutions was first studied at low temperatures, around −5 to +5 °C to check if the sudden decrease of gelation time with temperature decrease below −0 °C, reported for cellulose–7 % NaOH–12 % urea (Cai and Zhang 2006) system, was seen in our solutions. We did not observe any decrease of gelation time with temperature decrease in this temperature range (Table 5.1). Gelation cannot be studied below −5 °C as far as water in 8 % NaOH–water starts crystallising at this temperature (the end of water melting peak was recorded at −6 to −4 °C when solutions were heated from −60 °C to room temperature, Egal et al. 2007).

As shown above (Roy et al. 2003; Gavillon and Budtova 2008), gelation time is exponentially temperature dependent at given cellulose concentration $t_{gel} = D \exp(-aT)$ where D and a are adjustable constants. In the range of cellulose and ZnO concentrations studied, the exponent a varies from 0.28 to 0.46. Similar results were obtained for the same microcrystalline cellulose dissolved in (7.6–9) % NaOH–water (a = 0.35–0.4) (Gavillon and Budtova 2008; Roy et al. 2003). In the presence of ZnO, solutions are gelling with temperature with the same physical mechanisms (with the same exponent) as without ZnO.

Influence of Cellulose Concentration on Gelation: Increasing polymer concentration facilitates gelation, since chains are closer to each other, allowing an easier formation of a network. The influence of cellulose concentration on gelation time in the presence of ZnO is given in Fig. 5.4. The inverse of t_{gel} as a function of cellulose concentration C_{cell} presented in a double logarithmic plot shows that there is a power law relation between these two parameters. A power law dependence of gelation time on carrageenan concentration was suggested by Ross-Murphy (1991):

$$t_{gel} \approx \frac{K}{[(\frac{C}{C_0})^{n'} - 1)]^p}$$

where C_0 is the critical concentration above which gelation could happen, n' is the number of polymer chains involved in a junction zone (assumed to be 2), K is a rate constant and p is a percolation exponent estimated to be around 2. For cellulose–NaOH–ZnO–water solutions, the overlap concentration C^*, which could be considered as critical concentration for gelation, is around 1 %, depending on temperature (Liu et al. 2011). C^* of the same microcrystalline Avicel cellulose in 9 % NaOH–water was reported to be 0.83 % below 20 °C and 1.25 % at 40 °C (Roy et al. 2003). Considering these values, (5.3) becomes $t_{gel} \approx k\, C_{cell}^{-n}$, where k is a rate constant and n is a kinetic exponent. The exponent n is related to the gelation kinetics as well as to the organisation of the junction zones in gel. The values of n obtained here show that they are more or less constant (within the experimental errors) and equal to 9 ± 2, with or without ZnO, strongly suggesting that junction characteristics are not strongly influenced by the addition of ZnO.

Influence of ZnO Concentration on Gelation: Table 5.1 shows that the increase of ZnO concentration delays gelation. We measured the gelation time of 6 % cellulose–8 % NaOH–water at various ZnO contents, 0, 0.7 and 1.5 % from 10 to 45 °C. A master plot can be obtained by shifting two data sets, for 0.7 and 1.5 % ZnO, towards the one without ZnO (Fig. 5.16). The same shape of the initial curves indicates that despite ZnO efficiency in delaying gelation, the increase in ZnO concentration does not change gelation kinetics.

We demonstrated that ZnO is quite efficient in delaying gelation of cellulose–NaOH–water solutions and it does not have much influence on either gelation kinetic order or the junction zones in cellulose gels. In addition, the addition

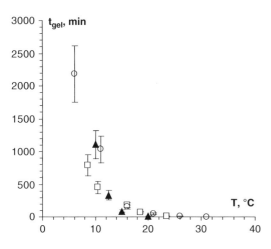

Fig. 5.16 Gelation time versus temperature for 6 % cellulose–8 % NaOH–water solutions with 0, 0.7 and 1.5 % ZnO shifted towards 0 % ZnO by 6.5 °C for 0.7 % ZnO and by 14 °C for 1.5 % ZnO. *Filled triangle*: without ZnO; *open square* with 0.7 % ZnO; *open circle* with 1.5 % ZnO

of ZnO does not change the properties of cellulose at the molecular level and does not improve the thermodynamic property of solvent towards cellulose. To dissolve cellulose, at least four NaOH molecules per one anhydroglucose unit are needed (Egal et al. 2007), with NaOH hydrates breaking cellulose intra-chain hydrogen bonds and linking to cellulose chains. The same is valid for cellulose–NaOH–water solutions containing urea (Egal et al. 2008). As shown many times, cellulose–NaOH–water solutions are not stable and cellulose chains tend to aggregate, leading to gelation. We are suggesting that when ZnO is suspended in NaOH–water solutions, a "network" of tiny particles is formed. The surface of particles is hydrolysed and a layer of hydroxide is formed ($\equiv$Zn–OH) attracting water molecules. When dissolved, $Zn(OH)_3^-$ and $Zn(OH)_4^{2-}$ ions are formed (Reichle et al. 1975; Degen and Kosec 2000) that also trap water. Thus, ZnO may play the role of a water "binder", strongly decreasing the amount of free water around the cellulose chains that may drive chain aggregation. ZnO stabilises the solution by keeping water far from the cellulose chains, as urea does (Egal et al. 2008). This could explain why ZnO only delays the gelation but does not change cellulose dissolution, solvent quality and gelation kinetics in NaOH–water solutions.

5.3.2.5 Cellulose Coagulation Kinetics from Cellulose–NaOH–Water Solutions (Gavillon and Budtova 2008; Sescousse and Budtova 2009; Sescousse et al. 2011a)

Shaping cellulose from NaOH–water-based solvents requires coagulation of cellulose, a process driven by diffusion coefficient of NaOH from cellulose solution or gel into the regenerating bath. The time of reaching complete coagulation is controlled by kinetics of cellulose solvent release into the bath which in turn depends on coagulation conditions and sample geometry. Obviously, coagulation time and rate are important parameters for cellulose processing. The understanding of the mechanisms governing coagulation kinetics opens the ways in controlling final cellulose morphology.

Coagulation kinetics of cellulose from cellulose–8 % NaOH–water gelled solutions wad studied by Gavillon and Budtova (Gavillon and Budtova 2008; Sescousse et al. 2011a). Coagulation of cellulose was performed in a bath of cellulose nonsolvent, water or alcohol, with a controlled volume and temperature. The diffusion coefficient of cellulose solvent (NaOH) from a sample towards the coagulation bath was studied as follows. A sample of Avicel–NaOH–water gel or Solucell–NMMO solution of a given weight and volume was placed in the regenerating bath at a fixed temperature. The proportion between sample/bath weights was kept constant and equal to ten. The amount of NaOH released into the bath was measured as a function of time by potentiometry and titration

The experimental data, i.e. the increase of NaOH amount in the coagulation bath as a function of time and in different conditions (various cellulose concentration, regenerating bath liquid, bath temperature), was analysed using Fick approach. It is widely applied in drug release field and formation of membranes due to phase separation and was already used to describe the kinetics of cellulose coagulation from cellulose–NMMO–water solutions (Biganska and Navard 2005). The applicability of Fick approach was checked by plotting the cumulative amount of substance $M(t)$ (NaOH) released in time t as a function of $\sqrt{t}$. The amount of a substance released in time from a semi-infinite plane can be described as follows:

$$\frac{M(t)}{M} = 1 - \sum_{n=0}^{\infty} \frac{8}{(2n+1)^2 \pi^2} \exp\left(\frac{-D\pi^2 t (2n+1)^2}{l^2}\right) \quad (5.4)$$

where M is the amount of substance released at $t = \infty$ (here M coincides with the amount of substance in the initial sample), D is diffusion coefficient and l is half sample thickness because diffusion takes place from its both sides.

If the diffusion coefficient is constant, several simplifications are used to determine D from the slope of $\frac{M(t)}{M} = f\left(\sqrt{\frac{t}{l^2}}\right)$ curves:

(a) Early time approximation ($0 \leq \frac{M(t)}{M} \leq 0.4$):

$$\frac{M(t)}{M} = 4\left(\frac{Dt}{\pi l^2}\right)^{1/2}$$

(b) Late time approximation ($0.4 \leq \frac{M(t)}{M} \leq 1$):

$$\frac{M(t)}{M} = 1 - \frac{8}{\pi^2} \exp\left(\frac{-\pi^2 Dt}{l^2}\right)$$

(c) Half time ($\frac{M(t)}{M} = 1/2$) approximation: Here the diffusion coefficient is calculated at the point where $\frac{M(t)}{M} = 1/2$; it is equal to $D = \frac{0.049}{(t/l^2)_{1/2}}$ where $(t/l^2)_{1/2}$ is the abscissa when $\frac{M(t)}{M} = 1/2$. The experimental data can be then fitted with formula 1 with $n = 0$.

The initial state of the Avicel–NaOH–water samples is a gel and the final state is coagulated swollen cellulose. During coagulation, the state of cellulose changes: A phase separation takes place. Different types of models explaining the transport behaviour of solutes in hydrogels or in porous media were tested (Gavillon and Budtova 2008). These models mainly include free-volume theory, hydrodynamic and obstruction approaches and their combinations as well.

The analysis of the influence of cellulose concentration on NaOH diffusion coefficients shows that "porous membrane" (i.e. free-volume or hydrodynamic approach) and not "hydrogel-obstruction" approach must be used for the understanding and interpretation of cellulose coagulation from cellulose–NaOH–water gels. The applicability of a free-volume or hydrodynamic approach is caused by the phase separation process occurring during coagulation of cellulose from cellulose–NaOH gels placed in a nonsolvent liquid. The interpretation of NMMO diffusion and thus of cellulose coagulation should also be with the "porous membrane" approach, but it is complicated by the high concentration of NMMO in the sample, by the presence of NMMO not linked to cellulose at low cellulose concentrations and the change of NMMO phase state during coagulation. An increase of the size of the diffusing entity due to NMMO dragging water molecules during diffusion should also be considered. NaOH activation energy obtained from diffusion experiments coincides with the activation energy calculated from the rheological experiments for cellulose–NaOH–water and pure NaOH–water solutions. This result confirms the suggestion made in Roy et al. (2003) that cellulose–NaOH–water solutions are in fact suspensions of NaOH hydrates with or without cellulose.

The influence of cellulose coagulation conditions on NaOH diffusion coefficient was studied by Sescousse and Budtova (2009). Gel discs were placed in 100 mL acetic acid regenerating bath of 0.1 mol L^{-1}, and pH evolution in time was recorded. The measured diffusion coefficients do not depend on the gelation mode (increasing temperature or time) within the experimental errors.

It was noted that gels were different in terms of hardness after "gentle" gelation (50 °C—2 h; 80 °C—15 min) as compared to "hard" gelation conditions (50 °C—20 h; 80 °C—90 min). Long heating times coupled with elevated temperatures lead to at least two phenomena: sample slight weight loss (syneresis due to microphase separation and probably cellulose degradation) and increase of cellulose–cellulose bondings. Both phenomena induce gel "densification" and are due the decrease of solvent thermodynamic quality with heating (Roy et al. 2003; Egal 2006). The obtained independence of the diffusion coefficient on gelation conditions shows that as well as for cellulose coagulation in water from cellulose–NaOH–water gel (Gavillon and Budtova 2007), the coagulation kinetics in an acid bath should be described with "membrane" approaches, like hydrodynamic or free-volume theories, that are not sensitive to gel structure on the molecular level. The diffusion coefficient of NaOH from non-gelled solution is three times lower than that from cellulose–NaOH–water gels, due to lower local cellulose concentration in gel "pores" and thus lower medium viscosity, as compared with the homogeneous cellulose concentration in solution. This result means that cellulose coagulation from cellulose–NaOH–water solution will take longer time than from the gel of the same concentration and geometry.

Diffusion coefficients were measured as function of acetic acid concentration in coagulation bath. They do not depend on acetic acid concentration within experimental errors. This means that coagulation of cellulose in water or in an acid bath from cellulose–NaOH gels of the same size and in the same temperature conditions will take the same time.

NaOH diffusion coefficient from 5 % cellulose–8 % NaOH–water gel to 100 mL of 0.1 mol L^{-1} acetic acid regenerating bath was determined for five bath temperatures: 22, 28, 37, 45 and 65 °C. Solution gelation was performed at 80 °C during 15 min. As expected, the diffusion coefficient decreases with temperature increase, varying from 3.2×10^{-4} mm^2 s^{-1} at 22 °C to 7×10^{-4} mm^2 s^{-1} at 65 °C. The diffusion coefficient values obtained were plotted versus inverse temperature and an activation energy of 16 ± 3 kJ mol^{-1} was calculated using Arrhenius law.

5.3.2.6 Influence of Cellulose Pulp Enzymatic Activation on Solubility in NaOH–Water (LeMoigne et al. 2010b)

The enzymatic treatment is a way to modify the composition of cellulose fibre and/or cellulose molecular weight in order to increase the accessibility of cellulose chains to chemical reagents. Enzymatic treatments were performed on three dissolving sulphite wood pulp samples named VHV-S (intrinsic viscosity 1,430 ml/g), SA (intrinsic viscosity 880 ml/g) and sample called LV-U (bleached sulphite pulp, intrinsic viscosity 350 ml/g) provided by Borregaard (Norway). A mixture of two commercial enzymes was used, Celluclast 1.5 L (cellulase derived from *Trichoderma reesei*, purchased by Novozymes, Denmark, used classically for the hydrolysis of lignocellulosic biomass feedstocks. This enzyme contains a broad spectrum of cellulolytic enzyme activities, like cellobiohydrolases and endo-1,4-glucanases) and Econase HC400 (highly concentrated liquid formulation of endo-1.4-beta-xylanase produced by *Trichoderma reesei* and standardised to minimum enzyme activity of 400,000 BXU/g as well as side activities like beta-glucanase, purchased by AB enzymes, Finland).

The non-treated and treated pulps were dissolved in an 8 % NaOH–water solution to test their alkaline solubility. The solutions were prepared as follows: 132 g of 12 % NaOH 12 %–water was stored at -6 °C. 2 g of pulp was added to 66 g of distilled water and stored 1 h at 4 °C. Then, 12 % NaOH–water and cellulose–water solutions were mixed together during 2 h at -6 °C and 1,000 rpm giving 200 g of a solution of 1 % cellulose in 8 % NaOH–water. These 1 % cellulose solutions were directly centrifuged to isolate the insoluble fraction.

The enzyme mixture used has two effects at short peeling times: (1) a digestion of the primary wall which is seen by the near absence of ballooning and (2) a destructuration action in the inside of the fibre which is seen by the large

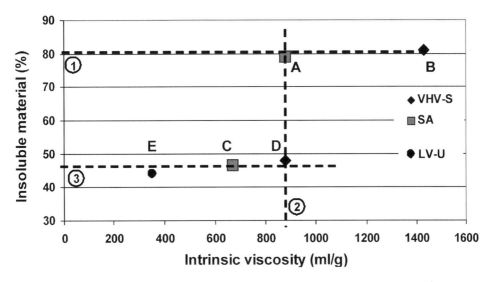

Fig. 5.17 Final amount of insoluble material versus intrinsic viscosity for the non-treated SA and VHV-S pulps (A, B), the 2 min treated SA and VHV-S pulps (C, D) and a non-treated LV-U pulp (E). From LeMoigne (2008)

decrease of DP. At long peeling time, the external walls are totally digested and the fibre structure is totally destructured, as seen by the absence of birefringence.

The observation by optical microscopy of the insoluble material clearly shows that the enzyme mixture used has two effects at short peeling times: (1) a digestion of the primary wall which is seen by the near absence of ballooning and (2) a destructuration action in the inside of the fibre which is seen by the large decrease of DP. At long peeling times, the external walls are totally digested and the fibre structure is totally destructured, as seen by the absence of birefringence. Non-treated samples mostly stay in a low swelling or ballooned state even after 2 h of mixing.

The SA sample contains fibres with high swelling ratio which can be related to the more severe digestion of the fibres during the original sulphite pretreatment. Several fibres are cut in sections due to the shearing involved by the mixing, and few fragments can also be observed. After 2 min of peeling, the shape of the insoluble parts is changing drastically. Only highly swollen sections and flat rings are observed. The removal of the primary wall by the enzymatic peeling gives a large swelling of the fibres and a subsequent cutting in flat rings. This cutting in flat rings was observed when putting and submitting fibres to shearing in 8 % NaOH–water, independently of the origin and the preparation mode of the wood pulp. With fibres without primary wall, the large swelling was shown to be sufficient to dismantle the fibres in highly swollen sections and flat rings. At longer peeling times, the insoluble material is made of small fragments implying that the fibres were totally destructured by the enzyme treatment.

The amount of insoluble material follows a linear relationship as a function of intrinsic viscosity and crystallinity. A simple way to explain the evolution of the alkaline solubility after the enzymatic treatment is thus to associate the increase of solubility to the decrease of the DP and of the crystallinity. However, the morphological study of the insoluble parts gives a more comprehensive interpretation on the influence of the structural changes due to the enzymatic treatment on the alkaline solubility. Figure 5.17 gives the amount of insoluble material of the non-treated (A, B) and 2 min enzymatically treated (C, D) SA and VHV-S pulps and of the LV-U pulp (E). By following the dotted line number 1 of Fig. 5.10, an interesting result is found. Despite the large difference of intrinsic viscosity of the two original SA and VHV-S pulps (880 ml/g and 1,430 ml/g, respectively), the amount of insoluble material is nearly the same. The DP is thus not the only parameter that governs the alkaline solubility after enzymatic treatment. This result was already found in the early work of the Asahi group (Kamide et al. 1984, 1992). The dotted line number 2 of Fig. 5.17 shows that the amount of insoluble material of the treated VHV-S sample (D, black diamond) is nearly two times lower than the original SA sample (A, grey square) whilst keeping the same intrinsic viscosity. This observation has to be related with the removal of the external walls that occur during the first minutes of treatment. The external walls are still present in the original SA sample, while they have been removed in the 2-min treated VHV-S sample. The removal of the external walls allows a higher swelling and thus eases the dismantlement and the fragmentation of the fibres during the solution preparation. The dotted line 3 shows that the LV-U sample (E) which was not enzymatically treated presents similar amount of insoluble material as compared to the treated SA and VHV-S samples (C, D), while its intrinsic viscosity is more than two times lower. The enzyme destructuration allows having a better solubility by favouring the fragmentation of the fibres during the dissolution.

The removal of the external walls and the macrostructural destructuration of the fibres must thus be considered as key factors in the dissolution of wood fibres in 8 % NaOH–water.

The enzymatic treatment is thus leading to a fast and large decrease of degree of polymerisation and of crystallinity, showing that the enzymes do not act simply on the fibre surface. The alkaline solubility of the different treated samples was investigated in a NaOH 8 %–water solution. As expected from thermodynamic considerations, there is a direct correlation between the solubility and the degree of polymerisation. However, aside thermodynamics, the removal of the external walls and the macrostructural destructuration of the fibres are key factors in the improvement of the dissolution of wood cellulose fibres. At constant intrinsic viscosities of the cellulose materials, the alkaline solubility is almost two times higher when the external walls are removed. The macrostructural destructuration of the fibre by the enzymes allows preserving a high degree of polymerisation while keeping a good alkaline solubility.

5.4 Cellulose–Ionic Liquid Solutions

As described above, several solvents have been successfully used for cellulose dissolution for making films and fibres and as homogeneous reaction media for synthesising cellulose derivatives. However, most of them are either toxic, or volatile, or with limited thermal stability, or needing rather high temperatures for the dissolution, or not allowing cellulose dissolution with reasonably high concentrations. There is thus a continuous search for new "green" cellulose solvents both for making objects from coagulated cellulose and also for the chemical derivatisation under homogeneous conditions. In this context, ionic liquids (IL) have been suggested as promising cellulose solvents. The first mention of cellulose dissolution in IL, quaternary ammonium salts, was published in US patent by Graenacher (1934). At the beginning of the twenty-first century, it was reported that cellulose can be dissolved at rather high concentrations (up to 15–20 %) without any pre-activation, in imidazolium-based ionic liquids, such as 1-butyl-3-methylimidazolium chloride (BMIMCl), 1-ethyl-3-methylimidazolium acetate (EMIMAc) and 1-allyl-3-methylimidazolium chloride (AMIMCl) (Swatloski et al. 2002; Zhang et al. 2005). Due to their ionic structure, IL possesses several advantages over common solvents: They are non-volatile, with high thermal stability, and have good dissolution properties for a large variety of chemical compounds. An interesting advantage is that the properties of IL may be easily tuned by varying anions or cations.

The dissolution of cellulose in ionic liquids is simple and occurs without a derivatisation step, which makes ionic liquids very attractive. Wet-spinning process can be used to obtain cellulose fibres; their properties were reported to be very similar to the ones known for lyocell fibres (Kosan et al. 2008a, b). Various novel cellulose materials can be prepared from cellulose–ionic liquid solutions, such as all-cellulose composites (Zhao et al. 2009), porous aerogel-like cellulose materials (Tsioptsias et al. 2008; Sescousse et al. 2011b), bioactive cellulose films (Turner et al. 2004) and functionalised cellulose microparticles (Lin et al. 2009). It is also possible to perform chemical reactions in homogeneous conditions (El Seoud et al. 2007): Cellulose sulphates, acetates and laurates and furoates were obtained (Gericke et al. 2009a; Barthel and Heinze 2006; Kohler and Heinze 2007). Characterisation of cellulose/IL solutions and understanding their properties and various practical applications have been summarised in recent reviews (Pinkert et al. 2009; Zhu et al. 2006).

Besides using IL as a reaction medium for synthesising cellulose derivatives, the research in cellulose–IL area can be divided into the following directions: understanding of (micro) structure of cellulose–IL solutions including molecular modelling, studying of solution properties including use of co-solvents and making materials for various applications. For example, using ^{1}H and ^{13}C NMR-technique cellobiose solvation in EMIMAc was recently reported (Zhang et al. 2010). Based on the analysis of the concentration dependences of chemical shifts for cellobiose and EMIMAc, authors concluded on hydrogen bonding between the anhydroglucose hydroxyl groups and both the [EMIM]$^+$ and [Ac]$^-$ ions (Zhang et al. 2010). A screening of more than 2000 ionic liquids with COSMO-RS approach (conductor-like screening model for realistic solvation) points to the anion as mainly being responsible for the respective dissolving power (Kahlem et al. 2010). Molecular dynamic simulations predict that 1-n-butyl-3-methylimidazolium acetate (BMIMAc) should have the strongest capability to break cellulose intra- and intermolecular hydrogen bonds (Gupta et al. 2011).

The understanding of cellulose–IL solutions rheological properties and cellulose hydrodynamic properties in these solvents is a need for a successful processing and for chemical derivatisation as well. Solution viscoelastic shear and extensional flow properties as a function of cellulose concentration and solution temperature were reported (Kosan et al. 2008a, b; Sammons et al. 2008; Collier et al. 2009; Haward et al. 2012). An extended study of the shear rheological properties of cellulose–AMIMCl (Kuang et al. 2008), cellulose–EMIMAc (Gericke et al. 2009b) and cellulose–BMIMCl solutions (Sescousse et al. 2010a, b) in dilute and semi-dilute states was performed. Cellulose intrinsic viscosity in EMIMAc was obtained for cellulose of various molecular weights in a large temperature range, from 0° to 100 °C, and the first attempt to determine Mark–Kuhn–Houwink constants was made (Gericke et al. 2009b).

It is known that the viscosity of imidazolium-based ionic liquids used to dissolve cellulose is much higher than the one of conventional organic solvents. This is a disadvantage for their use for cellulose processing and also as a reaction medium. For the latter case, dimethylformamide (DMF) and dimethylsulfoxide (DMSO) were used as co-solvents to decrease the overall viscosity (Gericke et al. 2009a, 2011). No decrease of solvent dissolution power was observed. These co-solvents and dimethylacetamide were mixed with EMIMAc, and cellulose solution's spinnability via electro-spinning was studied (Hairdelin et al. 2012). Shear rheological properties of cellulose solutions in AMIMCl and BMIMCl mixed with DMSO were investigated (Le et al. 2012). Cellulose concentration regimes and viscosities in these binary solvents were compared with the same parameters in pure EMIMAc. It was shown that IL-DMSO seems to be cellulose θ-solvent as suggested for EMIMAc in Gericke et al. 2009b, and the conformation of cellulose is not changed with the addition of DMSO not only in the dilute regime but also in the entanglement regime. The use of co-solvents opens new ways of tuning cellulose–IL solution viscosity. Still a lot of research has to be done to characterise and understand cellulose–IL-co-solvent solution properties.

Below we present results on the properties of cellulose–ionic liquid solutions performed in CEMEF/Mines ParisTech in collaboration with EPNOE partners and with School of Physics and Astronomy, University of Leeds, UK. Most of the results described below have been published in Gericke et al. (2009b), Sescousse et al. (2010a, b), Lovell et al. (2010) and Le et al. (2012).

5.4.1 Materials and Methods

Four celluloses of different molecular weight were used: microcrystalline cellulose with DP 180 and 300 (cell 170 and cell 300 in the following), spruce sulphite pulp with DP 1000 (cell 1000) and bacterial cellulose with DP 4420 (cell 4420). Ionic liquids were EMIMAc and BMIMCl. All products were purchased from Sigma-Aldrich except bacterial cellulose which was produced in Research Centre for Medical Technology and Biotechnology GmbH, Germany. Cellulose was dried at 50 °C in vacuum prior to use. Solvent and cellulose were mixed under nitrogen and stirred in a sealed vessel at 80 °C for at least 24 h to ensure complete dissolution. Clear solutions were obtained; they were stored at room temperature and protected against moisture absorption.

Rheological measurements on cellulose–EMIMAc solutions were performed on a Bohlin Gemini rheometer equipped with plate–plate geometry and a Peltier temperature control system. Steady-state viscosity of cellulose–BMIMCl solutions was measured using ARES rheometer from TA Instruments, with plate–plate geometry. In order to prevent moisture uptake and evaporation at elevated temperatures for both systems, a small quantity of low viscosity silicon oil was placed around the borders of the measuring cell. Because BMIMCl and cellulose–BMIMCl solutions are solid at room temperature (melting temperature of BMIMCl is around 70 °C as given by the manufacturer), the solutions were heated prior to measurements up to 140 °C and kept for 10 min, and viscosity–shear rate dependences were recorded at fixed temperatures, from 130° down to 70 °C.

Pulsed-Field Gradient ^{1}H NMR Spectroscopy. These experiments were performed in the group of Dr. M. E. Ries, School of Physics and Astronomy, University of Leeds, UK. Self-diffusion coefficients of both the [EMIM]$^+$ and [Ac]$^-$ ions were determined by a pulsed-field gradient ^{1}H NMR technique using a widebore Avance II NMR spectrometer (Bruker BioSpin) operating at a proton resonant frequency of 400 MHz. A specialised Diff60 diffusion probe (Bruker BioSpin) capable of producing a maximum field gradient of 24 T m^{-1} was employed in the experiments. More details are given in Lovell et al. (2010).

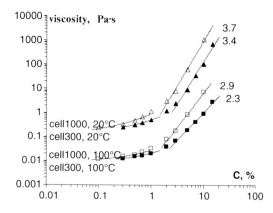

Fig. 5.18 Viscosity as a function of cellulose concentration for solutions of cell 300 and cell 1,000 in EMIMAc at 20° and 100 °C; the slopes for the concentrated region and indicated at the corresponding data. Adapted with permission from Gericke et al. (2009b). Copyright 2009 American Chemical Society

5.4.2 Steady-State and Intrinsic Viscosity of Cellulose–EMIMAc and Cellulose–BMIMCl Solutions

Steady-state viscosity of cellulose–EMIMAc and cellulose–BMIMCl solutions was measured at various concentrations and temperatures. For all systems a Newtonian plateau was recorded for at least two decades of shear rates (Gericke et al. 2009b; Sescousse et al. 2010a, b). Viscosity mean values were calculated for each concentration and temperature and were used in the following analysis. An example of viscosity η versus cellulose concentration for cell 300 and cell 1,000 dissolved in EMIMAc at 20° and 100 °C is presented in Fig. 5.18. Solution viscosity decreases with increasing temperature, which is typical for classical polymer solutions and for cellulose dissolved in other solvents. Two regions on each viscosity–concentration dependence can be seen:

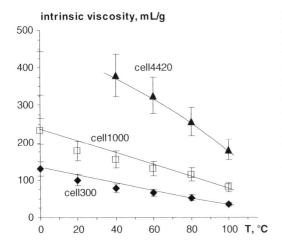

Fig. 5.19 Intrinsic viscosity of three cellulose samples dissolved in EMIMAc as a function of temperature. Lines are to guide the eye. Adapted with permission from Gericke et al. (2009b). Copyright 2009 American Chemical Society

a linear one in dilute regime and power law $\eta \sim C^n$ above the overlap concentration C^*. The exponent n was calculated for all cellulose–EMIMAc solutions; the values ranged from about 4 at low temperatures (0–40 °C) to 2.5–3 at high temperatures (60–100 °C). Similar values were reported for cellulose–LiCl/DMAc ($n = 3$ for bacterial cellulose and $n = 4$ for cotton linters and dissolving pulp) (Matsumoto et al. 2001).

Cellulose intrinsic viscosity $[\eta]$ was determined for solutions studied at different temperatures (Fig. 5.19). The intrinsic viscosity strongly decreases with temperature increase, for all samples investigated. This is a direct indication of a decrease in the thermodynamic quality of EMIMAc. A similar temperature influence on the size of cellulose macromolecules was observed for cellulose dissolved in 9 % NaOH–water (Roy et al. 2003). Cellulose–NaOH–water solutions are known to gel with temperature increase when polymer concentration exceeds the overlap concentration; no gelation or phase separation was observed for celluloses dissolved in imidazolium-based ionic liquids, unless they contain water.

Cellulose intrinsic viscosities obtained in EMIMAc were compared with the ones in other solvents, such as in cupriethylenediamine hydroxide (Cuen), in LiCl–DMAc, in 9 % NaOH–water and in cadoxene (Gericke et al.

2009b). In Cuen cellulose intrinsic viscosity was higher than in EMIMAc for the same cellulose sample, and in NaOH–water it was very similar. The overall conclusion is that EMIMAc is thermodynamically not a better solvent for cellulose as compared with the other ones. The exponent in Mark–Houwink equation relating intrinsic viscosity and molecular weight varied between 0.4 and 0.6 indicating that EMIMAc seems to be cellulose theta solvent.

The viscosity of cellulose–BMIMCl solutions was compared with the one of cellulose–EMIMAc (Sescousse et al. (2010a, b)). The properties of these imidazolium-based solvents are very different: EMIMAc is a room-temperature liquid with no glass transition or melting/crystallisation temperatures reported, while BMIMCl is solid at room temperature and melts around 60–70 °C. Because of the difference in solvent viscosity, the viscosity of cellulose–BMIMCl solution is 5–6 times higher than the one of cellulose–EMIMAc solution in the same conditions. If plotting the relative viscosities for both types of cellulose solutions in dilute regime, this difference disappears (Fig. 5.20).

The comparison between cellulose intrinsic viscosity in EMIMAc and in BMIMCl for the same cell 180 is demonstrated in Fig. 5.21. BMIMCl solvent behaves in an extremely similar way to EMIMAc. In overall, dilute regime of the studied solutions is at polymer concentrations below 1 % for room temperatures (only for cellulose–EMIMAc solution, cellulose–BMIMCl is solid) and slightly lower than 2 % for the temperatures above 80 °C for both solutions.

5.4.3 Influence of Water on Cellulose–EMIMAc Viscosity

It is known that imidazolium-based ionic liquids are extremely hygroscopic: They absorb humidity from the air not only modifying solvent viscosity but completely changing solution thermodynamics as far as water is cellulose nonsolvent. Cellulose processing on the pilot scale, as it is described in Kosan et al. (2008a, b), involves water evaporation from cellulose–IL–water slurry in a close-loop cycle of cellulose dissolution/coagulation.

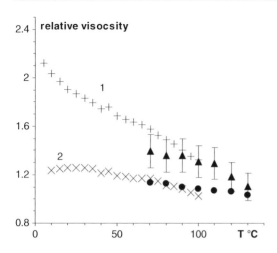
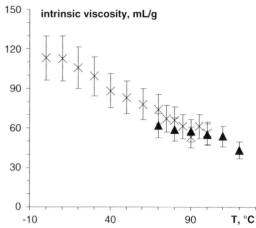

Fig. 5.20 Relative viscosities as a function of temperature for cell 180-EMIMAc (*crosses*) and cell 180-BMIMCl (*dark points*) solutions with 0.5 % (1) and 0.25 % (2) cellulose. The error bars of 10 % are shown only for 0.5 % cellulose–BMIMCl solutions in order not to overload the graph. Adapted with permission from Sescousse et al. (2010a). Copyright 2010 American Chemical Society

Fig. 5.21 Cell 180 intrinsic viscosity as a function of temperature for EMIMAc (*crosses*) and BMIMCl (*triangles*) solvents. Adapted with permission from Sescousse et al. (2010a). Copyright 2010 American Chemical Society

Water is removed from the slurry up to the moment when cellulose is dissolved in the ionic liquid. It is thus extremely important, both for cellulose shaping and chemical derivatisation as well, to understand how water influences cellulose–IL solution properties on the molecular level.

First, EMIMAc–water was prepared as a "new solvent" by mixing the components in different proportions. The viscosity of EMIMAc–water mixtures was analysed. Viscosity–shear rate dependence for EMIMAc–water at various water concentrations and temperatures was measured. In all cases studied, a Newtonian flow was recorded within at least two decades of the shear rates. Mean Newtonian viscosity values were determined, plotted as a function of water content and compared with viscosity calculated according to the logarithmic additive mixing rule. A noticeable difference between calculated and measured viscosity was obtained (Le et al. 2012). It should also be noted that heat is produced during mixing of EMIMAc and water (Le et al. 2012); it is undoubtfully an exothermal reaction suggesting, together with viscosity, special interactions between water and EMIMAc. This is an important result to keep in mind for the analysis of cellulose behaviour in ionic liquids in the presence of water.

An example of steady-state viscosity of a dilute cellulose–EMIMAc solution for various water contents is shown in Fig. 5.22. At water concentrations below 15 wt%, a Newtonian plateau was always found for 1 % cellulose solutions and lower concentrations. However, with 15 wt% water content and above the flow of dilute cellulose solution is shear thinning, clearly demonstrating that the properties of solutions changed despite the fact that they are transparent and no non-dissolved particles were seen with the optical microscope. We hypothesise that above 15 % water cellulose is not completely dissolved in EMIMAc; it exists as a sort of swollen agglomerates or "micro-gels" that are flowing as a "suspension". These soft species consisting of highly swollen cellulose chains are deforming and structuring under shear leading to viscosity decrease with the increase of shear rate. If increasing cellulose concentration in EMIMAc–15 wt% water mixture to 3 % cellulose, non-dissolved crystals can be seen with polarised optical microscopy (Le et al. 2012). 15 % of water in EMIMAc thus seems to be the

5 Preparation and Properties of Cellulose Solutions

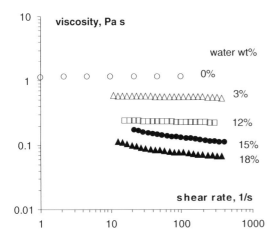

Fig. 5.22 Flow curves of 1 % cell 180–EMIMAc–water solutions with water concentrations of 0, 3, 12, 15 and 18 wt%, at 10 °C. The error bars are smaller or equal to the point size. With kind permission from Springer Science + Business Media: Le et al. (2012), Fig. 5.2

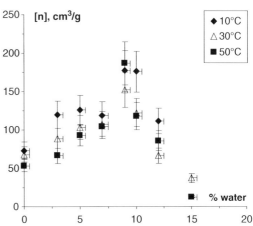

Fig. 5.23 Cell 180 intrinsic viscosity as a function of water concentration in wt% in EMIMAc–water at various temperatures. The error bars for [*filled square*] are 20 %. With kind permission from Springer Science + Business Media: Le et al. (2012), Fig. 5.5

maximal limit of water allowing the dissolution of 1 % cellulose in these conditions.

Cellulose intrinsic viscosity as a function of water content in the solvent EMIMAc–water is presented in Fig. 5.23 for some selected temperatures. The intrinsic viscosity first increases with water concentration, goes through a maximum at about 8–10 wt% water and then decreases. The decrease of cellulose intrinsic viscosity at high water content, above 10 wt%, was expectable: Polymer coil should contract when a large amount of nonsolvent is added. The increase of polymer intrinsic viscosity with the addition of nonsolvent is more complicated to interpret. When a polymer is dissolved in a mixed solvent of different thermodynamic quality towards the polymer, various cases may occur. The presence of a bad or a nonsolvent does not necessarily decrease binary solvent thermodynamic quality. For example, a mixture composed of a poor and nonsolvent may become a good solvent: The second viral coefficient changes the sign from negative to positive and goes through a maximum as a function of solvent composition (Masegosa et al. 1984). A convex shape of the intrinsic viscosity versus solvent composition, with solvent being a mixture of good + poor, poor + θ-solvent or good + nonsolvent, had already been reported for synthetic polymers (Pouchly and Patterson 1976; Pingping et al. 2006; Budtov et al. 2010). Such a deviation from the classical Flory lattice theory was interpreted by the preferential solvation introduced as "local" solvent composition being different from the "bulk" composition (Dondos and Benoit 1973; Hong and Huang 2000).

Considering the results obtained on the flow of dilute cellulose–EMIMAc–water solutions (Fig. 5.22), the presence of non-dissolved cellulose in 3 wt% cellulose–EMIMAc–15 wt% water and the fact that water and EMIMAc are interacting (exothermal mixing, deviation from the additive viscosity), the following interpretation of the increase of cellulose intrinsic viscosity with the increase of water content (left-hand part of Fig. 5.23) was suggested (Le et al. 2012). Cellulose and water are competing for EMIMAc. Because of strong EMIMAc–water interactions, cellulose is self-associating, probably via inter- and intramolecular hydrogen bonding, and aggregates are formed. It is the hydrodynamic size of these aggregates and not

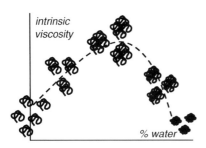

Fig. 5.24 A schematic illustration of the influence of water on cellulose intrinsic viscosity in EMIMAc–water: Formation of cellulose–cellulose aggregates with the increase of water content. With kind permission from Springer Science + Business Media: Le et al. (2012), Fig. 5.6

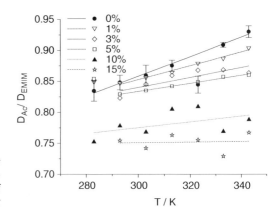

Fig. 5.25 The ratio of [Ac]⁻ to [EMIM]⁺ diffusion coefficients at various cell 180 concentrations as a function of temperature. Lines correspond to linear approximations. Error bars are shown only for EMIMAc data in order not to overload the graph. Reprinted with permission from Lovell et al. (2010). Copyright 2010 American Chemical Society

of one molecule that is determined with the intrinsic viscosity, as suggested in the schematic presentation in Fig. 5.24. The higher the water concentration, the larger the size of cellulose aggregates up to such water content at which cellulose becomes hardly soluble leading to a further strong decrease of the intrinsic viscosity.

5.4.4 Influence of Cellulose on the Diffusion of [EMIM]⁺ and [Ac]⁻ Ions in Cellulose–EMIMAc Solutions

In order to better understand cellulose–ionic liquid solution microscopic properties (transport phenomena, polymer–solvent interactions), diffusion-ordered two-dimensional NMR spectroscopy (DOSY) was used to measure ions' self-diffusion coefficients and correlate their mobility with cellulose concentration and solution temperature (Lovell et al. 2010). First, the change in the chemical shifts of each EMIMAc spectral band on cellulose concentration was recorded and analysed. The obtained cellulose concentration dependence of chemical shifts in cellulose–EMIMAc solutions was similar to the ones reported in Zhang et al. (2010) for cellobiose dissolved in EMIMAc. It was thus confirmed that cellulose dissolution in EMIMAc proceeds via hydrogen bonding and coordination of [EMIM]⁺ and [Ac]⁻ ions with cellulose hydroxyl groups.

The diffusion coefficients of [EMIM]⁺ and [Ac]⁻ ions at different cellulose concentrations and solution temperatures were calculated. The smaller anion was observed to diffuse slower than the larger cation (Lovell et al. 2010), a result already reported for other imidazolium-based ionic liquids (Noda et al. 2001; Tokuda et al. 2006) and also confirmed by molecular dynamic simulations (Tsuzuki et al. 2009, Urahata and Ribeiro 2005). The faster diffusion of a larger cation was explained by anisotropic diffusion of imidazolium ring in the directions of the ring plane as compared with the direction orthogonal to it (Urahata and Ribeiro 2005). For the case of cellulose–EMIMAc solution, the ratio between the diffusion coefficients of anion to cation is smaller than one, and it increases with temperature (Fig. 5.25) (Lovell et al. 2010). It is interesting to note that the presence of cellulose increases the difference between the diffusivity of anion and cation, and this ratio becomes less sensitive to temperature (Fig. 5.25).

The influence of cellulose concentration on the diffusion of ions was analysed with different theoretical approaches developed for polymer solutions. Obstruction, hydrodynamic and free-volume models were tested for fitting experimental data (Lovell et al. 2010). For example, by applying the obstruction model we realised that the notion of bound solvent and hydration shell around cellulose macromolecule should be considered to match theoretical predictions with experimental data. The proportion [EMIM]$^+$: [Ac]$^-$:hydroxyl = 1:1:1 was taken as a background approach. It gives the ratio of 5 between the volume fraction of cellulose and volume fraction of cellulose with bound solvent. With this ratio, different variations of the obstruction model approach fit, more or less, the experimental data (Lovell et al. 2010). The result obtained predicts that when cellulose concentration in EMIMAc reaches ≈ 27 wt%, all solvent molecules are bound in the primary solvation shell. This cellulose concentration seems to be the maximal possible to be dissolved in EMIMAc (Lovell et al. 2010). In the next section it will be shown that this prediction is confirmed by cellulose–EMIMAc–DMSO phase diagram.

The free-volume model was found not to give a satisfactory agreement with the experimental data for both ions' diffusion dependence on cellulose concentration. On the contrary, the hydrodynamic model as developed by Phillies (1986) showed a good agreement with the obtained experimental data (Lovell et al. 2010).

5.4.5 Phase Diagram and Solubility Limit of Cellulose in EMIMAc in the Presence of DMSO

DMSO (as well as DMF) is used to perform chemical reactions on cellulose in the homogeneous conditions. The main reason is that these co-solvents strongly decrease ionic liquid viscosity. They can also be interesting candidates for cellulose processing, allowing matching solution viscosity to the desired values. However, as far as DMSO is not cellulose solvent, the key question for processing is how the maximal cellulose concentration soluble in EMIMAc–DMSO depends on DMSO content. The goal of this work was to answer this question and build a phase diagram of cellulose in EMIMAc–DMSO.

Cellulose–EMIMAc–DMSO solutions were prepared at various EMIMAc–DMSO propor-

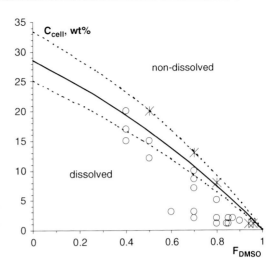

Fig. 5.26 Phase diagram for cell 180 in EMIMAc–DMSO. *Solid line* corresponds to the maximal cellulose concentration soluble in EMIMAc–DMSO calculated supposing 1 anhydroglucose unit binding 2.4 mol of EMIMAc; *dashed lines* correspond to AGU binding 3 mol (*lower*) or 2 mol (*higher*) of EMIMAc

tions and different cellulose concentrations and carefully analysed with optical microscopy using polarised and nonpolarised light in the view of the presence of non-dissolved cellulose. The magnification used was such that 5–10 μm objects were easily detectable. With this method, it is not possible to judge if the dissolution occurred on the molecular level; however, a phase diagram can be built in the first rough approximation. The results are presented in Fig. 5.26 as cellulose concentration in the system versus weight fraction of DMSO in the mixture EMIMAc–DMSO. Open points correspond to completely transparent (in nonpolarised light) or completely dark (in polarised light) solutions; crosses correspond to the cases when even a few small non-dissolved particles were detected.

The proportion between the maximal cellulose concentration dissolved at a given co-solvent fraction and the corresponding EMIMAc concentration in the solution was estimated. We supposed that neither cellulose nor EMIMAc interact with DMSO. The amount of EMIMAc moles per anhydroglucose unit (AGU) turned out to be a constant value around 2.4, not depending on DMSO concentration in solution, within the experimental errors. It seems that indeed EMIMAc and DMSO as well as cellulose and DMSO are not interacting. The theoretical maximal concentration of cellulose soluble in EMIMAc–DMSO was calculated for various cases: $C_{cell\ max} = 2$, 2.4 and 3 mol of EMIMAc. The results are shown in Fig. 5.26: solid line for $C_{cell\ max} = 2.4$ mol EMIMAc and upper and lower dashed lines for $C_{cell\ max} = 2$ and 3 mol EMIMAc, respectively. $C_{cell\ max} = 3$ mol EMIMAc overestimates the amount of EMIMAc needed to dissolve cellulose as far as the corresponding dashed line is passing through the well-dissolved region. 2 mol of EMIMAc per one AGU underestimates the amount if EMIMAc needed to dissolve cellulose; the corresponding line intersects the region that contains non-dissolved cellulose. We can thus conclude that one AGU needs about 2.3–2.5 mol of EMIMAc to be dissolved. This means that maximal cellulose concentration possible to dissolve in EMIMAc is around 28–30 wt%, the values predicted by NMR (Lovell et al. 2010).

5.4.6 Conclusions

Ionic liquids are a class of powerful cellulose solvents; they open new opportunities in cellulose processing and performing homogeneous chemical reactions. In our research we are trying to understand the fundamental properties of cellulose–ionic liquid solutions (polymer–solvent interaction, solvent thermodynamic quality, cellulose dissolution limit), and we are using ionic liquids for making highly porous cellulose materials, aerocellulose, as described in Chap. 5.

5.5 Stabilisation of Cellulose: NMMO Solutions (Lyocell Process)

Quality of the cellulose pulp and dissolution conditions determine the spinning behaviour and the performance of the finished material. Unfortunately, it is not possible to evaluate the suitability of pulps for the cellulose–NMMO process called lyocell only by analysis data such as intrinsic viscosity, molecular weight and their distribution or the content of heavy metals, ash or carbonyl and carboxyl groups. Although cellulose is more or less chemically identical in all pulps, there are significant differences in their physical states, mainly influenced by origin and extraction process of the pulps. Therefore, standard conditions are required to maintain dissolution and spinning quality, especially with the aim of upscaling (Michels and Kosan 2000). Solution quality, solution state and thermostability are important factors for characterising cellulose solutions for using them in the lyocell process.

5.5.1 Solution Quality and Solution State

Solution quality refers to the "spinning stability", which guarantees continuous fibre shaping of large spinning masses under technical conditions. Possible inhomogeneities like undissolved pulp residues can lead to clumps in the jet holes followed by permanent breaks of the filament. Consequently, the maximum diameter of particles is in direct relation to the spinnable fineness. To ensure a fineness of 1.3 dtex (B-type fibre), the spinning dope should not contain any undissolved particles >15 μm. A visual observation of the solution quality gives a with a detection threshold down to ~20 μm. Therefore, particle analysis by laser deflection must be used to provide information about the content and the distribution of particles in the range from 0.5 to 175 μm. Kosan and Michels (1999) have shown that the particle content depends on pulp origin and pretreatment and that the undissolved particle distribution is determined by the activation and solution conditions such as shear flow present in the mixing reactor, dissolution temperature and time.

Whereas solution quality describes the microscopic and submicroscopic range of undissolved particles, the solution state characterises its molecular range. Using the relatively easy accessible rheological parameters zero shear viscosity η_0, storage G' and loss modulus G'' and other parameters derived from the ones mentioned

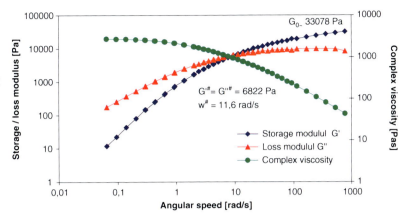

Fig. 5.27 Master curve of a deformation function for a cotton pulp as a function of angular speed (frequency) (DP 445, 13.2 % in NMMO) at 85 °C

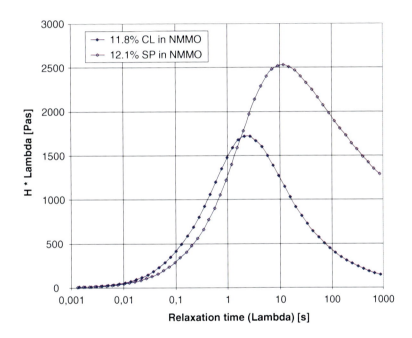

Fig. 5.28 Relaxation time spectra weighted of different wood pulp/ NMMO solutions with varying DP and inhomogeneities

above, width of the molecular weight distribution of linear polymers in concentrated solutions can be determined (Michels 1998). Zero shear viscosity is a function of the molecular weight, concentration and temperature. By monitoring G' and G'' as a function of angular speed ω at a reference temperature (usually 85 °C), master curves of a deformation function can be calculated (Fig. 5.27). The quotient of zero shear viscosity and viscosity at the "crossover" point where $\eta_\# = G'/\omega_\# = G''/\omega_\#$ characterise the vis-

coelastic behaviour of the dope and correspond to the molecular inhomogeneity parameter $U_\eta = \eta_0/\eta_\# - 1$. Measurements of different pulp origins give inhomogeneity parameter U_η of 2.5–4.2 for cotton, 4.9–8.0 for wood and 6.8–9.5 for cellulose mixtures.

The relaxation time, calculated from the deformation function, is proportional to the molecular weight. Figure 5.28 displays the weighted relaxation time spectra of different wood pulp/NMMO solutions with varying DP and inhomogeneities.

Lower molar mass fraction of the pulp causes increased inhomogeneities connected with low relaxation times and, consequently, leads to worse fibre performances. Short-chain fractions accelerate the relaxation and do not contribute to the entangled cellulose network. Both inhomogeneity and relaxation time have a direct effect on the range of suitable spinning conditions and on the mechanical parameters of the spun fibres or filaments because they determine the kinetics of fibre shaping. Temperature dependence of the relaxation time explains the significant influence of the spinning temperature on spinning stability and fibre properties.

5.5.2 Thermostability of Cellulose Solutions

Cellulose dissolution and shaping of the solutions are always connected with an alteration of the morphological structure accompanied by chain length reduction. Origin of the pulp, the nature of the solvent and the process conditions determine the decrease of the degree of polymerisation (DP). Whereas the DP reduction in systems like DMAc/LiCl is quite low, ionic solvents such as NMMO or ionic liquids (ILs) are involved in chemical reactions provoking degradation of both polymer and solvent. Therefore, a technical upscaling of a fibre-shaping technology is only possible if the process is stabilised, as for the lyocell process, which breakthrough arose from the introduction of NaOH/propyl gallate system, which suppresses the radical formation of NMMO and chain scission of cellulose (Buijtenhuijs et al. 1986).

Although the dissolution of cellulose in NMMO, being a non-derivatising solvent, and its fibre spinning are entirely physical processes, chemical alterations may appear under industrial conditions. High temperatures during cellulose dissolution and long residence times of the solutions during the spinning process result in discoloration and degradation of NMMO and cellulose. These reactions lower the recovery rate of the solvent and decrease the product performance (Taeger et al. 1985; Rosenau et al. 2001). A mass-related accumulation of the reaction mixture in the course of production increases the risk of exothermic reactions, so-called thermal runaway reactions, with an erratic temperature and pressure rise. Both chemical stabilisation and special technological features may prevent exothermic reaction and deflagrations (Firgo et al. 1994).

NMMO/cellulose segregation occurs as soon as the hydrogen bond system, which is considered as being responsible for cellulose solubility, is deteriorated (Novoselov et al. 1997; Cibik 2003). Local enrichment of NMMO and generated boundary surfaces of crystallised cellulose represent potential sites for subsequent reactions. NMMO is well known as an oxidising reagent in organic chemistry. Further, high amounts of many additives are able to disrupt hydrogen bonds by changing the polarity of the cellulose/NMMO system. In Fig. 5.29 the factors influencing the thermal stability are summarised.

Detailed ESR spectroscopy analysis of degradation processes in cellulose/NMMO solutions and, hence, formation of radical types have been carried out by Konkin et al. (2006) and Wendler et al. (2008). It shows that nitroxyl-type radicals R–NO·R' and oxymethylene species ·O–CH_2~ and ·CH_2–O~ (R11) at the first step were detected under irradiation as a result of ring degradation by a first-order reaction pathway (Scheme 5.1). Proposed NMMO ring degradation mechanism includes three variants (Ia, Ib, Ic) of di-radical types. The second step as a second-order reaction is invisible by ESR because the products are not radicals. However, they have to be potential precursors of further radical formations by step III under the continued laser illumination and were studied by high-performance liquid chromatography (HPLC). Further on, methyl ·CH_3 and formyl ·CHO radicals are formed at the third step.

Analytical investigations of lyocell dopes with regard to the thermostability are commonly provided by rheological, calorimetric, chromatographic and spectroscopic methods (Taeger et al. 1985, Rosenau et al. 2001). Especially, measurements of the pressure gradient and the generation of chromophores give information about the time dependence behaviour of the sample when kept at a constant temperature (Wendler et al. 2005a). Beside the type of cellulose, its concentration plays an important role. Higher cellulose

5 Preparation and Properties of Cellulose Solutions

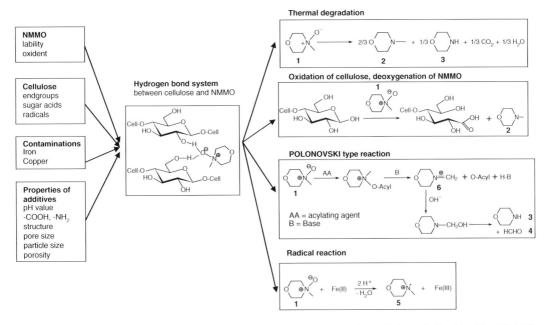

Fig. 5.29 Main actions and degradation reactions in the system cellulose/*N*-methylmorpholine-*N*-oxide (NMMO) adapted from Rosenau et al. (2001) and Taeger et al. (1985)

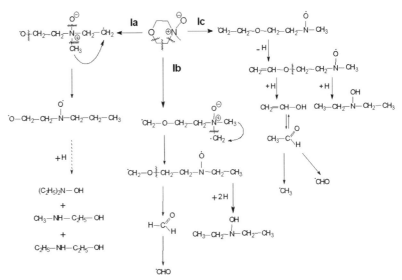

Scheme 5.1 Suggested reaction pathway leading to degradation of NMMO–cellulose mixtures

concentrations in the solution mean more reactive end groups and, as a consequence, higher concentrations of degradation products measured. Chromatographic methods revealed morpholine, *N*-methylmorpholine, *N*-formylmorpholine and formaldehyde as the main products resulting from reactions of the N–O bond. The activation of the most labile structure in the molecule occurs by protonation, complex formation with metals or O-alkylation. The cleavage of the NMMO ring

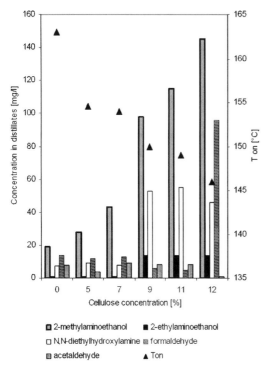

Fig. 5.30 Ring degradation products and the corresponding onset temperature (T_{on}) of cellulose/NMMO solutions with varying cellulose concentrations

structure as a consequence leads to resonance-stabilised nitroxyl radicals and stable ring degradation products. Acetaldehyde, 2-methylaminoethanol, 2-ethylaminoethanol and N,N-diethylhydroxylamine were detected by HPLC and attributed to cleavage products. Figure 5.30 displays the individual fingerprints of ring degradation products for solutions with different cellulose concentrations and the corresponding onset temperature.

The first thermal activity of a solution indicated as the onset temperature (T_{on}) is preferably measured by calorimetric methods and determined by plotting the deviation of pressure with respect to time (dp/dt) versus temperature. Cellulose/NMMO solutions give onset temperatures in the range 130–160 °C. Together with DSC mini-autoclave measurements with a sample weight of 2 g and analysis of the slope of pressure, it gives information about gas evolution even at lower temperatures and shows larger differentiations resulting in a more accurate characterisation of the influences of stabilisers and additives.

In order to prevent segregation and degradation of cellulose and NMMO, a stabiliser has to adjust the pH value, to scavenge by-products and to complex heavy metals. Propyl gallate commonly used as phenolic antioxidant in plastics, elastomers, fuels, etc. (Klemchuk 1985) was first introduced as a standard stabiliser preferably applied for trapping of radicals by forming relatively stable phenoxy radicals (Brandner and Zengel 1980). Further patents proposed reductants, e.g. amines, sulphites, aldehydes, sugars (Lukanoff and Schleicher 1981), bases (Franz et al. 1983), phosphonates (Laity 1983), sterically hindered phenols (Michels and Mertel 1984) and radical scavengers like 3,3′-thiodipropanol or carbon tetrabromide (Guthrie and Manning 1990). The combination of alkaline and antioxidant stabilisation (NaOH, propyl gallate) is generally accepted as the most efficient mode (Kalt et al. 1993). However, propyl gallate may interact with cellulose end groups and form coloured reaction products.

Up to now, no stabilisation system can be assessed as an ideal practical solution. The mentioned stabilisation procedures exclusively concern cellulose/NMMO solutions without any additive. Substances with functional groups or surface-active materials additionally cause degradation processes. TITK investigations of solutions modified with an IER and stabilised with NaOH and propyl gallate show decreased thermal stability leading to exothermic reactions during extrusion. Similar results were obtained in case of the presence of activated charcoal (Wendler et al. 2005b). Finally, it turned out that low thermal stability and large reduced viscosity will result in unstable spinning performance.

Table 5.4 summarises different bases, amines, complexing agents and phenolic oxidants employed as stabilisers in TITK under real time conditions. The measurements of the onset temperature (T_{on}) were taken as indicator for thermal stability. Commonly used stabilisers like propyl gallate; Irganox, a phenolic hydrazine stabiliser in plastics processing or the morpholine/HCHO system show high T_{on} for unmodified solutions

Table 5.4 Comparison of onset temperatures obtained from dynamic mini-autoclave measurements for 9 % cellulose/N-methylmorpholine-N-oxide solutions stabilised with different stabilisers

Stabiliser type	Amount of stabiliser (%)	T_{on} (°C)
None	–	146
NaOH	0.04	149
NaOH, propyl gallate	0.04; 0.06	155
NaOH, NH$_2$OH, propyl gallate	0.04; 0.1; 0.06	160
Benzyl amine	5	149
Morpholine	0.1	153
N-Methylmorpholine	0.33	150
N-Formylmorpholine	0.42	149
Morpholine, HCHO	0.32; 0.1	160
Polyethylene imine	5	153
Luvitec	5	149
Irganox	5	156
Ba(OH)$_2$	0.14	152
BHT	0.1	150
ISDB/BSDB	0.21; 0.21	152
NaOH, propyl gallate[a]	0.04; 0.06	131
Irganox[a]	5	135
Morpholine, HCHO[b]	0.32; 0.1	140

[a] 9% cellulose/N-methylmorpholine-N-oxide solution modified with activated charcoal (95 % with respect to cellulose)
[b] 9% cellulose/N-methylmorpholine-N-oxide solution modified with weak acidic cation exchange resin (95 % with respect to cellulose)

but fail to work for solutions modified with reactive charcoal or ion exchange resins. Therefore new stabilisers able to weaken surface reactions must be found. Additives such as activated charcoal or carbon black may have a serious effect as heterogeneous catalysts increasing degradation.

TITK investigations using a novel stabilising system with the combination of iminodiacetic acid sodium salt (ISDB) each covalently bound to a styrene/divinylbenzene copolymer exhibit favourable stabilising properties of cellulose/NMMO solutions modified with carboxyl group containing surface-active additives (Büttner et al. 2003). To estimate the effects of the polymeric stabilising system, the chelating ionogenic groups of ISDB in combination with the amine reaction potential of BSDB have to be taken into account. The iminodiacetic acid group forms coordination bonds between its donor atom nitrogen and metals. On the other hand, the partly sodium-neutralised carboxylic groups act as a buffer keeping constant pH values and thus maintaining the hydrogen bond system between cellulose and NMMO. Benzylamine of BSDB as a primary amine is able to bind carbon dioxide, acids and aldehydes. Additionally, BSDB may deactivate highly reactive intermediates, such as the aminiumyl radical, by scavenging these compounds within the polymeric skeleton.

Calorimetric, spectroscopic and rheological measurements were used to describe the stabilisation effects of ISDB/BSDB on modified solutions compared to solutions without stabilisers and the "state-of-the-art" stabiliser NaOH/PG by Wendler et al. (Wendler et al. 2006, 2008). Table 5.5 summarises an extract of the investigated additives. Since it is not possible to present here detailed examinations, only T_{on} is used to evaluate the thermal stability when the new stabiliser system is applied. Generally, a shift of T_{on} to higher values is registered on an average increase of 4.5 °C in a range from 0 to 8 °C.

5.6 Cellulose–Other Polysaccharide Blends

Multicomponent systems (polymer blends, alloys, composites, interpenetrating networks) have been receiving more and more attention owing to their scientific interest and practical applications. Blending is an efficient technique for the development of novel materials with improved properties (Guo 1999; Cazacu and Popa 2004). Especially cellulose materials offer several advantages when combined with synthetic or natural polymers due to their low density, high modulus, strength and high stiffness, little damage during processing, small requirements of processing equipment, biodegradability and relatively low price (Zadorecki and Michell 1989; Joly et al. 1996). Hydrogen bonding between –OH groups are recognised as an important factor in providing driving forces for the attainment of thermodynamic miscibility in many polymer blend systems where

Table 5.5 Comparison of onset temperatures obtained from dynamic mini-autoclave measurements for modified cellulose/NMMO solutions (9 % cellulose) without and with ISDB/BSDB stabiliser system

Additive		T_{on} (°C)	
Type	Amount of additive (%)[a]	Without stabiliser	With stabiliser
–	0	146	152
Aluminium oxide	33.3	150	153
Aluminium/silicon oxide	75	147	153
Titanium oxide	100	155	155
Yttrium oxide	15	152	154
Zinc oxide	50	158	160
Zeolite	50	144	150
Lead zircon titanate	50	158	158
Barium hexaferrite	500	132	137
Nano-silver	0.05	148	155
Graphite	100	143	147
Paraffin	70	153	160
PVP/MA copolymer	20	147	151
Weak acidic cation exchange resin	95	148	156
Strong acidic, gel-type cation exchange resin	100	150	154
Strong basic anion exchange resin	100	157	160
Superabsorbing polymer (polyacrylic acid)	50	151	157
Carbon black	100	144	149
Activated charcoal, strong reactive	95	131	134
Activated charcoal, strong reactive	50	137	141
Activated charcoal, reactive	50	140	144
Activated charcoal, medium reactive	50	147	152
Activated charcoal, weakly reactive	50	149	156

[a]With respect to cellulose

polysaccharides are present (Nisho 1994). Desired coupling of different polysaccharides' functionalities by joint or separate dissolving allows the shaping of the arising solution and solidification towards the polysaccharide blend (Taeger et al. 1997).

Most reports in literature concern the combination of a polysaccharide (including derivatives) and a synthetic polymer. There exist only few reports about polysaccharide/polysaccharide blends from different R&D groups, e.g. combination of pure cellulose with chitosan (Hasegawa et al. 1992), starch, xanthan, locust bean gum (LBG), guar gum, tragacanth gum, chitosan (Taeger et al. 1997), chitin (Pang et al. 2003; Kondo et al. 2004), sodium alginate (Kim et al. 2007) or carrageenan (Prasad et al. 2009). Some interesting studies dealt with hemicelluloses as structure regulators of native cellulose (Atalla et al. 1993; Zhang and Tong 2007). Recently, the incorporation of 15 different polysaccharides into a cellulose matrix using three solvents was studied to evaluate the interactions of polysaccharides or mixtures of those in solutions and the solid state after shaping (Wendler et al. 2010). In the following main results of that comprehensive study are inserted.

Various polysaccharides were chosen to find new structured biopolymer blends bearing adjustable properties, as shown in (Fig. 5.31). Three solvent routes for dissolving were followed: (1) sodium hydroxide–water (NaOH), (2) N-methylmorpholine-N-oxide monohydrate (NMMO) and (3) ionic liquids (ILs). By studying structure–property relations with a variety of analytical methods, new possibilities for the design of interesting cellulose-based fibre materials can be opened.

5 Preparation and Properties of Cellulose Solutions

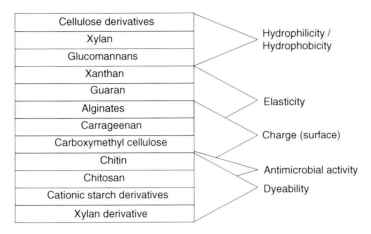

Fig. 5.31 Expected effects of selected blend polysaccharides on cellulose properties

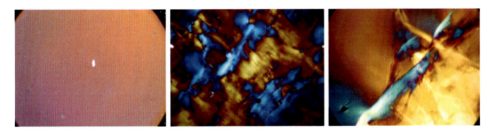

Fig. 5.32 Microscopic images of cellulose/blend polysaccharide solutions in EMIMac. *Left*: cellulose/LBG (97.5/2.5); *middle*: cellulose/LBG (95/5); *right*: cellulose/tragacanth gum (95/5)

Dissolution screenings yielded promising polysaccharides which were used for the preparation of cellulose/polysaccharide solutions and, consequently, for shaping of blends with cellulose. Chitin and chitosan were not soluble in any of the three solvents. Dissolution was possible in the case of carboxymethylcellulose, cellulose carbamate, xylan, xanthan and the mannans, guar gum, locust bean gum and tragacanth gum. In opposite to NaOH and NMMO, the EMIMac ionic liquid is not able to dissolve polysaccharides with functional groups like carrageenan or methylhydroxyethylcellulose above only about 1 %.

Solubility and miscibility with cellulose were evaluated by microscopy, DSC, particle analysis and rheology. To evaluate the states of solutions, microscopic images between cross polarisers were taken at room temperature. Most of the pure polysaccharide solutions were transparent and clear. Interestingly, only one polysaccharide solution—cellulose carbamate (15 %) in EMIMAc—showed an anisotropic behaviour which is of special importance for shaping of carbamate in ionic liquids (Rademacher et al. 1986). In the case of combination with cellulose and EMIMAc, xanthan, tragacanth gum and LBG, anisotropic structures could be observed as shown in Fig. 5.32 and interpreted as liquid crystalline (Kosan et al. 2008a, b; Chanzy et al. 1982). Usually, cellulose solutions in NMMO with polymer concentration higher than 20 % and molar ratios water/NMMO lower than one were found to be anisotropic.

Different types of the polysaccharide blends can also be shown for NMMO solutions.

Fig. 5.33 Microscopic images of cellulose/xylan solutions at 90 °C with a concentration ratio of 75/25 (**a**) 50/50 (**b**), 25/75 (**c**) and 0/100 (**d**) after image analysing (xylan particles are surrounded)

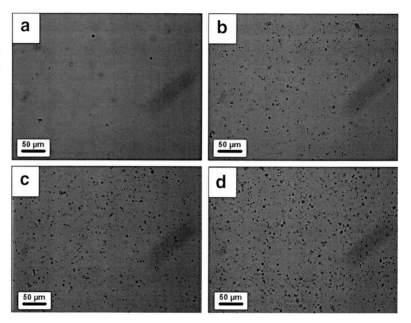

Figure 5.33 shows cellulose/xylan solutions and reveals the presence of particles of xylan in the range of 2–4 μm suggesting immiscibility with cellulose. However, it has to be noticed that these xylan particles are still present at the cellulose/xylan ratio 0:100, meaning that the presence of these particles is not due to the immiscibility between xylan and cellulose chains but to the interaction of xylan with the NMMO monohydrate. A possible explanation for the observed formation of xylan particles is the hydrophobic character of xylan chains related to the amount of side chains such as arabinose (Sternemalm et al. 2008) that can involve aggregation of the chains in aqueous solvent such as NMMO monohydrate. In comparison, cellulose/xanthan solutions reveal a continuous phase for all concentration ratios. It was not possible to distinguish the xanthan phase from the cellulose phase in the solutions.

Rheological investigations of cellulose/xylan solutions reveal no entanglement or interactions between cellulose and xylan chains. This is confirmed by a decay of zero shear viscosity accompanied by a decreasing relaxation time with increasing xylan amount. The rheological study of cellulose/xanthan solutions shows the inverse phenomenon compared to xylan, exhibiting a large increase of zero shear viscosity and relaxation time with xanthan concentration, mainly due to the high molar mass of xanthan chains (up to two millions).

Fibres in form of a multifilament yarn were manufactured from cellulose–other polysaccharide solutions in NaOH, NMMO and EMIMac using a wet-spinning method. As expected, textile physical properties are lower than unmodified fibres with slightly better properties for fibres spun in NMMO and EMIMac (Table 5.6). Residue analysis was carried out to determine the uptake rates of various polysaccharides which were added to the cellulose spinning dope during fibre production. The contents of arabinose, galactose, mannose, xylose and glucose in pure polysaccharide and in the blended fibre were measured. The content of the polysaccharide blended with cellulose in the fibres increases in the order NaOH–NMMO–EMIMac. This content depends rather on solvent than on type of polysaccharide and technology. Even though NMMO and EMIMac fibres were spun at the same equipment, the latter case favours higher amounts of the blended polysaccharide. It can be concluded that the coagulation of polymers from NMMO is

Table 5.6 Mechanical properties and sugar composition of fibres spun with the three solvent systems

Blend PS	Property	Fibre type NaOH	EMIMac	NMMO
Pure cellulose	Tenacity (cN/tex)	17.1	46.5	44.2
Carrageenan		17.3		38.2
CC		17.0		
CMC		14.3[a]	42.7	
Starch		13.4	36.7	
Xylan			42.8	37.1
LBG		12.9	43.6	
Guar gum			42.9	
Xanthan		14.2	43.8	41.3
Tragacanth gum		15.7	41.6	41.3
Pure cellulose	Elongation (%)	11.0	14.7	13.1
Carrageenan		7.6		12.3
CC		9.4		
CMC		3.5[a]	12.4	
Starch		9.0	11.2	
Xylan			13.2	13.5
LBG		17.2	12.9	
Guar gum			12.5	
Xanthan		11.1	13.4	11.6
Tragacanth gum		12.0	11.7	11.1

[a] Yarn

different, meaning that the hydrogen bond system between cellulose chains and the blended polysaccharide is weaker than inside the cellulose skeleton. The release of the blended polysaccharide out off the skeleton might be easier.

Spinning of fibres from all the solvents was possible even when using the higher molar polysaccharides. It has to be pointed out that the insertion of fine-milled polysaccharides into the cellulose is the crucial factor as previously described for the modified lyocell process where organic or inorganic compounds are inserted (see Chap. 5). Nevertheless, the morphology of blended solutions is of great interest to prevent clogging of filters and to control the final morphology and properties of the regenerated fibres.

X-ray diffraction measurements of the coagulated cellulose/polysaccharide blend fibres were performed to evaluate crystal orientation. Tenacity of cellulose fibres depends on chain orientation in the amorphous regions, whereas modulus depends on crystallinity and crystallite orientation. All examined fibre types give a cellulose II crystal structure. Main lattice planes are (1–10) and (110/020), which appear at the equator as discrete spots with small arcs of varying amplitude indicating different crystallite orientation. Figure 5.34 shows, for example, flat film diagrams of a cellulose/gum tragacanth fibre. Among the polysaccharide blends of the same fibre type, there are less significant differences between solvents, with the exception of the EMIMac fibres, whilst between polysaccharide blends, differences are larger. NMMO fibres have the highest orientation compared to EMIMac and NaOH fibre types.

From AFM analysis, it can be concluded that immiscibility state is not changed by the spinning, whereas the fibres exhibit distinctive differences in surface topography with regard to the solvent. Intertwining areas of different components can be differentiated. In general, on 1×1 μm sample size, surface roughness of fibres is increasing in the following sequence: EMIMac $\leq$ NMMO < NaOH (Fig. 5.35). Most

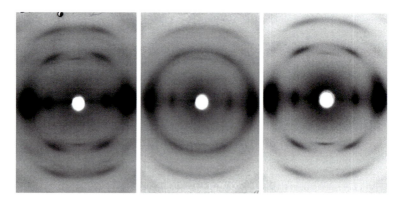

Fig. 5.34 X-ray flat film diagrams of cellulose/gum tragacanth fibre: *left* EMIMac, *middle* NaOH, *right* NMMO

Fig. 5.35 AFM of surface topography on 1 × 1 μm sample size for pure cellulose (*above*) and cellulose/xanthan (*below*) fibres from EMIMac (*left*, roughness 2.0 nm), NMMO (*middle*, 4.8 nm) and NaOH (*right*, 20.6)

fibres show irregular surfaces with fibrillar bundles or grainy structures of two phases as in the case of xanthan or tragacanth.

Contact angle measurements revealed only minor changes compared to the unmodified cellulose fibres. It has to be pointed out that the standard deviation of all samples is relatively high; the reason could be the sample non-homogeneity. The water contact angles of cellulose/polysaccharide fibres from EMIMac and NaOH are almost the same, i.e. 83–88°, while blend fibres from NMMO are slightly more hydrophilic, i.e. values are ranging from 73 to 76°. However, water contact angles are comparatively higher than viscose (68°) or modal fibre types (77°) (Persin et al. 2002).

Electrokinetic investigations (zeta potential measurements, ZP) were carried out to study the surface charges of the blended fibres. Apart from some exceptions, NMMO and EMIMac fibres show plateau values ranged from −7 to −10 mV, whereas the ZP values of NaOH fibres are in regions between −7 and −2 mV. The latter values can be seen in connection with the distinctive higher water retention, lower crystallite orientation and even much higher roughness as AFM measurements revealed. Indeed, water retention values of NMMO and EMIMac fibres are ranging between 40 and 54 % in comparison to main higher values obtained for the NaOH fibres (77–93 %).

Even though X-ray measurements revealed only little changes in the crystalline orientation of the blended fibres, further experiments were done to find out how the disturbed morphology of cellulose in blended fibres affects the accessibility, reactivity and adsorption properties. Mainly, the volume and inner surface areas of the voids play the decisive role in all cellulose heterogeneous chemical reactions (Fink et al. 1998). Consequently, sorption behaviour studied after special treatment of the fibres with distilled water buffered to varying pH value is enhanced for fibres from NaOH compared to EMIMac and NMMO fibres, which can be related to the higher surface roughness of the surface and lower crystallinity. Furthermore, dyeability tests show the same trends regard to the fibre types. Compared to the reference pure cellulose fibres, dye affinity of the blend fibres is clearly enhanced. As expected, fibres with carboxymethylcellulose and xanthan exhibit the highest distribution coefficient and monolayer sorption capacity for the cationic dye used. It appears that the solvent and/or the spinning process exerted an influence on the accessibility of carboxyl groups.

As was shown, a variety of parameters have an impact on the performance of the blend fibres, mainly the type of spinning technology used, the nature of the solvent and the type of polysaccharide mixed with cellulose. It can be concluded that what matters is the type of polymer added to cellulose more than the miscibility or not character of the mixture.

5.7 Cellulose–Lignin Mixtures in Solution (Sescousse et al. 2010b)

Cellulose and lignin are the main components of plant fibres. The synergy of their properties, together with hemicellulose, results in all varieties and special properties of natural fibres. For most of applications, cellulose and lignin have to be separated. During the separation, cellulose and lignin are in contact being in solution or suspension state (e.g. in papermaking); it is thus very important to know if they interact or not. Some applications suggest using lignin as a component in blends with polysaccharides. If cellulose or its derivatives are mixed with lignin, will the final material demonstrate the synergetic properties as it occurs in a natural fibre? The answer depends on the type of lignin, common solvent and mixing conditions used.

Literature reports some interactions between cellulose and lignin when they coexist in a natural fibre. It was shown that lignin is oriented preferentially parallel to the surface of the cell wall (Terashima and Seguchi 1988; Atalla and Agarwal 1985). Carbohydrates seem to form complexes with lignin as a unit of secondary cell wall structure (Chesson 1993). Molecular dynamics studied the growth and deposition of lignin on a cellulose surface in a cell wall. These studies supported the above-mentioned experimental finding and demonstrated not only a certain spatial organisation of lignin around cellulose microfibrils but also gave some evidences of links between two polymers (Houtman and Atalla 1995; Perez et al. 2004; Besombes and Mazeau 2005). However, the nature of these links was not specified.

In papermaking, lignin should be separated from cellulose. Possible redeposition of lignin onto cellulose fibres during cooking, washing and bleaching is of great importance, and thus, the compatibility and interactions between cellulose and lignin must be known. Maximova et al. (2001, 2004) studied the absorption of lignin by cellulose fibres and showed that it is absorbed due to capillary forces; lignin can be then easily removed by washing with water. They concluded that there was no true molecular adsorption of lignin onto fibres under various pH and ionic strengths investigated.

Blending of lignin with cellulose and cellulose derivatives has been extensively studied in the nineties (see, e.g. Rials and Glasser 1989, 1990; Dave et al. 1992; Dave and Glasser

1997). Fibres and films were usually prepared by mixing polymers in a common solvent followed either by solvent evaporative drying (film casting) or washing (fibres wet spinning). In the case of organosolv lignin–hydroxypropyl cellulose (HPC) films, miscible amorphous blends at low organosolv lignin concentrations were prepared and phase separation above 50 % of lignin was observed (Rials and Glasser 1990). For the mixtures of organosolv lignin with ethyl cellulose or with cellulose acetate butyrate (CAB), similar results were reported, but lignin content at which phase separation occurred was lower, 5–10 %. Fibres were spun from organosolv lignin–CAB mixtures and regenerated in water (Glasser et al. 1998). SEM and TEM results showed pore-size increase with the increase of organosolv lignin content. Analysing the above-mentioned results obtained on organosolv lignin–cellulose derivatives mixtures, Glasser et al. (1998) claim that there is an evidence of strong intermolecular interactions between the two components. Cellulose and lignin were mixed in dimethylacetamide/LiCL solvent and mixture viscosity was studied by Glasser et al. (1998): An increase of mixture dynamic elastic modulus with the addition of lignin was recorded, and it was explained by strong "secondary interactions" between the components.

Work performed in Centre de Mise en Forme des Matériaux, France, had the objective to study how the addition of lignin influences the properties of cellulose–8 % NaOH–water solutions. Certain simplifications, as compared with the naturally occurring systems, were made: We used microcrystalline cellulose and organosolv lignin. Mixture viscosity was compared to the result obtained with various cases known for polymer pairs forming interpolymer complexes or making an immiscible blend. Gelation of cellulose–8 % NaOH solution in the presence of lignin was investigated. Finally, dry lignocellulose materials were obtained according to the procedure used to prepare aerocellulose (Gavillon and Budtova 2008; Sescousse and Budtova 2009). Samples were analysed in terms of porosity and amount of lignin.

5.7.1 Cellulose–Lignin Mixtures in Aqueous 8 % NaOH

Microcrystalline cellulose Avicel PH101 was of DP 180. Organosolv lignin ("lignin" in the following, from vTI-Institute for Wood Chemistry, Hamburg, Germany) was from Norway spruce pulp. The molecular weight is $M_w = 7,300$ g mol^{-1} with polydispersity of 4.1, as determined with GPC using polyethylene glycol calibration. This lignin is soluble in aqueous solutions of neutral and high pH. Cellulose–NaOH solutions were prepared as described in Gavillon and Budtova (2008) and Sescousse and Budtova (2009). Briefly, an aqueous solution of 12 % wt NaOH was cooled down to −6 °C. Cellulose was swollen in distilled water in a certain proportion and kept at 5 °C. 12 % NaOH–water and swollen-in-water cellulose were mixed at −6 °C with a stirring rate of 1,000 rpm for 2 h. The ready solutions of various cellulose concentrations in 8 % NaOH–water were stored at 5 °C to avoid aging. Lignin was dissolved in aqueous 8 % NaOH by stirring for a few hours at room temperature. Cellulose–lignin mixtures in 8 % NaOH–water were prepared by simply mixing ready solutions in different proportions. All mixtures prepared were visually homogeneous and did not show any phase separation within 1 week storage time in refrigerator. Two types of mixtures were prepared: (a) series of mixtures with the total polymer concentration being constant and composition varied and (b) mixtures with cellulose concentration being constant and lignin concentration (and thus total polymer concentration) varied.

To prepare dry samples, cellulose–lignin–8 % NaOH–water mixtures were poured into cylindrical moulds with dimensions of about 3–5 cm height and 2 cm diameter and kept at 65 °C for 2 h. These conditions were chosen to ensure cellulose gelation (Roy et al. 2003; Gavillon and Budtova 2008). The gels were then placed in coagulation bath of either distilled water or 0.1 and 1 mol L^{-1} acetic acid. Bath liquid was regularly changed until pH did not vary anymore indicating that all NaOH is washed out. The samples were then rinsed with water to remove acid and

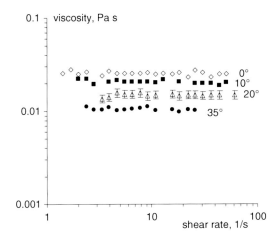

Fig. 5.36 Flow of 20 % lignin–8 % NaOH–water solution at various temperatures. With kind permission from Springer Science + Business Media: Sescousse et al. (2010b), Fig. 5.1

then washed in acetone to remove water which is not compatible with CO_2. Swollen-in-acetone samples were dried in supercritical CO_2 conditions as described earlier (1 L autoclave, 80 bar, 35 °C) by (Gavillon and Budtova 2008; Sescousse and Budtova 2009). After depressurisation (4 bar per hour at 37 °C), light brown samples of aerolignocellulose were extracted and analysed.

The flow of cellulose–(7–9 %)NaOH–water solutions and viscosity–concentration–temperature dependences have been studied extensively (Roy et al. 2003; Egal 2006; Gavillon and Budtova 2008). It was shown that above the overlap concentration solutions are gelling with time and temperature increase. Gelation time is exponentially temperature dependent, as shown by Roy et al. (2003) and Gavillon and Budtova (2008). Viscosity of cellulose solutions in conditions far from gel point is Newtonian below the overlap concentration and slightly shear thinning above it. The flow of lignin–8 % NaOH–water was investigated for 6–20 % lignin concentrations in the temperature interval from 0° to 35 °C. In all cases studied lignin solutions behave like Newtonian fluids; an example for 20 % lignin–8 % NaOH–water is shown in Fig. 5.36. This is what should be expected due to lignin low molecular weight. The activation energies Ea were calculated from viscosity–temperature dependences: $Ea_{15\% \; lign} = 16 \pm 2$ kJ mol^{-1} and $Ea_{20\% \; lign} = 18 \pm 2$ kJ mol^{-1}.

One way to estimate the interactions between the components in a common solvent is to study how mixture viscosity varies with its composition. The experimental data η_{exp} obtained should be then compared with a theoretical dependence of mixture additive viscosity η_{add} calculated at different compositions which assumes the absence of special interactions between the components (i.e. polymer1–polymer2 interactions are the same as the interactions between the macromolecules of the same type). For the mixtures where the concentration of at least one of the components is above the overlap concentration, the additive viscosity η_{add} must be calculated according to the logarithmic rule below:

$$\ln \eta_{add} = \phi_1 \ln \eta_1 + \phi_2 \ln \eta_2$$

where ϕ_1 and ϕ_2 are the weight fractions at which the components are present in the mixture ($\phi_1 + \phi_2 = 1$); η_1 and η_2 are the components' viscosities at $\phi_1 = 1$ and $\phi_2 = 1$, respectively.

In order to plot the experimental dependence of cellulose–lignin mixture viscosity as a function of mixture composition, the flow of cellulose–lignin–8 % NaOH–water of various compositions with total polymer concentration of 6 % was studied at 15 °C. These conditions were chosen as a compromise between not too low viscosity of lignin solution and not too fast gelation of cellulose solutions (e.g. 7–8 % cellulose solutions are gelling irreversibly within a few minutes as soon as they are extracted from the thermobath where they were prepared at −6 °C). Gelation of cellulose and cellulose–lignin mixtures was studied by measuring G' and G'' at 15 °C at frequency 1 Hz and stress 1 Pa. As it will be shown in the following, gelation of all 6 % mixtures studied at 15 °C is slow enough to be neglected for this type of measurements: The duration of one flow experiment was less than 30–40 min, while gelation takes more than 10 h. Because the mixtures showed a slightly shear-thinning flow, like observed previously for cellulose–(7–9)% NaOH–water solutions (Roy et al. 2003; Egal

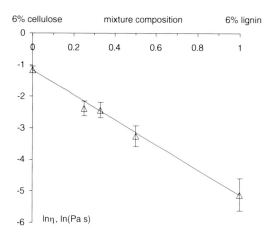 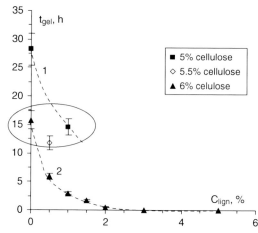

Fig. 5.37 Viscosity versus mixture composition for mixtures with 6 % total polymer concentration. *Points* are experimental data; *line* corresponds to the additive dependence calculated according to (5.1) (see details in the text). With kind permission from Springer Science + Business Media: Sescousse et al. (2010b), Fig. 5.3

Fig. 5.38 Gelation time as a function of lignin concentration in the mixtures containing 5 % (1), 5.5 % (*open point*) and 6 % (2) cellulose. *Lines* are given to guide the eye. The *circle* shows the mixtures with the same total polymer concentration, 6 %. With kind permission from Springer Science + Business Media: Sescousse et al. (2010b), Fig. 5.5

2006; Gavillon and Budtova 2008), the viscosity values to be plotted as a mixture composition were taken at the shear rate of 10 s^{-1}.

The viscosity of 6 % mixtures at various compositions is shown in Fig. 5.37 together with the calculated additive dependence. The experimental points coincide with the additive dependence demonstrating the evidence of the absence of any special interaction between cellulose and lignin in 8 % NaOH–water that could lead to the formation of new bonds. This important finding means that not only there are no bonds built between cellulose and lignin molecules but that these polymers coexist in the common solvent and may phase separate as soon as their total concentration becomes high enough. Electrostatic repulsive interactions between cellulose and lignin could be one of the reasons of the absence bonds built between the components as far as both polymers should be negatively charged in 8 % NaOH–water.

It was shown (Roy et al. 2003; Egal 2006; Gavillon and Budtova 2008) that cellulose–(7–9)%NaOH–water solutions are irreversibly gelling with the increase of temperature, time and concentration. It was interesting to check if and how the presence of lignin influences gelation of cellulose–8 % NaOH–water

solutions. Gelation time, t_{gel}, was chosen as one of the important parameters characterising gelation. It was determined from the evolution of elastic G' and viscous G'' moduli at a fixed temperature, 15 °C, for various mixture compositions and taken at the moment when $G' = G''$.

Gelation time of cellulose–lignin mixtures is shown in Fig. 5.38 for three mixtures containing 5, 5.5 and 6 % of cellulose. The amount of lignin was varied. Despite that low molecular weight lignin solutions in 8 % NaOH–water are not gelling, lignin presence in the mixture, even in low amounts, accelerates cellulose–8 % NaOH gelation (see a drastic decrease in gelation time for 5 % cellulose–1 % lignin as compared to 5 % cellulose–0 % lignin or with the increase of lignin content in the mixture of 6 % cellulose with 0.5, 1, 1.5 and 2 % lignin, Fig. 5.38).

Another way of looking at the same data gives a key in the understanding of what is happening in cellulose–lignin mixtures. Let us consider mixtures with the same total polymer concentration, 6 %: 6 % cellulose, 5.5 % cellulose–0.5 % lignin and 5 % cellulose–1 % lignin (shown with a circle in Fig. 5.38). On one hand, a decrease of cellulose concentration from 6 to 5 % in solutions without lignin leads to a strong increase in gelation time, as expected. On the other hand, gelation of all 6 % mixtures, pure 6 % cellulose, 5.5 % cellulose–0.5 % lignin and 5 % cellulose–1 % lignin takes practically the same time which is surprising because 0.5 and 1 % lignin solution has a very low viscosity and cannot be considered as a gelling component. A most probable explanation is that due to possible electrostatic repulsion between the polymers, lignin "helps" the formation of cellulose-rich and cellulose-poor domains, thus leading to cellulose–lignin microphase separation. The increase of local cellulose concentration facilitates gelation; lignin can thus be considered as gelation promoter for cellulose–8 % NaOH solutions. The result obtained on gelation also suggests lignin and cellulose incompatibility in 8 % NaOH–water. Cellulose–lignin gels should thus be much more heterogeneous as compared with pure cellulose–8 % NaOH–water gels.

5.7.2 Coagulation of Cellulose from Cellulose–Lignin–8 % NaOH–Water Gels and Aero-Lignocellulose Morphology

Cellulose from cellulose–NaOH–water gels can be coagulated if placing a gel into a liquid which is a nonsolvent for cellulose (water, aqueous acid solutions, alcohols). A 3D object made of cellulose "network" with nonsolvent liquid filling the pores is formed. Because lignin is not gelling and is not bound to cellulose, the question is what will happen with lignin during cellulose coagulation. The answer depends on the type of liquid used in coagulation bath. Qualitative observations are easy as far as lignin gives a dark brown colour either to the samples or to the coagulation bath. When water bath was used, it became brown when cellulose–lignin gel was placed in it. This is because lignin was washed out from cellulose as far as lignin used is water-soluble. The higher the bath acidity, the lower the amount of lignin washed out from the sample during coagulation because lignin is not soluble in acidic media and it coagulates inside the cellulose network. In the latter case, samples remain dark: The higher the bath acidity, the darker the samples.

If cellulose–lignin–8 % NaOH–water gels are very heterogeneous, coagulated samples should also be of very heterogeneous porosity. In order to study the porosity of swollen-in-nonsolvent coagulated cellulose, two ways of drying that preserve the pores against collapse can be used: either freeze drying or drying under supercritical conditions. We used the same way of drying in supercritical CO_2 as reported for the preparation of ultralight and highly porous pure cellulose, aerocellulose (Gavillon and Budtova 2008; Sescousse and Budtova 2009). It was shown that aerocellulose can be obtained from cellulose–8 % NaOH–water solutions and gels after cellulose coagulation in water, acid or alcohol bath, washing in acetone and drying in supercritical CO_2.

New aero-lignocellulose with various lignin contents and of various porosities were obtained using the same approach as described above for aerocellulose preparation (see Chap. 5). SEM images of dry ligno-aerocellulose coagulated in the baths of two different acid concentrations are shown in Fig. 5.39. All the other preparation parameters were identical. The lower is acid bath concentration (Fig. 5.39a), the more lignin was washed out and thus larger pores are obtained.

The proportion between cellulose and lignin in two dry samples was determined in vTI-Institute for Wood Chemistry, Hamburg, Germany, for the aero-lignocelluloses made from the same initial mixture with high initial lignin content but coagulated in different baths: one of 0.1 mol L^{-1} acetic acid and the other of 1 mol L^{-1}. The initial mixture was 4 % cellulose–8.6 % lignin, giving the initial proportion lignin/cellulose = 2.15. Sample preparation was identical except acetic acid concentration in coagulation bath. The amount of lignin in the aero-lignocelluloses was determined as a Klason and as acid-soluble lignin using UV spectrometer. The amount of acid-soluble fraction was very low, below 5–3 % of Klason lignin. The

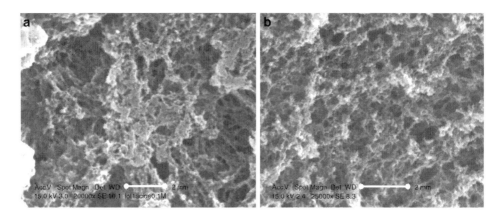

Fig. 5.39 SEM images of aero-lignocellulose from cellulose–lignin mixtures: regenerated in 0.1 mol L^{-1} (**a**) and in 1 mol L^{-1} acetic acid bath. The scale bar is 2 μm. Concentration of cellulose and lignin in the initial mixture was 4 and 4.3 %, respectively. Gelation was at 65 °C for 2 h. With kind permission from Springer Science + Business Media: Sescousse et al. (2010b), Fig. 5.7

proportion between lignin and cellulose obtained was 0.14 for the dilute acid bath and 0.33 for the concentrated one. In other words, keeping in mind that the amount of cellulose in wet and dry samples does no change, 82 % of lignin that was mixed with cellulose was then washed out during coagulation in 0.1 mol L^{-1} acetic acid bath and 65 % was washed out in 1 mol L^{-1} bath. It is a direct proof that cellulose and lignin macromolecules are not bound together. The reason could be, as mentioned above, the electrostatic repulsive interactions.

The influence of lignin concentration on aero-lignocellulose is reflected by sample density: 0.135 and 0.1 g cm^{-3} for sample with the initial cellulose/lignin proportion in the mixture 4 % cellulose–3.3 % lignin and 4 % cellulose–8.6 % lignin, respectively. Coagulation bath was 0.1 mol L^{-1} acetic acid. Different densities are due to large voids that are formed because of washed-out lignin during coagulation: The higher the lignin concentration in the mixture, the more are there large pores. However, the specific surface obtained with BET analysis for these two and other samples is practically the same within the experimental errors, around 200 ± 20 m^2 g^{-1}. The reason is that large pores do not make any significant contribution to pores' surface. Similar values were obtained for aerocelluloses prepared from pure cellulose solutions in 8 % NaOH–water (Gavillon and Budtova 2008) and in N-methylmorpholine-N-oxide (Innerlohinger et al. 2006).

Conclusions

Mixtures of cellulose and organosolv lignin in a common solvent, 8 % NaOH–water, were prepared. Mixture viscosity coincides with the one calculated according to the mixing rule, thus suggesting the absence of any bonds made between cellulose and lignin. The presence of lignin accelerates cellulose gelation, possibly due to the formation of cellulose-rich domains. Highly heterogeneous porous samples were obtained via cellulose coagulation and drying under CO_2 supercritical conditions. A significant part of lignin was washed out during coagulation confirming the repulsion between cellulose and lignin in NaOH–water solvent. The higher the lignin concentration in the mixture, the larger pores were obtained in the dry material.

References

Adusumali RB, Reifferscheid M, Weber H, Roeder T, Sixta H, Gindl W (2006) Mechanical properties of regenerated cellulose fibres for composites. Macromol Symp 244:119–125

Atalla RH, Agarwal UP (1985) Raman microprobe evidence for lignin orientation in the cell wall of native tissue. Science 227:636–638

Atalla RH, Hackney JM, Uhlin I, Thompson NS (1993) Hemicelluloses as structure regulators in the aggregation of native cellulose. Int J Biol Macromol 15:109–112

Barthel S, Heinze T (2006) Acylation and carbanilation of cellulose in ionic liquids. Green Chem 8:301–306

Benoît H (1948) Calcul de l'écart quadratique moyen entre les extrémités de diverses chaînes moléculaires de type usuel. J Polym Sci 3:376–388

Besombes S, Mazeau K (2005) The cellulose/lignin assembly assessed by molecular modeling. Part 2: seeking for evidence of organization of lignin molecules at the interface with cellulose. Plant Physiol Biochem 43:277–286

Biganska O, Navard P (2005) Kinetics of precipitation of cellulose from cellulose-NMMO-water solutions. Biomacromolecules 6:1949–1953

Blachot JF, Brunet N, Cavaille JY, Navard P (1998) Rheological behaviour of cellulose/(N-methylmorpholine N-oxyde-water) solutions. Rheol Acta 37:107–114

Boerstel H, Maatman H, Westerink JB, Koenders BM (2001) Liquid crystalline solutions of cellulose in phosphoric acid. Polymer 42:7371–7379

Brandner A, Zengel HG (1980) German Patent 303468

British Celanese (1925) GB 263810

Budtov VP, Bel'nikevich NG, Litvinova LS (2010) Thermodynamics and viscosities of dilute polymer solutions in binary solvents. Polym Sci Ser A 52:362–367

Buijtenhuijs FA, Abbas M, Witteveen AJ (1986) The degradation and stabilization of cellulose dissolved in N-methylmorpholine-N-oxide (NMMO). Papier 40:615–619

Büttner T, Graneß G, Wendler F, Meister F, Dohrn W (2003) German Patent 2003 (TITK) DE 10331342

Cai J, Zang L (2005) Rapid dissolution of cellulose in LiOH/urea and NaOH:urea aqueous solutions. Macromol Biosci 5:539–548

Cai J, Zhang LN (2006) Unique gelation behavior of cellulose in NaOH/Urea aqueous solution. Biomacromolecules 7:183–189

Cai J, Zhang L, Zhou J, Li H, Chen H, Jin H (2004) Novel fibres prepared from cellulose in NaOH:urea aqueous solutions. Macromol Rapid Commun 25:1558–1562

Cai J, Zhang L, Zhou J, Qi H, Chen H, Kondo T, Chen X, Chu B (2007) Multifilament fibers based on dissolution of cellulose in NaOH/urea aqueous solution: structure and properties. Adv Mater 19:821–825

Cai J, Zhang L, Liu S, Liu Y, Xu X, Chen X, Chu B, Guo X, Xu J, Cheng H, Han C, Kuga S (2008) Dynamic self-assembly induced rapid dissolution of cellulose at low temperatures. Macromolecules 41:9345–9351

Cazacu G, Popa VI (2004) Blends and composites based on cellulose materials. In: Dumitriu S (ed) Polysaccharide: structural diversity and functional versatility, vol 2. Dekker, New York, pp 1141–1177

Chanzy H, Roche E (1976) Fibrous transformation of Valonia cellulose I into cellulose II. J Appl Polym Symp 28:701

Chanzy H, Nawrot S, Peguy A, Smith P, Chevalier J (1982) Phase behavior of the quasiternary system N-methylmorpholine-N-oxide, water, and cellulose. J Polym Sci 20:1909–1924

Chanzy H, Noe P, Paillet M, Smith P (1983) Swelling and dissolution of cellulose in amine oxide/water systems. J Appl Polym Sci 37:239–259

Chaudemanche C, Navard P (2011) Influence of fibre morphology on the swelling and dissolution mechanisms of Lyocell regenerated cellulose fibres. Cellulose 18:1–15

Chen X, Burger C, Wan F, Zhang J, Rong L, Hsiao B, Chu B, Cai J, Zhang L (2007) Structure study of cellulose fibers wet-spun from environmentally friendly NaOH-urea aqueous solutions. Biomacromolecules 8:1918–1926

Chesson A (1993) Mechanistic model of forage cell wall degradation. In: Jung HG, Buxton DR, Hatfield RD, Ralph J (eds) Forage cell wall structure and digestibility. UAS Wisconsin, Madison, p 358

Cibik T (2003) Untersuchungen am System NMMO/H$_2$O/Cellulose. PhD Thesis, Technical University of Berlin

Ciechańska D, Wawro D, Stęplewski W, Wesolowska E, Vehvilonen M, Nousiainen P, Kamppuri T, Hroch Z, Sandak, Janicki J, Włochowicz A, Rom M, Kovalainen A (2007) Ecological method of manufacture of the cellulose fibres for advanced technical products. In Edana conference, Nonwovens Research Academy, 29–30 Mar 2007, University of Leeds, UK

Clark AH, Ross-Murphy SB (1987) Structural and mechanical properties of biopolymer gels. Adv Polym Sci 83:57–192

Cohen-Adad R, Tranquard A, Peronne R, Negri P, Rollet AP (1960) Le système eau-hydroxyde de sodium. Comptes Rendus de l'Académie des Sciences, Paris, France, 251 (part 3), pp 2035–2037

Cross CF, Bevan EJ (1892) Improvements in Dissolving Cellulose and Allied Compounds. British patent no. 8,700

Collier JR, Watson JL, Collier BJ, Petrovan S (2009) Rheology of 1-butyl-methylimidazolium chloride cellulose solutions. II. Solution character and preparation. J Appl Polym Sci 111:1019–1027

Cuissinat C, Navard P (2006a) Swelling and dissolution of cellulose, Part I: free floating cotton and wood fibres in N-methylmorpholine-N-oxide – water mixtures. Macromol Symp 244:1–18

Cuissinat C, Navard P (2006b) Swelling and dissolution of cellulose, Part II: free floating cotton and wood fibres in NaOH water-additives systems. Macromol Symp 244:19–30

Cuissinat C, Navard P (2008) Swelling and dissolution of cellulose, Part III: Plant fibres in aqueous systems. Cellulose 15:67–74

Cuissinat C, Navard P, Heinze T (2008a) Swelling and dissolution of cellulose, Part IV: Free floating cotton and wood fibres in ionic liquids. Carbohydr Polym 72:590–596

Cuissinat C, Navard P, Heinze T (2008b) Swelling and dissolution of cellulose, Part V: Cellulose derivatives fibres in aqueous systems and ionic liquids. Cellulose 15:75–80

Danilov SN, Samsonova TI, Bolotnikova LS (1970) Investigation of solutions of cellulose. Russ Chem Rev 39:156–168

Dave V, Glasser WG (1997) Cellulose-based fibres from liquid crystalline solutions: 5. Processing and morphology of CAB blends with lignin. Polymer 38:2121–2126

Dave V, Glasser WG, Wilkies GL (1992) Evidence of cholesteric morphology in films of cellulose acetate butyrate by transmission electron microscopy. Polym Bull 29:565–570

Davidson GF (1934) The dissolution of chemically modified cotton cellulose in alkaline solutions. Part I: In solutions of NaOH, particularly at T°C below the normal. J Text Inst 25:T174–196

Davidson GF (1936) The dissolution of chemically modified cotton cellulose inalkaline solutions. Part II: A comparison of the solvent action of solutions of Lithium, Sodium, Potassium and tetramethylammonium hydroxides. J Text Inst 27:T112–T130

Davidson GF (1937) The solution of chemically modified cotton cellulose in alkaline solutions. III. In solutions of sodium and potassium hydroxide containing dissolved zinc, beryllium and aluminum oxides. J Text Inst 28:2

Degen A, Kosec M (2000) Effect of pH and impurities on the surface charge of zinc oxide in aqueous solution. J Eur Ceram Soc 20:667–673

Dondos A, Benoit H (1973) The relationship between the unperturbed dimensions of polymers in mixed solvents and the thermodynamic properties of the solvent mixture. Macromolecules 6:242–245

Drechsler U, Radosta S, Vorwerg W (2000) Characterization of cellulose in solvent mixtures with N-methylmorpholine N-oxide by static light scattering. Macromol Chem Phys 201:2023–2030

Ducos F, Biganska O, Schuster KS, Navard P (2006) Influence of the Lyocell fibre structure on their fibrillation. Cell Chem Technol 40(5):299–311

Egal M (2006) Structure and properties of cellulose/NaOH aqueous solutions, gels and regenerated objects. PhD thesis, Ecole des Mines de Paris/Cemef, Sophia-Antipolis, France

Egal M, Budtova T, Navard P (2007) Structure of aqueous solutions of microcrystalline cellulose/sodium hydroxide below 0 °C and the limit of cellulose dissolution. Biomacromolecules 8:2282–2287

Egal M, Budtova T, Navard P (2008) The dissolution of microcrystalline cellulose in sodium hydroxide-urea aqueous solutions. Cellulose 15:361–370

El Seoud OA, Koschella A, Fidale LC, Dorn S, Heinze T (2007) Applications of ionic liquids in carbohydrate chemistry: a window of opportunities. Biomacromolecules 8:2629–2647

Fink H-P, Weigel P, Purz H-J (1998) Formation of lyocell-type fibres with skin-core structure. Lenz Ber 78:41–44

Fink HP, Gensrich J, Rihm R (2001) Structure and properties of CarbaCell-type cellulosic fibres. In: Proceedings of the 6th Asian textile conference, Hong-Kong, 22–24 Aug 2001

Fink H-P, Weigel P, Purz HJ, Ganster J (2001b) Structure formation of regenerated cellulose materials from NMMO-solutions. Prog Polym Sci 26(9):1473–1524

Firgo H, Eibl K, Kalt W, Meister G (1994) Kritishe fragen zur zukunft der NMMO-technologie. Lenz Ber 9:81–90

Flemming N, Thaysen AC (1919) On the deterioration of cotton on wet storage. Biochem J 14:25–29

Franks NA, Varga JK (1979) Process for making precipitated cellulose. US Patent 4,145,532

Franz H, Reusche P, Schoen W, Wiesener E, Taeger E, Schleicher H, Lukanoff B (1983) (AdW Teltow, TITK) German Patent 218104, 17 Oct 1983

Gavillon R, Budtova T (2007) Kinetics of cellulose regeneration from cellulose-NaOH-water gels and comparison with cellulose-N-methylmorpholine-N-oxide-water solutions. Biomacromolecules 8:424–432

Gavillon R, Budtova T (2008) Aerocellulose: new highly porous cellulose prepared from cellulose-NaOH aqueous solutions. Biomacromolecules 9:269–277

Gericke M, Liebert T, El Seoud O, Heinze T (2011) Tailored media for homogeneous cellulose chemistry: ionic liquid/co-solvent mixtures. Macromol Mater Eng 296:83–493

Gericke M, Liebert T, Heinze T (2009a) Interaction of ionic liquids with polysaccharides, 8 – synthesis of cellulose sulfates for polyelectrolyte complex formation. Macromol Biosci 9:343–353

Gericke M, Schlufter K, Liebert T, Heinze T, Budtova T (2009b) Rheological properties of cellulose/ionic liquid solutions: from dilute to concentrated states. Biomacromolecules 10:1188–1194

Glasser WG, Rials TG, Kelley SS, Dave V (1998) Studies of the molecular interaction between cellulose and lignin as a model for the hierarchical structure of wood. In: Heinze TJ, Glasser WG, Rojas O (eds) Cellulose derivatives. Modification, characterization and nanostructures. ACS symposium series 688 Chapter 19, pp 265–282

Glasser WG, Atalla RH, Blackwell J, Brown R Jr, Burchard W, French AD, Klemm DO, Nishiyama Y (2012a) About the structure of cellulose: debating the Lindman hypothesis. Cellulose. doi:10.1007/s10570-012-9691-7

Glasser WG, Atalla RH, Blackwell J, Brown R Jr, Burchard W, French AD, Klemm DO, Navard P, Nishiyama Y (2012b) Erratum to: about the structure of cellulose: debating the Lindman hypothesis. Cellulose. doi:10.1007/s10570-012-9702-8

Graenacher C (1934) Cellulose solution, US patent 1943176, 9 Jan 1934

Graenacher C, Sallman R (1939) Cellulose solutions. US Patent 2179181

Guo Q (1999) Thermosetting Polymer Blends: Miscibility, Crystallization, and Related Properties. In: Shonaike GO, Simon G (eds) Polymer blends and alloys, Marcel Dekker: New York, Chap. 6, pp 155–187

Gupta AK, Cotton JP, Marchal E, Burchard W, Benoit H (1976) Persistence length of cellulose tricarbanilate by small angle neutron scattering. Polymer 17:363–366

Gupta KM, Hu Z, Jiang J (2011) Mechanistic understanding of interactions between cellulose and ionic liquids: A molecular simulation study. Polymer 52:5904–5911

Guthrie JT, Manning CS (1990) The cellulose/N-methyl-morpholine-N-oxide/H$_2$O system: degradation aspects. In: Kennedy JF, Phillips GO, Williams PA (eds) Degradation Aspects, Cellulose Sources and Exploitation. Ellis Horwood, New York, pp 49–57

Hairdelin L, Thunberg J, Perzon E, Westman G, Walkenstrom P, Gatenholm P (2012) Electrospinning of cellulose nanofibers from ionic liquids: the effect of different cosolvents. J Appl Polym Sci 125:1901–1909

Haque A, Morris E (1993) Thermogelation of methylcellulose. Part I: Molecular structures and processes. Carbohydr Polym 22:161–173

Harrison W (1928) Manufacture of carbohydrate derivatives. US Patent 1,684, 732

Hasegawa M, Isogai A, Onabe T, Usada M, Atalla RH (1992) Characterization of cellulose–chitosan blend films. J Appl Polym Sci 45:1873–1879

Hattori K, Abe E, Yoshide T, Cuculo JA (2004) New solvents for cellulose. II Ethylenediamine/thiocyanate salt system. Polym J 36:123–130

Haward SJ, Sharma V, Butts CP, McKinley GH, Rahatekar SS (2012) Shear and extensional rheology of cellulose/ionic liquid solutions. Biomacromolecules 13:1688–1699

Hill JW, Jacobsen RA (1938) US patent 2,134,825

Hock CW (1950) Degradation of cellulose as revealed microscopically. Text Res J 20:141–151

Holt C, Mackie W, Sezllen DB (1976) Configuration of cellulose trinitrate in solution. Polymer 17:1027–1034

Hong PD, Huang HT (2000) Effect of co-solvent complex on preferential absorption phenomenon in polyvinyl alcohol ternary solutions. Polymer 41:6195–6204

Houtman CJ, Atalla RH (1995) Cellulose-lignin interactions. A computational study. Plant Physiol 107:997–984

Innerlohinger J, Weber HK, Kraft G (2006) Aerocellulose: aerogels and aerogel-like materials made from cellulose. Macromol Symp 244:126–138

Isogai A, Atalla RH (1998) Dissolution of cellulose in aqueous NaOH solutions. Cellulose 5:309–319

Jin H, Zha C, Gu L (2007) Direct dissolution of cellulose in NaOH:thiourea/urea aqueous solutions. Carbohydr Polym 342:851–858

Johnson DL (1969) Compounds dissolved in cyclic amine oxides. US Patent 3,447,939

Joly C, Kofman M, Gauthier RJ (1996) Polypropylene/cellulose fiber composites chemical treatment of the cellulose assuming compatibilization between the two materials. J Macromol Sci Pure Appl Chem A33 (12):1981–1996

Kahlem J, Masuch K, Leonhard K (2010) Modelling cellulose solubilities in ionic liquids using CPSMO-RS. Green Chem 12:2172–2187

Kalt W, Männer J, Firgo H (1993) (Lenzing) PCT Int. Appl. 9,508,010, 14 Sep1993

Kamide K, Saito M (1983) Persistence length of cellulose and cellulose derivatives in solution. Makromol Chem Rapid Commun 4:33–39

Kamide K, Saito M (1987) Cellulose and cellulose derivatives: recent advances in physical chemistry. Adv Polym Sci 83:1–56

Kamide K, Okajima K, Matsui T, Kowsaka K (1984) Study on the solubility of cellulose in aqueous alkali solution by deuteration IR and ^{13}C NMR. Polym J 16–12:857–866

Kamide K, Saito M, Kowsaka K (1987) Temperature dependence of limiting viscosity number and radius of gyration for cellulose dissolved in aqueous 8 % sodium hydroxide solution. Polym J 19:1173–1181

Kamide K, Yasuda K, Matsui T, Okajima K, Yamashiki T (1990) Structural change in alkali-soluble cellulose solid during its dissolution into alkaline solutions. Cellulose Chem Technol 24:23–31

Kamide K, Okajima K, Kowsaka K (1992) Dissolution of natural cellulose into aqueous alkali solution: role of super-molecular structure of cellulose. Polym J 24–1:71–96

Kasaai M (2002) Comparison of various solvents for determination of intrinsic viscosity and viscometric constants for cellulose. J Appl Polym Sci 86:2189–2193

Kihlman M, Wallberg O, Stigsson L, Germgard U (2011) Dissolution of dissolving pulp in alkaline solvents after stream explosion pretreatments. Holzforschung 65:613–617

Kim IS, Kim JP, Kwak SY, Ko YS, Kwon YK (2006) Novel regenerated cellulosic material prepared by an environmentally-friendly process. Polymer 47:1333–1339

Kim J, Wang N, Chen Y, Lee S-K, Yun G-Y (2007) Electroactive-paper actuator made with cellulose/NaOH/urea and sodium alginate. Cellulose 14:217–223

Klemchuk PP (1985) Antioxydants. In: Gerhartz W, Yamamoto YS (eds) Ullmann's encyclopedia of industrial chemistry, vol A3. Weinheim, VCH, pp 91–111

Kohler S, Heinze T (2007) Efficient synthesis of cellulose fuorates in 1-N-butyl-3-methylimidazolium chloride. Cellulose 14:489–495

Kondo T, Kasai W, Brown RM (2004) Formation of nematic ordered cellulose and chitin. Cellulose 11:463–474

Konkin A, Wendler F, Roth H-K, Schroedner M, Bauer R-U, Meister F, Heinze T, Aganov A, Garipov R (2006) Electron spin resonance study of radicals generated in cellulose/N-methylmorpholine solutions after flash photolysis at 77 K. Magn Reson Chem 44:594–605

Kosan B, Michels C (1999) Chem Fibers Int 49:50–54

Kosan B, Michels C, Meister F (2008a) Dissolution and forming of cellulose with ionic liquids. Cellulose 15:59–66

Kosan B, Schwikal K, Hesse-Ertelt S, Meister F (2008) In: Proceedings 8th international symposium alternative cellulose – manufacturing, forming, properties, Rudolstadt, Germany, 03–04 Sept 2008

Kuang QL, Zhao JC, Niu YH, Zhang J, Wang ZG (2008) Celluloses in an ionic liquid: the rheological properties of the solutions spanning the dilute and semidilute regimes. J Phys Chem B 112:10234–10240

Kunze J, Fink HP (2005) Structural changes and activation of cellulose by caustic soda solution with urea. Macromol Symp 223:175–187

Kuo Y-N, Hong J (2005) Investigation of solubility of microcrystalline cellulose in aqueous NaOH. Polym Adv Technol 16:425–428

Laity PR (1983) (Courtaulds) PCT International Application 8,304,415, 7 June 1983

Laszkiewicz B (1998) Solubility of bacterial cellulose and its structural properties. J Appl Polym Sci 67:1871–1876

Laszkiewicz B, Cuculo JA (1993) Solubility of cellulose III in sodium hydroxide solution. J Appl Polym Sci 50:27–34

Laszkiewicz B, Wcislo P (1990) Sodium cellulose formation by activation process. J Appl Polym Sci 39:415–425

Le KA, Sescousse R, Budtova T (2012) Influence of water on cellulose-EMIMAc solution properties: a viscometric study. Cellulose 19:45–54

LeMoigne N (2008) Mécanismes de gonflement et de dissolution des fibres de cellulose. Thèse de doctorat. Ecole Nationale Supérieure des Mines de Paris. Sophia Antipolis, France

LeMoigne N, Montes E, Pannetier C, Höfte H, Navard P (2008) Gradient in dissolution capacity of successively deposited cell wall layers in cotton fibers. Macromol Symp 262:65–71

LeMoigne N, Bikard J, Navard P (2010a) Contraction and rotation and contraction of native and regenerated cellulose fibres upon swelling and dissolution: the role of stress unbalance. Cellulose 17(3):507–519

LeMoigne N, Jardeby K, Navard P (2010b) Structural changes and alkaline solubility of wood cellulose fibers after enzymatic peeling treatment. Carbohydr Polym 79:325–332

Liebert TF (2010) Cellulose solvents-remarkable history, bright future. In: Liebert TF, Heinze TJ, Edgazr KJ (eds) Cellulose solvents: for analysis, shaping and chemical modification, ACS Symposium Series 1033, Oxford Press University, pp 3–54

Lin C-X, Zhan H-Y, Liu M-H, Fu S-Y, Lucia LA (2009) Novel preparation and characterisation of cellulose microparticles functionalised in ionic liquids. Langmuir 25:10116–10120

Lindman B, Karlström G, Stigsson L (2010) On the mechanism of dissolution of cellulose. J Mol Liq 156(1):76–81

Liu Y, Piron DL (1998) Study of tin cementation in alkaline solution. J Electrochem Soc 145:186–190

Liu W, Budtova T, Navard P (2011) Influence of ZnO on the properties of dilute and semi-dilute cellulose-NaOH-water solutions. Cellulose 18:911–920

Lovell PA (1989) Dilute solution viscometry. In: Colin B, Colin C (eds) Comprehensive polymer science, the synthesis, characterization, reactions and applications of polymers, vol I, Polymer characterization. Pergamon Press, Oxford

Lovell CS, Walker A, Damion RA, Radhi A, Tanner SF, Budtova T, Ries ME (2010) Influence of cellulose on ion diffusivity in 1-ethyl-3-methyl-imidazolium acetate cellulose solutions. Biomacromolecules 11:2927–2935

Lu A, Liu Y, Zhang L, Potthast A (2011) J Phys Chem B 115:12801–12808

Lue A, Liu Y, Zhang L, Potthast A (2011) Light scattering study on the dynamic behaviour of cellulose inclusion complex in LiOH/urea aqueous solution. Polymer 52:3857–3864

Lukanoff B, Schleicher H (1981) (AdW Teltow) German Patent 158656, 27 Apr 1981

Marsh JT (1941) The growth and structure of cotton, Mercerising. Chapman & Hall, London

Masegosa RM, Prolongo MG, Hernandez-Fuentes I (1984) Preferential and total sorption of poly(methyl methacrylate) in the cosolvent formed by acetonitrile with pentyl acetate and with alcohols (1-butanol, 1-propanol, and methanol). Macromolecules 17:1181–1187

Matsui T, Sano T, Yamane C, Kamide K, Okajima K (1995) Structure and morphology of cellulose films coagulated from novel cellulose/aqueous sodium hydroxide solutions by using aqueous sulphuric acid with various concentrations. Polym J 27–8:797–812

Matsumoto T, Tatsumi D, Tamai N, Takaki T (2001) Solution properties of celluloses from different biological origins in LiCl-DMAc. Cellulose 8:275–282

Maximova N, Osterberg M, Koljonen K, Stenius P (2001) Lignin adsorption on cellulose fibre surfaces: effect on surface chemistry, surface morphology and paper strength. Cellulose 8:113–125

Maximova N, Stenius P, Salmi J (2004) Lignin uptake by cellulose fibres from aqueous solutions. Nord Pulp Pap Res J 19:135–145

McCormick CL, Lichatowich DK (1979) Homogeneous solution reactions of cellulose, chitin, and other polysaccharides to produce controlled-activity pesticide systems. J Polym Sci Polym Lett Ed 17(8):479–484

McCormick CL, Callais PA, Hutchinson BH (1985) Solution studies of cellulose in lithium chloride and N,N-dimethylacetamide. Macromolecules 18:2394–2401

Michels C (1998) Beitrag zur Bestimmung von Molmasseverteilungen in Cellulosen aus rheologischen Daten. Determination of the mole-mass distribution of cellulose, using rheological data. Das Papier 52(1):3–8

Michels C, Kosan B (2000) Chem Fibers Int 50:556–561

Michels C, Mertel H (1984) (TITK) German Patent 229708, 13 Dec 1984

Mikolajczyk W, Struszczyk H, Urbanowski A, Wawro D, Starostka P (2002) Process for producing fibres, film, casings and other products from modified soluble cellulose. Poland, Patent no. WO 02/22924 (21 mars 2002)

Miller-Chou B, Koenig JL (2003) A review of polymer dissolution. Prog Polym Sci 28:1223–1270

Morris ER (1990) Shear-thinning of "random coil" polysaccharides: characterisation by two parameters from a simple linear plot. Carbohydr Polym 13:85–96

Morris ER, Cutler AN, Ross-Murphy S, Rees DA, Price J (1981) Concentration and shear rate dependence of viscosity in random coil polysaccharide solutions. Carbohydr Polym 1:5–21

Musatova GN, Mogilevskii EM, Ginzberg MA, Arkhangelskii DN (1972) The dissolution temperature of cellulose xanthate. Fibre Chem 2:451–453

Nägeli C (1864) Ueber den inneren Bau der vegetabilischen Zellenmem- branen Sitzber. Bay. Akad. Wiss. Munchen 1:282–323

Nisho Y (1994) Hyperfine composites of cellulose with synthetic polymers. In: Gilbert RD (ed) Cellulosic polymers, blends and composites. Hanser Publishers, New York, pp 95–113

Noda A, Hayamizu K, Watanabe M (2001) Pulsed-gradient spin-echo H-1 and F-19 NMR ionic diffusion coefficient, viscosity, and ionic conductivity of non-chloroaluminate room-temperature ionic liquids. J Phys Chem B 105:4603–4610

Noordermeer JWM, Daryanani R, Janeschitz-Kriegl H (1975) Flow birefringence studies of polymer conformation: cellulose tricarbanilate in two characteristics solvents. Polymer 16:359–369

Northolt MG, Boerstel H, Maatman H, Huisman R, Veurink J, Elzerman H (2001) The structure and properties of cellulose fibres spun from an anisotropic phosphoric acid solution. Polymer 42:8249–8264

Novoselov NP, Tret'yak VM, Sinel'nikov EV, Saschina ES (1997) Russ J Gen Chem 67(3):430–434

Okajima K, Yamane C (1997) Cellulose filament spun from cellulose aqueous NaOH solution system. Cell Commun 4:7–12

Ott E, Spurlin HM, Grafflin MW (1954) In Cellulose and cellulose derivatives (Part 1). Interscience Publisher, New York, p 353

Pang F-J, He C-J, Wang Q-R (2003) Preparation and properties of cellulose/chitin blend fiber. J Appl Polym Sci 90:3430–3436

Pennetier G (1883) Note micrographique sur les altérations du cotton. Bull Soc Ind Rouen 11:235–237

Persin Z, Stana-Kleinschek K, Kreze T (2002) Hydrophilic/hydrophobic characteristics of different cellulose fibres monitored by tensiometry. Croatica chemica acta 75(1):271–280

Perez DDS, Ruggiero R, Morais LC, Machado AEH, Mazeau K (2004) Theoretical and experimental studies on the adsorption of aromatic compounds onto cellulose. Langmuir 20:3151–3158

Phillies GDJ (1986) Universal scaling equation for self-diffusion by macromolecules in solution. Macromolecules 19:2367–2376

Pickering SU (1893) The hydrates of sodium, potassium and lithium hydroxides. J Chem Soc 63:890–909

Pingping Z, Yuanli L, Haiyang Y, Xiaoming C (2006) Effect of non-ideal mixed solvents on dimensions of poly(N-vinylpyrrolidone) and poly(methyl methacrylate) coils. J Macromol Sci Part B Phys 45:1125–1134

Pinkert A, Marsh KN, Pang S, Staiger MP (2009) Ionic liquids and their interactions with cellulose. Chem Rev 109:6712–6728

Potthast A, Rosenau T, Buchner R, Röder T, Ebner G, Bruglachner H, Sixta H, Kosma P (2002) The cellulose solvent system/N,N-dimethylacetamide/lithium chloride revisited: the effect of water on physicochemical properties and chemical stability. Cellulose 9:41–53

Pouchly J, Patterson D (1976) Polymers in mixed solvents. Macromolecules 9:574–579

Prasad K, Kaneko Y, Kadokawa J (2009) Novel Gelling Systems of κ-, ι- and λ-Carrageenans and their composite gels with cellulose using ionic liquid. Macromol Biosci 9:376–382

Qi H, Chang C, Zhang L (2008a) Effects of temperature and molecular weight on dissolution of cellulose in NaOH/urea aqueous solutions. Cellulose 15:779–787

Qi H, Cai J, Zhang L, Nishiyama Y, Rattaz A (2008b) Influence of finishing oil on structure and properties of multifilament fibers from cellulose dope in NaOH/urea aqueous solution. Cellulose 15:81–89

Rademacher P, Bauch J, Puls J (1986) Investigations of the wood from pollution-affected spruce. Holzforschung 40:331–338

Ramos LA, Assaf JM, El Seoud OA, Frollini E (2005a) Influence of the supramolecular structure and physicochemical properties of cellulose on its dissolution in a lithium chloride/N,N-dimethylacetamide solvent system. Biomacromolecules 6:2638–2647

Ramos LA, Frollinni E, Heinze T (2005b) Carboxymethylation of cellulose in the new solvent dimethylsulfoxide/tetrabutylammonium fluoride. Carbohydr Polym 60:259–267

Reichle RA, McCurdy KG, Hepler LG (1975) Zinc hydroxide – solubility product and hydroxyl–complex stability-constants from 12.5-75 °C. Can J Chem 53:3841–3845

Rials TG, Glasser WG (1989) Multiphase materials with lignin. VI. Effect of cellulose derivative structure on blend morphology with lignin. Wood Fiber Sci 21:80–90

Rials TG, Glasser W (1990) Multiphase materials with lignin: 5. Effect of lignin structure on hydroxypropyl cellulose blend morphology. Polymer 31:1333–1338

Röder T, Morgenstern B (1999) The influence of activation on the solution state opf cellulose dissolved in N-methylmorpholine N-oxide-monohydrate. Polymer 40:4143–4147

Röder T, Morgenstern B, Glatter O (2000) Light scattering studies on solutions of cellulose in N,N-dimethylacetamide/lithium chloride. Lenz Ber 79:97–101

Rollet AP, Cohen-Adad R (1964) Les systèmes "eau-hydroxyde alcalin". Revue de Chimie Minérale 1:451

Rosenau T, Potthast A, Sixta H, Kosma P (2001) The chemistry of side reactions and by-product formation in the system NMMO/cellulose (Lyocell process). Progr Polym Sci 26:1763–1837

Ross-Murphy SB (1991) Concentration dependence of gelation time. In: Dickinson E (ed) Food polymers, gels and colloids. Royal Society of Chemistry, Cambridge, pp 357–368

Roy C, Budtova T, Navard P, Bedue O (2001) Structure of cellulose-soda solutions at low temperatures. Biomacromolecules 2:687–693

Roy C, Budtova T, Navard P (2003) Rheological properties and gelation of aqueous cellulose-NaOH solutions. Biomacromolecules 4:259–264

Ruan D, Zhang L, Zhou J, Jin H, Chen H (2004) Structure and properties of novel fibers spun from cellulose in NaOH/thiourea aqueous solution. Macromol Biosci 4(12):1105–1112

Ruan D, Lue A, Zhang L (2008) Gelation behaviours of cellulose solution dissolved in aqueous NaOH-thiourea at low temperature. Polymer 49:1027–1036

Russler A, Lange A, Potthast A, Rosenau T, Berger-Nicoletti E, Sixta H, Kosma P (2005) A novel method for analysis of xanthate group distribution in viscoses. Macromol Symp 223:189–200

Russler A, Potthast A, Rosenau T, Lange T, Saake B, Sixta H, Kosma P (2006) Determination of substituent distribution of viscoses by GPC. Holzforschung 60:467–473

Saito G (1939) Das verhalten der zellulose in alkalilösungen. I. Mitteilung. Kolloid-Beihefte 29:365–454

Sammons RJ, Collier JR, Rials TG, Petrovan S (2008) Rheology of 1-butyl-methylimidazolium chloride cellulose solutions. I. Shear rheology. J Appl Polym Sci 110:1175–1181

Schulz L, Seger B, Burchard W (2000) Structures of cellulose in solution. Macromol Chem Phys 201:2008–2022

Segal L, Eggerton F (1961) Some aspects of the reaction between urea and cellulose. Text Res J 31:460–471

Seger B, Aberle T, Burchard W (1996) Solution behaviour of cellulose and amylose in iron-sodium tartrate (FeTNa). Carbohydr Polym 31(1–2):105–112

Sescousse R, Budtova T (2009) Influence of processing parameters on regeneration kinetics and morphology of porous cellulose from cellulose-NaOH-water solutions. Cellulose 16:417–426

Sescousse R, Le KA, Ries ME, Budtova T (2010a) Viscosity of cellulose-imidazolium-based ionic liquid solutions. J Phys Chem B 114:7222–7228

Sescousse R, Smacchia A, Budtova T (2010b) Influence of lignin on cellulose-NaOH-water mixtures properties and on Aerocellulose morphology. Cellulose 17:1137

Sescousse R, Gavillon R, Budtova T (2011a) Wet and dry highly porous cellulose beads from cellulose-NaOH-water solutions: influence of the preparation conditions on beads shape and encapsulation of inorganic particles. J Mater Sci 46:759–765

Sescousse R, Gavillon R, Budtova T (2011b) Aerocellulose from cellulose-ionic liquid solutions: preparation, properties and comparison with cellulose-NaOH and cellulose-NMMO routes. Carbohydr Polym 83:1766–1774

Sobue H, Kiessig H, Hess K (1939) The cellulose-sodium hydroxide-water system as a function of the temperature. Z Physik Chem B 43:309–328

Sprague BS, Noether HD (1961) The relationship of fine structure to mechanical properties of stretched saponified acetate fibers. Text Res J 31:858–865

Sternemalm E, Höije A, Gatenholm P (2008) Effects of arabinose substitution on the material. Properties of arabinoxylan films. Carbohydr Res 343:753–757

Struszczyk H, Wawro D, Starostka P, Mikolajscyk W, Urbanowski A (2000) EP 1317573 B1 "Process for producing fibres, film, casings and other products from modified soluble cellulose", 13/09/2000

Swatloski RP, Spear SK, Holbrey JD, Rogers RD (2002) Dissolution of cellulose with ionic liquids. J Am Chem Soc 124:4974–4975

Taeger E, Franz H, Mertel H, Schleicher H, Lang H, Lukanoff B (1985) Formeln Fasern Fertigware 4:14–22

Taeger E, Berghof K, Maron R, Meister F (1997) Eignshaftsänderungen im Alceru-faden durch zweitpolymere. Lenz Ber 76:126–131

Tamai N, Oano H, Tatsumi D, Matsumoto T (2003) Differences in rheological properties of plant and bacrterial cellulose in LiCl//N,N-dimethylacetamide. J Soc Rheol Jap 31(3):119–130

Tasker S, Baadyal JPS, Backson SCE, Richards RW (1994) Hydroxyl accessibility in celluloses. Polymer 35(22):4717–4721

Terashima N, Seguchi Y (1988) Heterogeneity in formation of lignin. IX. Factors affecting the formation of condensed structures in lignin. Cell Chem Technol 22:147

The Editors of "Dyer and Calico Printer" (1903) Mercerisation: a practical and historical manual, vol I. Heywood and Company Ltd., London

Tokuda H, Ishii K, Susan M, Tsuzuki S, Hayamizu K, Watanabe M (2006) Physicochemical properties and structures of room-temperature ionic liquids. 3. Variation of cationic structures. J Phys Chem B 110:2833–2839

Tripp VW, Rollins ML (1952) Morphology and chemical composition of certain components of cotton fiber cell wall. Anal Chem 24:1721–1728

Tsioptsias C, Stefopoulos A, Kokkinomalis I, Papadopoulou L, Panayiotou C (2008) Development of micro- and nano-porous composite materials by processing of cellulose with ionic liquids and supercritical CO_2. Green Chem 10:965–971

Tsuzuki S, Shinoda W, Saito H, Mikami M, Tokuda H, Watanabe M (2009) Molecular dynamics simulations of ionic liquids: cation and anion dependence of self-diffusion coefficients of ions. J Phys Chem B 113:10641–10649

Turbak AF, Hammer RB, Davies RE, Hergert HL (1980) Cellulose solvents. Chem Tech 10:51–57

Turbak AF, El-Kafrawy A, Snyder FW, Auerbach AB (1981) Solvent system for cellulose, US Patent 4,302,252

Turner MB, Spear SK, Holbrey JD, Rogers RD (2004) Production of bioactive cellulose films reconstituted from ionic liquids. Biomacromolecules 5:1379–1384

Urahata SM, Ribeiro MCC (2005) Single particle dynamics in ionic liquids of 1-alkyl-3-methylimidazolium cations. J Chem Phys 122:024511–024520

Vehviläinen M, Kamppuri T, Rom M, Janicki J, Ciechanska D, Grönqvist S, Sioika-Aho M, Christoffersson K, Nousiainen P (2008) Effect of wet spinning parameters on the properties of novel cellulosic fibres. Cellulose 15:671–680

Warwicker JO, Jeffries R, Colbran RL, Robinson RN (1966) A review of the literature on the effect of caustic soda and other swelling agents on the fine structure of cotton. St Ann's Press, Manchester, 93

Wendler F, Graneß G, Heinze T (2005a) Characterization of autocatalytic reactions in modified cellulose/NMMO solutions by thermal analysis and UV/VIS spectroscopy. Cellulose 12(4):411–422

Wendler F, Kolbe A, Meister F, Heinze T (2005b) Thermostability of lyocell dopes modified with surface – active additives. Macromol Mater Eng 290:826–832

Wendler F, Graneß G, Büttner R, Meister F, Heinze T (2006) A novel polymeric stabilizing system for modified lyocell solutions. J Polym Sci Part B Polym Phys 44:1702

Wendler F, Konkin A, Heinze T (2008) Studies on the stabilization of modified lyocell solutions. Macromol Symp 262:72–84

Wendler F, Meister F, Wawro D, Wesolowska E, Ciechanska D, Saake B, Puls J, Le Moigne N, Navard P (2010) Polysaccharide Blend Fibres Formed from NaOH, N-Methylmorpholine-N-oxide and 1-Ethyl-3-methylimidazolium acetate. Fibers Text Eastern Eur 18(79):21–31

Weng L, Zhang L, Ruan D, Shi L, Xu J (2004) Thermal gelation of cellulose in a NaOH/thiourea aqueous solution. Langmuir 20(6):2086–2093

Winter HH, Chambon F (1986) Analysis of linear viscoelasticity of a crosslinking polymer at the gel point. J Rheol 30(2):367–382

Yamada H, Kowsaka K, Matsui T, Okajima K, Kamide K (1992) Nuclear magnetic study on the dissolution of natural and regenerated celluloses onto aqueous alkali solutions. Cell Chem Technol 26:141–150

Yamane C, Saito M, Kowsaka K, Kataoka N, Sagara K, Kamide K (1994) New cellulosic filament yarn spun from cellulose/aq NaOH solution. In: Proceedings of '94 cellulose R&D, 1st annual meeting of the Cellulose Society of Japan (Cellulose Society of Japan, ed.) Tokyo, pp 183–188

Yamane C, Saito M, Okajima K (1996a) Industrial preparation method of cellulose-alkali dope with high solubility. Sen'I Gakkaaishi 52–6:310–317

Yamane C, Saito M, Okajima K (1996b) Specification of alkali soluble pulp suitable for new cellulosic filament production. Sen'I Gakkaaishi 52–6:318–324

Yamane C, Saito M, Okajima K (1996c) Spinning of alkali soluble cellulose-caustic soda solution system using sulphuric acid as coagulant. Sen'I Gakkaaishi 52–6:369–377

Yamane C, Saito M, Okajima K (1996d) New spinning process of cellulose filament production from alkali soluble cellulose dope-net process. Sen'I Gakkaaishi 52–6:378–384

Yamashiki T, Kamide K, Okajima K, Kowsaka K, Matsui T, Fukase H (1988) Some characteristic features of dilute aqueous alkali solutions of specific alkali concentration (2.5 mol l-1) which possess maximum solubility power against cellulose. Polymer J 20(6):447–457

Yamashiki T, Kamide K, Okajima K (1990a) New cellulose fibres from aq. alkali cellulose solution. In: Kennedy JF, Phillips GO, Williams PA (eds) Cellulose sources and exploitation. Ellis Horwood Ltd., New York, pp 197–202

Yamashiki T, Matsui T, Saitoh M, Okajima K, Kamide K (1990b) Characterisation of cellulose treated by the steam explosion method. Part 1: Influence of cellulose resources on changes in morphology, degree of polymerisation, solubility and solid structure. Br Polym J 22:73–83

Yamashiki T, Matsui T, Saitoh M, Okajima K, Kamide K (1990c) Characterisation of cellulose treated by the steam explosion method. Part 2: Effect of treatment conditions on changes in morphology, degree of polymerisation, solubility in aqueous sodium hydroxide and supermolecular structure of soft wood pulp during steam explosion. Br Polym J 22:121–128

Yamashiki T, Saitoh M, Yasuda K, Okajima K, Kamide K (1990d) Cellulose fibre spun from gelatinized cellulose/aqueous sodium hydroxide system by the wet-spinning method. Cell Chem Technol 24:237–249

Yamashiki T, Matsui T, Kowsaka K, Saitoh M, Okajima K, Kamide K (1992) New class of cellulose fiber spun from the novel solution of cellulose by wet spinning method. J Appl Polym Sci 44:691–698

Zadorecki P, Michell AJ (1989) Future prospects for wood cellulose as reinforcement in organic polymer composites. Polym Compos 10:69–77

Zhang H, Tong M (2007) Influence of hemicelluloses on the structure and properties of Lyocell fibers. Polym Eng Sci 47:702–706

Zhang L, Ruan D, Zhou J (2001) Structure and properties of regenerated cellulose films prepared from cotton linters in NaOH/urea aqueous solution. Ind Eng Chem Res 40:5923–5928

Zhang L, Ruan D, Gao S (2002) Dissolution and regeneration of cellulose in NaOH/thiourea aqueous solution. J Polym Sci Part B 40:1521–1529

Zhang H, Wu J, Zhang J, He J (2005) 1-allyl-3-methylimidazolium chloride room temperature ionic liquid: a new and powerful nonderivatizing solvent for cellulose. Macromolecules 38:8272–8277

Zhang S, Li FX, Yu JY (2009) Preparation of cellulose/chitin blend bio-fibres via direct dissolution. Cell Chem Technol 43:393–398

Zhang JM, Zhang H, Wu J, Zhang J, He JS, Xiang JF (2010) NMR spectroscopic studies of cellobiose solvation in EmimAc aimed to understand the dissolution mechanism of cellulose in ionic liquids. Phys Chem Chem Phys 12:1941–1947

Zhang S, Li FX, Yu JY (2011) Kinetics of cellulose regeneration from cellulose-NaOH/thiourea/urea/H2O system. Cell Chem Technol 45:5

Zhao Q, Yam RCM, Zhang B, Yang Y, Cheng X, Li RKY (2009) Novel all-cellulose ecocomposites prepared in ionic liquids. Cellulose 16:217–226

Zhou J, Zhang L (2000) Solubility of cellulose in NaOH/Urea aqueous solution. Polym J 32(10):866–870

Zhou J, Zhang L, Cai J, Shu H (2002a) Cellulose microporous membranes prepared from NaOH/urea aqueous solution. J Memb Sci 210:77–90

Zhou J, Zhang L, Shu H, Chen F (2002b) Regenerated cellulose films from NaOH/urea aqueous solution by coagulating with sulphuric acid. J Macromol Sci Phys B41(1):1–15

Zhou J, Zhang L, Cai J (2004) Behaviour of cellulose in NaOH/urea aqueous solution characterized by light scattering and viscosimetry. J Polym Sci Part B Polym Phys 42:347–353

Zhu SD, Wu YX, Chen QM, Yu ZN, Wang CW, Jin SW, Ding YG, Wu G (2006) Dissolution of cellulose with ionic liquids and its application: a mini-review. Green Chem 8:325–327

Cellulose Products from Solutions: Film, Fibres and Aerogels

6

Frank Wendler, Thomas Schulze, Danuta Ciechanska,
Ewa Wesolowska, Dariusz Wawro, Frank Meister,
Tatiana Budtova, and Falk Liebner

Contents

6.1	Regeneration of Polysaccharides: Overview	154
6.2	**Viscose Rayon Process**	157
6.3	**CarbaCell Process**	159
6.4	**Lyocell Process**	159
6.4.1	Dissolution of Polysaccharides in NMMO	159
6.4.2	Shaping of Lyocell Solutions	160
6.4.3	New Ways	162
6.4.4	Modified Lyocell Process	163
6.4.5	Ionic Liquids	165
6.4.6	New Technologies	168
6.5	**CELSOL®/BIOCELSOL® Process**	170
6.5.1	Modification of Cellulose Pulp	170
6.5.2	Preparation of Alkaline Solutions from Bio-Modified Cellulose for Film and Fibre Manufacture	172
6.5.3	Technology Process for Film Forming and Fibre Spinning	172
6.5.4	Properties of CELSOL/BIOCELSOL Film and Fibres	173
6.5.5	Feasibility Studies	174
6.6	**Ultra-Lightweight and Highly Porous Cellulose Aerogels and Aerogel-Like Materials**	175
6.6.1	Cellulose I Aerogels	176
6.6.2	Cellulose II Aerogels	178
6.6.3	Concluding Remarks on Aerogels	181
References		182

Abstract

It belongs to one of the oldest experiences of mankind to use fibrous materials of plant or animal origin with different fibre lengths and fineness as basic materials of human clothes. Cellulose forms linear macromolecules, preferentially useful for an application as textile fibres. Common used pulp fibres are too short for those applications and have to be transformed into staple fibres like wool or endless fibres like silk by means of solution shaping procedures. Dissolution of cellulose in common protic solvents like water or alcohol is hindered due to intra- and intermolecular hydrogen bonds. For that reason cellulose derivatives like cellulose nitrate, cellulose xanthogenate or cellulose acetate were used first in order to manufacture cellulose man-made fibres. Situation changes after investigation of direct dissolving liquids like cyclic amine oxides, especially N-methylmorpholine-N-oxide, and large-scale manufacturing of it became reality. In the late 1970s up to the late 1990s, direct dissolution technology, nowadays well known as Lyocell process, was developed up to technical scales. Dry-wet shaping has recently developed to a powerful tool for manufacturing of fibres, films, non-woven or other shapes based on cellulose. The Lyocell process also offers new opportunities for chemical or physical functionalisation of cellulose shapes. Because of technological

F. Meister (✉)
Abt.-Ltr. Chemische Forschung, Thuringian Institute for Textile and Plastics Research, Breitscheidstraße 97, 07407 Rudolstadt, Germany
e-mail: meister@titk.de

difficulties caused by the thermal and chemical sensitivity of the NMMO monohydrate and the strong fibrillation behaviour of dry-wet shaped fibres, alternatives are to be found to overcome these problems. CELSOL® and BIOCELSOL® processes are two approaches which investigated for substitution of Lyocell process. Native pulp fibres have to be pretreated before they could be used for direct dissolution in aqueous NaOH, the common solvent of CELSOL and BIOCELSOL processes. Development on direct dissolution in caustic soda is still under investigation and scaled up into semi-technical and technical scales.

Besides fibres and films, the manufacture of highly porous materials is a very active field, offering the possibility to use them in a wide range of applications, from biomedical and cosmetics to insulation and electrochemistry.

A review of the different processes able to manufacture fibres, films and porous objects is given here.

6.1 Regeneration of Polysaccharides: Overview

It belongs to one of the oldest experiences of mankind to use fibrous materials of plant or animal origin with different fibre lengths and fineness as basic materials of human clothes. From ancient times, wools, especially in particular linen, cotton or other cellulose-based, native fibres have been used for that purpose. The reasons lie in the complex macromolecular structure, which is characterised by ß-1.4 glycosidic linkages of one anhydrous glucose unit (AGU) to the other and a strong hydrogen bonding system (Fig. 6.1). As a consequence, cellulose forms linear macromolecules, preferentially useful for an application as textile fibres. Dissolution in common protic solvents like water or alcohol is hindered due to intra- and intermolecular hydrogen bonds. Furthermore, cellulose forms a macromolecule which is marked by a polycrystalline structure (Klemm et al. 1998).

Since the amount of plant fibres is limited, their quality is varying and processing for textile uses is rather expensive, cellulose pulps extracted from wooden sources or fast growing plants are used. But a lot of technical knowledge and technological experiences had to be accumulated before the first man-made fibre became reality (Woodings 2004). As the length of pulp fibres is too short for a textile use, they have to be processed by means of a continuous spinning associated with a regenerating technology. Apart from the development of suitable pulp manufacturing technologies, efficient techniques for cellulose derivatisation and shaping had to be invented, too (Fig. 6.2).

The history of man-made cellulosic fibres started with the invention of thermo-formability of cellulose nitrate dissolved in ether-alcohol blends by Christian Friedrich Schönbein in 1846. These early studies were initiated by the need to substitute the filaments of light bulbs, which were made of carbonised bamboo fibres, until then only. In Audemars (1855) proved that blends of dissolved cellulose nitrate and rubber can be shaped into fibrous structures. After Swan had found that dissolved nitrocellulose can be extruded through a nozzle and then coagulated and denitrated by the action of ammonium sulphide in 1883, the way was open for the manufacturing of cellulose man-made fibres in large amounts. In 1884 Comte de Chardonnet de Grange (Chardonnet 1884) patented his 'Matière textile artificielle ressemblant à la soie' as the first 'artificial silk', and until 1925 Chardonnet silk was manufactured in France, Switzerland, Germany and Hungary at a large production scale.

Parallel to the development of cellulose nitrate silk in 1892, Cross and Bevan together with Beadle (Cross et al. 1892) invented the dissolvability of cellulose dispersed in caustic soda with the help of carbon disulphide. Furthermore, they discovered that dissolved cellulose can be regenerated in an acid spinning bath. For the exploitation of their invention, a Viscose Spinning Syndicate (VSS) was established in 1903. This syndicate granted royalty rights to several manufacturers in USA (General Artificial

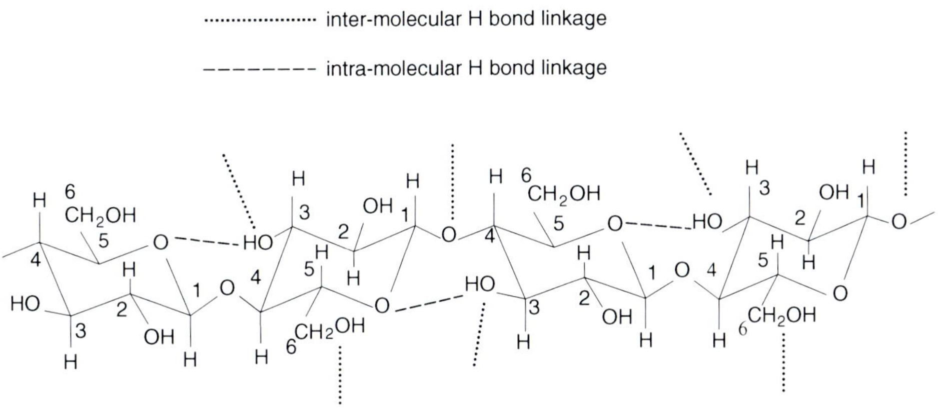

6 Cellulose Products from Solutions: Film, Fibres and Aerogels

Fig. 6.1 Cellulose molecule chain

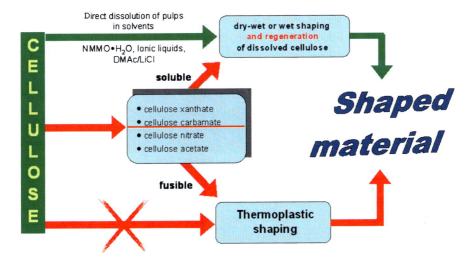

Fig. 6.2 Technological approaches for cellulose shaping

Silk Co.), France (Société Français de la Viscose) and Germany. A manufacturing licence for viscose silk in Great Britain was assigned to Courtaulds Ltd. in 1904. Based on the results of Cross, Bevan and Beadle, Courtaulds concentrated its own R&D activities at the new plant site at Coventry to increase the yield of first quality yarn by 40 % in 1907 and, after invention of Donnersmarck Müller spinning bath patent, to about 90 % in 1911. The basis for the commercial success of Courtaulds was the usage of Zn salts instead of costly ammonium salts before. After takeover of all business activities in America in 1909 and installation of its own viscose fibre plant at Marcus Hook in Pennsylvania in 1910, Courtaulds became the leading company in viscose technology.

In Germany Guido Prince Henckel of Donnersmarck acquired the licence for viscose fibre manufacturing. In 1904 his 'Kunstseiden und Acetat-Werke' started large-scale production at Sydowsaue nearby Stettin/Szczecin. Further pioneering inventions like the composition of a new coagulation bath containing sodium hydrogen sulphate (Koppe and Müller 1905) brought the xanthate technology into an economic efficient commercial scale. Kunstseiden and Acetat-Werke in Sydowsaue failed to achieve the top level of first quality yarn and were bought out by Vereinigte Glanzstoff-Fabriken in 1911. Viscose fibres are nowadays still the most important cellulose man-made fibres and are manufactured in a scale of about 3.1 million t/a, which is about a twentieth part of worldwide man-made fibre production.

Besides cellulose nitrate silk and rayon, another cellulosic fibre is Bemberg silk, which was developed by Schweitzer, Depaissis (1890), Fremery, Urban, Pauly (Pauly 1897, Anonymous 1997) and Thiele (Thiele 1901) from the dissolution of cellulose in tetra-amino copper(II) hydroxide, which is nowadays mainly used only in laboratories for the measurement of the degree of polymerization of cellulose. However, it was used for gentlewoman stockings and lingerie as well as for exclusive colourful dress fabrics. Today's worldwide Bemberg silk manufacturing capacity is very small, at about 18 kt/a.

Finally, Schützenberger (1865) invented cellulose triacetate, which was firstly used only for films and acetone lacquers. In 1901 the Americans Miles and Eichengrün, who were both working for Bayer Co., invented independently of each other the dissolvability of secondary acetate (cellulose 2.5 acetate) in acetone. From such dissolution, the Bayer Company developed Cellit safety films in 1906. One year later, Eichengrün investigated the spinnability of cellulose acetate dissolution to form fibres and filaments. Around 1910, Camille Dreyfus together with his brother Henri started the manufacturing of acetate man-made fibres in Basel. Also, at the plant Sydowsaue, acetate fibre spinning trials were initiated. Contrary to the wet shaping of rayon or Bemberg silk, the solvent acetone is evaporated passing a drying tunnel and can be recycled by a recondensation step. Because tests for the colouration of acetate fibres for the common usage of viscose and cotton failed, further investigations for an acetate technology development were stopped until 1920, when R. Clavel discovered disperse dyes. Afterwards, new cellulose acetate silk plants were erected all over the world. For instance, the Anglo-American Celanese in 1923, the Tennessee Eastman in 1929, the DuPont in 1929 and the Asahi in 1925 started acetate fibre manufacturing. In Germany the Aceta GmbH, Berlin, a joint venture of IG Farben and Vereinigte Glanzstoff-Fabriken, was installed (Pötsch 1997). Among others, Paul Schlack, who was a member of the R&D team at Aceta GmbH, studied acetate spinning before he dealt with synthetic man-made fibre spinning technologies (Pötsch 1999). Today, cellulose acetate fibres for textile usage are manufactured at a scale of about 65 kt/a.

A further shaping technology for cellulose developed by the Finnish companies Kemira Oy Saeteri, a well-known viscose fibre manufacturer, and Neste Oy is the CarbaCell process. Their activities are based on the original investigations of Hill and Jacobsen (1938) in the late 1930s showing that reaction of cellulose and urea led to a cellulose derivative, which is soluble in diluted aqueous sodium hydroxide. The resulted carbamate solution could be spun into a diluted acidic or sodium carbonate containing regenerating bath to form fibres exhibiting properties similar to viscose rayon fibres. However, in case of incomplete regeneration, carbamate fibres tend to self-bonding. The two companies stated that the CarbaCell technology is cheaper than the viscose rayon one. However, it has never been commercialised, as well as the further process modification developed by Fraunhofer Society and Zimmer Company in the late 1990s (Loth et al. 2002).

Recently, the direct dissolution of cellulose in amine oxides, well known as the Lyocell process, was considered as an alternative to the viscose rayon technology. Based on the invention of Swiss researchers Graenacher and Sallmann (1939) who discovered in the late 1930s that N-methylamine-N-oxide is able to dissolve cellulose directly without any chemical changes, Eastman Kodak and American Enka started further investigations in the late 1950s. In 1969, Johnson from Eastman Kodak found that monocyclic N-oxides, like N-methylmorpholine-N-oxide (NMMO), can be preferentially used as cellulose direct solvents. In the late 1970s Franks and Varga developed the way to increase the solid concentration of spinning dopes by carefully controlling water content (Franks and Varga 1977). Afterwards McCorsley (1979b) developed further key elements for a possible commercial process.

In the early 1980s Enka and Courtaulds set up pilot studies with the objectives to develop fibre spinning as well as the necessary solvent recovery procedure. However, in 1981, Enka decided to stop all research activities and commercialisation of Lyocell because of the latent danger of the exothermic solvent decomposition occurring with these solutions at temperatures close to the processing one. After having been taken over by Akzo Nobel, Enka restarted R&D activities again under the name 'Newcell', a process variant for manufacturing of cellulose filaments with the same process. In a common agreement Akzo Nobel and Courtaulds made a contract to exchange their rights.

Before acquiring Courtaulds in 1998, Akzo Nobel had granted a Lyocell licence to the Austrian viscose fibres manufacturer Lenzing, allowing it to enter the field with a process concept very similar to Courtaulds. In the second half of the 1980s up to the second half of the 1990s, pilot plants were installed in Coventry and later on in Grimsby, UK, as well as in Lenzing, Austria, and Rudolstadt, Germany.

In 1992 Courtaulds started its first full-scale production plant in Mobile, Alabama, USA, which was expanded to a double capacity in 1996. One year later, Lenzing started the full-scale production at Heiligenkreuz, Austria. The first European production plant of Courtaulds, which was merged with the fibre activities of Akzo Nobel into Acordis, started its production in 1998 at Grimsby. After the separation of Tencel Ltd. from Acordis group, Lenzing bought it in 2004 and consolidated all the Lyocell fibre production activities in one hand. At the present time, under the brand name Tencel, Lenzing produces around 120,000 t/a. Upcoming production lines are situated in China, India and South Korea. In Rudolstadt, Germany, Smartfiber as a spin-off company of the Thuringian Institute for Textile and Plastics Research produces and sells modified Lyocell fibres (known as ALCERU® process) with a capacity of 500 t/a.

Although further derivatising and/or cellulose direct dissolving liquid systems such as DMSO/paraformaldehyde, DMF/N_2O_4 and/or dimethylacetamide/LiCl, NaOH/H_2O, NaOH/H_2O/urea/thiourea, ferric tartaric acid complex (FeTNa), H_3PO_4 and ionic liquids have been developed, no other commercial fibre or film spinning process has been established so far.

6.2 Viscose Rayon Process

At present, viscose fibres (rayon) dominate the production of regenerated cellulose (man-made cellulosics) including also films, membranes and sponges. Based on the cellulose regeneration from cellulose xanthogenate—a metastable intermediate of cellulose and CS_2—viscose process exhibits a broad versatility arising from modification of different steps inside the process line. Variation of the degree of polymerization (DP) of the primary pulp, usage of additives in the xanthogenate solution or coagulation bath and stretching during fibre shaping are opportunities to widen the application range of the fibre from high wet modulus (HWM, Modal) to high-strength cord fibres.

Viscose technology (Fig. 6.3) starts from the alkalinisation suppressed (steeping state) with ca. 18 % NaOH at temperatures between 20 and 55 °C to convert cellulose I into alkali cellulose, which enhances the reactivity and permits improved penetration of CS_2. After a ripening step to decrease the DP from 750–850 to 270–350 and the removing of excess NaOH, sulphurisation provides the xanthogenate, which is dissolved in aqueous NaOH under high shearing at 10 °C.

This viscous solution pre-ripened for 24–48 h, filtered and degassed can be shaped into fibres or foils in a wet-spinning process. During the coagulation in a sulphuric acid Na_2SO_4/$ZnSO_4$ containing bath, high-purity cellulose is regenerated, followed by washing and after treatments. Parameters of the precipitation conditions determine the fibre structure and properties, respectively. In Fig. 6.4 cross section of a viscose fibre is displayed showing a lobed skin-core morphology compared to CarbaCell and Lyocell fibres (Klemm et al. 2005). The more 'softer' precipitation results in a lower crystallinity responsible for the low tendency to fibrillate. Admixture of amines and polyether glycols delays the

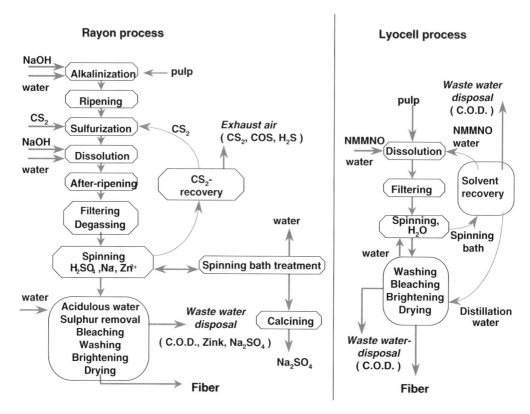

Fig. 6.3 Processes of cellulose regeneration: *left* derivatisation (viscose/rayon) and *right* direct dissolution (Lyocell)

Fig. 6.4 TEM micrographs of the cross section of different types of cellulosic fibres [Adapted from Klemm et al. (2005). Copyright Wiley-VCH Verlag GmbH & Co. KGaA. Reproduced with permission]

coagulation of the polymer resulting in an enhanced crystalline orientation connected with an improvement of the mechanical properties characterising Modal fibres.

Special developments are crimped fibres for bulkier, warmer fabrics using hot acid (Krässig 1982) or high-water-absorbing hollow fibres achieved by injection of gases. Further, incorporation of additives improves the dyeability as well as expands applications such as flame retardant, antibacterial or sun-protected fibres.

Owing to the very complex technology and the problematic environmental loads from the use of CS_2, heavy metals and by-products, research activities were intensified to decrease emissions and to meet the environmental

$$H_2N-CO-NH_2 \rightarrow HOCN + NH_3\uparrow$$
$$HOCN + HO-Cell \rightarrow H_2N-CO-O-Cell$$

Scheme 6.1 Reaction of cellulose with urea into cellulose carbamate

demands: optimisation of the chemical input, especially minimising the use of CS_2, purification of exhaust gas and wastewater and search for alternative derivatisation processes. Here, the conversion of cellulose to cellulose carbamate subsequently processed on existing viscose spinning systems could constitute an advanced alternative as described in next section.

6.3 CarbaCell Process

Cellulose carbamate is synthesised in a heterogeneous reaction using urea, which pyrolyses at 135 °C into ammonia and isocyanic acid. In the presence of hydroxyl groups, isocyanic acid can form carbamate (Scheme 6.1). Under special conditions all three hydroxyl groups of cellulose can react to give a product containing nitrogen in the range of 1–3.5 % (Ekman et al. 1986). Variation of reaction conditions generates by-products, e.g. biuret or cyanic acid that can be washed out in the end of the reaction path.

Cellulose carbamate can be easily dissolved in NaOH and spun as fibres via the CarbaCell process (Voges et al. 2000). Similar to the viscose process, cellulose pulp is alkalized, pre-ripened and subsequently derivatised. Advantageously, being stable enough at ambient temperature, cellulose carbamate can be produced at large scales and then shipped to viscose plants where it could be handled like xanthate, however without using of hazardous products like CS_2 and heavy metal contained in precipitation baths. Solutions are spun into an acidic precipitation bath (preferably sulphuric acid), and cellulose is regenerated by the decomposition of the carbamate groups in a salt-containing alkaline bath. ^{13}C NMR spectra revealed a cellulose II crystallite orientation accompanied by residues of unhydrolysed carbamate groups (Rihm 2003). Electron micrographs of ultrathin fibre sections in Fig.6.4 show that carbamate fibres are comparable with Lyocell (Klemm et al. 2005). Their homogeneous and oval cross-sectional shapes are very different from the lobed shape and skin-core-like morphology of viscose rayon confirming the distinctive higher crystallinity of carbamate fibres (Table 6.1).

Although regenerated fibres give similar textile physical properties compared to viscose rayon and can be applied for hollow fibres, nonwovens, sponges and highly porous materials (Pinnow et al. 2008), fibre production beyond lab scale has not been started, yet.

6.4 Lyocell Process

6.4.1 Dissolution of Polysaccharides in NMMO

Challenges regarding the production of a fibre in a more efficient way than viscose rayon in view of easy processing, higher performances and good environmental standards have strengthened the Lyocell process development. Direct dissolution of the cellulose favours this process, seen as a new generation of man-made cellulose fibre making. The solvent can be reprocessed almost completely. Among other amine oxides, N-methylmorpholine-N-oxide (NMMO) in its monohydrate form is able to dissolve cellulose and other polysaccharides due to its strong N–O dipole moment and molecule size breaking the hydrogen bond system (Franks and Varga 1977). Without further activation and no derivatisation of cellulose, dissolution is achieved only by the interaction of hydrogen bonds between NMMO, water and the polymer. Starting from a suspension of the polymer in a 50 % NMMO solution for swelling, water is removed at temperatures between 80 and 120 °C under vacuum up to ca.

Table 6.1 Degrees of crystallinity and tensile properties of cellulosic fibres

Fibre species	Degrees of crystallinity (%)	Tenacity, cond.[a] (cN/tex)	Elongation, cond.[g] (%)	Wet abrasion resistance (T)
Viscose	27–31 (Dohrn et al. 2001)	20–25 (Franks and Varga 1977)	18–23 (Franks and Varga 1977)	488 (Geyer et al. 1994)
Modal	39 (Fink et al. 2006)	34–36 (Franks and Varga 1977)	12–14 (Franks and Varga 1977)	1,426 (Geyer et al. 1994)
Lyocell	35 (Dohrn et al. 2001)	40–42 (Franks and Varga 1977)	15–17 (Franks and Varga 1977)	40 (Geyer et al. 1994)
Cuprammonium	52 (Fink et al. 2006)	15–20 (Franks and Varga 1977)	10–20 (Franks and Varga 1977)	
Cotton	78 (Fink et al. 2006)	24–36 (Franks and Varga 1977)	7–9 (Franks and Varga 1977)	
CarbaCell	34–43 (Dohrn et al. 2001)	13–26 (Fink et al. 1998)	8–25 (Fink et al. 1998)	
Cellulose carbamate from NMMO	38–41 (Fischer et al. 2006)	52–50 (Gavillon and Budtova 2008)	9–12 (Gavillon and Budtova 2008)	15 (Gavillon and Budtova 2008)

[a]Conditioned dry (65 % relative humidity, 20 °C)

13 % by mass (McCorsley and Varga 1979a). As it can be seen in Fig. 6.5, there is only a small solution process window, which is limited by NMMO monohydrate state of the solvent at the lowest dissolution concentration level and the strong increase of the cellulose–NMMO dope viscosity at higher concentration level. If going outside the window by excess of water, cellulose is precipitated, a process which is used during the spinning process. Up to 25 % of cellulose can be dissolved, but for technical purposes, typically concentrations of 12–14 % are used. Additives can be incorporated into the dope leading to modified shaped cellulosic bodies as described in Sect. 6.4.4. Lyocell solutions bear also thermal and shearing stress leading to degradation of both solvent and polymer. The development of Lyocell was always accompanied by the search for stabilisers such as propyl gallate (see Sect. 6.4.2).

6.4.2 Shaping of Lyocell Solutions

Compared to the viscose route, the Lyocell technology is some more demanding on pulp quality, does not have the restrictions caused by hazardous loads of CS_2, but does have strong restrictions by heavy metal contents and further by-products (Fig. 6.3). Although differences between Tencel, Lenzing and ALCERU production lines exist, the main principles are as follows. Typically, pulps of different origins (wood, eucalyptus, cotton, hemp, etc.) with degrees of polymerisation (DP) between 400 and 1,200 can be used. For wetting and swelling, the cut pulp is immersed in an aqueous solution of 50–75 % NMMO (pH 10–11) with small amount of propyl gallate. As can be seen in Fig. 6.5 and described later, in case of modified fibres, one or more additives can be incorporated at this step. To enhance swelling, the solution is sheared by ultra-turrax mixing. The resultant mixture containing 12–14 % cellulose is conveyed into a stirring reactor equipped with baffles to distil the excess water at temperatures from 80 to 120 °C under vacuum, homogenise the solution effectively, minimise undissolved particles and remove air bubbles. Another variant of water removing is the thin Filmtruder evaporation by means of steam-heating jackets around a rotator. Careful vacuum adjustment is necessary by reason of minimising the temperature to protect the solvent from exothermic degradation as described in Chapter 5.5. The resulting solution is then pumped through a transport tube system, pressures up to 150 bars being necessary for transportation of the viscous mass and passing the filters. Thermal load and high pressure connected with impurities may initiate exothermic reactions with rapidly increasing temperature and pressure. To prevent this, technical equipment is fitted with coolers and bursting discs. Filtration of the solution is needed to remove impurities, especially undissolved

6 Cellulose Products from Solutions: Film, Fibres and Aerogels

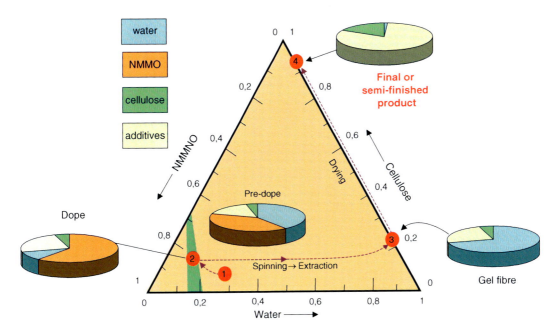

Fig. 6.5 Ternary phase diagram of the system NMMO–water–cellulose additive [Adapted from McCorsley and Varga (1979a) Copyright Wiley-VCH Verlag GmbH & Co. KGaA. Reproduced with permission]

polymer particles and inorganic materials from the used pulp.

Fibre shaping is carried out by extruding the solution through a nozzle jet consisting of thousands of holes and via an air gap with a length of 10–100 mm into an aqueous amine oxide precipitation bath. Here, the cellulose is coagulating instantaneously and NMMO is almost totally exchanged by water. After further washing steps, the fibres are dried. Precipitation baths are then rinsed through ion-exchange columns to recycle the NMMO up to 99 % purity. Fibre formation is a process of elongation deformation, which is influenced by the pressure building up in the die channel inside the nozzle and enhanced by passing the air gap (Michels et al. 1994, Michels 1998, Fink et al. 2001). The kinetics of fibre forming is caused by the geometry of die channel, length of the air gap, cellulose concentration, molecular weight distribution, temperature, injection velocity and drawing rate. These parameters have a significant impact on the orientation of cellulose macromolecules and, consequently, on the performance of the shaped fibres. Tensile properties such as tenacity, elongation, loop tenacity as well as humidity adsorption behaviour determine the application fields of the fibre. After washing, fibres can be bleached and handled with avivage (soft finishing) or antistatic materials depending of final uses. Dyeing behaviour is as comfortable as for cotton with a variety of dyes.

Lyocell fibres have distinguished features such as strength in both wet and dry states, high modulus of elasticity, good sorption behaviour, wearing properties, gloss and touch (Table 6.1). Besides these outstanding characteristics, Lyocell fibres suffer from the tendency of fibrillation, expressed by a decreased wet abrasion number (Table 6.1). Highly oriented crystallites, a high amorphous orientation and longitudinally stretched pores between the fibrils with a relatively large pore volume especially in the water-swollen state cause the splitting of fibrils into microfibrils leading to the so-called peach-skin touch (Bredereck and Hermanutz 2005). To avoid fibrillation, additional processing such as chemical cross-linking and precipitation in alcoholic baths or enzyme treatments (Nechwatal et al. 1996, Fink et al. 1998) is necessary and a topic of research.

Although the technological scaleup has reached a very high level, scientific activities are conducted to reach different requirements such increasing capacity per staple fibre production line, minimising the defects in filament spinning, making homogenous film without bubbles and producing nonwovens/microfibres. Recently, new developments of the kneader technology of LIST were directed to the production of higher concentrated cellulose solutions (Diener 2010). Continuous tests succeeded with up to 18 % cellulose with an expected future composition for industrial scale of 20–24 % where the dissolving concentration is not equal to the spinning concentration. The desired cellulose concentration is reached by dilution with NMMO with adjustable dope viscosity.

Since cellulose/NMMO solutions behave like a melt, cellulose films can be prepared in a very elegant way analogously to synthetic polymers. In regard to packaging applications, films were prepared via the viscose route (cellophane) and cuprammonium process (cuprophane) already in the early 1960s (Voss 1961; Bandel 1961). Blown-extrusion process is a way to produce films and spun fleeces (melt-blown nonwovens) from NMMO melt (Fink et al. 1997). Longitudinal and transverse properties can be adjusted influencing morphology and pore structure of the film by the subsequent precipitation process, which is not possible in melt-extruded films from synthetic polymers. Application fields range from packaging materials and food casings to dialysis membranes (Fink et al. 2001).

6.4.3 New Ways

Direct dissolution has opened from the beginning subsequently modified routes to exploit all degrees of freedom such as the polymer, the solvent or the admixture of functional components to target higher spinning capacities, improved thermal stability and recycling rates and innovative fibre performances.

An interesting approach is the dissolution of cellulose carbamate in NMMO with the aim to develop high-strength fibres and to improve economic aspects of Lyocell process. The advantages of both cellulose carbamate/NaOH and cellulose/NMMO systems are combined using the better soluble derivative with the more efficient solvent. An improved state of the solutions, higher concentrations of the polymer and liquid-crystalline states can be reached (Fink et al. 2004, 2006). As mentioned above, production of cellulose carbamate can be divided from dissolution and spinning process since the derivative is stable over long periods.

By variation of temperature, dwell time, raw material charge and cellulose concentration, conditions can be found in order to adjust carbamates of special DP and DS. Based on that, solutions of up to 30 % cellulose carbamate in NMMO can be prepared. Rheological and microscopic investigation revealed distinctive hints of liquid-crystalline solutions at higher DP and concentrations. Viscosity curves of solutions with different carbamate concentrations show a shear-thinning behaviour (Fig. 6.6). Higher concentrations are shifting the beginning of the nonlinear region to lower shear rates. This trend is also seen for cellulose carbamates when increasing DP.

Spinning was performed at a pilot plant with concentration up to 21 %. Carbamate fibres from NMMO are characterised by a tensile strength up to 60 cN/tex and a modulus of maximum 2,500 cN/tex. Similar to the CarbaCell fibre (see Chap. 3), cross section shows an oval, dense homogeneous morphology connected with pronounced crystallinity and crystallite orientation (Fig. 6.4). Further, carbamate groups are remaining in the fibre as confirmed by ^{13}C NMR. However, during the process line, nitrogen content is diminished from around 3 % in the primary product to 0.5 % in the spun fibre.

Since thermal stability of technical carbamate solutions is comparable to cellulose/NMMO solutions, stabilisation succeeded by using the system NaOH–NH$_2$OH–propyl gallate. Even above 130 °C, degradation of solvent and polymer occurred. Reaction calorimetry revealed a moderate pressure rise around this temperature, which is ascribed to the hydrolytic cleavage of the carbamate bond (Fig. 6.7). Generated gases

6 Cellulose Products from Solutions: Film, Fibres and Aerogels

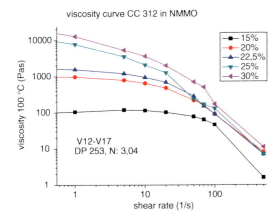

Fig. 6.6 Relationship between viscosity and shear rate for different cellulose carbamate concentrations (DP 253, N-content 3.04) (Fink et al. 2006)

CO_2 and NH_3 induce insecure spinning behaviour with breaking of the filaments and explain the nitrogen reduction.

Therefore, to reduce the loss of the carbamate group during spinning, a compromise must be found, which allows the thermal stabilisation of the spinning solution by means of complex reagents and pH adjustment and also the carbamate stabilisation in an acidic medium. Two ways were followed to solve the problem: setback of the pH value in the coagulation bath (1) and directly in the cellulose carbamate/NMMO solution (2).

As shown in Table 6.2, the lower the pH in the coagulation bath, the higher the nitrogen content in the fibre. Weak acidic stabilisation with propyl gallate and gallic acid increases the nitrogen content to 1.2 % and 1.5 %, respectively, while keeping good textile parameters, but thermostability is evidently deteriorated and the DP of the polymer is decreased, however. Consequently, the use of both alkaline/propyl gallate stabilisation of the solution and acetic acid adjustment of the bath bears the maintenance of the carbamate group.

Cellulosic fibres bearing a functional moiety inside the skeleton provide interesting features to be aimed at changing the wetting and interaction behaviour, e.g. sorption of dyes.

6.4.4 Modified Lyocell Process

Apart from manufacturing man-made fibres, the direct dissolution process of cellulose offers a broad diversity of technological opportunities. Using the ALCERU® process (ALCERU—brand name of the direct dissolution process developed by TITK), direct dissolution variants represent a unique and versatile spinning technology that has opened up new ways of cellulose material shaping. Additional to the modification of the fibre morphology, e.g. hollow fibres, fibrids and film stripes, a shape functionalisation may be achieved by means of chemical derivatisation or even by physical incorporation of additives with functional properties into cellulose solutions (Fig. 6.8) (Meister et al. 1998; Schulze et al. 2009). Especially, physical functionalisation is an active and important topic of research of the Lyocell fibre business beyond the textile sector. Due to the extraordinary high receiving capacity for secondary components as well as mild aftertreatment conditions, high amounts of insoluble additives can be homogeneously distributed into or around the cellulosic chain skeleton while keeping the shaped fibres structure.

Although basic research on functional Lyocell shapes already started in the middle of the 1990s, especially initiated by Vorbach and co-workers, the number of published papers remains rather low. However, the patent literature has undergone a remarkable growth at the same time. The physical dissolving process of cellulose without derivatisation and the high loading capacity of the spinning solution enable the incorporation of chemically different *types of additives*, e.g. ceramics (Vorbach and Taeger 1995), carbon black (Vorbach and Taeger 1995), activated charcoal (Vorbach et al. 1999), ion-exchange resins (IER) (Büttner et al. 1999), superabsorbent polymers (SAP) (Dohrn et al. 2001), etc., if the additive particle size is less or equal to 10 μm, which represents the typical fibre diameter. Furthermore, several functional cellulose fibres have been developed in the last decade exhibiting antimicrobial activities (incorporation

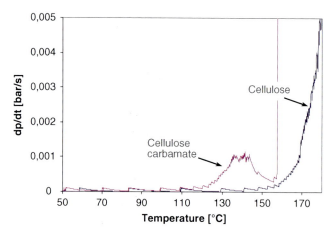

Fig. 6.7 Pressure rise of a 12 % cellulose/NMMO solution and 15 % cellulose carbamate/NMMO solution measured by means of reaction calorimeter. Cascade course is caused by scale spreading

Table 6.2 Selected properties of cellulose carbamate fibres spun from NMMO (concentration 21 %) by means of different systems

Parameter	Stabilisation					
	Propyl gallate, NaOH, NH$_2$OH				Propyl gallate	Gallic acid
	Coagulation pH value (acetic acid)					
	6.0	5.0	4.0	3.0		
Fineness (dtex)	2.21	2.27	1.95	2.12	2.34	2.17	2.41
Elongation, cond. (%)	12.0	9.1	9.0	9.1	8.8	10.9	7.13
Tenacity, cond. (cN/tex)	47.0	44.9	41.7	44.4	45.9	43.2	46.2
Loop tenacity (cN/tex)	8.86	20.1	16.3	15.9	16.9	11.7	9.81
Elasticity module (cN/tex)	2,000					1,887	1,757
Nitrogen content (%)	0.59	0.25	0.73	0.96	0.98	1.21	1.53

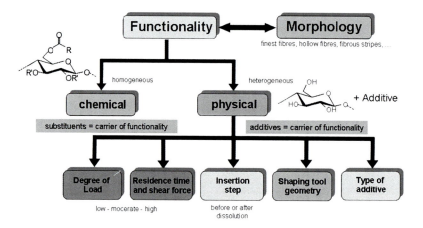

Fig. 6.8 Functionalisation of shaped bodies via the modified Lyocell process

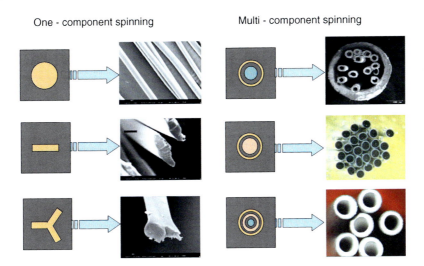

Fig. 6.9 Modification of fibres cross section by changing the shaping tool geometry

of zinc, silver or copper salts or even silver nanoparticles), electrical conductivity (incorporation of conductive carbon black) or high adsorption capacity using these diverse technological opportunities. Some of developed functional cellulose fibres have been transferred into a commercial manufacturing scale at Smartfiber Corp., Rudolstadt.

Recently, TITK found that applications of nanoscaled particles help to immobilise lipophilic additives being incompatible with the hydrophilic cellulose matrix. Deploying that approach, liquid, meltable or volatile hydrophobic compounds like cosmetic oils or insect repellents can be incorporated directly into cellulose fibres matrix by usage of well-dispersed silicic acid nanoparticle hydrocarbon composites together with organically modified layered silicate in order to stabilise the thermodynamic incompatibility of the organophile component and cellulose solution and to achieve a stable spinning process (Krieg et al. 2009). Investigation of bio-functionality as well as bio-compatibility of all the developed functional cellulose matrices proved its suitability as skin-compatible textile fibres according to Öko-Tex Standard 100®.

The *degree of load* with respect to the amount of cellulose can be varied from a fraction of few percent, e.g. in case of UV-stabilised textile fibres up to several hundred per cent, e.g. for piezoceramic green fibres. Such a broad load capacity range is nearly unparalleled in man-made fibre manufacturing.

Because of the chemical and thermal sensitivity of the N-methylmorpholine-N-oxide (NMMO) used (cf. Chap. 3) and/or of the compounds inserted into the cellulose dope, it is essential to decide at which *step* of the dissolving procedure they should be added.

Moreover, direct dissolution technology offers some more opportunities for cellulose shaping (Schulze et al. 2009). Round bulk fibres or fibres with flat or trilobal cross section and hollow structures could be moulded only by changing the shaping tool geometry (Fig. 6.9).

Furthermore, fibres, nonwoven, spherical and fibrillated particles might be manufactured by means of changing either shearing forces or residence time within a shearing field during the cellulose regeneration step only (Fig. 6.10).

6.4.5 Ionic Liquids

A promising approach to overcome the problematic thermostability of NMMO is the use of ionic liquids (ILs). Their outstanding material and solvent properties have raised an interest to these 'green' solvents (Wasserscheid and Welton 2008). Variation of the cationic base, anion and

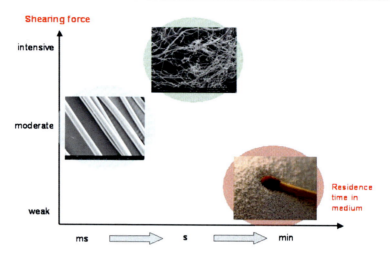

Fig. 6.10 Modification of Lyocell mould style by means of shearing forces and residence time

alkyl chain permits tunable adjustment of different properties such as melting point, viscosity, density or solvation. Even due to their ability to interact via hydrogen bonds and melting points below 100 °C, they exhibit ideal prerequisites to be solvents for cellulose and other polysaccharides (Swatloski et al. 2001). Preferably, cellulose can be regenerated from solutions of 1-alkyl-3-methylimidazolium ILs by precipitation into a variety of nonsolvents including water, alcohols or acetone allowing adjustment of topographic and morphologic features, especially in case of membranes (Liebert 2008). Activated charcoal and nanoscaled materials can also be incorporated into the solution for the adjustment of special features (Wendler et al. 2009).

Although the zero-shear viscosities at 85 °C of cellulose solutions with equal concentrations strongly depend on the type of IL, fibres with acceptable textile physical properties can be spun by means of both dry-wet and wet technologies (Kosan et al. 2008a, b). Employing 1-ethyl-3-methylimidazolium acetate allows up to 20 % of cellulose to be dissolved and regenerated (Table 6.3).

Another positive feature of ILs is their ability to dissolve, besides cellulose, many natural and synthetic polymers. Simultaneous dissolution of cellulose and polyacrylonitrile (PAN) is possible and functional fibres can be prepared. Surprisingly, a consecutive dissolution could be made: firstly cellulose and afterwards PAN are dissolved independent from kind and composition of the polymer mixture (Kosan et al. 2008a, b). The produced fibres exhibit a series of extraordinary properties. Table 6.4 displays the textile parameters and the clearly reduced water retention ability. Moreover, at a cellulose/PAN volume ratio of about 50:50, a considerable rise of fibrillation resistance of the blend fibres was ascertained. Improvement of fibrillation tendency goes with a drop of water retention value compared to pure cellulose fibres.

The spinning process with the addition of up to 0.1 % nanoscaled silver particles proceeds without any impact on textile physical parameters (Table 6.5) compared to unmodified fibres (Table 6.3). A crucial issue concerning the functional effect of the fibre is the proper distribution of silver particles in the spinning solution. Desired silver content is adjustable with a recovery rate of ~70 %. What is noteworthy in the application of colloidal silver in these concentration ranges is that it results in fibres of very high whiteness (Wendler et al. 2007).

Table 6.6 summarises the antimicrobial activity of fibres modified with nanoscaled silver particles. Fibres with porous MicroSilver (80–140 nm) show microbial effects only for higher concentrations (from 0.1 %). Because of the formation of aggregates, the antimicrobial performance of nanoscaled particles is

6 Cellulose Products from Solutions: Film, Fibres and Aerogels

Table 6.3 Textile physical properties of fibres spun from different ionic liquids (cellulose: eucalyptus pulp) (Adapted from Liebner et al. (2008). Copyright Wiley-VCH Verlag GmbH & Co. KGaA. Reproduced with permission)

	Solution				
	1	2	3	4	5
Cellulose concentration (%)	13.6	15.8	13.2	18.9	19.6
Solvent	BMIMCl	EMIMCl	BMIMAc	BMIMAc	EMIMAc
Fineness (dtex)	1.46	1.84	1.67	1.64	1.76
Tenacity, cond. (cN/tex)	53.4	53.1	44.1	48.6	45.6
Elongation, cond. (%)	13.1	12.9	15.5	12.9	11.2
Loop tenacity (cN/tex)	33.1	29.5	22.1	25.1	19.9
Wet abrasion test (T)	61	37	17	22	24
Water retention value (%)	64.6	68.2	79.3	71.4	68.1
DP of the fibre	514	493	486	479	515

Table 6.4 Textile physical properties of fibres modified with polyacrylonitrile (PAN) and spun from EMIMAc

	Solution					
	6	7	8	9	10	11
Mass relation cellulose/PAN	100/0	90/10	80/20	70/30	60/40	50/50
Volume relation cellulose/PAN	100/0	87/13	75/25	64/36	53/47	43/57
Solid content	11.2	12.1	13.9	15.2	17.2	19.2
Cellulose concentration	11.2	10.9	11.1	10.6	10.3	9.6
Fineness (dtex)	1.7	1.7	1.7	1.7	1.7	1.7
Tenacity, cond. (cN/tex)	50.3	44.6	35.1	30.4	23.9	19.4
Elongation, cond. (%)	11.7	10.6	9.2	15.5	10.3	12.9
Elongation, wet (%)	12.8	12.2	12.9	18.2	16.9	28.9
Loop tenacity (cN/tex)	22.2	19.5	18.6	16.6	11	13.1
Wet abrasion test (T)	28	36	59	114	618	4,466
Water retention value (%)	65.9	64.4	61.3	57.3	45.3	37.5

Cellulose: eucalyptus pulp [Adapted from Liebner et al. (2008). Copyright Wiley-VCH Verlag GmbH & Co. KGaA. Reproduced with permission]

Table 6.5 Textile physical properties of fibres modified with MicroSilver (80–140 nm) and NanoSilver (5–20 nm) particles spun from different ionic liquids [Adapted from Liebner et al. (2008). Copyright Wiley-VCH Verlag GmbH & Co. KGaA. Reproduced with permission]

	Fibre 1	Fibre 2	Fibre 3
Concentration of cellulose (%)	12	23.5	23.5
Ionic liquid	BMIMCl	EMIMAc	EMIMAc
Type of additive	MicroSilver 80–140 nm	MicroSilver 80–140 nm	NanoSilver 5–20 nm
Concentration of additive (%)	0.1	0.1	0.05
Fineness (dtex)	1.80	1.62	1.56
Tenacity, cond. (cN/tex)	42.3	55.9	50.8
Elongation, cond. (%)	10.6	9.1	10.0
Loop tenacity (cN/tex)	22.3	9.6	22.9

Table 6.6 Test results of antimicrobial activity of solvent fibres modified with MicroSilver (80–140 nm) and NanoSilver (5–20 nm) particles spun from NMMO and ionic liquids

Fibre	Type of additive	Solvent	Antimicrobial activity Staphylococcus aureus	Klebsiella pneumoniae
1	0.1 % MicroSilver	BMIMCl	Non-antibacterial	Non-antibacterial
2	0.1 % MicroSilver	EMIMAc	Strong antibacterial	Strong antibacterial
2[a]	0.1 % MicroSilver	EMIMAc	Significant antibacterial	Strong antibacterial
3	0.05 % NanoSilver	EMIMAc	Strong antibacterial	Strong antibacterial
3[a]	0.05 % NanoSilver	EMIMAc	Slight antibacterial	Strong antibacterial
4	–	EMIMAc	Slight antibacterial	Non-antibacterial
5	–	NMMO	Significant antibacterial	Non-antibacterial
Solvent	–	NMMO	Non-antibacterial	Non-antibacterial
Solvent	–	EMIMAc	Slight antibacterial	Non-antibacterial

[a]After washing [Adapted from Liebner et al. (2008). Copyright Wiley-VCH Verlag GmbH & Co. KGaA. Reproduced with permission]

achievable only by increasing the concentration. On the other hand, fibres with colloidal NanoSilver (5–20 nm) exhibit a higher performance (from a concentration of 0.05 %). Further, it can be drawn from Table 6.6 that when using chloride containing ILs as solvent, no antimicrobial effect is detectable. Even the application of acetate gives a strong antimicrobial effect for both test strains. After defined washing (40 times, 60 °C), a strong effect is measured for *Klebsiella pneumoniae* and a significant one for *Staphylococcus aureus* in the case of 0.1 % MicroSilver.

A composition of 50 % cellulose and 50 % adsorption materials, like activated charcoal, results in fibres which can be processed by conventional techniques to form nonwovens with sufficient mechanical stability and uniform charcoal loading. The charcoal particles are evenly distributed in the whole fibre. Additionally, the cellulose acts as a three-dimensional carrier that is responsible for the mechanical stability.

Nonwovens made with this technology have been investigated concerning its adsorption ability of organic compounds from their vapours as displayed in Fig. 6.11. In general, it should be noticed that carbon tetrachloride has a higher ability to adsorption on activated charcoal than toluene. The differences between carbon tetrachloride and toluene in its adsorption behaviour can be explained by specific properties of the compounds. An equilibrium state has been reached and stabilised after 1 day. However, the equilibration loading level is already reached after 8 h. The fibres have a lower affinity towards carbon tetrachloride and toluene than the pure active charcoal. Thus, a certain amount of the adsorbent must be inactive. First investigations of the produced fibres from ILs revealed that the adsorption effect does not reach the efficiency of the NMMO spun fibres. Reason might arise from the insufficient solvent extraction during fibre aftertreatment. Even an additional reaction time in the desiccator as well as Soxhlet extraction using acetone brought an improvement of the intake capacity, comparable to those from NMMO.

6.4.6 New Technologies

The basic idea behind all efforts to create new products is to make use of the interaction between functionality and morphology. Together with the use of technologies other than dry or wet spinning, the number of resulting products is almost unlimited.

In this way, thin fibre webs were prepared by simply blowing polymer solutions through spinnerets and initiate coagulation before placing onto sieve belts. This so-called Nanoval process makes it possible to get fibre webs consisting of fine fibres from almost any types of soluble polymers. The entire processing is not flawed with the problem of expensive spinning machines and a subsequent fibre processing anymore, but yields

6 Cellulose Products from Solutions: Film, Fibres and Aerogels

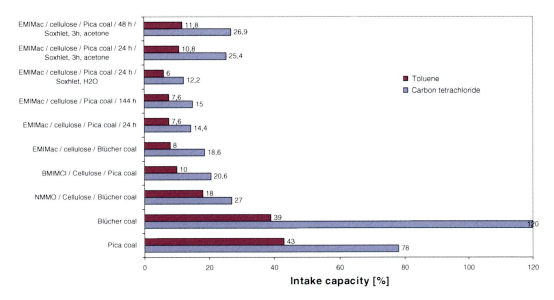

Fig. 6.11 Intake capacity of different activated charcoals (1—medium reactive, 2—strong reactive) and fibres spun from NMMO and ionic liquids. The adsorption time is 24 h except where denoted [Adapted from Wendler et al. (2007). Copyright Wiley-VCH Verlag GmbH & Co. KGaA. Reproduced with permission]

a final product of outstanding quality and performance by means of a comparatively simple process. This process is characterised by one-step processing of spinning solutions to produce fibre webs. Cold air blast and water fog are applied to spray spinning solutions through spinnerets and nozzles with comparatively large capillary diameter without having problems with regard to the dope viscosity. The main difficulty concerns the splitting of the extruded dope jets into fine fibrids, preferably into those being in the sub-micro-range (<1 μm), resulting in a very homogeneous web morphology with no droplets or other faulty structures. Also the formation of long fibres is possible. As a simple process, almost all types of additives and polymer/solvent combinations might be processed, e.g. cellulose/carbon black/NMMO, inorganic (ceramic) powders and man-made polymers dissolved in aprotic solvents.

All of the new technologies have one item in common, and this is replacing staple fibre or filament spinning by getting elongated shapes of, however, often nonuniform length and less defined shape, but enhanced diameter-to-length ratio. This can also be done by extruding polymer solutions or melts into a carrier matrix and removing the matrix afterwards. Such so-called bicomponent spinning might be the right choice when polymer or solution is sensitive to heat or shows poor forming abilities.

Following the basic conception of Lyocell processing, shear coagulation represents another interesting shaping opportunity called fibrid spinning. Instead of drawing fibres through steady coagulation baths, here the spinning solution is injected straight into a stream of the same coagulation bath resulting in long and very fine fibrids. The formed fibrids can easily be isolated from the slurry, where the process liquors are recycled completely (Fig. 6.12).

A much more elegant way to shape polymers is to apply an electrical field for spraying solutions (a process called electro-spinning). Providing that solution viscosity and elasticity are adjusted properly, a web consisting of nano-fibres can be prepared (Fig. 6.13). Thus, coating carrier materials with nano-webs becomes very easy in this way.

It is important to point out that all process material and compounds used in fibrid and fine fibre making as mentioned above are completely recyclable and that no hazardous by-products are formed.

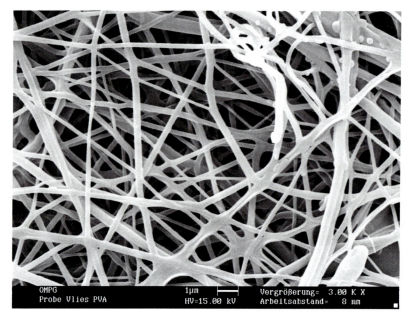

Fig. 6.12 Nano-fibres onto support web

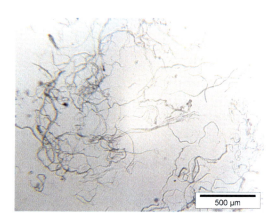

Fig. 6.13 Fine cellulose fibrids made by shear coagulation

6.5 CELSOL®/BIOCELSOL® Process

6.5.1 Modification of Cellulose Pulp

The literature shows many methods of cellulose modification to transform cellulose into an alkaline soluble form that is useful for film forming and fibre spinning. The most known methods of activation of cellulose pulp are alkalization, steam explosion, hydrothermal treatment and enzymatic modifications. The main goal of modification is the partial cellulose degradation decreasing polymerization degree (DP) as well as changes in molecular, supra molecular and morphological structure (Kennedy et al. 1990, 1992, 1995) followed by increase of cellulose reactivity and solubility in aqueous alkaline solutions, like NaOH/water.

6.5.1.1 Alkalization

The process is based on cellulose treatment with 18–19 wt% NaOH solution leading to the formation of a new cellulose derivative—alkali cellulose. During alkalization, swelling of cellulose pulp with lowering of its DP takes place. Introduction of some additives such as zinc oxide, urea and thiourea into NaOH solution causes an enhanced cellulose swelling ability and reactivity as well as offers the possibility to prepare more stable spinning solutions with higher concentration of cellulose. In combination of NaOH with thiourea and/or urea, cellulose is dissolved directly and quickly and fibres can be formed, characterised by good mechanical properties (Ruan et al. 2004; Chen et al. 2006; Huajin et al. 2007).

6.5.1.2 Steam Explosion

In order to increase cellulose reactivity and solubility in aqueous NaOH solution, the cellulose pulp can be subjected to modifications using steam explosive treatment. This process is based on short-time vapour-phase cooking at temperatures in the range of 180–250 °C followed by a fast decompression. Main effects of steam explosion conditions (pressure, temperature and time of treatment) on the changes in morphology, degree of polymerization, solubility in aqueous alkaline solution and supermolecular structure of pulp were widely investigated (Kamide et al. 1984; Yamashiki et al. 1990a, b). Modified cellulose under optimal conditions, i.e. 2.9 MPa and a treatment time of 30 s, can lead to the preparation of 9–10 % cellulose in aqueous NaOH solution with a yield of 99 wt%. Steam explosion is an interesting alternative method of cellulose activation for the preparation of dissolving cellulose and for the production of new textile fibres.

6.5.1.3 Hydrothermal Treatment

Method of cellulose activation by hydrothermal method is based on heating of cellulose pulp suspension in water or methanol at 100–170 °C under hyperbaric pressure. Appropriate conditions enable to lower the average DP of cellulose and polydispersity and to increase the solubility of the pulp in aqueous NaOH (Struszczyk et al. 1997, 2000a, b). Established at the Institute of Biopolymers and Chemical Fibres, IBWCh, Poland, the method was envisaged as an alternative to the viscose process and elaborated for manufacturers of cellulose carbamate (Keunecke et al. 1999). The controlled hydrothermal activation offers highly reactive cellulose pulp, which allows after reaction with urea the production of cellulose carbamate. The resulting pulp is characterised by good solubility in aqueous alkaline solution as well as excellent spinnability for producing fibres and films.

Usage of hydrothermally treated cellulose pulp for the manufacture of cellulose fibres was the focus of further studies at IBWCh (Wawro et al. 2009). The initial cellulose, hydrothermally treated with water at temperatures of about 100–200 °C under a pressure of about 0.1–1.5 MPa for 30–70 min, is directly soluble in NaOH solutions. Alkaline solutions with cellulose concentration of 5–7.5 wt% were used for the production of cellulose fibres characterised by tenacities of around 22 cN/tex and elongations at break of ca. 15 %.

6.5.1.4 Enzymatic Modification

Another method is the application of specific enzyme complexes (cellulases), which causes controlled degradation of polymer macromolecule and its activation. As a result of enzyme action, the average DP and the crystallinity degree of cellulose decrease and the number of hydrogen bonds is reduced, which are the reasons of improved solubility of cellulose in aqueous NaOH solutions. It has been found that accessibility of the pulp to enzymes is increased by the use of suitable pre-treatment methods, particularly mechanical disintegration (Ciechańska et al. 1996, 2007; Struszczyk et al. 2000c). Firstly, the dry cellulose pulp is swollen in demineralised water at ambient temperature and disintegrated with an agitator. Then, the surplus water is squeezed out and the pulp is fed in a Werner Pfleiderer tearing machine where it is mechanically torn for 90 min at 35 °C. Afterwards the wet pulp is subjected to the bio-modification procedure treatment with cellulolytic enzyme in a 0.05 M acetate buffer at pH 4.8. The cellulose pulp is separated from the enzyme liquor on a 'Nutsche' filter and washed with water, while the enzyme is inactivated (Ciechańska et al. 2005). Such prepared pulp is ready for the dissolution in NaOH.

Enzymatic modification of cellulose has been studied at IBWCh together with Tampere University of Technology (Finland) for several years, protected by Polish, Finnish and European patents (Struszczyk et al. 1991, 1995, European project undated, Ciechanska et al. 2006) and given the name CELSOL. The process was further developed within the 6th EU Framework Programme, in the project *Biotechnological Process for Manufacturing Cellulosic Products with Added Value*. The acronym BIOCELSOL was used as name of the process and the fibres and films obtained. The process was designed to replace the viscose method by a new environmentally friendly technology (mild conditions of biomodification, possibility of enzyme recirculation, low environmental hazard) for manufacturing of shaped cellulosic products.

6.5.2 Preparation of Alkaline Solutions from Bio-Modified Cellulose for Film and Fibre Manufacture

To prepare alkaline spinning solutions, the enzymatically treated cellulose pulp containing 75 wt % water is dissolved into 10.2 wt% aqueous sodium hydroxide containing a small amount of zinc oxide (Ciechańska et al. 2005). The dissolving process is carried out in a mixing tank equipped with a high-speed agitator and cooling jacket. Time and temperature of cellulose dissolving have a significant effect on the properties of cellulose solutions and should be adjustable depending on the characteristics of the starting material. Usually, the process is performed from 60 to 180 min at temperature of about 0 °C. After cellulose is dissolved, the solution is filtrated to remove undissolved particles and deaerated to remove air bubbles. Prepared solutions should be assessed on cellulose content, NaOH and ZnO content, falling ball viscosity, dynamic viscosity and filterability to assure proper quality of the solution for spinning trials. Ideal alkaline cellulose solutions designed for fibre spinning and film shaping should be characterised by 6–6.5 wt% cellulose, 7.8 wt% NaOH, 0.87 wt% ZnO, insoluble particle content below 1 wt% and alkali/cellulose ratio of 1.2.

6.5.3 Technology Process for Film Forming and Fibre Spinning

CELSOL/BIOCELSOL films from alkaline cellulose solutions can be formed by a batch method with the use of a hand coater, which allows manufacturing film sheets with adequate size (Ciechanska et al. 2006), or by a continuous method using a spinning line for film forming. Device designed for film forming using batch method is equipped with movable, smooth glass plate with size 60 × 80 cm. A filtered and deaerated cellulose solution at temperature of about 10 °C is poured out on the plate and exactly distributed using knife coater with a gap of 0.5 mm. Then, the plate with solution is immersed into coagulation bath containing 150 g/l H_2SO_4 and 150 g/l Na_2SO_4 at room temperature for 10 min. Next, cellulose film is placed in tank and subjected to the flow of demineralised water at temperature of 30 °C for about 2 h and dried using plate dryer.

Cellulose films can be also manufactured using laboratory spinning line. After deaeration and filtration, the spinning solution is placed in pressure tank with a capacity of 30 dm^3. The solution is transferred to gear pump using pressure 0.5 MPa and next to the spinneret head equipped with a crack slot with length of 100 or 35 mm and width of 0.5 mm. A solution of sulphuric acid and sodium sulphate constitutes the coagulation bath. The length of the coagulation path is changeable depending on the immersion depth of the film, usually 200–300 cm. Forming of the films occurs in a vertical/horizontal mode. The formed film passes a godet and runs to the washing bath containing sulphuric acid allowing a speed of 5–15 m/min with a width range of up to 100 mm. Washing of the films is accomplished in troughs equipped with two godet sets, of which one is immersed in an acidic coagulation bath while the other is fixed 100 cm above the bath. Washing water moves at 45–50 °C in counter flow to the film. After collecting on a winder in pieces max of 300 m long, the films are once more rinsed with water in a container; then, finishing agents are applied and dried.

CELSOL/BIOCELSOL fibre spinning from bio-modified cellulose pulp has been developed at IBWCh, too (Ciechańska et al. 2005, 2006). The process was investigated using a wet spinning method on a large laboratory spinning line. As widely investigated, spinning conditions, composition of the spinning bath, type of spinneret, draw ratio and spinning speed influence the

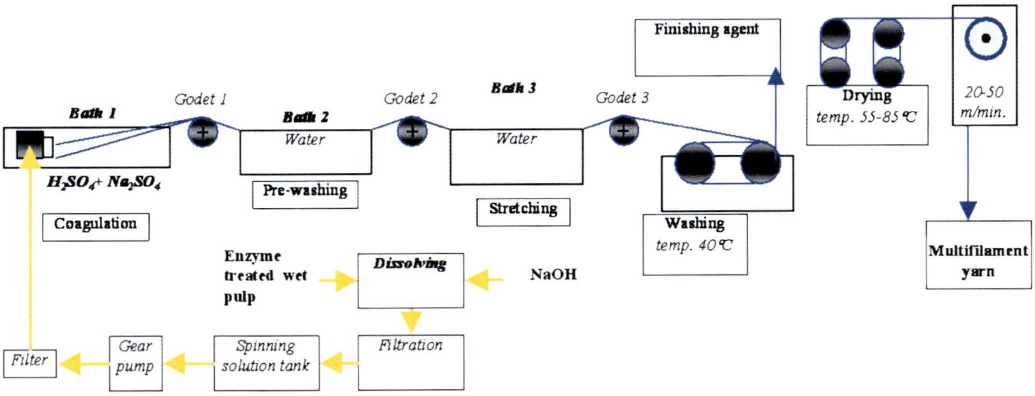

Fig. 6.14 Technological scheme of CELSOL/BIOCELSOl fibre spinning

mechanical properties of the shaped fibres. The spinning solution is conveyed to the spinning bath by means of a 6.0-cm^3/revolution metering pump through a Pt-Rh spinneret with 300 holes of a single-hole diameter of 0.08 mm or, alternatively, 1,000 holes of 0.06 mm in diameter. A solution containing 100–150 g/dm^3 of sulphuric acid and 150 g/dm^3 of sodium sulphate serves as a coagulation bath, which temperature is maintained at about 15 °C. The spun fibres are stretched ($R_{max} = 30\%$) in water at 85 °C, rinsed with water at 40 °C, passed through a spin-finishing solution containing 7 g/dm^3 of Berol Fintex, dried on cylinders at 55–85 °C and finally tension-collected on spools as multifilament yarn. The applied spinning speed amounts to 20–50 m/min. Figure 6.14 presents the scheme of cellulose fibre spinning.

6.5.4 Properties of CELSOL/BIOCELSOL Film and Fibres

Cellulose films produced according to viscose method are commonly named cellophane. Next to paper, it was the most popular packaging material in the first half of the twentieth century. Nowadays, cellophane is seen more as a niche product. Only a few companies are manufacturers of cellophane, like in Europe Innovia Film Ltd., UK, with a cellophane production capacity of about 18,000 t/a in Wigton, England. Additionally, Innovia operates a cellophane plant in Tecumseh, Kansas, USA, which represents a capacity of around 12,000 t/a. There are several smaller producers in the Far East, most of them in China. Despite the enormous decline in the consumption of cellulose film in the last three decades, the future of the material seems rather bright and an increasing demand is expected in the market for the coming years. Tenacity and elongation of BIOCELSOL films were tested and compared with commercially available films as shown in Table 6.7. It can be concluded that the properties of BIOCELSOL films are comparable to quality indices of the film offered by the European producer.

When the idea was conceived to convert cellulose enzymatically into a material for direct dissolution, first and main goal for an industrial application was fibre manufacture. CELSOL/BIOCELSOL fibres resemble those made according to the viscose route. Table 6.8 displays quality indices of BIOCELSOL fibres in comparison to several viscose fibres, which are shown for benchmarking.

BIOCELSOL offers a technology, which is capable of yielding cellulose fibres in a wide titre range with tenacity in the range of 1.8–1.9 cN/dtex, while elongation, Young modulus and crystallinity are comparable to viscose fibres. Such quality parameters dedicate the fibre to standard textile techniques and many further applications. Additionally, the use as nonwovens for medical devices and hygiene products is possible. It is assumed that other physical properties

Table 6.7 Quality indices of BIOCELSOL films (Sescousse and Budtova 2009)

Film type	Thickness (mm)	Tenacity (MPa)	Elongation at break (%)
Commercial (Innovia)	0.02–0.04	125	22
BIOCELSOL hand cast	0.03	ca. 105	ca. 30
BIOCELSOL continuous cast	0.01–0.06	100–105	15–30

When the idea was conceived to convert cellulose enzymatically into a material for direct dissolution, first and main goal for an industrial application was fibre manufacture. CELSOL/BIOCELSOL fibres resample those made according to the viscose route. Table 6.8 displays quality indices of BIOCELSOL fibres (Sescousse and Budtova 2009) in comparison to several viscose fibres, which are shown for benchmarking

Table 6.8 Quality indices of BIOCELSOL fibres in comparison to viscose fibres (Sescousse and Budtova 2009)

Fibre	Titre (dtex)	Tenacity (cN/dtex)	Elongation (%)	Young modulus (cN/dtex)	Crystallinity (%)
Viscose 1[a]	1.36	2.33	18.2	38.3	38.3
Viscose 2[a]	1.95	2.05	17.4	33.1	27.9
Viscose 3[a]	1.73	2.15	18.4	39.4	39.2
BIOCELSOL[b]	1.4–4.2	1.8–1.9	14–17	22–50	35–42

[a]Viscose fibres from different manufacturers were taken as reference
[b]Range of quality indices concerns the variety of fibres prepared

Fig. 6.15 BIOCELSOL film sheets and multifilament yarn

like dye affinity and absorption ability are similar to viscose fibres since there is a close relationship in chemical and molecular structure. BIOCELSOL film and fibre production lines are shown in Fig. 6.15.

6.5.5 Feasibility Studies

CELSOL/BIOCELSOL is an attempt towards introducing an environmentally friendly technology to the manufacture of various cellulose products like fibres and film and replaces the well-established but hazardous viscose technology. In this context, the following conspicuous features of the new technology deserve highlighting:

- The technology bears the chance to eliminate carbon disulphide (CS_2) from the manufacturing technique of cellulose fibres and films.
- With CELSOL/BIOCELSOL, the possibility is given to increase the use of renewable raw materials in the manufacture of valuable market articles. Virtually no crude oil-derived materials appear in these technologies, and moreover, its products are biodegradable.
- In contrast to the viscose process, CELSOL/BIOCELSOL is a clean technology. Throughout the entire production routes, only mild and save conditions are employed like moderate temperatures (up to 50 °C) and atmospheric pressure operations. There is

no emission of gases to the atmosphere, nor are the wastewaters contaminated with sulphuric compounds other than sodium sulphate.

Further, raw material, energy and labour costs of the BIOCELSOL and viscose fibre processes were compared. Raw material consumption was estimated to be 7–8 % higher than of the viscose process by reason of higher alkali consumption. Energy costs are somewhat difficult to compare because relevant information from the viscose process is not available, and there is no pilot-scale fibre process line for BIOCELSOL. Labour costs were estimated to be the same for both processes, although the employees' comfort is considerably advantageous in the BIOCELSOL process due to the elimination of carbon disulphide.

It should also be stressed that environmental taxes due to emissions from the use of carbon disulphide in the viscose process are not included because they depend on the location of the factory. Costs related to the handling, storage, safety procedures and recovery of carbon disulphide are not either included, but they are turning the figures in favour of the BIOCELSOL process. However, the BIOCELSOL process must be further optimised to decrease alkali consumption and improve end product properties. Pilot plant experiments should also be carried out to evaluate the final feasibility of the BIOCELSOL process.

6.6 Ultra-Lightweight and Highly Porous Cellulose Aerogels and Aerogel-Like Materials

Novel added-value cellulosic materials were developed in the first decade of the twenty-first century: ultra-light, highly porous aerogels and aerogel-like materials based on cellulose or cellulose derivatives. This new class of promising materials offers a wide range of potential applications, from biomedical and cosmetic (delivery systems, scaffolds) to insulation and electrochemical (when pyrolysed). Despite the fact that these highly porous celluloses often contain large micron-size pores and are not prepared via classical sol–gel techniques, the term 'aerogel' has been widely accepted and will be therefore used for simplification.

Cellulose aerogels can be divided into two main classes: (1) those based on cellulose I such as bacterial cellulose (BC) (Maeda et al. 2006; Liebner et al. 2010) and nano- or microfibrillated cellulose prepared via mechanical disintegration of native cellulose sometimes followed by an enzymatic treatment (Pääkko et al. 2008) and (2) those based on cellulose II that can be obtained via cellulose dissolution-coagulation (Jin et al. 2004; Innerlohinger et al. 2006; Gavillon and Budtova 2008; Liebner et al. 2008, 2009; Tsioptsias et al. 2008; Sescousse and Budtova 2009; Deng et al. 2009; Aaltonen and Jauhiainen 2009; Duchemin et al. 2010; Sescousse et al. 2011). In all cases wet cellulose aerogel precursor is dried in such a way that the porosity is largely retained and the pores remain open, i.e. via freeze drying or drying under supercritical conditions. In the latter case carbon dioxide is used; $scCO_2$ is one of the most commonly used supercritical fluids in polymer chemistry and technology due to its low cost, non-inflammability and low critical point temperature and pressure.

The precursor of cellulose I-based aerogels is a 'network' of nano-fibres filled with water. If using supercritical carbon dioxide for drying, water has to be replaced by a $scCO_2$-miscible organic solvent such as ethanol or acetone beforehand. The final $scCO_2$-dried material is also a network of cellulose I nano-fibres. Even though it is similar to cellulose II aerogels, lower densities and higher porosities can be obtained at almost zero contraction during drying due to higher skeleton crystallinity and molecular weight.

The preparation of cellulose II-based aerogels, so-called aerocellulose, is inspired by the synthetic routes of making classical aerogels: sol–gel process followed by drying in supercritical CO_2. Cellulose is dissolved in a direct solvent, for example, N-methylmorpholine-N-oxide (NMMO) monohydrate (Innerlohinger et al. 2006; Liebner et al. 2008, 2009), 8 % NaOH-water (Gavillon and Budtova 2008; Sescousse

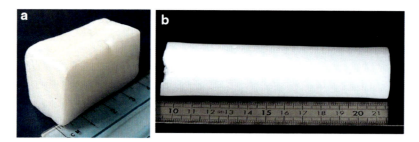

Fig. 6.16 Cellulose aerogels: (**a**) from bacterial cellulose, reproduced from Liebner et al. (2010) with kind permission from Wiley-VCH, Weinheim, and (**b**) from microcrystalline cellulose via dissolution in 8%NaOH–water

and Budtova 2009), LiCl/DMAc (Duchemin et al. 2010), calcium thiocyanate (Jin et al. 2004) or ionic liquid (Tsioptsias et al. 2008; Deng et al. 2009; Aaltonen and Jauhiainen 2009; Sescousse et al. 2011), coagulated in a nonsolvent (water, alcohols) and then dried either via freeze drying (Jin et al. 2004; Deng et al. 2009; Duchemin et al. 2010) or with scCO$_2$. (Innerlohinger et al. 2006; Liebner et al. 2008, 2009; Gavillon and Budtova 2008; Tsioptsias et al. 2008; Sescousse and Budtova 2009; Aaltonen and Jauhiainen, 2009; Sescousse et al. 2011). When coagulated in water, the latter is replaced by a scCO$_2$-miscible organic solvent before supercritical drying as described above. In all cases cited no chemical cross-linking is used to 'stabilise' the cellulose 'network'; it is formed during cellulose coagulation either from a solution (Jin et al. 2004; Innerlohinger et al. 2006; Liebner et al. 2008, 2009; Deng et al. 2009; Aaltonen and Jauhiainen 2009; Duchemin et al. 2010; Sescousse et al. 2011) or from a physical gel (Gavillon and Budtova 2008; Sescousse and Budtova 2009). The density of cellulose II aerogels depends on the initial cellulose concentration in solution and on the accuracy and dimensional stability throughout drying. Foaming agents can be added to increase the porosity of cellulose II aerogels (Gavillon and Budtova 2008).

A brief review of research on aerogels from both underivatised cellulose I (Fig. 6.16a) and II (Fig. 6.16b) conducted in two EPNOE laboratories is given below.

6.6.1 Cellulose I Aerogels

Aerogels that contain a considerable portion of the crystal phase I can be obtained from both highly crystalline bacterial cellulose (Iα, triclinic space group P1) and plant cellulose after mechanical disintegration of fibres into microfibrils (Iβ, monoclinic P2$_1$). In contrast to cellulose II aerogels, their cellulose I counterparts have to be prepared directly from respective cellulose I lyogels, i.e. cellulose 'networks' whose pores are filled with a cellulose antisolvent. This measure is necessary due to their thermodynamically lower stability and hence inevitable conversion into the more stable cellulose II polymorph upon dissolution and regeneration.

High purity, average molecular weight, crystallinity, mechanical stability (Bodin et al. 2007) and bio-compatibility (Helenius et al. 2006; Jonas and Farah 1998) are some of the features that render bacterial cellulose an interesting source for highly porous scaffolds that would have a considerable application potential in medicine and health care (Barud et al. 2007; Sheridan et al. 2002).

Bacterial cellulose can be grown (semi-)continuously (Kralisch et al. 2008, 2010) or discontinuously in batch mode (Geyer et al. 1994). In batch mode both static and dynamic cultivation techniques have been established. Static cultivation of the cellulose-producing strain *Acetobacter xylinum* in Hestrin-Schramm culture medium using glass tanks is probably the most frequently applied technique to produce bacterial cellulose.

The growth of the bacterial cellulose takes place at the phase boundary between air and culture medium surface. Throughout a growth period of 30 days at 30 °C, the bacterial cellulose layer deposited below the culture medium surface reaches a thickness of about 3–4 cm with the most recent formed cellulose on top. After harvesting, the BC aquogels are repeatedly subjected to alkaline treatment (0.1 M aq. NaOH, 90 °C, 20 min, three times, e.g.) and subsequent thorough washing to eliminate growth medium residues and the protein fraction. Typically, bacterial cellulose sheets are sterilised prior to further use in medical or cosmetic applications.

Aerogels from bacterial cellulose were first described by Maeda (2006) who dried the respective aquogels with supercritical ethanol at 243 °C and 6.38 MPa. The obtained materials had densities of 6 mg cm^{-3} and porosities of more than 99 % and consisted of 20–60-nm-thick microfibrils forming a network. The harsh drying conditions used are, however, supposed to effect considerable chemical alteration of the cellulose network structure as evident from the strong increase of the stress measured at 60 % strain ($\sigma 60\%$) from 20 kPa (hydrogel) to 54 kPa (aerogel) (Maeda et al. 2006). In particular hornification that takes place in (ligno)cellulosic materials upon thermal drying beyond 180–200 °C (Cheng et al. 2009) leads to considerable stiffening of the polymer structure (Diniz et al. 2004) and negatively affects both hydrophilicity and rewettability of the aerogels. Drying with supercritical carbon dioxide at 40 °C and 10 MPa turned out to be a good alternative to fully preserve the chemical integrity of bacterial cellulose (Liebner et al. 2010).

As the presence of solvents with high surface tension such as water—even in traces—would inevitably lead to serious pore collapsing during the drying procedure, bacterial cellulose aerogels are subjected to a thorough solvent exchange prior to scCO$_2$ drying. It has been confirmed that the highly porous network structure of BC aerogels can be fully preserved at almost zero shrinking if ethanol is used to replace water before drying (Liebner et al. 2010). This is due to the good miscibility of ethanol with scCO$_2$ and the almost negligible difference between the densities of scCO$_2$ and scCO$_2$-expanded ethanolic phase at 40 °C and 10 MPa. The obtained ultralightweight materials have densities of down to 5 mg cm^{-3} and feature a surprisingly high resistance to shrinking when stored at elevated relative humidity (see Fig. 6.17). In contrast to aerogels obtained from the triple amount of cotton linters via dissolution and regeneration following the Lyocell approach which shrunk at 65 % r.h. by about 50 %, BC aerogels almost fully retain their shape over a time period of 84 days.

Nitrogen sorption at 77 K and small-angle X-ray scattering experiments along with scanning electron micrographs evidence that BC aerogels have a spread porosity comprising micro-, meso- and macropores. Similarly to other aerogels such as from cellulose II (see next section), cellulose acetate (Fischer et al. 2006), polyurethane (Pirard et al. 2003) and silica (Scherer et al. 1995), mercury intrusion cannot be applied due to the fragility of the materials. However, thermoporosimetry using xylene as confined medium has been confirmed to be a useful tool to obtain valuable information about the pore characteristics of soft matter. It was shown that the peak of pore size distribution of BC aerogels is at around 60–100 nm; however, larger micron-size pores of up to $\varnothing = 2$ μm are present too (Liebner F et al., unpublished results, 2011).

The mechanical strength and stiffness of BC aerogels have been confirmed to be dependent on the sample orientation. The observed anisotropy is due to the nearly two-dimensional bacterial growth and hence cellulose synthesis at the phase boundary air/culture medium. Correspondingly, a higher Young's modulus of about 145 kPa was measured for the two orthogonal directions spanning the plain of microbial cellulose synthesis and only of 57 kPa in downward 'growth' direction where the once synthesised BC slowly veers from the surface into the culture medium (Liebner F et al., unpublished results, 2011).

Bacterial cellulose aerogels were shown to be fully rewettable when the corresponding alcogels obtained by replacing water by ethanol were dried with scCO$_2$ (Liebner F et al., unpublished

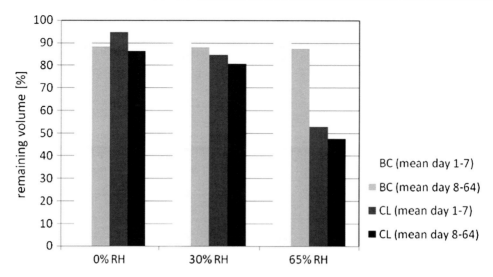

Fig. 6.17 Dimensional stability of scCO$_2$-dried cellulosic aerogels obtained from bacterial cellulose alcogels (BC, theoretical density: 10 mg cm^{-3}) and Lyocell dopes from cotton linters (CL, theoretical density: 30 mg cm^{-3}) in dependence on time and relative humidity during storage.

results, 2011) which is not the case for freeze drying which allows for 88–96 % rewetting only, depending on the applied technique. The repeatable cycle of drying and rewetting under full retention of the pore features along with the earlier mentioned properties renders BC scaffolds promising materials for controlled slow release applications in wound treatment, health care or dehabituation.

Microfibrillated plant cellulose has been recently demonstrated to be an interesting alternative source for cellulose I aerogels (Pääkko et al. 2008). It has been confirmed to have a high robustness towards mechanical stress which can overmatch that of comparable aquogels from acidic hydrolysed cellulose by up to two orders of magnitude (Pääkkö et al. 2007).

6.6.2 Cellulose II Aerogels

Cellulose II aerogels can be obtained by dissolving cellulose in any direct solvent. A general preparation scheme is given in Fig. 6.18. If dissolving cellulose in aqueous NaOH, gelation of cellulose–NaOH–water solution with time and temperature (Roy et al. 2003; Gavillon and Budtova 2008) can be used to make a three-dimensional object: the solution is poured into a mould and gelled and the shape of the mould determines the final aerogel shape. In a similar way 3D cellulose aerogels of various shapes can be obtained using solvents that solidify at room temperature, such as NMMO monohydrate (Liebner et al. 2008, 2009) or certain ionic liquids (for example, 1-butyl-3-methylimidazolium chloride) (Sescousse et al. 2011a). Another way is to form cellulose aerogel precursors directly from solutions: for example, cellulose beads of various shapes, from flat plates to spherical, were prepared by dripping cellulose–NaOH–water solutions into water; they were then dried (after water exchange by acetone) in scCO$_2$ to obtain aerogels (Sescousse et al. 2011b).

The morphologies of cellulose II aerogels prepared with different solvents are shown in Fig. 6.19.

Two types of structures can be distinguished: a 'network'-like and a 'globular'-like. A network-like morphology is obtained from gelled (in NaOH–water) and solidified (NMMO monohydrate) cellulose solutions (Fig. 6.19a, b, respectively). A globular morphology is obtained when solutions are coagulated directly from liquid state

Fig. 6.18 Main steps in cellulose II aerogel preparation

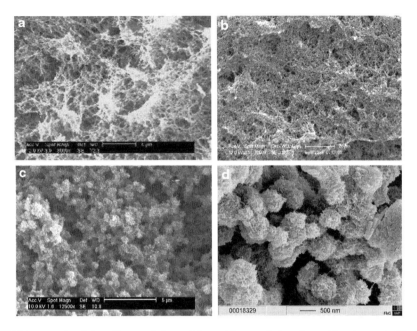

Fig. 6.19 SEM images of cellulose II aerogels from: (**a**) gelled 5 % microcrystalline cellulose–8 % NaOH–water; (**b**) solid 3 % bleached hardwood sulfite pulp-NMMO monohydrate solution, reproduced from Liebner et al. (2009) with kind permission from De Gruyter, Munich; (**c**) 3 % microcrystalline cellulose–EMIMAc solution, reprinted from Sescousse et al. (2011b), Copyright 2011, with permission from Elsevier and (**d**) 3%cellulose (DP = 950)-NMMO monohydrate solution, published with permission from Gavillon and Budtova (2008), Copyright 2008 American Chemical Society and reprinted from Sescousse et al. (2011b), Copyright 2011, with permission from Elsevier

such as for molten NMMO monohydrate or the ionic liquid, 1-ethyl-3-methylimidazolium acetate (EMIMAc) (Fig. 6.19c, d). The development of different morphologies is due to different mechanisms of phase separation. If solutions are gelled or solidified before coagulation (network-like morphology), a preceding phase separation splits the free solvent phase (NaOH hydrates, NMMO·H$_2$O crystals) apart from a second phase consisting of cellulose and bound solvent. Coagulation then takes place in two steps: the nonsolvent first dilutes regions with 'free' solvent and then subsequently removes the rest of solvent bound to cellulose. In EMIMAc and molten NMMO monohydrate, cellulose is homogeneously distributed all over the solution and phase separation occurs in one step, via spinodal decomposition, creating regular small spheres (Fig. 6.19c, d).

The density of cellulose II aerogels strongly depends on the initial cellulose concentration in solution and on the applied drying method. The impact of the cellulose content on the aerogel density is shown in Fig. 6.19. For cellulose of a similar molecular weight (around DP ≈ 200) and the same drying conditions, density is

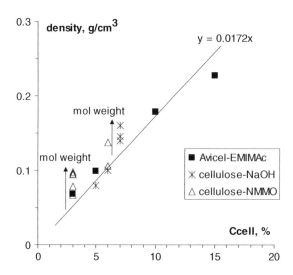

Fig. 6.20 Aerocellulose density as a function of the initial cellulose concentration. Drying in supercritical conditions of microcrystalline cellulose–EMIMAc and cellulose 8 %NaOH–water was performed in Centre Energétique et Procédés, Mines ParisTech, France, see details in Gavillon and Budtova (2008), Sescousse et al. (2011a) and Sescousse and Budtova (2009). Data on cellulose–NMMO are taken from Liebner et al. (2009)

directly proportional to the initial cellulose concentration. The density of aerogels was confirmed to increase with the molecular weight of the used cellulose (shown by arrows in Fig. 6.20).

Density can be strongly reduced if the solution is 'foamed' with a surfactant (Gavillon and Budtova 2008): for example, the addition of 1 % Simulsol SL 8 (alkyl polyglucoside) to an aqueous solution containing 8 w% NaOH and 5 w% microcrystalline cellulose reduced the final density of the aerogel from 0.14 g cm^{-3} (without surfactant) to 0.06 g cm^{-3}.

The size of the pores in cellulose II aerogels as obtained with BET analysis is around 10–15 nm (Liebner et al. 2009; Sescousse et al. 2011a). The overall size distribution is very wide, from few nanometers to several microns, as shown in Fig. 6.18. The determination of the pore size distribution in the macropore range is, however, a very challenging task due to the fragility of the cellulose pore network structure. Mercury intrusion, a commonly used method for investigating macroporosity, was shown to fail for cellulose aerogels as pore walls collapsed under the applied pressure preventing mercury to penetrate the pores. The BET surface area obtained from nitrogen sorption experiments at 77 K varied from 100 to 300 m^2 g^{-1} for all systems studied.

The mechanical properties of cellulose II aerogels were determined from compression experiments. The Young modulus E of all aerogels (obtained from microcrystalline cellulose that was dissolved either in EMIMAc or aqueous NaOH (8 w%) and of various celluloses of higher molecular weights dissolved in NMMO monohydrate) was found to be proportional to the aerogel density in power three (Fig. 6.21) (Sescousse et al. 2011a). This is a typical scaling for classical aerogels but not for the open-cell regular foam model which predicts square-law dependence. Structural defects appearing during cellulose coagulation can be the reason of the similarity between cellulose II aerogels and classical aerogels formed via sol–gel process.

New nano-structured carbons with interesting properties for electrochemical applications were obtained by pyrolysis of cellulose II aerogels (Guilminot et al. 2008; Rooke et al. accepted). By varying cellulose concentration, coagulation conditions and pyrolysis parameters, monolithic carbon aerogels with controlled shape and porosity were obtained (see Fig. 6.22). As compared

6 Cellulose Products from Solutions: Film, Fibres and Aerogels

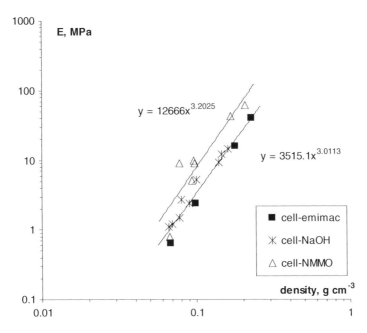

Fig. 6.21 Young modulus as a function of cellulose aerogel density; data on cellulose–EMIMAc and cellulose–8 % NaOH are taken from Sescousse et al. (2011a), data on cellulose–NMMO are taken from Liebner 2009

Fig. 6.22 Aerocellulose from microcrystalline cellulose–8 % NaOH–water and its carbon counterpart

6.6.3 Concluding Remarks on Aerogels

A very brief review of results obtained in two EPNOE laboratories on cellulose I and cellulose II aerogels has been presented. New cellulose materials with a high added value have been developed. Ultra-lightweight cellulose aerogels are very promising materials and open a wide range of various applications: as scaffolds and carriers, potential insulating materials and being pyrolysed in batteries and fuel cells.

with the non-carbonised cellulose II aerogel, the respective carbon counterparts have a higher density of 0.2–0.4 g cm^{-3} and a mesopore volume of 2–4 cm^3 g^{-1}. The obtained carbon aerogels show a 30–50 % higher capacity in button cell batteries than the industrial reference as tested by SAFT (France). In single PEM fuel cell test bench, the texture of the new carbon aerogels was furthermore demonstrated to allow an improved mass transfer at the cathode (Rooke et al. 2011).

Conclusions

Modern times and an insatiable mankind require new and advanced polymeric products. But how if the natural resources are limited and the variety of polymers keep abreast of growing demand and desired performance? A possible answer could be to take well-known polymers and bring out more interesting properties of them having in mind that new practical values not always have to be based on complicated structural features. In many cases, it might be just a matter of morphology of the new product that may give it outstanding performances in everyday use. On the other hand, polymer producers are continuously seeking alternatives to common uses and are increasingly being forced to reduce production costs due to resource exhaustion and economic pressure. These are some of the facts driving further developments. Being in the vanguard of progress and creating ideas how to make use of renewable resources means to make an important contribution to overcome the future's serious raw material exhaustion. The basic idea behind all efforts to create new products is to make use of the interaction between functionality and morphology.

Considering that cellulose and other polysaccharides are huge biopolymer resources, varieties of regeneration processes were intensively investigated. Still, viscose fibre is dominating the market, and various activities were initiated to improve the viscose process with the aim of reducing air and water pollution. R&D has emphasised the production figures concerning quality, versatility and economics on such a level that is quite far to reach for alternative processes. Usage of cellulose carbamate as alternative derivatisation sulphur-free method could not prevail against so far, despite the possibility of running on existing viscose plants. Also CELSOL/BIOCELSOL is a safe technology employing throughout the entire production routes only mild and safe conditions like moderate temperatures and atmospheric pressure operations. There is no emission of gases to the atmosphere, and the wastewaters are contaminated with sulphuric compounds other than sodium sulphate. However, the BIOCELSOL process must be further optimised to decrease the alkali consumption and to improve the end product properties.

Direct dissolution processes when combined with an advanced textile processing could be a further approach. Yet, experiments with H_3PO_4, $ZnCl_2/H_2O$, $NaOH/H_2O$/water, DMA/LiCl, DMF/LiCl or DMSO/LiCl as solvents have not succeeded outside the lab scale. Either the textile physical parameters like tenacity and elongation were not adequate enough or the process optimization was not satisfying or the costs for waste management were too high. Only Cu^{2+}/NH_3 system is used for preparing cuprammonium membranes for blood dialysis as the standard material worldwide.

Solely NMMO surely offers well progress as solvent in a commercial way for staple fibre and filament production. Although variations in the technology are limited and the produced fibre species are restricted in their variety in the textile sector, adding secondary, functionalizing agents to a polysaccharide, for instance, changes its properties in a very interesting and sometimes unexpected manner. Together with the use of technologies other than dry or wet spinning, the number of resulting products is almost unlimited.

References

Aaltonen O, Jauhiainen O (2009) The preparation of lignocellulosic aerogels from ionic liquid solutions. Carbohydr Polym 75:125–129

Anonymous (1997) 100th anniversary of the Pauly Patent. Chem Fiber Intern 47:431

Audemars G (1855) British Patent 283

Bandel W (1961) Svensk Papperstidn 64:893–897

Barud HS, Barrios C, Regiani T, Marques RFC, Verelst M, Dexpert-Ghys J, Messaddeq Y, Bodin A, Bäckdahl H, Fink H, Gustafsson L, Risberg B, Gatenholm P (2007) Influence of cultivation conditions on mechanical and morphological properties of bacterial cellulose tubes. Biotechnol Bioeng 97(2):425–434

Bodin A, Bäckdahl H, Fink H, Gustafsson L, Risberg B, Gatenholm P (2007) Influence of cultivation conditions on mechanical and morphological properties of bacterial cellulose tubes. Biotechnol Bioeng 97(2):425–434

Bredereck K, Hermanutz F (2005) Rev Prog Color 35:59–75

Büttner R, Claussen F, Knobelsdorf C, Krieg M, Taeger E (1999) DE 19917614 A1. Production of highly adsorbent cellulose fibre for use in filters to remove ions or polar substances, 19/04/1999

Chardonnet AM (1884) French Patent 165 349

Chen X, Burger C, Fang D, Ruan D, Zhang L, Hsiao BS, Chu B (2006) Polymer 47:2839–2848

Cheng K-C, Catchmark JM, Demirci A (2009) Enhanced production of bacterial cellulose by using a biofilm reactor and its material property analysis. J Biol Eng 3:12

Ciechańska D, Galas E, Struszczyk H (1996) Biotransformation of cellulose. Fibres Text East Eur 4(3–4):148

Ciechańska D, Wawro D, Stęplewski W, Kazimierczak J, Struszczyk H (2005) Formation of fibres from bio-modified cellulose pulp. Fibres Text East Eur 13(6(54)):19–23

Ciechanska D, Wawro D, Stęplewski W, Wesolowska E (2006) Ecological method of manufacture of the multi-filament cellulose fibres for knitted and woven products. In: Proceedings of the 7th International Symposium "Alternative cellulose – manufacturing, forming, properties", Sept 6–7, 2006, Rudolstadt, Germany

Ciechańska D, Wawro D, Stęplewski W, Wesolowska E, Vehvilonen M, Nousiainen P, Kamppuri T, Hroch Z, Sandak, Janicki J, Włochowicz A, Rom M, Kovalainen A (2007) Ecological method of manufacture of the cellulose fibres for advanced technical products. In: Edana conference, Nonwovens Research Academy, 29–30 March 2007, University of Leeds, UK

Cross CF, Bevan EJ, Beadle C (1892) British Patent 8700

Deng M, Zhou Q, Du A, Kasteren JMN, Wang W (2009) Preparation of nanoporous cellulose foams from cellulose-ionic liquid solutions. Mater Lett 63 (21):1851–1854

Depaissis LH (1890) French Patent 203 741

Diener A (2010) How to maximize cellulose fiber capacity? In: Proceedings of the 9th International Symposium "Alternative cellulose – manufacturing, forming, properties", Sept 09–10, 2010, Erfurt, Germany

Diniz JMBF, Gil MH, Castro JAAM (2004) Hornification - its origin and interpretation in wood pulps. Wood Sci Technol 37:489–494

Dohrn W, Büttner R, Knobelsdorf C, Notz I, Schuemann M, Herrmann E, Werner G (2001) DE 10137171 A1, "Verfahren zur Herstellung von cellulosischen Formkörpern mit superabsorbierenden Eigenschaften", 31/07/2001

Duchemin BCJ, Staiger MP, Ticker N, Newman RH (2010) Aerocellulose based on all-cellulose composites. J Appl Polym Sci 115:216–221

Ekman K, Eklund V, Fors J, Huttunen JI, Selin JF, Turunen OT (1986) Cellulose carbamate. In: Young RA, Rowell RM (eds) Cellulose structure, modification and hydrolysis. Wiley, New York, pp 131–148

European project (undated): BIOCELSOL – biotechnological process for manufacturing cellulosic products with added value. Contract number: NMP2–CT–2003-505567

Fink H-P, Weigel P, Bohn A (1997) Lenz Ber 76:119–125

Fink H-P, Weigel P, Purz H-J (1998) Formation of lyocell-type fibres with skin-core structure. Lenz Ber 78:41–44

Fink H-P, Weigel P, Purz H-J, Ganster J (2001) Structure formation from regenerated cellulose materials from NMMO solutions. Prog Polym Sci 26:1473–1524

Fink H-P, Weigel P, Rihm R (2004) DE 10 2004 007 616 B4 "Verfahren zur Herstellung von Fasern und anderen Formkörpern aus Cellulosecarbamat und/oder regenerierter Cellulose", 17/02/2004

Fink H-P, Ebeling H, Rihm R (2006) Fibre formation from liquid crystalline solutions of cellulose carbamate in N-methylmorholine-N-oxide. In: Proceedings of the 7th international symposium "Alternative cellulose – manufacturing, forming, properties", Sept 6–7, 2006, Rudolstadt, Germany

Fischer F, Rigacci A, Pirard R, Berthon-Fabry S, Achard P (2006) Cellulose-based aerogels. Polymer 47:7636–7645

Franks NA, Varga JK (1977) US 4145532

Gavillon R, Budtova T (2008) Aerocellulose: new highly porous cellulose prepared from cellulose-NaOH aqueous solutions. Biomacromolecules 9:269–277

Geyer U, Heinze T, Stein A, Klemm D, Marsch S, Schumann D, Schmauder HP (1994) Formation, derivatization and applications of bacterial cellulose. Int J Biol Macromol 16 (6):343–347

Graenacher C, Sallman R (1939) US Patent 2179181

Guilminot E, Gavillon R, Chatenet M, Berthon-Fabry S, Rigacci A, Budtova T (2008) New nanostructured carbons based on porous cellulose: elaboration, pyrolysis and subsequent use as substrate for proton exchange membrane fuel cell electrocatalyst particles. J Power Sources 185:717–726

Helenius G, Bäckdahl H, Bodin A, Nannmark U, Gatenholm P, Risberg B (2006) In vivo biocompatibility of bacterial cellulose. J Biomed Mater Res A 76A(2):431–438

Hill JW, Jacobson RA (1938) US Patent 2134825

Huajin J, Chunxi Z, Lixia G (2007) Carbohydr Polym 342 (6):851–858

Innerlohinger J, Weber HK, Kraft G (2006) Aerocellulose: aerogels and aerogel-like materials made from cellulose. Macromol Symp 244:126–135

Jin H, Nishiyama T, Wada M, Kuga S (2004) Nanofibrillar cellulose aerogels. Coll Surf A Physicochem Eng Aspects 240:63–67

Jonas R, Farah LF (1998) Production and application of microbial cellulose. Polym Degrad Stab 59 (1–3):101–106

Kamide K, Okajima K, Matsui T, Kowsaka K (1984) Polym J 16:857–866

Kennedy F, Phillips GO, Williams PA (1990) Cellulose sources and exploitation: industrial utilization, biotechnology and physico-chemical properties. Ellis Horwood, New York

Kennedy F, Phillips GO, Williams PA (1992) Lignocellulosics: science, technology, development and use. Ellis Horwood, New York

Kennedy F, Phillips GO, Williams PA (1995) Cellulose and cellulose derivatives: physico-chemical aspects and industrial applications. Woodhead Publishing, Cambridge

Keunecke G, Struszczyk H, Starostka P, Mikolajscyk W, Urbanowski A (1999) US 5,906,926 A "Method for modified manufacture of cellulose carbamate", 25/05/1999

Klemm D, Philipp B, Heinze T, Heinze U, Wagenknecht W (1998) Comprehensive cellulose chemistry. Wiley, Weinheim

Klemm D, Heublein B, Fink H-P, Bohn A (2005) Angew Chem Int Ed 117:3422–3458

Koppe P, Müller M (1905) German Patent 187 947

Kosan B, Michels C, Meister F (2008a) Cellulose 15:59–66

Kosan B, Nechwatal A, Meister F (2008b) Chem Fiber Intern 4:234–236

Kralisch D, Hessler N, Klemm D (2008) Kontinuierliches Verfahren zur Darstellung von bakteriell synthetisierter Cellulose in flächiger Form, DE 10 2008 046 644.1

Kralisch D, Hessler N, Klemm D, Erdmann R, Schmidt W (2010) White biotechnology for cellulose manufacturing—the HoLiR concept. Biotechnol Bioeng 105(4):740–747

Krässig H (1982) Textilveredlung 17:333–343

Krieg M, Mooz M, Rauch C, Meister F (2009) Optimization of a new cellulosic fiber with a depot-function for insecticide substances. In: Proceedings of the Xth international jena symposium on tick-born diseases, Mar 19–21, Weimar, Germany

Liebert T (2008) Macromol Symp 262:28–38

Liebner F, Potthast A, Rosenau T, Haimer E, Wendland M (2008) Cellulose aerogels: highly porous, ultralightweight materials. Holzforschung 62:129–135

Liebner F, Haimer E, Potthast A, Loidl D, Tschegg S, Neouze M-A, Wendland M, Rosenau T (2009) Cellulosic aerogels as ultra-lightweight materials. Part 2: synthesis and properties. Holzforschung 63:3–11

Liebner F, Haimer E, Wendland M, Neouze M-A, Schlufter K, Miethe P, Heinze T, Potthast A, Rosenau T (2010) Aerogels from unaltered bacterial cellulose: application of scCO$_2$ drying for the preparation of shaped, ultralightweight cellulosic aerogels. Macromol Biosci 10(4):349–352

Loth F, Schaaf E, Weigel P, Fink -P, Gensrich HJ (2002) WO 03/099 871

Maeda H (2006) Preparation and properties of bacterial cellulose aerogel and its application. Cell Commun 13(4):169–172

Maeda H, Nakajima M, Hagiwara T, Sawaguchi T, Yano S (2006) Preparation and properties of bacterial cellulose aerogel. Jpn J Polym Sci Technol 63:135–137

McCorsley CC (1979b) US 4144080

McCorsley III CC, Varga JK (1979a) US 4142913 A "Process for making a precursor of a solution of cellulose", 06/03/1979

Meister F, Vorbach D, Michels C, Maron R, Berghof K, Taeger E (1998) Chem Fiber Intern 48:32–35

Michels C (1998) Das Papier 52:3–8

Michels C, Maron R, Taeger E (1994) Lenz Ber 9:57–60

Nechwatal A, Nicolai M, Mieck K-P (1996) Text Chem Color 28:24–27

Pääkkö M, Ankerfors M, Kosonen H, Nykänen A, Ahola S, Österberg M, Ruokolainen J, Laine J, Larsson PT, Ikkala O (2007) Enzymatic hydrolysis combined with mechanical shearing and high-pressure homogenization for nanoscale cellulose fibrils and strong gels. Biomacromolecules 8(6):1934–1941

Pääkko M, Vapaavuori J, Silvennoinen R, Kosonen H, Ankerfors M, Lindstrom T, Berglund LA, Ikkala O (2008) Long and entangled native cellulose I nanofibers allow flexible aerogels and hierarchically porous templates for functionalities. Soft Matter 4(12):2492–2499

Pauly H (1897) German Patent 98 642

Pinnow M, Fink H-P, Fanter C, Kunze J (2008) Macromol Symp 262:129–139

Pirard R, Rigacci A, Maréchal JC, Quenard D, Chevalier B, Achard P, Pirard JP (2003) Characterization of hyperporous polyurethane-based gels by non-intrusive mercury porosimetry. Polymer 44(17):4881–4887

Pötsch WR (1997) Chem Fiber Intern 47:432

Pötsch WR (1999) Paul Schlack - der Erfinder des Perlons - und die Filmfabrik Wolfen. In: Zur Industriegeschichte der Bitterfelder Region, vol 7, pp 70–76 (ISSN 1432–7406)

Rihm R (2003) Röntgen-Strukturuntersuchungen an Celluloseregeneratfasern. Dissertation, Technical University of Berlin

Rooke J, de Matos Passos C, Chatenet M, Sescousse R, Budtova T, Berthon-Fabry S, Mosdale R, Maillard F (2011) Synthesis and properties of platinum nanocatalyst supported on cellulose-based carbon aerogel for applications in PEMFCs. J Electrochem Soc 158(7): B779–789

Roy C, Budtova T, Navard P (2003) Rheological properties and gelation of aqueous cellulose-NaOH solutions. Biomacromolecules 4:259–264

Ruan D, Zhang L, Zhou J, Jin H, Chen H (2004) Macromol Biosci 4:1105–1112

Scherer GW, Smith DM, Qiu X, Anderson J (1995) Compression of aerogels. J Non-Cryst Solids 186:316–320

Schulze T, Kosan B, Niemz F-G, Mooz M, Krieg M, Meister F (2009) Technical opportunities for shaping natural polymers by environmentally friendly technologies. In: Proceedings European Polymer Congress, July, 12–17, Graz, Austria

Schützenberger S (1865) Comp Rend 61:485–486

Sescousse R, Budtova T (2009) Influence of processing parameters on regeneration kinetics and morphology of porous cellulose from cellulose-NaOH-water solutions. Cellulose 16:417–426

Sescousse R, Gavillon R, Budtova T (2011) Aerocellulose from cellulose-ionic liquid solutions: preparation, properties and comparison with cellulose-NaOH and cellulose-NMMO routes. Carbohydr Polym 83:1766–1774

Sescousse R, Gavillon R, Budtova T (2011a) Aerocellulose from cellulose-ionic liquid solutions: preparation, properties and comparison with cellulose-NaOH and cellulose-NMMO routes. Carbohydr Polym 83:1766–1774

Sescousse R, Gavillon R, Budtova T (2011b) Wet and dry highly porous cellulose beads from cellulose-NaOH-water solutions: influence of the preparation conditions on beads shape and encapsulation of inorganic particles. J Mater Sci 46:759–765

Sheridan RL, Morgan JR, Mohamed R (2002) Biomaterials in burn, wound dressings. In: Severian D (ed) Handbook of polymeric biomaterials. Marcel Dekker, New York

Struszczyk H, Wawro D, Ciechańska D, Nousiainen P, Dolk M (1991) Direct dissolving cellulose: behaviour and properties. In: Cellulose '91 conference, New Orleans

Struszczyk H, Ciechańska D, Wawro D (1995) Fibres Text East Eur 3(1):47–49

Struszczyk H, Wawro D, Starostka P, Mikolajscyk W, Urbanowski (1997) A PL 188788 A "Method for fabrication of fibres, films and other products from cellulose", 20/11/1997

Struszczyk H, Wawro D, Starostka P, Mikolajscyk W, Urbanowski A (2000a) US 6,106,763 A "Process for producing cellulosic mouldings", 22/08/2000

Struszczyk H, Wawro D, Starostka P, Mikolajscyk W, Urbanowski A (2000b) EP 1317573 B1 "Process for producing fibres, film, casings and other products from modified soluble cellulose", 13/09/2000

Struszczyk H, Wesołowska E, Ciechańska D (2000c) Development of the biological utilisation of textile wastes. I. Scaling up of the biodegradation process. Fibres Text East Eur 8(2 (29)):74–78

Swatloski RP, Rogers RD, Holbrey JD (2001) WO 03/029329 A3, "Dissolution and processing of cellulose using ionic liquids", 31/07/2001

Thiele E (1901) German Patent 154 507

Tsioptsias C, Stefopoulos A, Kokkinomalis I, Papapdoupoulou L, Panayiotou C (2008) Development of micro- and nano-porous composite materials by processing cellulose with ionic liquids and supercritical CO_2. Green Chem 10:965–971

Voges M, Brück M, Fink H-P, Gensrich J (2000) The CarbaCell process – an environmentally friendly alternative for cellulose man-made fibre production. In: Proceedings of the Akzo-Nobel Cellulosic Man-made Fibre Seminar, Stenungsund

Vorbach D, Taeger E (1995) DE 19542533 "Highly shape-sensitive sensor material production method for humidity, stress or temperature sensor", 15/11/1995

Vorbach D, Taeger E, Schulze T (1999) DE 19910012 "Verfahren zur Herstellung von Formkörpern", 08/03/1999

Voss J (1961) Svensk Papperstidn 64:863–871

Wasserscheid P, Welton T (2008) Ionic liquids in synthesis, 2nd edn. Wiley-VCH, Weinheim

Wawro D, Stęplewski W, Bodek A (2009) Manufacture of cellulose fibres from alkaline solutions of hydrothermally-treated cellulose pulp. Fibres Text East Eur 17(3 (74)):18–22

Wendler F, Meister F, Montigny R, Wagener M (2007) Fibres Text East Eur 64–65:41–45

Wendler F, Kosan B, Krieg M, Meister F (2009) Macromol Symp 280:112–122

Woodings C (2004) Regenerated cellulose fibres. Woodhead Publishing Limited, Cambridge

Yamashiki T, Matsui T, Saitoh M, Okajima K, Kamide K, Sawada T (1990a) Br Polym J 22(2):121–128

Yamashiki T, Matsui T, Saitoh M, Okajima K, Kamide K, Sawada T (1990b) Br Polym J 22(1):73–83

Polysaccharide Fibres in Textiles

Lidija Fras Zemljic, Silvo Hribernik, Avinash P. Manian,
Hale B. Öztürk, Zdenka Peršin, Majda Sfiligoj Smole,
Karin Stana Kleinscheck, Thomas Bechtold,
Barbora Široká, and Ján Široký

Contents

7.1 Introduction 188
7.2 Characterisation of Cellulose Structure by Sorption Behaviour: Changes During Processing 192
7.3 Effects of Alkali Type on Cellulose Fibre Properties 192
7.4 Accessibility and Reactivity 194
7.5 Structural Assembly of Textiles and Its Influence in Processing Swollen State 195
7.6 Plasma Activated Regenerated Cellulose Fibres and Their Properties 196
7.7 Antimicrobial Functionalisation of Cellulose Fibres Using Polysaccharides 199
7.8 Functionalisation of Regenerated Cellulose Fibres by Magnetic Nanoparticles 203
7.8.1 Modification of Regenerated Cellulose Fibres by Magnetic Coatings 204
7.9 Spun-Dyed Cellulosics 205
7.9.1 Mass Colouration of Viscose or Cuprammonium Rayon 205
7.9.2 Mass Colouration of Lyocell 207
7.10 Outlook 207
References 208

T. Bechtold (✉)
Research Institute for Textile Chemistry and Textile Physics, University Innsbruck, Hoechsterstrasse 73, 6850 Dornbirn, Austria
e-mail: Thomas.Bechtold@uibk.ac.at

Abstract

Besides naturally grown cellulose fibres like cotton, hemp or flax, interest in textile fibres made up from regenerated cellulose is growing. By sure the use of a polymer material, which is provided by nature in huge amounts, favours its use as more sustainable material compared to oil-based products. However, a much stronger argument is the high variability of the properties that can be achieved, which allows design an extremely wide range of products.

In this chapter the main characteristics of textile fibres from regenerated cellulose are highlighted. Dependent on the production process, pore characteristics, accessibility and surface can be shaped. The chemical reactivity of the cellulose polymer and the swelling behaviour of the fibrous structure permit many chemical conversion processes towards specialised products.

Representative examples for fibre modifications are highlighted in this chapter, among them are fibre reorganisation during swelling processes, accessibility-controlled reactivity, plasma treatment for surface modification, antimicrobial functionalisation, deposition of magnetic nanoparticles and incorporation of pigments. The given examples demonstrate the diversity of processing strategies which all lead to unique products with specific functionality.

7.1 Introduction

Textiles belong to the group of consumer goods which are produced to fulfil a fundamental human need. Thus, the production of textile fibres represents a huge volume, with a strong tendency of growth. In 2009, the total volume of textile fibres produced per year can be estimated to 70.5 Mio tons. A schematic presentation of the production volumes of important textile fibres is given in Fig. 7.1. Different fibres contribute to the total volume, among them polyester (polyethylene terephthalate, PET) being the most important fibre in volume with 32.5 Mio tons out of the 41.5 Mio tons big segment of man-made fibres (anon 2010a; anon 2010b).

The most important natural fibre is cotton, with a total amount of 22–23 Mio tons harvested in 2009. Besides cotton, other cellulose-based natural fibres are used for textile clothing production and technical products, among them being flax, hemp, jute and sisal, however with much smaller volume. The reason for the comparable low production volumes of flax or hemp can be seen in their more complex and laborious production resulting in higher costs for flax and in the different properties of the stem fibre bundles.

Hemp has been tested for technical applications, for example, as reinforcing fibres in polymer extrusion. Besides the costs, a major drawback for a wider use of these natural cellulose fibres results from the existing limitation to modify and optimise fibre characteristics for a certain application.

At present the farmland capacity for growth of cotton faces more and more limitations, and the supply with high-quality water for irrigation of cotton plantations becomes difficult. Thus, increased attention has been directed to substitute cotton as textile raw material. The major part of growth in future fibre production can be expected to be covered by man-made fibres, at present mainly synthetic fibres, in particular polyester fibres.

While polyester fibres can be understood as cost-effective high-performance material, the unique moisture and water-related properties of cotton only can be substituted by use of cellulose-based fibres, so-called man-made cellulosics. Regenerated cellulose fibres are formed from α-cellulose-containing raw materials, for example, pulp or cotton linters. The total production of cellulosics in 2009 can be estimated with 2.95 Mio tons, which is ~13 % of the volume of cotton. While the present share seems rather small, the future potential of this class of fibres is enormous.

Several reasons can be formulated to demonstrate the future importance of these fibres:

– The plants (e.g. beech, eucalyptus) grow under less demanding climate conditions without the very large water consumption for irrigation and pesticide uses as required for cotton.
– Clean technologies for production of regenerated cellulose fibres have been developed during the past, among them the NMMO process, ionic liquids, and including the environmentally optimised viscose process (Schuster et al. 2003).
– The limited growth of cotton production only can be compensated by growth in regenerated cellulose fibres as farmland capacity is limited.
– The fibre-spinning technology delivers a wide range of modified fibres; thus, variability of products is extremely wide, and a high number of specialised products are already on the market or are expected to appear in the near future.

The excellent position of regenerated cellulose fibres for textile applications can be demonstrated by the comparison of basic fibre properties with cotton and polyester as competing materials (Table 7.1). While only limited space for variation in the polymer chain itself is possible for polyester and cotton cellulose can only be modified in the later textile chemical processing, there is an extremely high number of options to modify polysaccharide materials as raw materials or during fibre production.

Due to their porous structure, cellulose fibres are described best by the so-called pore model. Chemicals for modification of structure and derivatisation can penetrate the accessible parts of the fibres, which enables the textile chemists to produced numerous variations during textile

7 Polysaccharide Fibres in Textiles

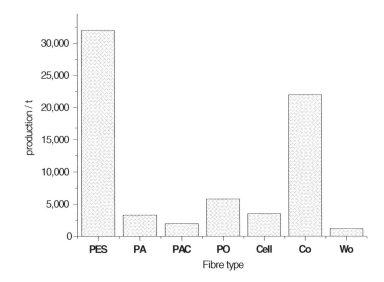

Fig. 7.1 Estimated total production volume for textile fibres in 2009. *PES* polyester, *PA* polyamide, *PAC* polyacrylonitrile, *PO* polyolefine, *cell* cellulosics, *Co* cotton, *Wo* wool (anon 2010b)

Table 7.1 Characteristics of polysaccharide (PS)-based fibres, compared to cotton and polyester fibres

Aspect	Fibres		
	Cotton	Polysaccharide fibres	Polyester fibres
Fibre polymer	Cellulose	Cellulose Alginate Chitosan Derivates (silyl ethers, acetates) Carbamate	Polyethylene-glycol-terephthalate Polylactic acid
Fibre spinning	Native	Viscose process Direct dissolution Dry spinning Melt spinning	Melt spinning
Fibre structure	Pore volume model (accessible pores in the fibre)	Pore volume model (accessible pores in the fibre)	Free volume model (glass transition temperature)
Characteristics	Hydrophil Swelling water Water sorption	Hydrophil/hydrophob Swelling in water Water sorption	Hydrophob Non swelling Only surface sorption
Reactivity	High Structure reorganisation Chemical reactions possible (dyeing finishing)	High Structural reorganisation Chemical reactions possible (dyeing finishing)	Low High structural stability
Variability	Medium	Extremely high	Low

chemical processing, like reorganisation by swelling operations, derivatisation or cross-linking (Jaturapiree et al. 2008; Bui et al. 2008). Besides chemical reactivity of the cellulose polymer, the swelling behaviour of cellulose-based fibres introduces another degree of freedom to achieve fibre structure modification and shaping.

Polyester fibres are described best by the free volume model. Chemicals, for example, dyes, can access into the fibre above the glass transition temperature where chain mobility is high. Chemical and structural modification is rather limited and not comparable to the highly reactive cellulose-based fibres.

For comparison of main types of fibres, their basic characteristics are shown in Table 7.2.

A common feature of all regenerated cellulose fibres is their hydrophilicity, which enables regenerated cellulose fibres to hold a unique position in the class of man-made fibres. At 60 % relative humidity and 20 °C, man-made cellulosics can absorb up to 10 % of water

Table 7.2 Comparison of important characteristics of cellulose fibres (Schuster et al. 2003)

Parameter	Unit	State	Viscose	Modal	Lyocell	Cotton
Degree of polymerisation	DP_w	–	200–250	400–600	600–800	2,000
Crystallinity IO	1	–	0.39	0.39	0.62	0.74
Water retention value	%	Wet	70–80	50–53	65–70	38–45
Void volume	ml/g	Wet	0.68	0.49	0.60	0.30
Inner surface	m^2/g	Wet	444	407	418	207
BET surface	m^2/g	Dry	276	277	246	132
Mean pore diameter	Å	Wet	31	23	28–30	28

while still providing the sensation of dryness, while synthetic fibres, for example, PET or PP, typically absorb less than 1 % moisture in such environmental conditions (Okubayashi et al. 2005). This high moisture buffering capacity of cellulose-based fibres makes them of interest to adjust sorption properties of a garment in order to improve wear comfort.

The relative high uptake of liquid water is of relevance for the production of textile structures designed to provide rapid sorbency and high-capacity intake of water (e.g. underwear, bed-sheets, tablecloth, hygiene applications).

Unfortunately, the high swelling of regenerated cellulose fibres also results in low dimensional stability of fabric and undesirable shrinkage during wet processing, for example, household laundry. In addition rather high amounts of water have to be removed during the drying step after a wet processing/household washing step, which results in comparably high energy costs for drying steps (Allwood et al. 2006).

In the field of technical textiles, cellulose fibres are useful for a wide range of applications. For the application as reinforcement material for tyres and other mechanical rubber goods like conveyor belts, the high uniformity, dry tenacity and dry modulus in combination with good temperature resistance makes the cellulose fibres an ideal material for these applications. An important further aspect is the good adhesion between the fibre surface and the rubber matrix after vulcanisation (Horrocks and Anand 2004).

Good absorbency properties of regenerated cellulose fibres make them of interest in the field of hygiene and medical applications. In particular, their suitability for processing by spunlace (hydroentanglement) techniques makes these fibres favourable for application in disposable cleaning and hygiene products (Wadsworth 2004). Examples for such products are caps, masks, cloths and wipes (Horrocks and Anand 2004).

At present, a number of cellulose fibre production technologies have reached technical scale. Besides the viscose technology, the NMMO process has reached substantial market volume. The need for replacement of petrol-based fibres by more sustainable materials using biopolymers, for example, polysaccharides, causes increasing activity to develop new routes for shaping of polysaccharides into fibres, with the most prominent processes at present being:

– The Celsol process which uses the solubility of cellulose in NaOH solutions (Struszczyk and Ciechanska 1998; Vehviläinen et al. 2008)
– The use of cellulose carbamate for fibre production from an aqueous alkaline solution (Ebeling and Fink 2009)
– The replacement of NMMO by ionic liquid systems
– The use of urea containing alkaline or non-alkaline solutions as cellulose solvents (Zhou and Zhang 2002; Tatarova et al. 2010)

The use of ionic liquids, NaOH solutions or modified cellulose derivatives like cellulose carbamate technology still is in an early stage of development. However, the existing technologies for cellulose fibre shaping already deliver fibres with specialised properties for particular textile applications.

Viscose fibres are produced by the xanthogenate route. The fibres show core-shell structure, high water absorbency and inner surface. These fibres are widely used both as staple yarns and filament yarn (viscose rayon) in home textiles, embroidery, clothing and underwear. The disadvantage is the low wet tensile strength of the fibres. The advantageous high water retention value makes such fibres of interest for the production of hygiene products, where also specialty fibres, for example, trilobal fibres, are used to achieve maximum water absorbency.

Flammability of viscose fibres could be overcome by incorporation of flame retardant like bis (2-thio-5,5-dimethyl-1,3,2-dioxaphosphorinyl) oxide (Sandoflam 5060, Sandoz) (Kampl and Six 1996) or polysilic acid (Link et al. 2005). The high water absorbency of viscose fibres is also advantageous for their use in geotextiles, where nonwoven cellulose fabric is used to entrap water for seed growth during dry periods (Greentex 2012).

Modal fibres are also produced via the xanthogenate route; however, retarded cellulose regeneration leads to all skin cellulose fibre with increased wet tenacity and lower swelling in alkaline solution. Modal and MicroModal fibres have a strong position in the field of textiles for apparel; in particular the softness and high water absorbency make them favourable for use in clothing worn next to the skin.

During the last 25 years, the production of textile fibres via the NMMO route (*N*-methylmorpholine *N*-oxide) gained importance. Wood pulp dissolves in NMMO monohydrate, and the fibres are then spun by extrusion of the cellulose solution via an airgap into a water/NMMO solution. This process is called lyocell. The lyocell fibres exhibit highly ordered cellulose chains and distinct fibrillar structure. Being a new fibre technology with a wide range of application, considerable research activity is still focussing on studies to modify fibre properties and develop new products. Activities include reorganisation of the fibres by swelling operations, chemical modification and cross-linking. Considerable amount of work was also addressed to study fibrillation, control of fibrillation during textile processing and prevention of fibrillation during consumer use.

The extraordinary, present and future potential of polysaccharides, in particular cellulose, as material for textile fibre production, is due to a number of unique properties:

- The bio-based material is available in huge amount and can be extracted by rather simple procedures.
- Chemical reactivity of polysaccharide polymers enables to synthesise fibres with a wide range of modifications.
- The accessibility and porosity of the three-dimensional polymer chain arrangement allows numerous structural modifications.
- The hydrophilic properties can be modified and varied to adjust the range of water absorbency.
- A wide range of processing routes already is available, providing materials with different properties.
- The combination with other fibres multiplies the variety of products.
- Additional modifications are possible during textile chemical treatments (e.g. alkalisation, cross-linking, specialty products).

Summarising, textile fibres based on regenerated cellulose offer an extremely wide range of product variations which can be accessed by selective processing strategies. Important examples for possible modification of a polysaccharide-based fibre product to optimise textile performance can be sorted following the material processing chain:

- Appropriated selection of polysaccharide materials
- Derivatisation of polymer chain, for example, carbamate
- Shaping of fibres
- Incorporation of functional materials
- Surface modification
- Material blending during textile processing
- Textile chemical modification, for example, alkalisation and cross-linking
- Structural assemblies/product design

In the following part of this Chap. 8, selected results of specialised research to modify cellulose fibre-based textiles are presented, to demonstrate the potential of this new 'old material' and to indicate directions of future progress both in research and technical application.

7.2 Characterisation of Cellulose Structure by Sorption Behaviour: Changes During Processing

During the manufacturing of cellulosic fibres into fabrics, the material passes through stages of pretreatments, colouration and finishing. These procedures entail a succession of wetting and drying steps. It was found that a repetition of wetting-drying treatments affect the pore characteristics of regenerated cellulosic fibres (Crawshaw and Cameron 2000; Kongdee et al. 2004). This phenomenon is well known as 'hornification', commonly defined as the relative reduction in water retention (WRV) compared to the original value due to drying (Diniz et al. 2004; Weise et al. 1996). Any structural changes in fibres due to cycles of wetting and drying will impact the accessibility of reagents and may also alter the reactivity of fibres. The effect of repeated wetting-drying cycles on the sorption properties of regenerated cellulosic fibres can be evaluated by measuring of the water retention, water vapour sorption and sorption of iodine. Water vapour and iodine sorptions characterise the accessible surface, while the interactions with liquid water represent the total swelling of substrates indicated by WRV. Wetting-drying treatment has been found to affect the sorption properties of lyocell, viscose and modal fibres and knit fabrics. Generally, the extent of changes in accessibility determined by different methods was found to be different for particular fibre type and to depend on substrate as well as sorbate type. There was evident influence of molecular and supramolecular structure of substrates through all experiments (Siroka et al. 2008).

Pretreatments such as desizing, scouring, bleaching, mercerising and ammoniating are applied to cotton fabrics in order to remove sizing from warp yarns, eliminate natural and foreign impurities, improve whiteness and increase tensile strength, dimensional stability, affinity for dyes and lustre (Corbman 1985; Lacasse and Baumann 2004). The sorption properties and complexation with iron of cotton fabrics treated at laboratory and technical scales by alkali and ammonia followed by bleaching and drying have been also studied. The obtained results show that although swelling treatments generally improve the level of accessibility in substrates, such improvements will not always translate into increased levels of sorption for all reagents (Jaturapiree et al. 2008).

7.3 Effects of Alkali Type on Cellulose Fibre Properties

In the textile industry, mercerisation is commonly defined as conversion of cellulose I structure of cotton to cellulose II, mainly achieved using concentrated sodium hydroxide solutions.

Mercerisation consists of three steps: swelling, decrystallisation during swelling and then recrystallisation by washing the mercerising agent from the cellulosic material (Vigo et al. 1969, 1970). Swelling of fibres was attributed to two effects. One is the electrostatic repulsion between the negatively charged carboxylate anions of fibres (Scallan and Grignon 1979, 1983), and the other is Donnan equilibrium theory. The latter theory tells that in concentrated alkali solutions, cellulose is expected to behave as a weak acid such that some of the OH groups of the solution would be deprotonated. The concentration of cations in the cellulosic structure would be higher than in the bulk solution to preserve electrical neutrality. The surplus amount of cations in the cellulosic material promotes an osmotic pressure differential and thus swells the cellulose (Donnan and Harris 1911). But Donnan theory was discarded by other researchers (Chedin and Marsaudon 1955, 1956, 1959; Warwicker 1967; Jeffries and Warwicker 1969; Warwicker and Wright 1967; Warwicker 1969). They mentioned that hydration of ionic species swells the cotton cellulose up to a threshold concentration, that is, interfibrillar swelling takes place, beyond which amount of water for hydration decreases resulting in decrease of degree of swelling. In this region the ionic species would then substitute three hydroxyl groups from cellulose chains to complete their hydration, that is, intrafibrillar swelling takes place. Denoyelle found that alkali mercerisation of native cellulose involves the formation of alcoholate, that is, –CHONa (7.1), during alkali treatment (Denoyelle 1959).

$$Cell-OH + NaOH \rightarrow Cell-O^-Na^+ + H_2O \quad (7.1)$$

Ranby explained mercerisation through thermodynamic analysis stating that transition from cellulose I to cellulose II is favoured through both entropy and enthalpy terms (Ranby 1952a). The effect of temperature on mercerisation was also found to be profound in that transition from cellulose I to cellulose II became less at elevated temperatures of rinsing step at even higher amounts of transitions happening through swelling at low temperatures (Jeffries and Warwicker 1969).

The conversion of cellulose I structure of hydrocellulose into cellulose II starts at 2.8 M and completed at 4.6 M NaOH. These threshold values for KOH were 2.7 M and 4.9 M, while they were 3.3 M and 4.9 M for LiOH (Dimick 1976). The threshold values of mercerisation depend on the origin of the cellulose samples in that cellulose samples with lower crystallinity, such as wood cellulose, would be mercerised at lower concentrations of alkali solutions than cellulose samples of higher crystallinity such as cotton linters (Dimick 1976; Ranby 1952b, c; Ranby and Mark 1955). Crystal lattice conversion from cellulose I to cellulose II was found to be in the order of LiOH < CsOH < RbOH < KOH < NaOH, while the swelling order was determined in the following order: CsOH < RbOH < KOH < NaOH < LiOH for cotton fibres (Vigo et al. 1969). By other researchers, the ranking for lattice conversion was proposed to be in the following order of: LiOH < NaOH < KOH analogous to the dissociation constants of the ions (Dimick 1976; Zeronian and Cabradilla 1972). KOH is known to dissociate completely in aqueous solution, NaOH is slightly associated, and LiOH dissociates the least among them (Dimick 1976; Gimblett and Monk 1954; Douglas and McDaniel 1965).

The presence of OH^- ions was reported to be responsible for the decrystallisation step of mercerisation. At equal hydroxyl concentrations, the lattice conversion was in the order of Li > Na > K analogous to the order of their hydration numbers which indicates that Donnan mechanism can also explain the decrystallisation step in addition to the presence of OH^- ions. As the hydrated cation size gets bigger, greater osmotic pressure exerts disrupting the crystal lattice. The cations were expected to retain their number of hydration spheres even though they may associate or form ion pairs with OH on cellulose (Tan et al. 1975). Donnan theory explains the swelling of cellulose but does not predict the threshold values for mercerisation (Dimick 1976). In conclusion, the most important parameter in the decrystallising process is said to be the activity of the OH^- ion since OH^- ions have the ability to penetrate and disrupt cellulose I only after their hydration spheres are depleted. Cations are said to play a second role which can be explained by Donnan theory since decrystallising is explained by both hydrogen bond disruption and swelling pressure.

During the alkali treatment of regenerated cellulosic fibres, fibre diameter, alkali retention value (ARV) and splitting test were conducted. Depending on alkali type, ARV, fibre diameter and split number decreased in the following way: TMAH > LiOH ≥ NaOH > KOH for lyocell and TMAH > NaOH ≥ LiOH > KOH for viscose and modal fibres (Öztürk et al. 2006a, b; Öztürk and Bechtold 2008).

After alkali treatment of fibres, water retention value, accessible pore volume, carboxyl content, colour strength after dyeing with C.I. Direct Red 81, tensile properties and abrasion resistance of fibres were measured. The threshold concentrations for intrafibrillar swelling of lyocell fibres were suggested as 1.5 M TMAH, 2 M KOH, 2.5–3 M NaOH and 5 M LiOH (Öztürk and Bechtold 2007; Öztürk et al. 2009, 2010a, b).

LiOH, NaOH and TMAH treatments increased the accessibility of water, moisture, C.I. Direct Red dye into the viscose and modal fibres by not only increasing the pore volume of the fibres but also modifying the lobal fibre cross-section into rounder cross-section. Loosening of the fibre structure after alkali treatments caused an increase in elongation at break but a decrease in the tensile strength for viscose and modal fibres. The effect of alkali types (LiOH, NaOH, TMAH) on properties of viscose and modal fibres was observed only for water retention value and colour strength (K/S) (Öztürk 2008).

7.4 Accessibility and Reactivity

Cellulosic substrates are subjected to swelling treatments in alkali solutions to improve their lustre, dimensional stability, wrinkle resistance and/or dyeability. The most common method is mercerisation with sodium hydroxide (Mercer 1850), although mercerisation with liquid ammonia (Bredereck and Commarmot 1998; Coats and Gailey 1968) is also practised to limited extents. Cellulosics swell strongly in alkali solutions (Zhang et al. 2005a) leading to changes in substrate properties such as pore structure (Kasahara et al. 2004; Bredereck et al. 2003), fibrillation tendency (Zhang et al. 2005b), crystallinity (Colom and Carrillo 2002) and surface characteristics (Persin et al. 2004) in fibres and yarn crimp and stiffness in fabrics (Ibbett and Hsieh 2001). Such swelling treatments lead to alterations in the accessibility and reactivity of cellulosic substrates to chemicals and reagents in subsequent treatments (Tatarova et al. 2011).

The swelling process and resulting changes in sorption and accessibility of cellulosics can be monitored with a number of methods. The degrees of swelling may be estimated from measurements of fibre diameter or from the amount of water retained in wet substrates after centrifugation at high speeds, termed water retention value (WRV). Changes of the pore structure in substrates may be estimated with inverse size exclusion chromatography (ISEC) and by monitoring the sorption and diffusion of fluorescent agents through fibre structures. The degrees of reagent sorption in substrates (salts, alkali, ions) may also be used as measures of substrate accessibility. In addition, the impact of prior swelling treatments on subsequent treatment steps such as enzymatic hydrolysis, cross-linking treatments also provide insights into the changes in substrate properties.

Distinct differences are observed between cotton and viscose fibre swelling in solutions of LiSCN, NaOH and their mixtures (Ehrhardt et al. 2007). The swelling extents, estimated from fibre diameters, are generally lower in Co compared to viscose and increase with alkali concentration in mixtures for both fibre types. Increasing salt concentration in mixtures, however, promotes swelling in Co fibres but exerts no influence on viscose fibre swelling. The greater swelling in viscose may be attributed to their lower degree of polymerisation and crystallinity/order as compared to Co. The differences in the effect of salt on swelling between the fibre types perhaps result from the differences in their supramolecular structure. The regenerated viscose fibre, being of the cellulose II type, is characterised by a higher density of intermolecular hydrogen bonds compared to Co, which is of the cellulose I type.

The swelling of another regenerated cellulosic, lyocell, in alkali solutions can cause fibre splitting. Interesting differences are observed when comparing the splitting propensities of the fibre with different alkalis. The fibre exhibits little to no splitting with any alkali above concentrations of 5 M, but at concentrations below that level, the fibre splitting decreases in the order of alkalis: tetramethyl ammonium hydroxide (TMAH) > NaOH $\geq$ LiOH > KOH. The high fibre-splitting propensities in alkali solutions of concentrations less than 5 M are attributed to the build-up of localised stresses within fibre structures as a result of nonhomogenous fibre swelling due to non-uniform alkali distribution within fibres (Öztürk and Bechtold 2008). It is believed that the strong sorption of alkali in fibres from solutions of concentrations greater than 5 M leads to more uniform distribution of alkali within fibres and therefore to more homogenous fibre swelling (Öztürk et al. 2006b, 2011). The greater homogeneity in fibre swelling results in reduced stress within fibres, which leads to reductions in fibre-splitting propensities.

In correlating the measured WRV with total pore volumes obtained with ISEC measurements on lyocell fibres treated with solutions of different alkalis over a range of concentrations, the WRV was consistently greater than the corresponding total pore volumes by ~10 % (Jaturapiree et al. 2006). The WRV represents the total volume of water in water-swollen fibres, while the total pore volume represents the volume of water-swollen fibres accessible to reagents. Hence, it appears that 10 % of the volume in alkali-treated water-swollen fibres may be regarded as a 'nonreactive' fraction.

Significant differences were observed in the accessibility of salts in lyocell and viscose fibres immersed in salt-alkali mixtures (Jaturapiree et al. 2006). In comparisons between NaCl and KCl, the Na^+ salt exhibited greater accessibility than the K^+ salt. Such differences may play a significant role in chemical reactions involving cellulosic fibres and other chemical reagents, such as cross-linkers, by influencing the overall extent and efficiency of reactions. Evidence of this is observed when lyocell-woven fabrics pretreated with NaOH and KOH solutions are subsequently subjected to a pad-dry-cure cross-linking treatment (Manian et al. 2006, 2008a; Jaturapiree et al. 2011). The alkali pretreated fabrics exhibit greater strength losses after cross-linking as compared to samples cross-linked without pretreatment. Differences were also observed in cross-linker distribution between samples cross-linked after and without alkali pretreatment, where the pretreated samples exhibited a greater surface concentration of the cross-linker, while the non-pretreated samples exhibited a more uniform distribution of cross-linker. It is believed that the differences in performance between substrates cross-linked after and without alkali pretreatments could be attributed to enhanced substrate accessibility of catalyst after pretreatments.

When the internal pore structure and accessibility of lyocell, viscose and modal fibres are assessed from the diffusion and distribution of a fluorescent whitening agent across their cross-sections, it is found that the accessibility/porosity at fibre centres is significantly lower than at the peripheral regions (Abu-Rous et al. 2007). Lyocell fibres appear to have greater accessibility/porosity through the fibre bulk as compared to viscose and modal, while viscose fibres exhibited greater porosity/accessibility at peripheral regions than modal fibres. In another study (Jaturapiree et al. 2011), it was found that the swelling treatments of lyocell fabrics with NaOH and KOH solutions cause increases in their porosity/accessibility, but there are qualitative differences in the changes caused by the two different alkali types. The diffusion of NaOH appears to reach greater depths within fibre structures thereby resulting in porosity/accessibility changes at greater depths within fibres. In comparison, the diffusion of KOH within fibres appears limited to shallower depths thereby limiting changes in porosity/accessibility to fibre peripheral regions.

The pretreatment of lyocell fabrics with NaOH solutions is observed to increase the degree of substrate degradation in subsequent enzymatic hydrolysis steps (Schimper et al. 2009). The increased degree of degradation is most evident on substrates pretreated with NaOH solutions at concentrations in the region where cellulosics exhibit the maximum degree of swelling and solubility. Enzymatic hydrolysis of cellulosics involves the sorption of high-molecular-weight proteins (the enzymes) onto accessible regions in cellulose. Hence, pretreatments of cellulosics with alkali concentrations in the region of their swelling/solubility maxima will significantly enhance accessibility in fibre structures and thereby promote the sorption of enzymes onto the substrate.

7.5 Structural Assembly of Textiles and Its Influence in Processing Swollen State

One of the basic research scopes in the cellulose chemistry and physics is the study of interactions of cellulose with alkali aqueous solutions. Effect of alkali treatment is complex and has a substantial influence on morphological, molecular and supramolecular level, causing changes in crystallinity, pore structure, accessibility, stiffness, unit cell structure and orientation of fibrils in cellulosic fibres. Factors such as the concentration of NaOH, treatment temperature, applied tension, residence time, source of cellulose, physical state of cellulose (fibril, fibre, yarn or fabric) and degree of polymerisation have an effect on the properties and degree of change upon treatment (Manian et al. 2008; Jaturapiree et al. 2006; Colom and Carrillo 2002; Heinze and Wagenknecht 1998). Therefore, pilot-scale experiments were conducted that simulated full-scale treatments to study the influence of various process parameters on the properties of lyocell fabrics treated with sodium hydroxide in continuous processes (Široký et al. 2009, 2010a, b). The process parameters of interest were alkali concentration (varied from 0.0, 2.53, 3.33, 4.48, 4.65 to 7.15 M), temperature (25 °C or 40 °C), applied tension on the fabric (49 N/m or 147 N/m), fabric type (plain-, twill- or sateen-woven) and duration of the treatment. All of these variables were adjusted and controlled in treatment compartment of the continuous process which consists of four different stages, treatment, stabilisation, washing and neutralisation.

When lyocell fabric (cellulose II-based polymer) undergoes treatment with aqueous sodium hydroxide, the most substantial changes in properties occur under the same process conditions, depending upon alkali concentration, temperature and applied tension (Široký et al. 2009, 2010a, b). This was evidenced by physicochemical properties of plain-woven fabrics as follows (Široký et al. 2009): peaks of fabric shrinkage were observed in samples treated with 3.33 M NaOH at 25 °C and with 4.48 M NaOH at 40 °C. The shrinkage peaks corresponded to peaks in flexural rigidity (in dry and wet state) and minima in water retention, crease recovery and breaking force, leading to the conclusion that the greatest molecular reorganisation of cellulose II fibres corresponds to the conditions where the maximum of crystalline conversion occurs.

Later (Široký et al. 2010a), it was proposed that at the maximum of swelling (crystalline conversion to amorphous phase) at certain alkali concentrations and temperatures (peaks with 3.33 M NaOH at 25 °C and with 4.48 M NaOH at 40 °C), the greatest reorganisation in the amorphous and quasi-crystalline structure is observed. Also, the conversion of cellulose II to Na-cellulose II takes place and causes changes in number and size of voids, cracks and pores within the fibre as well as size of crystals within the crystalline structure and is influenced by the alkali treatment parameters, such as alkali concentration, temperature and tension applied.

Study of dye sorption by hydrolysed reactive dyes onto lyocell fabrics (Široký et al. 2010b) revealed that it is very thermodynamically favourable for substrates treated with 2.53 to 3.33 M NaOH. Also within the same research, it was concluded that pores in the fibre are significantly affected by alkali treatment (<20 Å diameter) and the accessibility of dye (14 Å) sorption into those pores accounts for the differences observed herein; maximum equilibrium concentration of sorbate on the sorbent (q_e), theoretical monolayer capacity (q_0) and adsorption energy (ΔG^0) are observed for cellulose II fibre treated with 2.53 to 3.33 M NaOH as this concentration range affects the greatest increase in accessible pore volume (APV).

7.6 Plasma Activated Regenerated Cellulose Fibres and Their Properties

The pretreatment and finishing of textiles by nonthermal plasma technologies becomes more and more popular as a surface modification technique (Shishoo 2007). Furthermore, they are particularly suited to textile processing because most textile materials are heat-sensitive polymers. Moreover, nonthermal plasma surface modifications can be achieved over large textile areas without affecting the bulk properties.

Physically, plasma modification is an ionised gas treatment. It is a high energy state of matter, in which a gas is partially ionised into charged particles, electrons and neutral, often highly excited molecules and atoms. The surface activation of polymers is carried out by exposure to non-polymer-forming plasma, such as O_2, N_2, NH_3 and the inert gases (Mozetič 2007). The treatment by energetic particles breaks the

7 Polysaccharide Fibres in Textiles

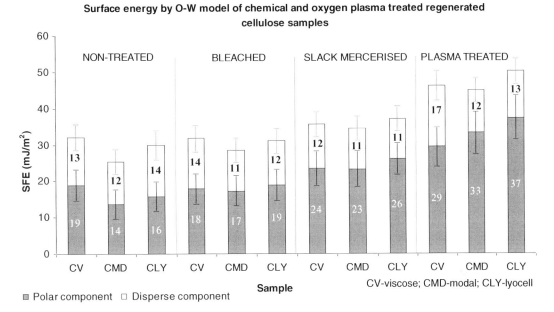

Fig. 7.2 Overall, polar and disperse components of the surface energies (SFE) of standard chemical and oxygen plasma-treated regenerated cellulose fabric samples. The absolute values were determined by Owens–Wendt–Rabel–Kaeble (O–W) calculation method

covalent bonds at the surface, leading to the formation of surface radicals. The latter can react with the active plasma species to form different chemically active functional groups at the surface. Surface contaminants and weakly bound polymer layers can dissociate into volatile side products, which can be pumped away.

Plasma technology provides activation with the required quality standards in terms of speed, homogeneity, process stability and efficiency (Holländera et al. 2004). Depending on the type and nature of the gases used, a variety of surface modifications can be achieved.

Of all plasma effects, the increasing of the hydrophilic character is without a doubt the most studied (Holländera et al. 2004; Akishev et al. 2003; Brüser et al. 2004; Yaman et al. 2009; Fras Zemljič et al. 2009a, b) because it influences many plasma applications. The treatment aims at the introduction of water compatible functional groups such as –COOH, –OH and –NH$_2$ using oxygen (O$_2$) and/or nitrogen (N$_2$) gas. Oxygen plasma active species attack the natural polymer surfaces and cause the incorporation of hydrophilic groups, that is, polar groups like carbonyl, carboxyl and hydroxyl.

The reactivity increases not only due to the formation of new reactive functional groups but also due to the activation of remaining hydroxyl groups, if present. Therefore, the moisture regain (Hwang et al. 2005), as well as the surface energy, are the parameters that are clearly affected by this plasma treatment. In Fig. 7.2, the effects of standard chemical and oxygen plasma treatments, as applied to regenerated cellulose fabrics, are presented. The effects were evaluated by means of surface energy using Owens–Wendt–Rabel–Kaeble (O–W) calculation method. The results showed that bleaching had little effect on the surface energy; treatment with alkali increased the surface energy, while oxygen plasma treatment resulted in a very significant effect. The hydroxyl groups were those that were affected by different treatments, the most by plasma activation (Fras Zemljič et al. 2009a). Their increased interaction capability to form hydrogen bond with liquids used increased the absolute values of polar components resulting in augment of the samples surface energy.

Highly reactive surface functional groups as created by plasma treatment are unstable; therefore, for an effective wettability enhancement,

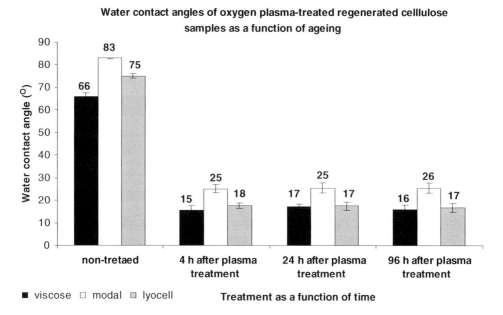

Fig. 7.3 Water contact angles for non-treated and oxygen plasma-treated fabric samples as a function of ageing (Peršin et al. 2008)

increased attention should be given to the permanence of the plasma effects (Akishev et al. 2003, 2000; Temmerman and Leys 2005; De Geyter et al. 2006; Morent et al. 2007). Figure 7.3 shows the water contact angles measured on regenerated cellulose fabrics, as a function of plasma-activation-effect time dependency. Results indicated no significant changes in water contact angles in 24 and 96 h after plasma activation (Peršin et al. 2008).

Nonequilibrium gaseous plasma could be used in order to introduce the desired type and quantity of reactive functional groups at the fibre surface (Morent et al. 2007; Bullett et al. 2004; Vesel et al. 2009) and to change, to a certain extent, the structural and surface properties. These changes are responsible for better mechanical bonding in composites (Tissington et al. 1992; Sheu and Shyu 1994; Morales et al. 2006).

Two types of plasma processing can be distinguished for improving dyeing and printing properties and used on all fibre types: depositing plasmas or nondepositing plasmas. In the former, the required functional groups are part of a coating that is plasma polymerised in situ at the fibre surface. Nondepositing plasmas (e.g. O_2 plasma) introduce functional groups that interact with an (ionic) dye molecule directly at the fibre polymer surface by chemical reaction or alter the hydrophobic character of the fibre surface to improve diffusion of (ionic) dye molecules (Sarmadi and Kwon 1993; Vesel et al. 2007).

The plasma treatment may well be used for introducing certain functional groups via a coating or a graft copolymer, for removing hydrophilic functional groups or for changing hydrophilic groups into non-hydrophilic ones. Therefore, the most straightforward way for achieving the hydrophobicity and/or oleophobicity is the treatment of a fabric in a nondepositing gas which grafts (or exchanges) single-fibre polymer atoms with hydrophobic groups such as fluorine groups (Krentsel et al. 1994, 1995; Yasuda et al. 1994; Mukhopadhyay et al. 2002a, b; Molina et al. 2005; Ho et al. 2007; Bongiovanni et al. 2007; Shen and Dai 2007).

Plasma may possibly also be used as a means to deposit nanometre thick layers of (in) organic or metallic material at a fibre surface. Several methods of plasma-assisted deposition on textiles have been proposed, for example, thermal evaporation, sputtering or anodic arc evaporation at low pressure, monomers in gaseous or aerosol phase deposited as a three-dimensional and cross-linked polymer network at the substrate surface. The latter, also called plasma polymerisation, could replace the traditional wet chemical treatment (Goodwin et al. 2002; Castelvetro et al. 2006).

When using a plasma treatment for surface cleaning, the proper selectivity of the plasma for the material to be removed from the fibre surface is required since the fibre structure itself should not be damaged (Riccobono 1973).

Plasma can be an efficient sterilising tool via vacuum pressure, highly energetic UV light and/or reactive plasma species. Plasma technology on textiles offers different biomedical applications, for example, as a barrier against microbes (Fras Zemljič et al. 2009a, b; Sodhi et al. 2001; Yuranova et al. 2006; Gorjanc et al. 2010).

Plasma treatment with efficient physical removal of fibre surface polymers could be specifically used for the creation of microscopic pits in order to, for example, decrease light reflection at the dyed fibre surface (Byrne and Brown 1972) and to achieve an acceptable whiteness (Prabaharan and Carneiro 2005). Using physical and chemical etching process conditions, the frictional properties of fibres may well change (Thorsen and Kodani 1966; Thorsen 1971, 1974). Changing the physical properties of yarns and fabrics (Abbott and Robinson 1977a) and in particular their increase in tensile strength (Sun and Stylios 2004, 2005, 2006), bursting strength (Sun and Stylios 2005; Abbott and Robinson 1977b) and wear resistance were among the first positive effects caused by a plasma treatment on a textile structure. Nonequilibrium gaseous plasma could introduce functional groups at the fibre surface which, by hydrogen bonding with atmospheric water, decrease the resistivity of the treated textile (Rashidi et al. 2004).

Plasma application for improving the efficiency of wet-finishing processes shows advantage on two levels (a) due to improved wetting properties of the textile product (Yoon et al. 1996; Wong et al. 2000) where less or no chemical wetting agent is needed and (b) because of an improved interaction between finishing product and fibre surface (Rakowski 1989) resulting in less finishing product used for identical properties (better bath exhaustion) (Zuchairah et al. 1997). Nevertheless, apart from the working gas and treatment conditions (Wakida et al. 1989), the fibre type also plays an important role in plasma modification processes (Malek and Holme 2003; Chen 1991).

The shift to high-functional, high-added value and technical fibrous materials (http://www.suschem.org, 5.4.2012; http://www.forestplatform.org 5.4.2012; Persin et al. 2011; http://etp.ciaa.eu, 5.4.2012; http://www.planttp.com, 5.4.2012; http://www.etcgroup.org, 5.4.2012; http://www.qub.ac.uk/pprc/, 5.4.2012) is deemed to be essential for substantial growth of textile industry in Europe, North America and other developed countries. The growing environmental and energy-saving concerns (anon 2008) will also lead to the gradual replacement of much traditional wet-chemistry-based processing, using large amounts of water, energy and effluents, by various forms of low-liquor and dry-finishing processes. Therefore, the anticipated message of this chapter is to give reader and/or potential users an overview of the advantages related to plasma technologies, with particular emphasis on their potential uses in the textile industry. It is in this context that plasma technology for textiles, when developed at a commercially viable level, will have a strong potential to bring in an attractive way new functionalities in fibrous materials in an ecologically friendly way.

7.7 Antimicrobial Functionalisation of Cellulose Fibres Using Polysaccharides

Over the past decade, the cationisation of polysaccharides in order to obtain antimicrobial properties has become recognised as a very important contribution to novel antimicrobial substances, due to the biodegradability and biocompatibility of such materials (Baumann et al. 1991; Culler et al. 1979; Hattori et al. 1998; Berscht et al. 1994; Berger et al. 2004; Kaputskii et al. 2005; Hoenich 2006; Muzzarelli 2009). Thus, cationic functional polysaccharides (CAT-PS) are very promising since they possess a wide antimicrobial activity useful for many applications, of which cosmetic, pharmacy, health, packaging products and medical textile are only a few examples (Lim and Hudson 2004; Muzzarelli et al. 1994; Wu and Kuga 2006). Another example is the interaction of positively charged polysaccharides with negatively charged proteins, which can be used for inactivating those enzymes responsible for the generation of malodorous compounds, thus formulating consumer products (Pecse et al. 2005; Ravi Kumar 2000).

It is clear that the antimicrobial activity of cationic polysaccharides is correlated with the density of the cationic amino groups, which can be, therefore, used as a guideline for designing antimicrobial substances (Baumann et al. 1991; Culler et al. 1979; Hattori et al. 1998; Berscht et al. 1994; Berger et al. 2004; Kaputskii et al. 2005; Hoenich 2006; Muzzarelli 2009; Lim and Hudson 2004; Muzzarelli et al. 1994; Wu and Kuga 2006; Pecse et al. 2005). Another example is the interaction of positively charged polysaccharides with negatively charged proteins, which can be used for inactivating those enzymes responsible for the generation of malodorous compounds, thus formulating consumer products (Pecse et al. 2005; Ravi Kumar 2000).

In the field of developing new medical textiles and devices from natural cellulose fibres, consumers and producers have extensively expressed the need for incorporating a biofunctional product within cellulose fibres, in order to increase their safety and to obtain bioactive properties. Thus, the use of cationic polysaccharides for cellulose fibre functionalisation when developing medical materials has recently gained considerable attention, as emphasised by numerous reviews on the topic (Kaputskii et al. 2005; Hoenich 2006; Kotel'nikova et al. 2003; Czaja et al. 2006). Among the various cationic polysaccharide products, chitin and its derivate chitosan are currently the most promising and useful antimicrobial coatings for cellulose fibres. Chitosan is a natural, renewable resource, a polycationic biopolymer (Fig. 7.4) that possesses a well-documented and wide spectrum of biological activity, including antibacterial and antifungal activities (Muzzarelli et al. 1994).

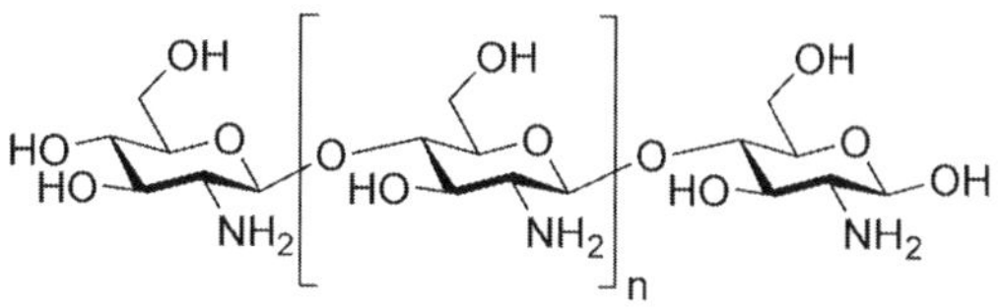

Fig. 7.4 The structure of chitosan (Muzzarelli et al. 1994)

Therefore, chitosan coatings offer many advantages over the traditional treatments of cotton and regenerated cellulose fibres because of their nontoxicity, biodegradability and biocompatibility.

As already mentioned, it is clear that for biomedical application, the charged cationic groups of functionalised cellulose materials are of high importance. Therefore, it is essential to understand their protonation behaviour in detail (positive-charge density as a function of pH and the ionic strength of the aqueous medium). It is only in this way that the influence of the amount of amino groups onto final fibre antimicrobial properties may be understood. For achieving this aim, it is extremely important to establish a combination of different precise analytical techniques in order to detect amino groups in such a complex, heterogeneous system as cellulose fibres coated with chitosan. In previous research from Fras Zemljič et al. (2009b), it was shown that combining wet techniques for the determination of accessible amino groups with XPS results gives a consistent picture of the amount of amino groups responsible for the antimicrobial character of cotton fibres functionalised with chitosan. Good agreement was obtained between the amount of amino groups determined by the spectrophotometric Acid Orange method and polyelectrolyte titration. This work also showed that a combination of these two techniques with XPS allows for the obtaining of a better and more detailed understanding of the content and distribution of amino groups onto cellulose (cotton) functionalised by chitosan. In addition to the above-mentioned techniques, it has become clear that conductometric titration and Kjeldahl analysis are also very suitable for determining the content of amino groups present in viscose cellulose fibres coated by chitosan (Fras Zemljič et al. 2009a).

However, in order to examine in detail the influence of amino groups on fibre antimicrobial activity, a more precise study by means of determining the pK value is of great importance. pK is not only important for classifying acid strength but it determines also the properties of a substance in nature or its possible use as an antimicrobial agent or drug. Determination of the pK value is, therefore, of great importance for biomedical applications as, for example, medical textiles in wound healing, gynaecological treatment, skin infection treatment, etc. Čakara et al. studied (Čakara et al. 2009) the protonation of cotton fabric with irreversibly adsorbed medical chitosan (cotton-chitosan) in an aqueous medium at 0.1-M ionic strength by means of potentiometric titrations and compared these results with the results obtained for only pure cotton and chitosan. For cotton-chitosan, the charging isotherm exhibits a charge reversal around pH $\cong$ 6.0, which is identified as the point of zero charge (PZC). The pure chitosan and the acid fraction that is present in cotton protonate according to the one-pK model, with $pK_{CT} = 6.3$ and $pK_{CO} = 4.7$, respectively. At pH > PZC, the charge of the acid fraction in cotton-chitosan is negative and constant, and the proton binding is attributed purely to the adsorbed chitosan. On the other hand, the cotton-bound acid exhibits a more complex protonation mechanism in cotton-chitosan than in the pure fabric, which is evidenced as an excess positive charge at pH < PZC and a deviation from the one-pK behaviour. Čakara et al. also showed that the irreversible adsorption of chitosan onto weakly acidic cotton fabric is predominately driven by a non-electrostatic attraction. Myllyte et al. (2009) also evidenced a non-electrostatic interaction between chitosan and cellulose at low pH. This may be attributed to specific structural interaction between chitosan and cellulose (H-bonds and hydrophobic interactions).

Establishing appropriate methods which are precise and repeatable enough for monitoring the influence of amino group quantity and quality (acidic strength) onto final cellulose fibre antimicrobial properties is extremely vital for creating future potentially advanced applications of fibres functionalised by chitosan or any other cationic polysaccharide.

Fras et al. examined the influence of chitosan concentration as a viscose fibre coating onto amino group content by conductometric titration and Kjeldahl analysis. It has been shown that viscose fibres treated with 0.5 % chitosan solution in lactic acid and further dried in air chamber contain 20 % less amino groups than the same fibres treated with 0.8 % of chitosan solution. When chitosan concentration in the solution is increased to 1 %, the fibre amino group content increased by around 40 % when compared to the sample treated with 0.8 % chitosan solution and even more than 80 % when compared to the sample treated with 0.5 % chitosan solution. Moreover, the antimicrobial activity against the microorganisms of viscose fibres treated with chitosan-lactic acid solution linearly increased with increasing amounts of amino groups in the fibres. The same trend was obtained in previous research from Fras Zemljič et al. (2009a) where the influence of low-pressure oxygen plasma treatment on the functionalisation of cellulose (viscose) material using chitosan (1 % solution) was investigated by conductometric titration and XPS. The effect of plasma treatment on the antimicrobial activity of chitosan-treated (impregnated) fabrics was also investigated. Chitosan adsorbed onto plasma-treated fabric was substantially more active as an antimicrobial agent for the investigated pathogen microorganisms. The plasma treatment resulted in a higher degree of chitosan adsorption onto fabric surfaces and consequently a higher amount of amino groups which are responsible for antimicrobial activity. Because the number of amino groups in plasma chitosan-treated samples is higher (conductometric titration, XPS) than for non-plasma chitosan-treated samples, the probability that a protonated amino group meets the bioplasm of bacteria would increase, resulting in a greater reduction capacity (Fig. 7.5).

In spite of chitosan excellent antimicrobial activity, it shows poor antioxidant properties (Ravi Kumar 2000). However, such properties

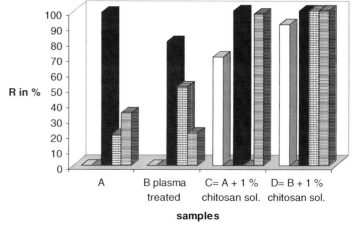

Fig. 7.5 Antimicrobial properties of viscose non-treated and chitosan-treated (impregnated) fabrics, previously nonactivated and previously plasma activated, respectively. *R* in % is the reduction of the organism

are significant when developing innovative biomaterials for medical devices, such as bioactive dressings and wound-healing isolation materials, or a bioactive material to be used in, for example, food packaging (Lim and Hudson 2004). For these applications it is essential to provide, besides antimicrobial inhibition, a reduction in those reactive oxygen species that are strongly implicated in the pathogenesis of, for example, wounds (Moseley et al. 2004), causing injury with biomolecules such as lipids, proteins and nucleic acids, as well as a depleting of mitochondrial DNA from human skin (Briganti and Picardo 2003), and/or, for example, retarding the degradation reactions of fats and pigments in food products (Lund et al. 2007). Kokol et al. studied the protonation behaviour of chitosan-cellulose (electrospunned) fibres grafted by two chemically similar phenolics, fisetin and quercetin (Fras Zemljič et al. 2011a).

Potentiometric titration was used to determine fibre-dissociable weak acids as a consequence of chitosan presence and deposited flavonoids. In addition, a conventional spectrophotometric method using C.I. Acid Orange 7 dye was used for determining amino groups only. Finally, and most importantly, the antioxidant and antimicrobial properties of fibres were estimated, respectively. It was clearly shown that the introduction of antioxidant onto chitosan-based fibres introduces a significant amount of anionic phenolic OH groups, leading to fibre antioxidant activity. The latter was also very much influenced by the determined pK values of weak phenolic acidic OH groups. Moreover, the attachment of antioxidants onto the chitosan cellulose-based fibres blocked accessible chitosan-amino groups, and, therefore, the inhibition of gram-positive bacteria *S. aureus* and fungus *C. albicans* is reduced, but still present. The reduction of *C. glabrata* is slightly increased after fibre treatment with quercetin, indicating that the synergistic activities of antioxidants and polysaccharide chitosan positively influenced the inhibition of this microorganism. The obtained knowledge may point towards further research activities, combining chitosan and flavonoid biomolecules, in order to develop new materials with targeted antioxidant and antimicrobial properties.

Unfortunately, the application of chitosan as a cellulose fibre coating is still limited, due to its action and solubility under acidic conditions, as well as its difficulties during purification being vital for high-value products and high-level (measuring precision and repeatability) analytical support. Therefore, an appropriate substitute

Table 7.3 The amount of dissociable groups in viscose fibres impregnated by chitosan and by aminocellulose, respectively

Sample	Positive charge [mmol/kg]	Negative charge [mmol/kg]
Reference material	0	30
Viscose fibres impregnated by chitosan	7.1	21.7
Viscose fibres impregnated by aminocellulose	57.5	4.0

for chitosan is required for specific applications, from among which novel soluble aminocellulose derivatives may be ideal. The synthesis of aminocellulose was investigated systematically, and different procedures may be used giving different quantities and qualities (primary, secondary, tertiary, etc.) of amines (Klemm et al. 1998).

Fras et al. applied water soluble 6-deoxy-6-(2-aminoethyl)aminocellulose (DAEAC) synthesised by Heinze et al. (Fras Zemljič et al. 2011b) onto the cellulose viscose fibres. This product has a well-defined structure and purity degree and contains primary and secondary amines on the same glucose unit. Fibres functionalised by DAEAC (in 1 % solution) were compared to fibres impregnated by chitosan solutions under the same conditions. The amount of weak dissociable groups (carboxyl groups and amines) was determined by potentiometric titration (see Table 7.3).

Positive charge is a consequence of protonated amino groups, while negative charge is a consequence of deprotonated carboxyl groups. It is obvious that aminocellulose coatings introduce a much higher amount of amino groups to the fibres. A lower amount of carboxyl groups was detected in samples impregnated by aminocellulose due to more expansive ionic bond formation between the carboxyls of the fibres and the amino groups of aminocellulose.

Moreover, because the number of amino groups in aminocellulose-treated samples is higher than for chitosan-treated samples, the probability that a protonated amino group meets the bioplasm of bacteria would increase, resulting in a greater reduction capacity for pathogenic bacteria and fungi. Thus, the antimicrobial properties of aminocellulose-treated fibres were much higher than for chitosan-treated fibres.

7.8 Functionalisation of Regenerated Cellulose Fibres by Magnetic Nanoparticles

Addition of magnetic particles to fibres results in fibrous material that possesses magnetic properties, magnitude of which depends on the properties of magnetic component used. Magnetic layers can impart conductivity and magnetic character to a starting material, hence making it useful for protection against static electricity charge, for protection against electromagnetic radiation and for filtration of magnetic particulate matter.

There has been some research on manufacturing of magnetic paper by incorporation of magnetic particles into pulp fibres, used in paper-making processes. Marchessault et al. (1992) patented a procedure for preparation of fibres with magnetic properties. Fibres with incorporated magnetic particles can be either obtained by inclusion of an inorganic magnetic component (Fe_2O_3, Fe_3O_4) into cellulose fibres (e.g. lignocellulosic fibres) or by generating magnetic particles in situ. Carrazana-Garcia et al. (1997) have prepared cobalt ferrite/pulp fibre composites for manufacturing of magnetic paper via alkali precipitation of hydrated metal chlorides. Zakaria et al. (2005) developed a procedure for loading lignocellulosic fibres, used in paper-making process, with preformed magnetite particles. Mancosky et al. (2005) have combined the bleaching process of fibres with magnetic nanomodification procedure. Iron particles were adsorbed to the surface of fibres via interaction between iron (II) oxide and cellulose hydroxyl groups. Magnetic properties were imparted to the modified fibres, and their electrical resistivity was decreased. Sourty et al. (1998) prepared two different cellulose gels, loaded with iron ferrite particles. Bacterial cellulose membrane and a regenerated cellulose film, made from a solution of cellulose pulp in N-methylmorpholine-N-oxide, were dipped in an iron chloride solution, followed by an addition of alkali catalyst and oxidation with H_2O_2 at an elevated temperature. In situ synthesised magnetite or maghemite particles provoked superparamagnetic properties of cellulose membranes. Sun et al. (2008) dissolved different cellulose substrates in an ionic liquid and then added magnetite particles to the solution. Cellulose solution with dispersed magnetic particles was spinned, and resultant magnetite-embedded fibres were coagulated in a water bath.

7.8.1 Modification of Regenerated Cellulose Fibres by Magnetic Coatings

7.8.1.1 Synthesis of Magnetite Particles

Magnetite particles can be synthesised using various procedures. To a great extent, particular procedure defines the properties of the product, especially its crystalline structure, specific surface area and water content (Schwertmann and Cornell 2000). Methods for magnetite synthesis can be roughly divided into procedures with thermal decomposition of the precursors, sonochemical ones and different types of precipitation. The most common and frequently encountered procedure for the preparation of monodisperse colloidal particles of magnetite is precipitation. We can differentiate between (1) partial reduction of Fe^{3+} and (2) partial oxidation of Fe^{2+}, both followed by precipitation and (3) co-precipitation of Fe^{3+} and Fe^{2+}.

Since properties of magnetite particles (size, crystalline structure) influence their magnetic characteristics (e.g. saturation magnetisation), a thorough study of particle synthesis alone was undertaken by Hribernik (2010). To optimise the procedure, molar concentrations of iron chloride solutions, their ratios and additions of reactants into the reaction mixture were varied. Monitoring of the synthesis and, consequently, determination of the magnetite formation in certain stages of reactant addition, that is, in specific pH regions, served as a tool for selection of subsequent coating procedures for cellulose fibres.

Results of particle crystallite size determination, their electrophoretic mobility and size of particle agglomerates in an aqueous dispersions and magnetic properties of dried powders, showed that addition of a catalyst solution (ammonium hydroxide) in a controlled manner into a precursor solution (iron chlorides) yields particles with larger crystallite sizes, in contrast to the procedure with a reversed reactant addition. These larger magnetite particles also exhibit higher values of saturation magnetisation and are more stable in aqueous dispersions. Molar ratio between bivalent and trivalent iron ions also influences the size of the particle crystalline domains. In case of nonstoichiometric ratios ($Fe^{3+}/Fe^{2+} \neq 2$), where an excess of bivalent iron ions exists, larger particles are produced.

Co-precipitation procedure with the controlled addition of precursor into a catalyst solution or instant mixing of all components into a reaction system appears less suitable for coatings of fibres, from a viewpoint of reduced control over particle formation.

7.8.1.2 Preparation of Magnetic Nanocoatings on Regenerated Cellulose Fibres

Regenerated cellulose fibres were coated with magnetic layers using different procedures: in situ co-precipitation of magnetite particles in the presence of fibres and adsorption of preformed magnetite particles onto fibre surface from a stable aqueous dispersion. Synthesis and adsorption conditions influenced the morphology of particle layers and their properties.

In a procedure where a suspension of fibres in ammonium hydroxide was used and a mixture of iron chloride solutions was added subsequently, fibres with uneven layers of particles resulted. Larger aggregates were formed on fibre surface. Due to the pH value of the reaction, which was never below value of 10, meaning that both species (fibres and particles) carried a negative surface potential (zeta potential), the deposition of formed particles onto the fibres in an ordered manner, due to the same charge, was hindered. Impregnation of fibres with the precursor solution (mixture of iron chlorides) and subsequent addition of ammonium hydroxide results in coverage of fibres, which is very homogenous especially in the case of pre-swollen fibres (using aqueous solutions of sodium hydroxide). Obvious disadvantage of this procedure is the low pH value of the starting precursor solution, which decreases the mechanical properties of coated fibres.

Modification of the procedure with the impregnation of fibres with iron chloride solutions, where the pH value of the precursor solution was increased beforehand, yields the most homogenously covered fibre surfaces. Elevation of pH value of the precursor solution to approx. 3.5 allows impregnation of fibres, since reduction of fibre mechanical properties is avoided, while at the same time precipitation of particles does not start. Only after subsequent addition of ammonium hydroxide is the rapid precipitation of particles occurring, which then form a dense surface layer on fibre surface. The particles are also expected to grow in the fibre inner structure. This procedure gives the highest value of the saturation magnetisation. Depending on the synthesis procedure and whether a fibre was subjected to pretreatment before the application of particles, saturation magnetisation varies between 1.33 emu/g and 10.50 emu/g (Hribernik 2010). Figure 7.6 shows a magnetisation curve for precipitated magnetite particles (values of saturation magnetisation 65–70 emu/g) and viscose fibres, coated with magnetic particles, using the same synthesis procedure as for particles alone, showing values of saturation magnetisation around 10 emu/g (Hribernik et al. 2009).

Organic/inorganic hybrid superparamagnetic materials, based on the conventional regenerated cellulose fibres with an attached layer of magnetic particles, were prepared. Particles of magnetite (Fe_3O_4), which is a soft magnetic material, were obtained with a co-precipitation reaction of precursor iron salts, and they were grown on fibre surface. Magnetic layers were either in situ precipitated on the textile material surface or deposited from a previously prepared ferrofluid.

Results have shown that the technique of nanoparticle formation influences particle morphology and, consequently, their functionality. However, the modification procedure yields new possibilities of modified textile applications.

7.9 Spun-Dyed Cellulosics

Mass colouration, spun dyeing or dope dyeing maybe defined as 'a method of coloring manufactured fibres by incorporation of the colorant in the spinning composition before extrusion into filaments' (anon 2000). One of the primary considerations in any mass colouration process is to ensure the chemical and physical stability of the polymer-colourant mixture. This is especially important with regard to regenerated cellulosics as their manufacturing processes involve treatment of the cellulose with strong oxidising/reducing agents, which may militate against the stability of colourants. But mass colouration of regenerated cellulosics has been found possible, and the different methods reported for its achievement are summarised below.

7.9.1 Mass Colouration of Viscose or Cuprammonium Rayon

7.9.1.1 Vat Dyes

Many of the techniques proposed for mass colouration of viscose or cuprammonium rayon involve the addition of vat dyes to the spinning dope, wherein the vat dye is reduced to its leuco form and oxidised back to its parent form in the course of manufacturing the substrate. Some techniques involve the addition of reduced vat dye to the spinning dope (I. G. Farbenindustrie A.-G. 1937; Kline and Helm 1939). In others, it is proposed that the vat dye be reduced in the spinning dope either by utilising the chemical reagents already present in the system (Lockhart 1932) or by the addition of reducing agents such as sodium hydrosulphite (Batt 1961). Yet others involve dispersing the vat dye in spinning dope as a pigment, forming the regenerated substrate, and treating the formed substrate with reagents to reduce the vat dye within (Ruesch and Schmidt 1936; I. G. Farbenindustrie A.G 1936; Maloney 1967). In

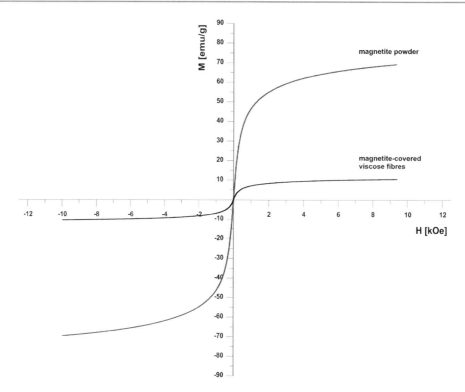

Fig. 7.6 Magnetisation *curve* for precipitated magnetite particles (values of saturation magnetisation 65–70 emu/g) and viscose fibres, coated with magnetic particles (values of saturation magnetisation 10 emu/g)

all these techniques, the oxidation of the vat dye back to its parent form is achieved, in general by treating the formed substrate with oxidising agents.

There are limitations to these techniques. Adding reduced vat dyes to the spinning dope may result in the stabilisation of the spinning dope (Ruesch and Schmidt 1936). Proper ageing and coagulation of the spinning dope is hindered, which affects the development of suitable viscosities for fibre/filament spinning. The reduced vat dyes are also susceptible to premature oxidation, which results in a non-uniform distribution of dye in the substrate (Dosne 1936). Many vat dyes are not reduced under the conditions that exist in the spinning dope (Lutgerhorst 1956), and the addition of reducing agents to the system renders the dope liable to gel formation (Batt 1961). The technique of dispersing vat dyes in the spinning dope and reducing them in the formed substrate is not without problems, firstly, because a uniform distribution of the dyestuff in the substrate is difficult to achieve, and secondly, because not all dyestuff in substrate may be reduced, causing visible specks of dyestuff particles to remain in the substrate (Batt 1961).

In some techniques, vat acids or the ester derivatives of the leuco compounds of vat dyes are added to the spinning dope (Dosne 1936; Lutgerhorst 1956). However, the leuco compounds are highly susceptible to oxidation, which may result in the formation of coarse dyestuff particles in dope thereby affecting the subsequent regeneration step (I. G. Farbenindustrie A.-G. 1937).

7.9.1.2 Dissolved Colourants

Some of the proposed mass colouration techniques involve adding to spinning dopes colourants dissolved in polar water-miscible solvents or dissolving colourants directly in the spinning dope (Soc. pour l'ind. chim. a Bale 1941; Phrix-Werke A.-G. 1965; Ciba Ltd. 1965, 1966, 1968; Wegmann and Booker 1966; Riehen and Reinach 1971a, b). The colourants used in these methods are selected dyes, dye derivatives or pigments. The choice of dyestuffs available for this technique is limited by the fact that not all dyestuffs can withstand the strong alkaline conditions present in the spinning dope or the strong acid treatments imparted to substrates during regeneration (Dosne 1936). Moreover, the use of water soluble dyestuffs in mass colouration of regenerated cellulosics has been observed to result in poor water fastness of the formed products (Batt 1961).

7.9.1.3 Dispersed Colourants

The dispersion of finely milled organic or inorganic pigments in spinning dopes has been suggested as a possible route for mass colouration (Novacel 1953; Manufactures de produits chimiques du Nord 1956; Jones 1959; Keil et al. 1961; Eskridge 1962; Gomm et al. 1964; Heinrich 1966; Schoenbach et al. 1967; Hama and Sakurai 1972), with additives being recommended in some cases to improve pigment dispersibility (Keil et al. 1961; Eskridge 1962; Gomm et al. 1964; Ono and Igase 1979). The process of milling pigments to obtain a suitable particle size is time intensive and is accompanied by the risks of recrystallisation and/or regrouping of dyestuff particles (Whitehead 1938). The possibility of poor pigment dispersion in spinning dope is an inherent risk in this technique, which may lead to problems both in the regeneration process and with the uniformity of colour in the formed substrates (Keil et al. 1961; Gomm et al. 1964). The dispersing agents added to pigment formulations sometimes cause undesirable foaming in the spinning mass (Schoenbach et al. 1967). The coloured substrates tend to be opaque (Phrix-Werke A.-G. 1965), exhibit dull shades (Dosne 1936) and may also exhibit dichroism (Eskridge 1962). This technique of mass colouration may also exert a deleterious effect on substrate strength (Ciba Ltd. 1968).

7.9.1.4 Other Methods

Other proposed techniques involve the suspension of sulphur dye intermediates in spinning dopes (Cassella Farbwerke Mainkur A.G. 1961) or utilising waste cotton textiles dyed with reactive dyes by mixing them with fresh cellulose, subjecting the mixture to xanthation and spinning coloured filaments therefrom (von der Eltz 1996).

7.9.2 Mass Colouration of Lyocell

Lyocell fibres are relatively new in comparison to the other regenerated cellulosics, with the first commercial samples being available in the mid-1980s and full-scale commercial production beginning in the early 1990s (White 2001). Hence, there are only a few methods reported for the mass colouration of lyocell.

In one of the first techniques reported (Bartsch and Ruef 1999), it is proposed that selected inorganic pigments, which contain small amounts of heavy metals and do not significantly decrease the decomposition temperature of the spinning mass, be mixed with the cellulose solution prior to fibre spinning. It has also been proposed that colourant or colourant precursors be mixed with the cellulose solution (Ruef 2001), with the caveat that the colourants be insoluble or sparingly soluble in amine oxide. In the most recent method reported (Bechtold and Manian 2005), it has been suggested that cellulose pulp be dyed with a vat dye and the dyed pulp optionally mixed with undyed pulp be used to spin lyocell fibres.

Like other processes, each method for the mass colouration of regenerated cellulosic fibres has its advantages and limitations. Mass colouration processes in general offer the advantages of high colour fastness in products and a high degree of colourant utilisation which make the process more ecofriendly, but are usually cost efficient only in large-scale production (Holme 2004). There exist niche markets for such products, and mass coloured regenerated cellulosic fibres are commonly used in technical textiles such as in medical applications and outdoor textiles.

7.10 Outlook

Polysaccharide-based fibres for textile applications are at the beginning of a new prosperity.

The growing awareness about substitution of petrol-based material directed attention of scientists to this wide field of research. As a result, substantial activity can be seen both in scientific literature and development activities of enterprises. Progress can be expected to occur in many fields of polysaccharide research:
- Solvents for polysaccharide shaping for new processes and products
- Modification of the polymer structure by chemical derivatisation to achieve new polymer and fibre properties
- Use of polymer blends in fibre production
- Incorporation of functional elements into the fibre structure
- Adjustment of material properties in textile processing (porosity, accessibility, surface)
- Chemical modification of the fibrous/textile structure (cross-linking, deposition, coating)

The examples for specialised textile fibres presented in this chapter thus can be understood as representative cases, to indicate the different options to modify polysaccharide-based materials for a certain textile application.

When we consider the high number of strategies to modify fibre properties in polysaccharide-based textiles, we can recognise that only a few possibilities have been explored and commercialised up to now. We understand that awareness of this future potential and the chances for development of new materials make researchers enthusiastic to work in this particular field of science.

References

Abbott GM, Robinson GA (1977a) The corona treatment of cotton, part I: silver cohesion. Text Res J 47:141–144

Abbott GM, Robinson GA (1977b) The corona treatment of cotton, part II: yarn and fabric properties. Text Res J 47:199–202

Abu-Rous M, Varga K, Bechtold T, Schuster KC (2007) A new method to visualise and characterise the pore structure of TENCEL® (Lyocell) and other mansmade cellulosic fibres using a fluorescent dye molecular probe. J Appl Polym Sci 106(3):2083–2091

Akishev Y, Kroepke S, Behnisch J, Hollander A, Napartovich A, Trushkin N (2000) Non-thermal plasma treatment of polymer films and fabrics based on a glow discharge at atmospheric pressure In: Wagner H, Behnke JF, Babucke G (eds) Proceedings of the international symposium on high pressure low temperature plasma chemistry, vol. 2. Greifswald, Germany, pp 481–485

Akishev YS, Grushin ME, Monich AE, Napartovich AP, Trushkin NI (2003) One-atmosphere argon dielectric-barrier corona discharge as an effective source of cold plasma for the treatment of polymer films and fabrics. High Energ Chem 37(5):286–291

Allwood JM, Laursen SE, Malvido de Rodriguez C, Boecken NMP (2006) Well dressed? The present and future sustainability of clothing and textiles in the United Kindgdom. University Cambridge Institute for Manufacturing, Cambridge. ISBN 1-902546-52-0

Anon (2000) A Glossary of AATCC Standard Terminology, AATCC Technical Manual 75, American Association of Textile Chemists and Colorists, Research Triangle Park, NC, USA, p 399

Anon (2008) ESFRI - European Road Map for Research Infrastructures; Roadmap 2008, ISBN 978-92-79-10117-5

Anon (2010a) PES-Weltproduktion +9 %. Melliand Textilberichte 91/1–2, pp 4–5

Anon (2010b) Globale Faserproduktion. Melliand Textilberichte 91/4-5, pp 149

Bartsch P, Ruef H (1999) Method for producing cellulosic moulded bodies. WO 99/46434, 16 Sep 1999

Batt IP (1961) Process for producing colored pellicular gel structures of regenerated cellulose. US 3005723, 24 Oct 1961

Baumann H, Keller B, Ruzicka E (1991) Partially cationized cellulose for non- thrombogenic membrane in the presence of heparin and endothelial-cell-surface-heparansulfate (ES-HS). J Membr Sci 61:253–268

Bechtold T, Manian AP (2005) Method of producing a dyed formed cellulosic article. WO 2007/070904, 19 Dec 2005

Berger J, Reist M, Mayer JM, Felt O, Peppas NA, Gurny R (2004) Structure and interactions in covalently and ionically crosslinked chitosan hydrogels for biomedical applications. Eur J Pharm Biopharm 57:19–34

Berscht PC, Nies B, Liebendorfer A, Kreuter J (1994) Incorporation of basic fibroblast growth factor into methylpyrrolidinone chitosan fleeces and determination of the in vitro release characteristics. Biomaterials 15:593–600

Bongiovanni R, Di Gianni A, Priola A, Pollicino A (2007) Surface modification of polyethylene for improving the adhesion of a highly fluorinated UV-cured coating. Eur Polym J 43:3787–3794

Bredereck K, Commarmot A (1998) Ammonia treatment of cellulosic fibers. Melliand Textilberichte 79(1–2): E19–E22

Bredereck K, Stefani HW, Beringer J, Schulz F (2003) Alkali- und Flüssigammoniakbehandlung von Lyocellfasern. Melliand Textilberichte 58(1–2):58–64

Briganti S, Picardo M (2003) Antioxidant activity, lipid peroxidation and skin diseases. What's new. J Eur Acad Dermatol Venereol 17(6):663–669

Brüser V, Heintze M, Brandl W, Marginean G, Bubert H (2004) Surface modification of carbon nanofibres in low temperature plasmas. Diamond Relat Mater 13:1177–1181

Bui HM, Lenninger M, Manian AP, Abu-Rous M, Schimper CB, Schuster KC, Bechtold T (2008) Treatment in swelling solutions modifying cellulose fibre reactivity – Part 2: Accessibility and reactivity. In: Macromolecular symposia: zellcheming 2007 conference proceedings, vol 262, pp 50–64

Bullett NA, Bullett DP, Truica-Marasescu FE, Lerouge S, Mwale F, Wertheimer MR (2004) Polymer surface micropatterning by plasma and VUV-photochemical modification for controlled cell culture. Appl Surf Sci 235:395–405

Byrne GA, Brown KC (1972) Modifications of textiles by glow-discharge reactions. J Soc Dyers Colour 88:113–117

Čakara D, Fras Zemljič L, Bračič M, Stana-Kleinschek K (2009) Protonation behavior of cotton fabric with irreversibly adsorbed chitosan: a potentiometric titration study. Carbohydr Polym 78:36–40

Carrazana-Garcia JA, López-Quintela MA, Rivas-Rey J (1997) Characterization of ferrite particles synthesized in presence of cellulose fibres. Colloids Surf A Physicochem Eng Asp 121:61–66

Cassella Farbwerke Mainkur A.G. (1961) Dyeing of regenerated cellulose in the spinning paste. BE 611323, 29 Dec 1961, vide Chemical Abstract No. 57:17664

Castelvetro V, Fatarella E, Corsi L, Giaiacopi S, Ciardelli G (2006) Graft polymerisation of functional acrylic monomers onto cotton fibres activated by continuous Ar plasma. Plasma Process Polym 3(1):48–57

Chedin J, Marsaudon A (1955) Action of caustic soda solutions on cellulose fibers. Equilibrium fixation of caustic soda. Mercerization. Die Makromolekulare Chemie 15:115–160

Chedin J, Marsaudon A (1956) The mechanism of the fixation of sodium hydroxide on the cellulose fiber-mercerization-structure of aqueous sodium hydroxide solutions II. Die Makromolekulare Chemie 20:57–82

Chedin J, Marsaudon A (1959) The adsorption and desorption mechanism of aqueous sodium hydroxide solutions in cellulose fibers III. Equilibriums of NaOH fixation. Die Makromolekulare Chemie 33:195–221

Chen JR (1991) Free radicals of fibers treated with low temperature plasma. J Appl Polym Sci 42(7):2035–2037

Ciba Ltd. (1965) Dyeing of regenerated cellulosic fibers and films. FR 1417575, 12 Nov 1965, vide Chemical Abstract No. 65:57365

Ciba Ltd. (1966) Transparent colored regenerated cellulose. NL 6514672, 13 May 1966, vide Chemical Abstract No. 65:82924.

Ciba Ltd. (1968) Process for the preparation of transparent colored shaped articles of regenerated cellulose with the aid of organic dyestuffs of low solubility in water. GB 1128158, 25 Sep 1968

Coats JP, Gailey RM (1968) Method of treating cellulosic materials. British Patent 1136417, 1968

Colom X, Carrillo F (2002) Crystallinity, crystallinity changes in lyocell and viscose-type fibres by caustic treatment. Eur Polym J 38(11):2225–2230

Corbman BP (1985) Textiles: fiber to fabric, 6th edn. McGraw-Hill, New York, NY. ISBN 0-07-Y66236-3

Crawshaw J, Cameron RE (2000) A small angle X-ray scattering study of pore structure in Tencel cellulose fibres and the effects of physical treatments. Polymer 41:4691–4698

Culler MD, Bitman J, Thompson MJ, Robbins WE, Dutky SR, Mastitis I (1979) In vitro antimicrobial activity of alkyl amines against mastitic bacteria. J Dairy Sci 62:584–595

Czaja W, Krystynowicz A, Bielecki Sand Brown RM (2006) The natural power to heal wounds. J Biomater 27(2):145–151

De Geyter N, Morent R, Leys C (2006) Surface modification of a polyester non-woven with a dielectric barrier discharge in air at medium pressure. Surf Coat Tech 201:2460–2466

Denoyelle G (1959) Action of soda solutions on native cellulose, formation of hydrated sodium cellulosate; swelling. Svensk Papperstid 62:390–406

Dimick BE (1976) The importance of the structure of alkali metal hydroxide solutions in decrystallizing cellulose I, PhD thesis, Lawrance University, The Institute of Paper Chemistry, Appleton, WI

Diniz JMBF, Gil MH, Castro JAAM (2004) Hornification-its origin and interpretation in wood pulps. Wood Sci Technol 37:489–494

Donnan FG, Harris AB (1911) Osmotic pressure and conductivity of aqueous solutions of congo red and reversible membrane equilibria. J Chem Soc Trans 99:1554–1577

Dosne H (1936) Colored cellulose material. US 2041907, 26 May 1936

Douglas BE, McDaniel DH (1965) Concepts and models of inorganic chemistry. Blaisdell, Waltham, MA, p 199

Ebeling H, Fink HP (2009) Method of producing cellulose carbamate blown film and use of the same, US Patent Application No. 2009/0259,032, 7 Apr 2009

Ehrhardt A, Groner S, Bechtold T (2007) Swelling behaviour of cellulosic fibers – Part I, fibers and textiles in Eastern Europe fibres. Text Eastern Eur 15(5–6):46–48

Eskridge BE (1962) Manufacture of pigmented viscose rayon. US 3033697, 8 May 1962

Fras Zemljič L, Peršin Z, Stenius P (2009a) Improvement of chitosan adsorption onto cellulosic fabrics by plasma treatment. Biomacromolecules 10(5):1181–1187

Fras Zemljič L, Strnad S, Šauperl O, Stana-Kleischek K (2009b) Characterization of amino groups for cotton fibers coated with chitosan. Text Res J 79(3):219–226

Fras Zemljič L, Čakara D, Kokol V (2011a) Antimicrobial and antioxidant properties of chitosan-based viscose fibres enzymatically functionalized with flavonoids. Text Res J 81(15):1532–1540

Fras Zemljič L, Čakara D, Michaelis N, Heinze T, Stana-Kleinschek K (2011b) Protonation behavior of

6-deoxy-6-(2-aminoethyl)amino cellulose: a potentiometric titration study. Cellulose 18:33–43

Gimblett FGR, Monk CB (1954) E.m.f studies of electrolytic dissociation. VII. Some alkali and alkaline earth metal hydroxides in water. Trans Faraday Soc 50:965–972

Gomm AS, Morgan LB and Wood L (1964) Process of incorporating aqueous pigment composition in viscose. US 3156574, 10 Nov 1964

Goodwin A, Herbert T, Leadley S, Swallow F (2002) In: Proceedings of 8th international symposium on high pressure low temperature plasma chemistry, vol 2. Pühajärve Estonia, pp P7.9–1

Gorjanc M, Bukošek V, Gorenšek M, Mozetič M (2010) CF4 plasma and silver functionalized cotton. Text Res J 80(20):2204–2213

Greentex international. http://www.greentex-international.com. Accessed 4.5.2012

Hama H, Sakurai H (1972) Dope dyeing viscose rayon having a good black luster. JP 47051968, 27 Dec 1972, vide Chemical Abstract No. 80:28345

Hattori K, Yoshida T, Nakashima H, Premanathan M, Aragaki R, Mimura T, Kaneko Y, Yamamo N, Uryu T (1998) Synthesis of sulfonated amino polysaccharides having anti-HIV and blood anticoagulant activities. Carbohydr Res 312:1–8

Heinrich E (1966) Spin-dyed regenerated cellulose products and process for their manufacture. GB 1046299, 19 Oct 1966

Heinze U, Wagenknecht W (1998) Comprehensive cellulose chemistry. functionalisation of cellulose, vol 2. Wiley, Weinheim. ISBN 10: 3527294899

Ho KKC, Lee AF, Bismarck A (2007) Fluorination of carbon fibres in atmospheric plasma. Carbon 45:775–784

Hoenich N (2006) Cellulose for medical applications: past, present, and future. Bioresources 1(2):270–280

Holländera A, Thome J, Keusgen M, I D, Klein W (2004) Polymer surface chemistry for biologically active materials. Appl Surf Sci 235:145–150

Holme I (2004) Coloration of technical textiles. In: Horrocks AR, Anand SC (eds) Handbook of technical textiles. Woodhead Publishing Ltd, Cambridge

Horrocks AR, Anand SC (2004) Handbook of technical textiles. The Textile Institute, Woodhead Publishing Ltd, Cambridge. ISBN 1 85573 385 4

Hribernik S, Sfiligoj-Smole M, Stana-Kleinschek K (2009) Formation of magnetic layers on regenerated cellulose fibres' surface, EPNOE Polysaccharides as a source of advanced materials: Turku/Åbo 2009, Book of abstracts

Hribernik S (2010) Study of pre-treatment and coating of regenerated cellulose fibres with nano particles, Doctoral thesis, University of Maribor, Slovenia

Hwang YJ, Mccord MG, An JS, Kang BC, Park SW, Kang BC (2005) The effects of helium atmospheric pressure plasma treatment on low-stress mechanical properties of polypropylene nonwoven fabrics. Text Res J 75:771–778

I.G. Farbenindustrie A.G. (1936) Process for the manufacture of dyed filaments and films. GB 448447, 8 Jun 1936

I.G. Farbenindustrie A.-G. (1937) The manufacture of dyed artificial masses from regenerated cellulose. GB 465606, 10 May 1937

Ibbett RN, Hsieh YL (2001) Effect of fiber swelling on the structure of lyocell fabrics. Text Res J 71(2):164–173

Jaturapiree A, Manian AP, Bechtold T (2006) Sorption studies on regenerated cellulosic fibres in salt-alkali mixtures. Cellulose 13(6):647–654

Jaturapiree A, Ehrhardt A, Groner S, Öztürk HB, Siroka B, Bechtold T (2008) Treatment in swelling solutions modifying cellulose fibre reactivity—Part 1: accessibility and sorption. In: Macromolecular symposia: Zellcheming 2007 conference proceedings, vol 262, pp 39–49

Jaturapiree A, Manian AP, Lenninger M, Bechtold T (2011) The influence of alkali pretreatments in lyocell resin finishing—Changes in fiber accessibility to cross-linker and catalyst. Carbohydr Polym 86:612–620

Jeffries R, Warwicker JO (1969) Function of swelling in the finishing of cotton. Text Res J 39(6):548–559

Jones FB (1959) Glossy, spun-dyed threads from aqueous cellulose solutions. DE 1067173, 15 Oct 1959

Kampl R, Six W (1996) Non-woven flame retardant textile fabric. WO 96/14461, 17 Mai 1996

Kaputskii FN, Gert EV, Torgashov VI, Zubets OV (2005) Hydrogels for medical applications fabricated by oxidative-hydrolytic modification of cellulose. Fibre Chem 37:485–489

Kasahara K, Sasaki H, Donkai N, Takagishi T (2004) Effect of processing and reactive dyeing on the swelling and pore structure of lyocell fibers. Text Res J 74(6):509–515

Keil A, Popp P, Krause E (1961) Process for the production of pigmented regenerated cellulosic fibers. GB 872207, 5 Jul 1961

Klemm D, Philipp B, Heinze T (1998) Comprehensive cellulose chemistry. Wiley, Winheim, pp 9–32

Kline HB, Helm EB (1939) Manufacture of artificial silk. US 2143883, 17 Jan 1939

Kongdee A, Bechtold T, Burtscher E, Scheinecker M (2004) The influence of wet/dry treatment on pore structure - the correlation of pore parameters, water retention and moisture regain values. Carbohydr Polym 57:39–44

Kotel'nikova NE, Wegener G, Paakkari T, Serimaa R, Demidov VN, Serebriakov AS, Shchukarevand AV, Gribanov AV (2003) Silver intercalation into cellulose matrix. An X-ray scattering, solid-state 13C NMR, IR, X-ray photoelectron, and Raman study. Russ J Gen Chem 73(3):418–426

Krentsel E, Fusselman S, Yasuda H, Yasuda T, Miyama M (1994) Penetration of plasma surface modification. II. CF4 and C2F4 low-temperature cascade arc torch. J Polym Sci A Polym Chem 32:1839–1845

Krentsel E, Yasuda H, Miyama M, Yasuda T (1995) Penetration of plasma surface modification. III.

Multiple samples exposed to CF4 and C2F4 low temperature cascade arc torch. J Polym Sci A Polym Chem 33:2887–2892

Lacasse K, Baumann W (2004) Textile chemicals. Environmental data and facts. Springer, Berlin. ISBN 978-3-540-40815-4

Lim SH, Hudson SM (2004) Synthesis and antimicrobial activity of a water-soluble chitosan derivative with a fiber-reactive group. Carbohydr Res 339(2):313–319

Link E, Mason CR, Tosti A, Karnik A (2005) Flame blocking liner materials. US 2005/0118919, 2 June 2005

Lockhart GR (1932) Manufacture of rayon. US 1865701, 5 July 1932

Lund MN, Hviid MS, Skibsted LH (2007) The combined effect of antioxidants and modified atmosphere packaging on protein and lipid oxidation in beef patties during chill storage. Meat Sci 76(2):226–233

Lutgerhorst AG (1956) Spundyed rayon. US 2738252, 13 Mar 1956

Malek RMA, Holme I (2003) The effect of plasma treatment on some properties of cotton. Iran Polym J 12 (4):271–280

Maloney MA (1967) Mass coloring of regenerated cellulose with vat dyes. DE 1253864, 9 Nov 1967

Mancosky DG, Lucia LA (2005) A novel and efficient approach for imparting magnetic susceptibility to lignocellulosic fibers. Carbohydr Polym 59:517–520

Manian AP, Abu-Rous M, Schuster KC, Bechtold T (2006) The influence of alkali pre-treatments in lyocell resin finishing. J Appl Polym Sci 100 (5):3596–3601

Manian AP, Abu-Rous M, Lenninger M, Roeder T, Schuster KC, Bechtold T (2008) The influence of alkali pretreatments in lyocell resin finishing – Substrate structure. Carbohydr Polym 71(4):664–671

Manufactures de produits chimiques du Nord (establissment Kuhlmann) (1956) Coloring viscose fibers. FR 1114803, 17 Apr 1956, vide Chemical Abstract No. 53:102818

Marchessault RH, Rioux P, Ricard S (1992) Preparation and synthesis of magnetic fibers. US Patent 5,143,583, 1 Sept 1992

Mercer J (1850) Improvements in the preparation of cotton and other fabrics and other fibrous materials. British Patent 13,296, 1850

Molina R, Espinós JP, Yubero F, Erra P, González-Elipe AR (2005) XPS analysis of down stream plasma treated wool: influence of the nature of the gas on the surface modification of wool. Appl Surf Sci 252:1417–1429

Morales J, Olayo MG, Cruz GJ, Herrera-Franco P, Olayo R (2006) Plasma modification of cellulose fibers for composite materials. J Appl Polym Sci 101 (6):3821–3828

Morent R, De Geyter N, Leys C, Gengembre L, Payen E (2007) Surface modification of non-woven textiles using a dielectric barrier discharge operating in air, helium and argon at medium pressure. Text Res J 77 (7):471–488

Moseley R, Hilton JR, Waddington RJ, Harding KG, Stephens P, Thomas DW (2004) Comparison of oxidative stress biomarker profiles between acute and chronic wound environments. Wound Repair Regen 12:419–429

Mozetič M (2007) Characterization of reactive plasmas with catalytic probes. Surf CoatTechnol 9–11 (201):4837–4842

Mukhopadhyay SM, Joshia P, Datta S, Macdaniel J (2002a) Plasma assisted surface coating of porous solids. Appl Surf Sci 201(1–4):219–226

Mukhopadhyay SM, Joshi P, Datta S, Zhao JG, France P (2002b) Plasma assisted hydrophobic coatings in porous materials. J Phys D: Appl Phys 35:1927–1933

Muzzarelli RAA, Mattioli-Belmonte M, Tietz C, Biagini R, Ferioli G, Brunelli MA, Fini M, Giardino R, Ilari P, Biagini G (1994) Stimulatory effect on bone formation exerted by a modified chitosan. Biomaterials 15:1075–1081

Muzzarelli RAA (2009) Chitins and chitosan for the repair of wound skin, nerve, cartilage and bone. Carbohydr Polym 76:167–182

Myllyte P, Salmi J, Laine J (2009) The influence of pH on the adsorption and interaction of chitosan with cellulose. Bioresources 4(4):1647–1662

Novacel SA (1953) Colored regenerated cellulose sponges. FR 1025296, 13 Apr 1953, vide Chemical Abstract No. 52:3968

Oeztuerk HB, MacNaughtan B, Mitchell J, Bechtold T (2011) What does LiOH treatment offer for lyocell fibers? Investigation of structural changes. Ind Eng Chem Res 50:9087–9094

Okubayashi S, Griesser UJ, Bechtold T (2005) Moisture sorption/desorption behaviour of various manmade cellulosic fibres. J Appl Polym Sci 9(4):1621–1625

Ono S, Igase T (1979) Dope dyeing of rayon. JP 54038919, 24 Mar 1979, vide Chemical Abstract No. 91:40853

Öztürk HB, Okubayashi S, Bechtold T (2006a) Splitting tendency of cellulosic fibers, part 1: the effect of shear force on mechanical stability of swollen lyocell fibers. Cellulose 13(4):393–402

Öztürk HB, Okubayashi S, Bechtold T (2006b) Splitting tendency of cellulosic fibers. part 2: effects of fiber swelling in alkali solutions. Cellulose 13(4):403–409

Öztürk HB (2008) Regenerated cellulosic fibers-Effect of alkali treatment on structure, chemical reactivity and fiber properties, PhD thesis, University of Innsbruck, Austria

Öztürk HB, Bechtold T (2007) Effect of NaOH treatment on the interfibrillar swelling and dyeing properties of lyocell (TENCEL) fibers. Fibers Text East Eur 15 (5–6):114–117

Öztürk HB, Bechtold T (2008) Splitting tendency of cellulosic fibers, part 3: splitting tendency of viscose and modal fibers. Cellulose 15(1):101–109

Öztürk HB, Potthast A, Rosenau T, Abu-Rous M, MacNaughtan B, Schuster KC, Mitchell J, Bechtold T (2009) Changes in the intra- and inter- fibrillar structure of lyocell (TENCEL) fibers caused by NaOH treatment. Cellulose 16(1):37–52

Öztürk HB, Abu-Rous M, MacNaughtan B, Schuster KC, Mitchell J, Bechtold T (2010a) Changes in the inter- and intra- fibrillar structure of lyocell (TENCEL) fibers after KOH treatment. Macromol Symp 294 (2):24–37

Öztürk HB, MacNaughtan B, Mitchell J, Bechtold T (2010b) Effects of tetramethylammonium hydroxide (TMAH) treatment on interfibrillar structure of lyocell (TENCEL) fibers. Mater Res Innov 14(3):224–230

Pecse A, Jordane AA, Carluci G, Cintio A (2005) Articles comprising cationic polysaccharides and acidic pH buffering means. US Patent Application 0124799 A1, 2005

Persin Z, Stana-Kleinschek K, Sfiligoj-Smole M, Kreze T (2004) Determining the surface free energy of cellulose materials with the powder contact angle method. Text Res J 74(&):55–62

Peršin Z, Vesel A Strnad S, Stana-Kleinschek K, Mozetič M (2008) XPS and sorption measurements of plasma-treated regenerated cellulose fabrics and ageing effects. In: Proceedings of the 24th annual meeting of the polymer processing society. PPS-24, Salerno, Italy, June 15–19, 2008 [COBISS.SI-ID 12397590]

Persin Z, Stana-Kleinschek K, Foster TJ, Van Dam JEG, Boeriu CG, Navard P (2011) Challenges and opportunities in polysaccharides research and technology: the EPNOE views for the next decade in the areas of materials, food and healthcare. Carbohydr Polym 84 (1):22–32

Phrix-Werke A.-G. (1965) Spun-dyed regenerated cellulose. NL 6407087, 18 Jan 1965, vide Chemical Abstract No. 63:63815

Prabaharan M, Carneiro N (2005) Effect of low-temperature plasma on cotton fabric and its application to bleaching and dyeing. Indian J Fibre Text 30 (1):68–74

Rakowski W (1989) Plasmamodifizierung der Wolle unter industriellen Bedingungen. Melliand Textilberichte 70:780–785

Ranby BG (1952a) The mercerization of cellulose. I. Thermodynamic discussion. Acta Chem Scand 6:101–115

Ranby BG (1952b) The mercerization of cellulose. II. A phase-transition study with X-ray diffraction. Acta Chem Scand 6:116–127

Ranby BG (1952c) The mercerization of cellulose. III. A phase-transition study with electron diffraction. Acta Chem Scand 6:128–138

Ranby BG, Mark HF (1955) The mercerization of cellulose. IV. Phase transition studies on technical wood pulps and cotton linters. Svensk Papperstid 58:374–382

Rashidi A, Moussavipourgharbi H, Mirjalili M, Ghoranneviss M (2004) Effect of low-temperature plasma treatment on surface modification of cotton and polyester fabrics. Indian J Fibre Text 29(1):74–78

Ravi Kumar MNV (2000) A review of chitin and chitosan applications. React Funct Polym 46:1–27

Riccobono PX (1973) Plasma treatment of textile: a novel approach to the environment problems of desizing. Text Chem Color 5:239–248

Riehen WM, Reinach FS (1971a) Difficulty soluble organic dye compositions for dyeing transparent, shaped, regenerated cellulose bodies. DE 1806199, 29 Apr 1971

Riehen WE, Reinach FS (1971b) Sparingly soluble organic dyestuffs. US 3620788, 16 Nov 1971

Ruef H (2001) Colored cellulosic shaped bodies. WO 01/11121, 15 Feb 2001

Ruesch R, Schmidt H (1936) Preparation of dyed filaments and films. US 2043069, 2 Jun 1936

Sarmadi AM, Kwon YA (1993) Improved water repellency and surface dyeing of polyester fabrics by plasma treatment. Text Chem Color 25(12):33–40

Scallan AM, Grignon J (1979) The effect of cations on pulp and paper properties. Svensk Papperstidning 82 (2):40–47

Scallan AM, Grignon J (1983) The effect of acidic groups on the swelling of pulps: a review. Tappi J 66 (11):73–75

Schimper CB, Ibanescu C, Bechtold T (2009) Effect of alkali pre-treatment on hydrolysis of regenerated cellulose fibers (part 1: viscose) by cellulases. Cellulose 16(6):1057–1068

Schoenbach V, Weissert J, Teige W (1967) Pigment dispersions for coloring viscose spinning masses. US 3337360, 22 Aug 1967

Schuster KC, Rohrer C, Eichinger D, Schmidtbauer J, Aldred P, Firgo H (2003) Environmentally friendly lyocell and rayon fibers. In: Wallenberger FT, Weston NE (eds) Natural fibers, polymers and composites—Recent advances. Kluwer Academic, Boston, MA, pp 123–146

Schwertmann U, Cornell RM (2000) Iron oxides in the laboratory. Preparation and characterization. Wiley, Weinheim. ISBN 978 3 527 29669 9

Shen L, Dai J (2007) Improvement of hydrophobic properties of silk and cotton by hexafluoropropene plasma treatment. Appl Surf Sci 253(11):5051–5055

Sheu GS, Shyu SS (1994) Surface properties and interfacial adhesion studies of aramid fibres modified by gas plasmas. Compos Sci Technol 52:489–497

Shishoo R (2007) Introduction – The potential of plasma technology in the textile industry. In: Shishoo R (ed) Plasma technologies for textiles. Woodhead Publishing Limited in association with The Textile Institute, Cambridge

Siroka B, Noisternig M, Griesser UJ, Bechtold T (2008) Characterisation of cellulosic fibers and fabrics by sorption and desorption. Carbohydr Res 343:2194–2199

Široký J, Manian AP, Široká B, Abu-Rous M, Schlangen J, Blackburn RS, Bechtold T (2009) Alkali treatments of lyocell in continuous processes. Part 1: effects of temperature and alkali concentration in treatments of plain-woven fabrics. J Appl Polym Sci 113 (6):3646–3655

Široký J, Blackburn RS, Bechtold T, Taylor J, White P (2010a) Attenuated total reflectance Fourier-transform Infrared spectroscopy analysis of crystallinity changes in lyocell following continuous treatment with sodium hydroxide. Cellulose 17(1):103–115

Široký J, Blackburn RS, Bechtold T, Taylor J, White P (2010b) Alkali treatment of cellulose II fibers and effect on dye sorption. Carbohydr Polym 84 (1):299–307

Soc. pour l'ind. chim. a Bale (1941) Pigment-containing spinning masses. CH 212386, 3 Mar 1941

Sodhi RNS, Sahi VP, Mittelman MW (2001) Application of electron spectroscopy and surface modification techniques in the development of anti-microbial coatings for medical devices. J Electron Spectros Relat Phenomena 121:249–264

Sourty E, Ryan DH, Marchessault RH (1998) Characterization of magnetic membranes based on bacterial and man-made cellulose. Cellulose 5:5–17

Struszczyk H, Ciechanska D (1998) Perspectives of Enzymes for Processing Cellulose for New Chemical Fibers. Enzyme Applications in Fiber Processing, ACS Symposium Series 687/25, pp 306–317

Sun D, Stylios GK (2004) The effect of low temperature plasma treatment on the scouring and dyeing processes of nature fabrics. Text Res J 74:751–756

Sun D, Stylios GK (2005) Investigating the plasma modification of natural fiber fabrics-the effect on fabric surface and mechanical properties. Text Res J 75:639–644

Sun D, Stylios GK (2006) Fabric surface properties affected by low temperature plasma treatment. J Mater Process Technol 173(2):172–177

Sun N, Swatloski RP, Maxim ML, Rahman M, Harland AG, Haque A, Spear SK, Daly DT, Rogers RD (2008) Magnetite-embedded cellulose fibers prepared from ionic liquid. J Mater Chem 18:283–290

Tan JS, Fisher LW, Marcus P (1975) 169th National Meeting of ACS, Philadelphia, PA, April

Tatarova I, Manian A, Siroka B, Bechtold T (2010) Non-alkali swelling solution for cellulose. Cellulose 17:913–922

Tatarova I, MacNaughtan W, Manian AP, Siroka B, Bechtold T (2011) Steam processing of regenerated cellulose fabric in concentrated LiCl/urea solution. Macromol Mater Eng. doi:= 10.1002/mame.201100272

Temmerman E, Leys C (2005) Surface modification of cotton yarn with a DC glow discharge in ambient air. Surf Coat Tech 200:686–689

Thorsen WJ, Kodani RY (1966) A corona discharge method of producing shrink-resistant wool and mohair. Text Res J 36(7):651–661

Thorsen WJ (1971) Improvement of cotton spinnability, strength, and abrasion resistance by corona treatment. Text Res J 41(5):455–458

Thorsen WJ (1974) Modification of the cuticle and primary wall of cotton by corona treatment. Text Res J 44 (6):422–428

Tissington B, Pollard G, Ward IM (1992) Study of the effects of oxygen plasma treatment on the adhesion behaviour of polyethylene fibres. Compos Sci Technol 44:185–195

Vehviläinen M, Taina T, Rom M, Janicki J, Ciechańska D, Grönqvist S, Siika-Aho M, Christoffersson KE, Nousiainen P (2008) Effect of wet spinning parameters on the properties of novel cellulosic fibres. Cellulose 15(5):671–680

Vesel A, Mozetič M, Hladnik A, Dolenc J, Zule J, Milošević S, Krstulović N, Klanjšek Gunde M, Hauptman N (2007) Modification of ink-jet paper by oxygen-plasma treatment. J Phys D Appl Phys 40:3689–3696

Vesel A, Mozetic M, Strnad S, Peršin Z, Stana-Kleinschek K, Hauptman N (2009) Plasma modification of viscose textile. Vacuum 84(1):79–82

Vigo TL, Wade RH, Mitcham O, Welch CM (1969) Synergistic effect of mixed bases in the conversion of cotton cellulose I to cellulose II. Role of cations as cocatalysts for crystal lattice rearrangement. Text Res J 39(4):305–316

Vigo TL, Mitcham O, Welch CM (1970) Decrystallization of cotton cellulose by benzyltrimethylammonium hydroxide followed by polar organic solvents. J Polym Sci (Polym Lett Ed) 8(6):385–393

von der Eltz A (1996) Recycling of dyed cellulosic wastes. EP 0717131, 19 Jun 1996

Wadsworth LC (2004) Nonwovens Science and Technology II In: Materials science and Engineering 554. http://www.engr.utk.edu/mse/Textiles/index.html. Accessed 4.5.2012

Wakida T, Takeda K, Tanaka I, Takagishi T (1989) Free radicals in cellulose fibers treated with low temperature plasma. Text Res J 59(1):49–53

Warwicker JO (1969) Cotton swelling in alkalis and acids. J Appl Polym Sci 13:41–54

Warwicker JO (1967) Effect of chemical reagents on the fine structure of cellulose. IV. Action of caustic soda on the fine structure of cotton and ramie. J Polym Sci A1 Polym Chem 5(10):2579–2593

Warwicker JO, Wright AC (1967) Function of sheets of cellulose chains in swelling reactions on cellulose. J Appl Polym Sci 11(5):659–671

Wegmann J, Booker C (1966) Colored viscose dope. DE 1220964, 14 Jul 1966

Weise U, Maloney T, Paulapuro H (1996) Quantification of water in different states of interaction with wood pulp fibres. Cellulose 3:189–202

White P (2001) Lyocell: the production process and market development. In: Woodings C (ed) Regenerated cellulose fibers. Woodhead Publishing Ltd, Cambridge

Whitehead W (1938) Colored organic derivatives of cellulose and method of making same. US 2128338, 30 Aug 1938

Wong KK, Tao XM, Yuen CWM, Yeung KW (2000) Effect of plasma and subsequent enzymatic treatments on linen fabrics. J Soc Dyers Colour 116(7–8):208–214

Wu M, Kuga S (2006) Cationization of cellulose fabrics by polyallylamine binding. J Appl Polym Sci 100:1668–1672

Yaman N, Özdoğan E, Seventekin N, Ayhan H (2009) Plasma treatment of polypropylene fabric for improved dyeability with soluble textile dyestuff. Appl Surf Sci 255:6764–6770

Yasuda T, Okuno T, Miyama M, Yasuda H (1994) Penetration of plasma surface modification. I. CF4 and C2F4 glow discharge plasma. J Polym Sci A Polym Chem 32:1829–1837

Yoon NS, Lim YJ, Tahara M, Takagishi T (1996) Mechanical and dyeing properties of wool and cotton fabrics treated with low temperature plasma and enzymes. Text Res J 66(5):329–336

Yuranova T, Rincon AG, Bozzi A, Parra S, Pulgarin C, Albers P, Kiwi J (2006) Performance and characterization of Ag–cotton and Ag/TiO2 loaded textiles during the abatement of E. coli. J Photoch Photobio A 181(2–3):363–369

Zakaria S, Ong BH, Ahmad SH, Abdullah M, Yamauchi T (2005) Preparation of lumen-loaded kenaf pulp with magnetite (Fe3O4). Mater Chem Phys 89:216–220

Zeronian SH, Cabradilla KE (1972) Action of alkali metal hydroxides on cotton. J Appl Polym Sci 16:113–128

Zhang W, Okubayashi S, Bechtold T (2005a) Fibrillation tendency of cellulosic fibers – Part 1 effects of swelling. Cellulose 12(3):267–273

Zhang W, Okubayashi S, Bechtold T (2005b) Fibrillation tendency of cellulosic fibres –part 3. Effects of alkali pretreatment of lyocell fibre. Carbohyd Polym 59(2):173–179

Zhou J, Zhang L (2002) Cellulose microporous membranes prepared from NaOH/urea aqueous solution. J Membr Sci 210(1):77–99

Zuchairah IM, Pailthorpe MT, David SK (1997) Effect of glow discharge-polymer treatments on the shrinkage behavior and physical properties of wool fabric. Text Res J 67(1):69–74

Cellulose and Other Polysaccharides Surface Properties and Their Characterisation

8

Karin Stana-Kleinschek, Heike M.A. Ehmann, Stefan Spirk, Aleš Doliška, Hubert Fasl, Lidija Fras-Zemljič, Rupert Kargl, Tamilselvan Mohan, Doris Breitwieser, and Volker Ribitsch

Contents

8.1 Introduction 216
8.1.1 Model Polysaccharide Surfaces: Cellulose 216
8.1.2 Cellulose Nanocrystals (CNC) 219
8.1.3 Textile Cellulose Fibres 219
8.1.4 Pulp Fibres 220

8.2 Methods 222
8.2.1 Polysaccharide Surface Chemical Composition 223
8.2.2 Polysaccharide Surface Morphology: Roughness and Layer Thickness 224
8.2.3 Polysaccharide Interaction Ability 226
8.2.4 Molecular Modelling 229

8.3 Interaction Ability and Surface Characterisation 230
8.3.1 Cellulose Model Films 230
8.3.2 Cellulose Nanocomposites and PS in Nanoparticle Synthesis 230
8.3.3 Textile and Pulp Cellulose Fibres and Foils ..232
8.3.4 Cellulose Charge Modification 235
8.3.5 Polysaccharide Hybrid Materials 238
8.3.6 Modified Polymer Surfaces with Bioactive Polysaccharides 241
8.3.7 Computations with Molecular Modelling 244

8.4 Conclusion and Outlook 246

References 247

K. Stana-Kleinschek (✉)
Laboratory of Characterization and Processing Polymers, University of Maribor, Smetanova ulica 17, 2000 Maribor, Slovenia
e-mail: karin.stana@uni-mb.si

Abstract

This chapter presents comprehensive information about surface phenomena of cellulose and polysaccharide surfaces. It comprises the necessary description of cellulose moieties, of measuring methods and recent results of polysaccharides surface modification and characterisation.

The first part describes different cellulose moieties starting with model cellulose surfaces, allowing basic studies of interface phenomena at well-defined surfaces and provides general information and deepens the understanding of interaction processes. It is extended to nanocrystalline cellulose and further to technological cellulose products.

The importance of structural information as the degree of crystallinity and amorphous regions and the voids size is briefly mentioned in the second part. The state-of-the-art measurement methods providing information about the chemical surface composition, the surface structure and roughness are discussed. The possibilities to measure thickness of cellulose layers using optical and other methods are presented. Particular attention is given to those methods providing information about polysaccharide surface at the solid/liquid interface, the surface energy, the quantification of surface charge, the interaction ability and the quantitative determination of adsorbed mass. A presentation of molecular modelling methods shows the ability of computational chemistry to describe such complex systems. Molecular mechanics force field, semiempirical and *ab initio* methods are described.

In the third part, recent results of surface modification, their characterisation and interaction abilities are presented. This comprises cellulose nanocrystals and nanocompounds, the stabilising effect of polysaccharides and the creation of functional groups on technical and biocompatible cellulose materials. Finally, results of molecular dynamics simulations of the polysaccharide–water interface are presented estimated by semiempirical and *ab initio* methods.

8.1 Introduction

Although there are plenty of known polysaccharides, we will focus mainly on the description and presentation of solid cellulose surfaces and their characterisation. However, some noticeable studies on the characterisation of other polysaccharide surfaces will be shown.

In order to understand and to study systematically solid cellulose surface and interface phenomena, it is important to distinguish between model cellulose surfaces (films and nanoparticles) and 'real' cellulose surfaces in the form of textile fibres, pulp fibres, different foils and other real structures or matrices. All of them differ in their supramolecular structures (fine structure), which define the surface character, interaction capacity and the accessibility of functional groups. For many industrially performed processes on cellulose matrices, accessibility in the water-swollen state is of even greater importance than accessibility in the dry or conditioned state (Stana-Kleinschek et al. 1999; Ribitsch et al. 2001), Fig. 8.1.

Some of the methods used today in modern surface characterisation can only be applied to model surfaces, rather than real systems. Because of the restrictions mentioned above, the differences between model surfaces and some real systems will be presented and discussed in the introduction.

Some specific phenomena in the surface characterisation of real as well as model cellulose matrices will be shown. For prediction of the molecular behaviour and thermodynamic properties, a general overview of the most common molecular modelling models will be briefly presented.

8.1.1 Model Polysaccharide Surfaces: Cellulose

Although there are several examples of polysaccharide model surfaces available in literature, we will focus in this chapter only on cellulose model surfaces due to their high importance in many large-scale industrial applications (Klemm et al. 2005).

In nature, several supramolecular structures of cellulose can be found that are all dominated by the formation of extended inter- and intramolecular hydrogen-bonded networks. One can distinguish between amorphous cellulose and several crystalline polymorphs (cellulose I_α, I_β, II, III, IV) that have been described and characterised recently (for further information, see Klemm et al. 2001, 2005). They differ in the overall organisation of the chain and thus in their hydrogen-bond structures. Only a few solvents are capable of penetrating and breaking up the extended hydrogen-bonded networks and dissolving cellulose on a molecular level. Common cellulose solvents are DMA/LiCl, ionic liquids such as EMIMAc and NMMO, which is used in the lyocell process (Liebert 2010).

The lack of a broad spectrum of solvents for cellulose, as well as the occurrence of several supramolecular structures, complicates the preparation of cellulose model films. Model films are commonly used to study the surface characteristics of natural polymers such as cellulose. The advantages of model films made from cellulose are obvious: a defined supramolecular structure, a defined surface morphology and a defined layer thickness (Kontturi and Tammelin 2006; Saarinen et al. 2008). In addition, model films can be prepared much faster and with higher reproducibility than macro-samples and enable surface interactions like adsorption/desorption to be studied in detail by means of QCM technology for instance (Kargl et al. 2012; Mohan et al. 2012a, b; Saarinen et al. 2008; Tammelin et al. 2006).

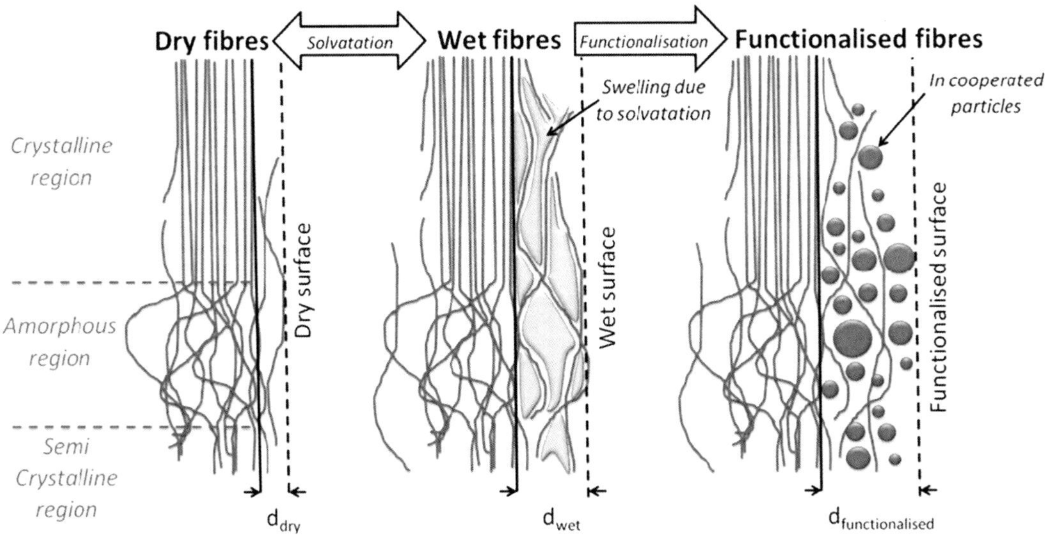

8 Cellulose and Other Polysaccharides Surface Properties and Their Characterisation

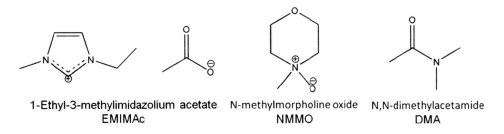

Fig. 8.1 Differences in dry, wet, swollen and functionalised fibres

Fig. 8.2 Chemical structures of some common cellulose solvents

Before we start to discuss the different possibilities for model film preparation, we have to define the demands on our model films. These demands can vary depending on the intended application. In most publications, the most important point is not only chemical but also morphological equivalence, because surface interactions such as adsorption or wetting are strongly dependant on the surface morphology and composition (Aulin et al. 2009; Eriksson et al. 2007; Notley et al. 2006; Notley and Wågberg 2005). Therefore, model films from cellulose in its native crystalline state (cellulose I) as dispersions in water have been prepared and characterised intensively during recent years (Ahola et al. 2008; Cranston and Gray 2008; Edgar and Gray 2003; Habibi et al. 2007).

However, as mentioned above, the preparation of model films is hampered by several difficulties, and a number of strategies have evolved. Besides the deposition method (usually spin coating or LB), the choice of solvent used for deposition is crucial. Unfortunately, some solvents (e.g. EMIMAc) react with cellulose to a certain extent; therefore, model surfaces gained from such solvents lack purity and no longer represent pure cellulose phases (Ebner et al. 2008). For other solvent systems (e.g. DMA/LiCl), salts added to break up the hydrogen-bonded network have to be removed after the model surface preparation, leading to rough model surfaces. Whilst the obvious strategy of using cellulose solutions is somewhat limited because of the limited number of commercially available solvents, the preparation of modified, more soluble cellulose compounds that may be converted to cellulose

Fig. 8.3 Schematic view of the preparation of cellulose model surfaces made from trimethylsilyl cellulose (TMSC) (Mohan et al. 2011)

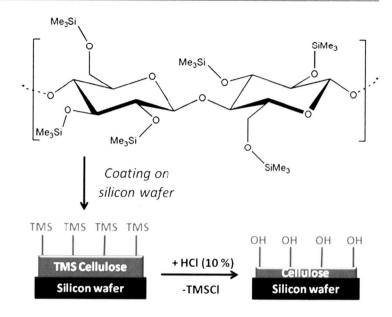

Fig. 8.4 Mechanisms of the acid hydrolysis of cellulose (*top*) and esterification of cellulose nanocrystals with sulphuric acid (*bottom*) (Habibi et al. 2010)

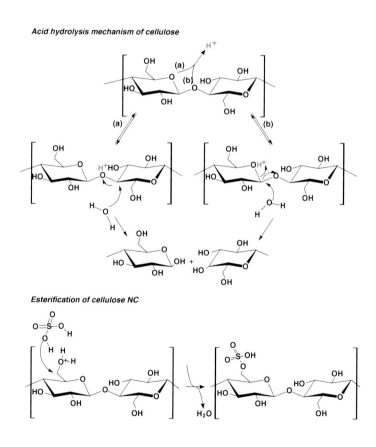

after the coating procedure offers many possibilities for the preparation of model surfaces. Here we will just refer to the latter method as the results are more reproducible. Several derivatives (xanthogenates, cellulose acetates) have already been considered; however, from our point of view, the most straightforward approach to prepare cellulose model surfaces is the use of trimethylsilyl cellulose (TMSC) (Schaub et al. 1993). Several detailed studies on the preparation, characterisation and utilisation of thin cellulose films regenerated from TMSC have been performed (Holmberg et al. 1997; Kontturi et al. 2003; Mohan et al. 2012; Rossetti et al. 2008; Wegner et al. 1996).

After coating on a flat substrate such as silicon wafers, the silyl groups can be removed very easily by simple exposure to an acidic atmosphere yielding cellulose model films, whose thickness can be tuned to a certain degree. The by-product, trimethylchlorosilane, is hydrolysed under these reaction conditions to yield silanol, which in a subsequent step should condense to give HMDS (hexamethyldisiloxane), which can be easily removed in vacuum.

In contrast to pulp and fibres, these model films provide a well-defined surface representing an ideal tool to study surface interactions (e.g. with electrolytes, surfactants, inorganic particles, biomolecules) such as adsorption (Penfold et al. 2007; Sczech and Riegler 2006; Tanaka et al. 2004). In addition, characterisation of these films is easier to perform due to homogeneity reasons. Therefore, analytical methods like AFM, XPS, ATR-IR and SEM are more meaningful than for "real" samples like fibres or pulp. However, the model films consist only of cellulose II, and therefore, any conclusions drawn (e.g. the adsorption behaviour on this substrate) cannot be transferred to other supramolecular cellulose structures without any constraints. Nevertheless, from an industrial point of view, cellulose II is the most important cellulose allotrope because it is widely present in manufactured films and also fibres.

8.1.2 Cellulose Nanocrystals (CNC)

In the 1950s, Ränby reported for the first time the preparation of colloidal suspensions of cellulose nanocrystals (CNC) by controlled acid hydrolysis of cellulose. Later, other starting products such as cotton (Favier et al. 1995), algae (Hanley et al. 1992), bacteria (Grunert and Winter 2002) and wood (Beck-Candanedo et al. 2005) were used for CNC preparation. Depending on the source and the hydrolysis conditions, the geometric dimensions of CNCs like length and width, can be varied. Commonly employed acids for CNC preparation include sulphuric, hydrochloric, phosphoric and hydrobromic acids (Filpponen 2009; Habibi et al. 2010; Usuda et al. 1967). Upon treatment of cellulose with strong acids (cooling is necessary to control the reaction), crystalline regions of the cellulose are degraded slower than amorphous regions. If hydrochloric acid is used as the hydrolysing agent, the ability to disperse the CNCs is limited and the aqueous suspensions tend to flocculate (Araki et al. 1998). If sulphuric acid is used, the surface hydroxyl groups of cellulose are esterified yielding charged surface sulphate esters. As a consequence, the CNC suspensions are stabilised electrostatically.

Due to the high surface-to-volume ratio, many hydroxyl groups are accessible for further modification including esterification, etherification, oxidation, silylation, polymer grafting, etc. In the field of nanocomposites, CNCs have become more and more important because of their appealing intrinsic properties such as nanoscale dimensions, high surface-to-volume ratio, unique morphology, low density and mechanical strength (Habibi et al. 2010).

8.1.3 Textile Cellulose Fibres

Different textiles and natural and regenerated cellulose fibres are good examples of 'real' cellulose matrices. The natural ones (cotton, flax, coco, ramie, etc.) exhibit excellent mechanical properties and may contain about 10–15 % of different noncellulosic compounds such as hemicellulose, pectic substances, proteins, waxes, pigments and mineral salts. Different production processes for regenerated fibres (like viscose, modal, lyocell) cause differences in their structure and in their reactivity (e.g. adsorption character), despite having the same chemical composition (Albrecht et al. 1997; Berger 1994; Brüger 1994; Cole 1996; Cook 1984; Krüger 1994). Not surprisingly, the resulting composition is also different from the naturally occurring ones.

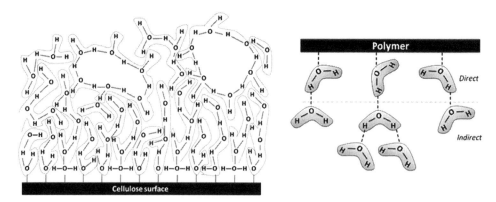

Fig. 8.5 Cellulose surfaces in the aqueous media (*left*), direct and indirect adsorption of water molecules on a polymer surface (*right*)

Cellulose textile fibres have a crystalline/amorphous microfibrillar structure (two-phase model), regardless of the transformation process of cellulose into the solution and further fibre-forming processes. The fibres consist of elementary fibrils, which consist of a succession of crystallites and intermediate less-ordered amorphous regions (Klemm and Philipp 1998; Krässig 1993; Schurz 1994; Schuldt et al. 1994). The amorphous regions and inner surface area of voids are decisive factors with regard to the fibre accessibility, reactivity and adsorption properties (Kreze et al. 2002). The determination of the correlation between structural differences and the adsorption abilities of fibres by determining the surface character of the fibres is especially important, since in this way the polymer ability to interact with the ingredients of the liquid phase, such as several kinds of ions, specific enzymes, surfactants and dyes, can be assessed. The formation of water clusters at the accessible surfaces due to the high cellulose sorption capacity is shown in Fig. 8.5. The swelling of the polymer surfaces may cause the immobilisation of water clusters including the counterions by polymer loops, which can be studied using different surface characterisation tools.

Cellulose-water interactions can be best understood as a competition in hydrogen-bond formation between the hydroxyl groups of the polymer and those between the polymer and a water molecule or water cluster (Klemm and Philipp 1998). The water penetrates into the fibre in the form of vapour or liquid water. It breaks down the secondary interactions between cellulose macromolecules and is absorbed into the fibre by hydrogen bonds, causing the fibres to swell. The amount of water vapour adsorbed at a certain temperature and humidity is one of the important criteria in describing the adsorption properties of regenerated cellulose.

Fibre sorption properties are influenced by the molecular (chemical structure, molecular mass, number of functional groups) as well as the supramolecular structure of fibres (molecular orientation, degree of crystallinity, crystallite dimensions, portion of amorphous regions, size and shape of voids, etc.). The amount of accessible hydroxyl and carboxyl groups and the portion of amorphous regions where the adsorption processes preferentially take place have a significant influence on the adsorption properties of fibres (Fras et al. 2004). Namely, neither water nor aqueous solutions of electrolytes, dyes or surfactants can penetrate into the crystalline regions of the fibre.

Nowadays, many different modern methods to elucidate the interaction ability and surface properties of textile cellulose fibre surfaces do exist (e.g. determination of the electrokinetic character of the fibre samples, contact angles and surface free energy determination, titration techniques). Some examples will be shown in the following chapters.

8.1.4 Pulp Fibres

Pulp fibres differ from man-made (regenerated) fibres and foils regarding their morphology,

Table 8.1 Characterisation possibilities of dry and wet polysaccharides in solid matrices (Zaera 2012)

Structure (Dry state)	Accessibility of functional surfaces/groups (Wet state)
SWAXS methods	Electrokinetic techniques
X-ray photoelectron spectroscopy	Titration techniques
FT-IR spectroscopy	UV–VIS spectroscopy, calorimetry, QCM
Scanning electron microscopy	Water retention/iodine sorption value
Atomic force microscopy	Size exclusion chromatography
	Tensiometry/goniometry

chemical composition and surface properties. This difference is a result of mechanical and chemical treatment and the sources used for pulp fibre production. The major source of pulp fibres is wood with some production based on annual non-wood plants with the use of recovered pulp fibres having steadily increased for decades. Wood as a raw material can be subdivided into two main sources, namely, softwood and hardwood. In general, softwood yields longer fibres with higher strength, whereas hardwood has better sheet-forming properties in paper production, shorter fibres and higher hemicellulose content. The structure and cell wall of a single fibre tracheid, the main constituent of softwood, is complex as it consists of several fibre walls that differ in the content of lignin, hemicelluloses, cellulose, pectin and waxes. In principle, each individual tracheid is separated from other fibres by the middle lamella. This middle lamella is rich in lignin, pectin and waxes and encloses the primary cell wall. The primary and secondary cell walls differ in the orientation and amount of cellulose microfibrils. Whereas the thinner primary wall (P) consists of randomly oriented cellulose fibrils, the next inner secondary wall (S1) consists of cellulose fibrils wound around the innermost secondary walls (S2 and S3). These two walls contain up to 85 % cellulose oriented along the fibre axis. Similar structures can be found in hardwood, even though there are other structural features like vessel elements that are responsible for water transport within the plant. Non-recovered pulp fibres are obtained either by chemical or mechanical disintegration of the above-mentioned sources of cellulose. Mechanical pulp can be thermally or chemically pretreated before mechanical refining. This process gives high yields because most of the lignin and hemicelluloses remain in the end product. On the other hand, chemical pulp is obtained through the digestion of lignin and parts of hemicelluloses in the alkaline sulphate or acidic sulphite process. The sulphate process (or kraft process) is, worldwide, by far the most important chemical pulping process and yields higher-strength paper than the sulphite process. In order to further reduce the amount of lignin in the pulp, the chemical pulp can be bleached additionally either with chlorine, elementary chlorine free (e.g. NaOCl) or chlorine free (with, e.g. oxygen or hydrogen peroxide). All these processes together with the cellulose source strongly determine the surface properties of the pulp fibres. Nevertheless, it is worth noting that most man-made fibres, as well as cellulose derivatives, are made from pulp as the primary source for cellulose that, depending on the purity, partly determines the properties of these materials. The properties of paper based on pulp fibres strongly depend on the lateral dimensions and interaction forces between single fibres. In a dry state, the fibres are bonded by hydrogen bonds, whereas in aqueous systems they can be easily dispersed. This makes cellulosic fibres ideal candidates for paper and pulp recycling. Besides the fibre dimensions, the existence of charges on the pulp surfaces is a dominating factor in its properties. Charges are either native (stemming from charged hemicellulose species) or are introduced during chemical or mechanical pulping. Usually kraft fibres possess fewer charges than sulphite pulp. This originates from sulphonated residual lignin, which bears sulphonic acid groups. In bleached pulps, the amount of charges is reduced due to the solubilisation of lignin and hemicelluloses during bleaching. Charges on the surfaces, the existence of hydrophobic components as well as the accessibility of the cellulose surface are the dominating factors when it comes to the adsorption of polyelectrolytes or water-soluble polymers on a pulp fibre, which are important phenomena in papermaking and the surface modification of cellulosic products from pulp (Zhou et al. 2006; Horvath et al. 2008; Biermann 1996; Groth and Wagenknecht 2001; Holik 2006). Therefore, a detailed characterisation of the elemental composition, charge, lignin content, amount of extractives and fibre morphology is a prerequisite to correlate the complex pulp fibre system with its surface properties.

8.2 Methods

The investigation of the structure and morphology of solid polysaccharide materials has been performed since many years. The characterisation of their surfaces, interfaces and interaction with compounds of the liquid phase is a less-investigated topic. Despite the development of new techniques and methods in the past years, it is still a challenging and often a time-consuming task. The difficulties in the analyses stem from the appearance of a variety of different structural features present in cellulose solid matrices (e.g. crystalline and amorphous regions, voids of different dimensions in the dry and swollen states, etc.) (Krässig 1993; Kreze and Malej 2003; Kreze et al. 2001, 2002), which results in different degrees of accessibility for solvents and reagents of the OH groups. Therefore, to describe the surface properties of a cellulose sample appropriately, several characterisation methods have to be applied. The choice of which characterisation methods are employed depends of course on the kind of samples (e.g. dry or swollen state, model film or real sample) as well as the type of parameters of interest (crystallinity, morphology, void area, etc.).

Several methods can be used to determine the degree of crystallinity, such as wide-angle X-ray scattering (WAXS) (Fink et al. 1985; Hermans and Weidinger 1949; Ruland 1961), ^{13}C-CP/MAS solid state NMR and FT-Raman spectroscopy (Schenzel et al. 2005). In Table 8.2, the crystallinities of different regenerated textile fibres determined by these methods are compared.

Other important factors in the description of cellulose solid matrices are the determination of the inner surface area, the pore volume and the pore size. Nitrogen adsorption, DSC–freezing point depression (Park et al. 2006), scattering methods like SAXS or SANS (Gray et al. 2010) and inverse SEC (Table 8.3) can be applied for this purpose.

A combination of different methods is needed to achieve a comprehensive material characterisation describing both processing as well as utilisation properties. Some surface and interface characterisations are particularly important in polysaccharide characterisation and are described briefly in the following:

- Chemical composition analyses (XPS, ToF-SIMS, MALDI).
- Morphology, roughness and layer thickness analyses (SEM, AFM, profilometry, ellipsometry, SARFUS).
- Interaction ability studies (charge determination, electrokinetic properties, QCM-D).

For further information on thermodynamics and interaction capacities among polysaccharides and

Table 8.2 Degree of crystallinity of different regenerated textile fibres determined by different methods (Fowkes 1962)

Method	Lyocell	Modal	Polynosic	Viscose
X_c (IR) (%)	46	34	29	34
X_c (Raman) (%)	40	30	25	29
X_c (NMR) (%)	34	22	13	19
X_c (WAXS) (%)	40	30	25	30

Table 8.3 Pore volume (V_p), specific inner surface (S_p), pore diameter (d) and surface per volume of cellulose textile fibres determined using SEC (Gruber 1998)

Parameter	Cotton	Lyocell	Modal	Viscose
Voids volume V_p (mL/g)	0.291	0.534	0.494	0.672
Specific inner surface S_p (m^2/g)	207	418	408	444
Void diameter d (nm)	2.9	2.7	2.3	3.1
Surface per volume (m^2/mL)	702	783	859	657

8 Cellulose and Other Polysaccharides Surface Properties and Their Characterisation

Table 8.4 The most common molecular mechanics force fields with their corresponding energy terms (Young 2001)

Name	Use	Energy term
Harmonic	Bond stretch	$k(l - l_0)^2$
	Angle bend	$k(\theta - \theta_0)^2$
Cosine	Torsion	$k[1 + \cos(n\theta)]$
Lennard-Jones 6–12	van der Waals	$\left[4k\left(\frac{A}{r}\right)^{12} - \left(\frac{B}{r}\right)^6 \right]$
Lennard-Jones 10–12		$\left[4k\left(\frac{A}{r}\right)^{12} - \left(\frac{B}{r}\right)^{10} \right]$
Coulomb	Electrostatic	$\dfrac{q_1 q_2}{4\pi\varepsilon_o r}$
Taylor	Stretch-bend	$k(\theta - \theta_0)[(l_1 - l_{1_0})(l_2 - l_{2_0})]$
Morse	Bond stretch	$D_e[1 - e^{-\alpha(l - l_0)}]^2$

Table 8.5 Some examples of semiempirical parameterisation models (Young 2001)

Name	Included terms	Use
CNDO (Complete neglect of differential overlap)	– Models valence orbitals with a minimal basis set of Slater-type orbitals	Some hydrocarbons
AM1 (Austin model 1)	– The atomic core terms are modified (addition of attractive & repulsive Gaussian functions) – More accurate for H-energies	For organic compounds
PM3 (Parameterisation method 3)	– Very similar equations to AM1 – More accurate for H-bond angles	For organic compounds

Table 8.6 *Ab initio* approximations (Young 2001)

Approximation	Included terms	Remarks
1. Central field approximation	– Coulombic electron–electron repulsion term by variational calculation ($E_{\text{(calculated)}} \geq E_{\text{(real)}}$) – The many-electron Schrödinger equation gets broken into many simple one-electron equations	Only the average effect, not the explicit repulsion Each one-electron equation is solved (= orbitals and orbital energies)
2. Gaussian-type orbitals (**GTO**)	– LCAO of basis functions – $E_{\text{(calculated)}} \geq E_{\text{(real)}}$ – Gaussian functions are multiplied by an angular function (for s-, p-, d- and f-orbital symmetry)	**s**: constant angular term **p**: x, y, z angular term **d**: xy, xz, yz, x^2-y^2, $4z^2-2x^2-2y^2$ angular term...

the component of the liquid phase, molecular modelling computations have been developed in the last few decades.

8.2.1 Polysaccharide Surface Chemical Composition

8.2.1.1 XPS: X-ray Photoelectron Spectroscopy

In polymer surface modification, it is of interest to identify the presence of specific functional groups. X-ray photoemission spectroscopy (XPS) or electron spectroscopy for chemical analysis (ESCA) can be used to determine the atomic composition of the top few nm solid surface. Upon exposure to X-ray photons, a surface will emit photoelectrons whose binding energies will be compared to known values in order to identify the elements and their oxidation state. The resulting spectrum is a plot of intensity (arbitrary units) versus binding energy (eV). The intensity of the emitted photoelectrons relates directly to the material surface atomic distribution and is therefore used to quantify the relative atomic composition and stoichiometric ratios (Gray et al. 2010; Zuwei et al. 2007).

The following information can be obtained:
- Elemental composition on the surface (1–8-nm depth).
- Thickness of one or more thin layers (1–8 nm) of different materials.
- Electronic state providing binding energy and material density.

The detection limits for most elements are in the range of parts per thousand; however, hydrogen cannot be detected.

8.2.1.2 Surface Analytical Methods Using Time-of-Flight Mass Spectrometry

Because XPS analysis is limited to the top few nm of the surface, polymer samples must be handled carefully as even minor surface contaminations could spoil the resulting spectrum. Time-of-flight secondary ion mass spectrometry (ToF-SIMS) is used to determine the type and quantity of ionisable chemical groups at the surface. When a surface is subjected to a beam of primary ions, it will emit secondary ions, which will be separated according to their mass to charge ratio (m/z) in a mass spectrometer. The resulting spectrum depicts signal intensity versus m/z and is used to gauge the relative intensities of chemical species. It also provides two-dimensional chemical maps for use in establishing surface homogeneity. Furthermore, etching of the sample with fullerene ions allows delicate depth profiling. Results obtained at different sample tilts, in combination with XPS, can provide an insight into the gradients of elements in surface layers (Belu1 et al. 2003).

In Maldi-ToF, the ions that originate from surfaces of macromolecular species are generated using matrix-assisted laser desorption/ionisation (MALDI). The ionisation is caused by a laser beam, and the energy is absorbed by a matrix molecule and transferred to the macromolecule; in this way, its fragmentation is avoided and vaporisation and ionisation are facilitated (Chensong et al. 2007).

8.2.2 Polysaccharide Surface Morphology: Roughness and Layer Thickness

8.2.2.1 SEM: Scanning Electron Microscopy

During SEM measurement a high-energy electron beam is scanned in a raster beam pattern, and the electrons interact with the surface atoms. Electrons (secondary electrons) are emitted from the specimen's surface and provide information about surface topography, composition and electrical conductivity. Specimens must be electrically conductive (at least at the surface and electrically grounded). Nonconducting specimens must be coated with an ultrathin coating of electrically conductive materials (Au, Au/Pd, graphite) by sputtering (Andrade et al. 2009; MacArtain et al. 2003; Van Riessen et al. 1994; Zuwei et al. 2007).

8.2.2.2 AFM: Atomic Force Microscopy

AFM is a method used to measure a surface's profile and to quantify its roughness in the sub nm range. The AFM scans the surface in three dimensions using a cantilever with a sharp tip with a curvature of several nm. Forces (mechanical contact, van der Waals, dipole-dipole interaction, electrostatic and magnetic) between the tip and the sample in close proximity deflect the cantilever, and this deflection is measured using a laser spot reflected from the top surface of the cantilever. Imaging by AFM is performed under static or dynamic conditions and different modes:
- Contact mode: The tip makes contact with the sample surface, and the tip deflection is used as a feedback signal.
- Non-contact mode: The oscillating cantilever's tip does not make contact with the sample surface (amplitude typically < 10 nm). The forces between tip and sample change the oscillation, and this is used to construct a topographic image.
- Tapping mode: The cantilever is oscillated at large amplitude (around 100 nm) and is

scanned over the sample. Tapping mode is gentle enough to visualise soft surfaces, adsorbed polymers and even under a liquid medium.
- Force spectroscopy: The tip-sample interaction forces are measured as a function of their distance, resulting in the so-called force-distance curve. This provides information about nanoscale interaction forces (van der Waals, dipole-dipole, dispersion, electrostatic).

The force resolution's detection limits are in the pN range, whilst the distance resolutions are better than 0.1 nm. Several advantages of this method are that it provides three-dimensional surface imaging at high resolution with no specific surface treatment required and it is possible to investigate biological species under in situ conditions (Abu-Lail and Camesano 2003; Kontturi et al. 2003a, b, c; Spirk et al. 2010).

8.2.2.3 Profilometry

Profilometers are instruments used to determine a surface profile and to quantify its roughness. There are two different types of profilometers (Jianshe 2007; Kocher et al. 2002):
- Contact profilometers: A diamond stylus (tip) is laterally moved across the sample at a specified contact force. Small vertical features ranging in height from 10 nm to 1 mm can be measured. This method is applicable on many kinds of surfaces, and impurities can be easily identified.
- Non-contact profilometers: An optical profilometer provides the information without coming into contact with the surface. Methods used are laser triangulation, confocal microscopy and coherent interferometry. The advantage of the non-contact method is that no damage occurs to the tips or surfaces. The profilometer vertical resolution is on a nm level (minimum 10 nm), and their horizontal resolution is in a range of 0.1–10 micrometres. The spot size ranges from a few μm to sub-μm.

8.2.2.4 Measurement of Surface Layer Thickness

Ellipsometry

Ellipsometry is an optical technique used to investigate the dielectric properties of thin films. It analyses the polarisation of light that is reflected from a sample and yields the refractive index or dielectric function tensor. It is commonly used to characterise the film thickness of single layers or complex multilayers ranging from sub nm to several micrometres. The advantages of ellipsometry are that specific surface modifications or a vacuum is not required and it can be applied to surface layers in their natural state. It is a disadvantage that in most cases a layer model must be established that considers the optical constants and thickness parameters. The model that best matches the experimental data is used to determine the optical constants and thickness parameters. The resolution is in the range of a few nm (Bergmair et al. 2009; Zuwei et al. 2007).

Sarfus

Sarfus is an optical quantitative imaging technique that analyses the polarisation state of reflected light from specific supports (surfs). These surfs provide contrast amplification and do not change the light polarisation after reflection. The polarisation is changed by samples deposited on the surf, and this is used to determine the layer thickness. The layer thickness is determined by comparing a colorimetric calibration standard with the analysed sample. The method visualises thin films (between 0.3 and 70 nm) and isolated nano-objects directly in real time and under environmental conditions (also in air or even in water). An additional advantage is the 3D representation of the analysed samples, and the sarfus method provides easy access to topological parameters such as thickness and roughness. A disadvantage is that the surface layer thickness measurement is limited to 70 nm thick layers and that the films of interest have to be deposited onto special surfaces, so called 'Surf' (Ausserre and Valignat 2006; Souplet et al. 2007; Valles et al. 2008).

8.2.3 Polysaccharide Interaction Ability

8.2.3.1 Contact Angle and Surface Energy Determination

By determining the static water contact angle (CA) on cellulose substrates with liquids of different polarity, information on the surface energy and, as a consequence, on the wettability of a surface is obtained (Chan 1994). This information is based on several models–(Fowkes 1962, 1964; Zisman 1964), equations of state (EOS) (Kwok and Neumann 1999), Owens, Wendt, Rabel, Kaeble (OWRK) (Kälble 1969; Kaelble 1972; Owens 1969; Rabel 1977) and Lifshitz-van der Waals acid base (LWAB) (van Oss et al. 1988)–that include different terms (e.g. polar, apolar, Lewis-acid–base interactions) contributing to the total surface energy. In the past decades, these models have been proven to predict surface properties quite well, and changes in the surface chemistry are reflected in a change in the total surface free energy, making contact angle measurements an important tool in the surface analysis of cellulose surfaces.

In contrast to the information derived from static water contact angle determinations, the determination of advancing and receding contact angles gives insight into the dynamic interactions between gas, liquid and solid phases. The advancing contact angle is determined by depositing a drop on a surface, measuring the contact angle and subsequently increasing the volume of the drop. Once the drop reaches a certain volume, the contact angle does not increase further; this is the advancing contact angle. The receding contact angle is determined by removing water from a certain drop volume until the contact angle does not decrease anymore; this is the receding contact angle. The difference between the advancing contact angle and the receding contact angle is called the contact angle hysteresis.

For fibrous materials, the powder contact angle method is used to determine the CA. The sample weight is measured as a function of time during the water adsorption phase. The contact angle between the solid (polymer sample) and water is calculated using a modified Washburn equation (Washburn 1921).

Contact angle determinations are simple and rapid measurements of surface hydrophilicity/hydrophobicity, but, purely on this basis, it is not possible to distinguish between different hydrophilic/hydrophobic functional groups present at the surface. Therefore, spectroscopic methods (i.e. XPS, ToF-SIMS) that are more precise and accurately describe changes in the surface chemistry must be applied as mentioned above.

The advantages of these methods include the rather low complexity and inexpensive equipment combined with the possibility of distinguishing between different contributions to the surface energy. The disadvantages are that the surface morphology plays an important role and that many repetitions are needed to obtain reliable results. The precision of CA depends on the method used and usually varies between 1 and 3 degrees.

8.2.3.2 Quantification of Surface Charge

The methods described below have been extensively applied in high-precision titrations aiming at determinations of complex equilibrium in solution (Fras Zemljič et al. 2004; Herrington and Petzold 1992; Laine and Stenius 1997; Laine et al. 1996; Wågberg et al. 1989; Zhang et al. 1994).

Potentiometric Titration

It is of interest to be able to quantify the immobile charges located at polysaccharide polymer surfaces resulting from weak acid groups. By applying a potentiometric titration to dispersed polymers (fibres), the concentration of free hydrogen ions [H^+] is determined after the stepwise addition of sodium hydroxyl (NaOH). This titration process is performed in an electrochemical cell whose potential is measured in order to determine the proton concentration. At high ionic strength, it can be assumed that the activity coefficients of hydrogen and hydroxyl ions are constant, and the cell potential is given by the Nernst equation. By determining the cell potential (proton concentration) as a function of the added NaOH volume of a blank solution and of a polymer dispersion, it is possible to calculate the amount of bound weak acids. The amount of acidic groups is calculated from the difference in the added NaOH volume between the blank and dispersed polymer sample at any given pH. Potentiometric titrations can also be used to evaluate the apparent dissociation constants of acids by applying Räsänen's model (Räsänen et al. 2001a, b). The dissociation constant of weak acids on polymer surfaces depends on the degree of dissociation due to increased electrostatic interaction between hydrogen ions and the surface as the degree of dissociation increases. The crossing point of the charge versus pH functions at different ionic strengths provides the point of zero charge (Biederman and Sillen 1952; Ciavatta 1963; Gran 1950; Laine et al. 1996; Lindgren et al. 2000; Räsänen et al. 2001a, b).

Conductometric Titration

The conductometric titration of weak dissociable groups on dispersed polymer surfaces is similar to that of soluble acids. The parameter measured is conductance, which is an additive function of the products of the concentration and equivalent conductance of each type of ion present in solution (Sjöström and Enström 1966; Katz and Beatson 1984). Increases or decreases in conductance are associated with changed concentrations of the two most highly conducting ions–hydrogen and hydroxyl ions. It is necessary to perform the measurement in the presence of a neutral salt (0.001M KCl) to avoid the Donnan equilibrium causing an unequal distribution in the mobile ions between the interior of the fibre wall and the external solution. Polymer surfaces are first converted to H^+–which means that all acidic groups are protonated. After this the polymer dispersion is titrated with NaOH, which leads to three sections in the titration function if the conductance is plotted against the amount of added NaOH (1) the neutralisation of liberated protons slightly lowers the conductivity, (2) neutralisation of carboxylic groups does not change the conductivity and (3) the increasing NaOH concentration in excess leads to increased conductivity. The total amount of acidic groups is determined from the intersection between the minimum conductance and the Sect. 3 linear extrapolation (Laine et al. 1994; Wågberg et al. 1989).

Polyelectrolyte Titration

When performing a charge titration using charged surfactants or polyelectrolytes as a titrant, the streaming potential is used as an indicator to determine the charge neutralisation. To quantify the charge of dispersed polymers, oppositely charged species are adsorbed on the polymer surface. The unadsorbed amount is determined using polyelectrolyte titration. To determine the total accessible charge, low-molecular polyelectrolytes or surfactants are added to the dispersed polymers. The charges on the surface as well as voids are compensated by the charged titrant molecules, and their concentration in the liquid phase after adsorption is determined. To determine the surface charge, high molecular weight polyelectrolytes are used that neutralise only charges at the outer surface. The experiments are performed via recording of the adsorption isotherms. This is done by determining the amount of titrant adsorbed as a function of the equilibrium concentration of titrant in the solution and extrapolation to zero solution concentration. A 1:1 charge stoichiometry is only achieved at low ionic strength (Laine 1994; Laine et al. 1996; Wågberg et al. 1989; Waltz and Taylor 1947; Wolf and Mobüs 1962; Scallan and Katz 1989).

8.2.3.3 Electrokinetic Phenomena: The ζ-Potential

The ζ-potential (ZP) can be used to estimate the effect of the surface charge on different processes such as aggregation, adsorption stability of dispersions, etc. The ZP is not the surface potential, but the potential at the border between the stationary and mobile liquid phases in the vicinity of a solid phase.

This represents the primary interaction area for molecules and particles diffusing from the liquid bulk phase towards the solid surface. It depends on the surface, as well as the liquid, chemical properties and is therefore a parameter that describes changes in a solid surface, as well as its interaction properties in contact with an aqueous (polar) liquid. Ideally, it would be best to measure the surface potential directly, but this is not experimentally accessible. In practice, the zeta potential is measured via different electrokinetic phenomena such as electrophoresis for small particles, streaming potential/streaming current for macroscopic surfaces and acoustophoresis for concentrated dispersions (Hunter 1981; Stana-Kleinschek and Ribitsch 1998).

The zeta potential is sensitive to changes in surface groups–surface modification by chemical and physical methods, dissociation of surface groups as a function of pH, changes in polarity of the surface, swelling (shift of the shear plane into the liquid phase) and the adsorption of components of the liquid phase, for example, surfactants, polyelectrolytes, dyes and specifically adsorbed ions. The electrokinetic experiments enable the determination of the zeta potential as a function of the pH, the determination of the isoelectric point as well as the determination of the affinity of components present in the liquid phase to the surface (adsorption kinetics) (Ribitsch et al. 2001; Stana-Kleinschek and Ribitsch 1998).

For streaming potential measurements, as usually applied for macroscopic surfaces, the ZP is calculated from the primary data streaming potential using the Smulochowski equation (Smulochowski 1903) and the approximation of Fairbrother and Mastin (Faibrother and Mastin 1924).

The advantage of this method is the great sensitivity to chemical changes on the surfaces and to adsorption processes, both of which are measurable in an aqueous environment. The disadvantages are the influence of surface swelling processes and the limits in determining directly the nature of dissociated groups from ZP measurements.

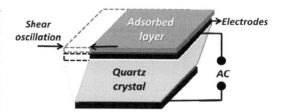

Fig. 8.6 Working principle of quartz crystal microbalance with dissipation (QCM-D) (Johannsmann D (2008) Phys Chem Chem Phys10:4516–4534)

8.2.3.4 QCM-D: Quartz Crystal Microbalance Dissipation

Quartz crystal microbalance (QCM) is a suitable method for adsorption/desorption studies at the solid/liquid interface. It can be used to monitor directly the adsorption of components of the liquid phase on the surface of polymer materials.

A QCM sensor consists of a thin quartz crystal plated with electrodes on the top and bottom as shown in Fig. 8.6. It is excited to oscillate in a transverse shear mode at its resonance frequency by applying an AC field across the electrodes. The resonance frequency is highly sensitive to the total oscillating mass; the damping characteristics are sensitive to the properties of the adsorbed mass. The resonant frequency ($f_0 \approx 5$ MHz) of the crystal decreases with additional mass adsorbed on its surface. The decrease in frequency is proportional to the adsorbed mass if the adsorbed layer is rigid, evenly distributed and much smaller than the mass of the quartz crystal. The adsorbed mass can be calculated using the Sauerbrey equation (Sauerbrey 1959). In the case of non-rigid adsorbed layers, the damping characteristics of the crystal oscillation (—the dissipation factor D—) provide information on the viscoelastic properties of the adsorbed layers (Marx 2003).

For adsorption measurement, the quartz crystal must be coated with the polymer material of interest. The adsorbing material is flushed along the surface, and changes in the total mass (adsorbed mass) as well as the viscoelastic properties of the adsorbed film are determined (Indest et al. 2008; Reimhulta et al. 2003; Zuwei et al. 2007).

The advantage of this method is that the smallest amounts of adsorbed material can be detected *in situ*. The disadvantage is that absolute values of adsorbed mass are not obtained. An approximation via the Sauerbrey equation is applied to calculate the adsorbed mass. The resolution of mass changes is around 20 ng/cm^2.

8.2.4 Molecular Modelling

Computational chemistry (molecular modelling) is the science of representing and manipulating molecular structures numerically by simulation of the molecular behaviour with quantum and Newtonian mechanic equations. Computational chemistry programmes allow scientists to generate and present molecular data including geometries (bond lengths, bond angles and torsion angles), energies (heat of formation, activation energy, reaction enthalpies, orbital energies, etc.), electronic properties (charges, ionisation potential, electron affinity), spectroscopic properties (vibrational modes, chemical shifts) and bulk properties (volumes, surface areas, diffusion, viscosity, etc.). Depending on the size of the investigated molecules, different methods can be used for the calculation of the above-mentioned properties. As in all scientific models, scientist's intuition and training are necessary to interpret the results critically and to compare them with experimental data. In the following, several approaches will be presented (Hyperchem 7 release 2002; Lipkowitz and Boyd 1990; Sax 2008).

8.2.4.1 Molecular Mechanics Force Field

This model includes the Newtonian mechanic approximations to get the optimal conformation of the considered molecule. The molecular mechanics energy expressions are based on a simple algebraic equation of the total energy of compounds containing constants obtained from spectroscopic data or *ab initio* calculations. Such a set with constants is called a force field, whereas the fundamental assumption is the transferability of parameters, which gives a simple calculation for large molecules (Berkert and Allinger 1982).

8.2.4.2 Semiempirical methods

Semiempirical calculations include a Hamiltonian and a wave function, whereas the core electrons are not included. This omission causes an error, which is corrected by introducing parameters into the method (e.g. from experimental data or *ab initio* calculations from similar molecules). The advantage of parameterisation is an enhancement in computational cost, but the molecules being computed must be similar enough to the ones used for parameterisation, otherwise the results can be very inaccurate. The computational requirements allow them to run conveniently on conventional computers. Vibrational spectra and other spectroscopic data may be predicted in principle, but in such cases it is more reliable to use *ab initio* methods (Clark 1985; Dewar 1969; Stewart 1990).

Semiempirical methods are parameterised to reproduce various results:
- Geometry and energy (usually heat of formation)
- Dipole moment
- Heat of reaction
- Ionisation potential
- Electronic spectra
- NMR shifts

8.2.4.3 *Ab Initio* Methods

The term '*ab initio*' means 'from the beginning' and signifies that the computations are derived directly from theoretical principles, without any experimental data (approximate quantum mechanical calculation) (Hehre et al. 1986).

The Hartree–Fock (HF) approximation is the most common type of *ab initio* calculation (HF). When applying large basis sets, this method usually yields accurate geometries and energies. However, for the calculation of reaction energies, it is less suitable because bond-breaking processes are not described very well within the HF routine.

In contrast, density functional theory (DFT) is able to describe such reaction energies properly. It is based on optimising the electron densities of all atoms present in a molecule. In most quantum

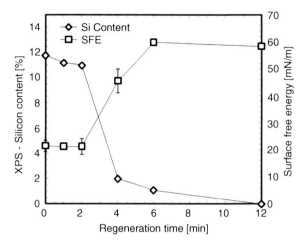

Fig. 8.7 Silicon content and surface free energies of cellulose model surfaces derived from different regeneration times from trimethylsilyl cellulose (Mohan et al. 2011)

chemical programmes, this theory is better implemented than HF, which reduces the computational costs significantly.

8.3 Interaction Ability and Surface Characterisation

8.3.1 Cellulose Model Films

As already mentioned in Chapter 8.1.1 well-defined model surfaces of pure cellulose can be prepared by spin coating a soluble cellulose derivative onto a substrate, followed by conversion back to cellulose via a simple regeneration process. One of these derivatives is trimethylsilyl cellulose (TMSC) with a high degree of substitution (typically more than 2.0) (Schaub et al. 1993). The spin coating parameters and the concentration of the polymer strongly determine the morphology and thickness of the resulting film. The trimethylsilyl groups are removed by being exposed to vapours of hydrochloric acid generating free OH groups. As a consequence the hydrophilicity of the surface is increased (compared to TMSC), as well as the surface free energies (Fig. 8.7). Carefully controlled hydrolysis reaction conditions allow the tailoring of the wettability properties, surface free energies and thickness of such films (Mohan et al. 2012a, b). For interaction studies of polymer and enzymes, for example such partly and fully regenerated surfaces of cellulose are of interest (Kontturi et al. 2003; Mohan et al. 2011).

8.3.2 Cellulose Nanocomposites and PS in Nanoparticle Synthesis

8.3.2.1 Chemical Modifications of Cellulose nanostructures

The incorporation of cellulose nanocrystals into nanocomposites is very attractive due to their nanoscale dimensions, high surface area, unique morphology, low density and mechanical strength. Cellulose nanocrystals have been successfully incorporated into many polymer matrices like polysiloxanes (Grunnert and Winter 2000), polysulfonates (Noorani et al. 2006), poly(caprolactone) (Morin and Dufresne 2002), poly(oxyethylene) (Azizi et al. 2004) or carboxymethyl cellulose (Choi and Simonsen 2006). In addition, many covalent and noncovalent chemical modifications have been attempted such as esterification, etherification, oxidation, silylation, polymer grafting, adsorption of surfactants and polymer coating. One main point of functionalisation is firstly the introduction of a stable positive or negative charge on the surface to increase the electrostatic stabilisation of the resulting dispersions. Furthermore, the tuning of the surface energy characteristics is an important factor; due to this, it is possible to improve compatibility with nonpolar or hydrophobic matrices in nanocomposites. In all modification processes,

Table 8.7 Determined dimensions of the different cellulose nanocrystal structures determined using SEM and TEM

Method	SEM		TEM	
Sample preparation	Au-coated		Redispersed in ethanol	
Spheres	$d = 20–50$ nm		$d = 10–100$ nm	
Rods	$w = 30–50$ nm	$l = \sim \mu m$	$w = < 10$ nm	$l = 200–400$ nm
Network	Not observed		$l, w = \sim \mu m$	

w width, l length, d diameter (Lu et al. 2010)

Table 8.8 Electrostatic fibre spinning with different crystallinities depending on the preparation method (Kim CW et al. 2006)

Degree of polymerisation	Solvent	Temperature (°C)	Degree of crystallinity (peak area)
210	NMMO/H_2O	50	52.9
210	NMMO/H_2O	70	56.8
1140	NMMO/H_2O	70	42.2
1140	LiCl/DMAc	25	~ 0.0

it is important to preserve the original morphology to avoid any polymorphic conversion and to maintain the integrity of the crystals (Habibi et al. 2010).

Preparation of Cellulose Nanocrystals

Cellulose nanocrystals are prepared via acid hydrolysis of cotton cellulose, followed by freeze drying in many protocols. Hydrolysis is performed using 64–65 wt.% sulphuric acid with 10 mL/g cellulose at 45 °C for around 60 min. After centrifugation and dialysis, the suspension is freeze dried. During characterisation rod-, sphere- and network-like structures can be identified, but cannot be separated by the common methods of centrifugation and filtration (Lu et al. 2010).

In Table 8.7, the dimensions of the different structural forms are summarised. The dimensions obtained by SEM are larger than those observed in TEM. A possible explanation may be the aggregation during freeze drying to form larger bundles, which are only loosely packed and which can be easily separated by the addition of ethanol in the course of TEM preparation.

Spheres are the most abundant structural motif; they are formed by self-assembled short cellulose rods via interfacial hydrogen bonding, which overcomes the repulsive interactions of the sulphate groups. Also the network formation occurred due to strong H-bonding interactions but was only observed in TEM. This could be due to the over irradiation of electron beams during TEM measurements.

Nanocomposites

Processing techniques have an important impact on the final nanocomposite properties, the intrinsic properties, the nature of the polymer matrix and also the interfacial characteristics and hence have to be taken into account:

- Casting-evaporation: Transfer of cellulose whiskers from an aqueous dispersion into an organic polymer matrix such as PVA (Ibrahim et al. 2010, Capadona et al. 2007).
- Sol–gel process: It has been reported that a three-dimensional template (due to cellulose nanocrystals' self-assembly) can be filled with a template polymer (Mathew et al. 2006).
- Extrusion: Pumping an aqueous dispersion of CNs coated with a surfactant (Park et al. 2007) into a melt polymer during extrusion.
- Electrostatic fibre spinning: Also known as electrospinning, this is a versatile method to manufacture fibres with diameters ranging from several micrometres down to 100 nm or less. This occurs due to electrostatic forces

and allows for further incorporation into different polymer matrices such as PEO or PVA (Kim et al. 2006).

8.3.2.2 Polysaccharides in Nanoparticle Synthesis

A rapidly growing field in nanoscience is the synthesis of well-dispersed and stable nanoparticles. In particular, silver nanoparticles (AgNP) are interesting targets for the synthesis of new materials. Their size-dependent optical, electronic and chemical properties make them interesting targets for the preparation of new materials (Evanoff and Chumanov 2005; Ferraria et al. 2010; Murray 2008; Schrand et al. 2008, Talapin et al. 2010, Twu et al. 2008). Moreover, it has been reported that silver products containing nanocrystalline silver kill microbes more efficiently than products where silver remains in the cationic form (Bhattacharya and Mukherjee 2008). Besides medical applications such as catheters, scaffolds and wound dressings (Chen and Schluesener 2008), they are also used in textile finishing, mainly in sportswear (El-Rafie et al. 2010; Klemencic et al. 2010). Although several wet chemical methods for the synthesis of AgNPs do exist, the preparation of well-dispersed, stable AgNPs still remains a challenge. In recent years, interest has focused on biopolymers, in particular polysaccharides, and their capability to act as stabilising and reducing agents for noble metal nanoparticles. In particular, biopolymers containing NH_2 groups coordinate metal ions efficiently, especially those with a soft character according to the HSAB-Pearson concept. An example therefore is the preparation of nanoparticles using chitosan, a deacetylated poly-1-4-glucosamine, as a reducing and stabilising agent (Twu et al. 2008). The only problem with using common chitosan as the stabilising and reducing agent is that high molecular weight chitosans are only soluble in acidic solutions, which influences the formation and aggregation of the nanoparticles. Therefore, water-soluble derivates like lactit-1-yl-chitosan are an improvement. This engineered polysaccharide is a weak polycation and shows water solubility

Table 8.9 Size distribution of AgNP stabilised and reduced with 3,6-O-sulfochitosan

Method	AFM	TEM	SAXS
Weighted	Number	Number	Volume
Diameter (nm)	27 ± 12	14 ± 7	19 ± 8

over a wide pH range, making it an interesting compound in the preparation of films and hydrogels with AgNPs for biomedical applications (Donati et al. 2009). 6-O-sulfochitosan is another highly water-soluble chitosan derivate. From a structural point of view, this derivate resembles heparin and less surprisingly, like heparin, it exhibits antithrombotic and anticoagulative effects. 6-O-sulfochitosan can be used as a stabilising and reducing agent in the preparation of AgNPs. These nanoparticles show a narrow size distribution (Table 8.9) and exhibit remarkable stability at neutral pH when stored at ambient atmosphere under exclusion of light for a period of three weeks. Aggregation and precipitation of silver is not observed, and particle sizes are only slightly increased after that time, which can be explained by electrostatic repulsion as well as the steric stabilisation that the 6-O-sulfochitosan provides (Breitwieser et al. 2012).

8.3.3 Textile and Pulp Cellulose Fibres and Foils

As mentioned in previous chapters, the correlation between the structural parameters and surface properties of cellulose solid matrices should be studied carefully. Table 8.10 shows some typical structural as well as surface characteristics of regenerated textile (Modal) and pulp fibres (mixed hardwood pulp). The differences in surface parameters between both types of fibres are significant despite the more or less identical chemical composition.

Modal fibres are more crystalline [and also have a different morphology (Krässig 1996)] than pulp ones, which are reflected in the higher contact angles and smaller capability to retain water. In contrast, pulp fibres show higher amounts of acidic groups, which correlates with

8 Cellulose and Other Polysaccharides Surface Properties and Their Characterisation

Table 8.10 Some structural as well as surface parameters of modal and pulp fibres

Characterisation method	Modal fibres		Pulp fibres	
Crystallinity (%)	65	±10	55	±8
Zeta potential (plateau) (mV)	−13	±1.15	−14*, −22 *	–
Total charge – weak acids (mmol/kg)	23.9	±0.3	49.7	±0.1
pK	3.7	–	3.4 and 5.6	–
Contact angle (°)	79.3	±1.1	67.13	±1.4
Water retention (%)	40.7	±0.7	67.13	±1.6

Crystallinity determined by wide-angle X-ray diffraction (WAXD). To determine crystallinity, peak areas were integrated and compared to the scattering curve of amorphous cellulose; *zeta potential* (ZP$_{plateau}$) was defined using oscillating streaming potential and streaming potential measurements, dependent on electrolyte pH (Fairbrother and Mastin 1924; Jacobasch and Bauböck GSchurz 1985); Reischl et al. 2006); *titrations*: the conductometric and potentiometric titration of the cellulose material suspension were carried out; total charge and the pK values were determined (Fras Zemljič et al. 2004, Myers 1999); *contact angle*: powder contact angle method determination for porous solid surfaces (fibres, powders ,etc.), contact angle calculation using a modified Washburn equation (Peršin et al. 2004); goniometric determination for porous solid surfaces (foils, films, etc.): measuring the force (F) between a wetting liquid and the surface of a solid; contact angle is determined using contact angle hysteresis: (fa)–(fr); *water retention* to standard DIN 53 814

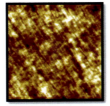

Fig. 8.8 Surface morphology of the cellulose foil (mean roughness defined using AFM) versus contact angle [defined using contact angle hysteresis and calculated using Young eq. (Myers 1999)]

Surface morphology of cellulose foil	
Topography range	24.52 nm
Average roughness	2.60 nm
Root mean square of roughness	3.25 nm
Contact Angle (H$_2$O)	17.6 °

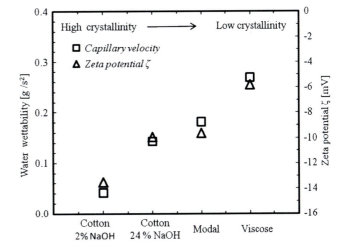

Fig. 8.9 Wettability (capillary velocity) versus zeta potential in plateau (at pH 9) of regenerated and natural cellulose fibres (alkaline modified; 2 % NaOH or 24 % NaOH)

the higher zeta potential in plateau values at pH 9 (Jacobasch and Schurz 1985). The pulps contain two types of acidic groups that dissociate in the pH range 2.5–7.5 (pK ≈ 3.4 and pK ≈ 5.5). The pK value of the stronger acid could be expected for uronic acids in hemicellulose (xylan), whilst the amount of the weaker acid correlates with the amount of lignin in the pulps

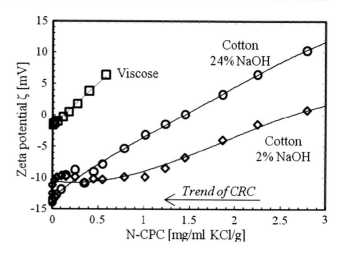

Fig. 8.10 Zeta potential as a function of cationic N-CPC adsorption onto natural and regenerated cellulose fibres

(Myers 1999). In regenerated fibres, there is only one type of acidic group (p$K \approx 3.5$; typical for carboxyl group in uronic acids) (Fras Zemljič et al. 2004).

The third class of 'real' solid cellulose matrices, the foils, exhibits extremely hydrophilic characteristics due to their surface morphology, despite having the same chemical composition (compared to the other two classes of cellulose solid matrices). The surface roughness, one of the most important parameters, defines the wettability of the solid matrices, which is very low (root mean square of roughness defined using AFM = 3.25 nm) (Fig. 8.8). The pK values of the foils found in our study were the same as in the textile and pulp fibres (p$K \approx 3.5$) typical for uronic acids.

When analysing only the textile fibres against each other (natural, cotton, vs. regenerated, viscose), it can be seen that they differ in hydrophilic/hydrophobic character and electrokinetic properties (Fig. 8.9) (Kreže 2005; Stana-Kleinschek and Ribitsch 1998; Stana-Kleinschek et al. 2003; Sfiligoj-Smole et al. 2003).

Compared to the hydrophilic character of regenerated fibres, the hydrophobic character of cotton is responsible for the high negative surface charge of cotton fibres. If cotton fibres are treated with 24 % NaOH, a decrease in ZP$_{plateau}$ is observed because of the changed cellulose structure (Krässig 1996). The transformation of cellulose fine structure from cell I to cell II causes an enlargement of the amorphous part of the fibre, which leads to cellulose having a higher reactivity. The ZP$_{plateau}$ values of structured cotton fibres become similar to those of the regenerated ones. An increased trend in the direction: cotton 2 % NaOH treated → cotton 24 % treated → regenerated cellulose fibres (viscose) is obtained by using a conventional water vapour sorption experiment, water retention value and determination of capillary velocities (Stana-Kleinschek 2002). There is an evident correlation between those parameters determined in the swollen state, that is, water vapour sorption, swelling in an aqueous medium and ZP versus time and pH; these are the parameters that describe the interaction capacity of the fibres.

The surfactant adsorption (N-acetylpyridiniumchloride (–N-CPC; $C_{21}H_{38}ClN$) with a molecular mass of 358.01 g/mol) shows a similar picture to the ZP-pH functions, see Fig. 8.10. The hydrophilic materials, with better accessibility obtained by fibre swelling and enlargement of the primary interfibrillar and/or intrafibrillar places, show very rapid adsorption of the cationic surfactant (critical reversal concentration–CRC of viscose = app. 0.01 mg N-CPC/mL KCl/g fibres).

An excellent correlation between solution exhaustion, concentration of the surfactant on the fibres and fibre void fraction can again be observed in the adsorption of N-CPC on the different kinds of regenerated cellulose fibres. The highest amount of N-CPC adsorbs on viscose fibres with the largest void fraction. In the case of water adsorption, as well as in the experiment where surfactant was adsorbed, the crystallinity does not predict the sorption character of the fibres. Although the crystallinity of modal fibres is not the highest, they are unable to adsorb the same amount of surfactant as lyocell, which has the highest crystallinity and largest voids. The Langmuir type of adsorption process can be observed if the adsorption process is determined using the electrokinetic approach. The same structure/accessibility/surface properties/interaction abilities correlations are found (Stana-Kleinschek and Ribitsch 1998).

8.3.4 Cellulose Charge Modification

It is well known that the charge of cellulosic fibres is an essential feature of their chemical and physical properties, and hence, it is of great importance to the final properties of textiles and paper pulp products manufactured from these fibres (Fras Zemljič et al. 2004, 2008a; Kreze et al. 2001, 2002).

8.3.4.1 Anionisation of Cellulose

The importance of acidic groups in cellulose fibres lies in their capability to ionise without changing the pH to extreme values. These dissociable groups can have a large impact on the final fibre properties, such as swelling, strength, hydrophilicity, adsorption capacity, bioactive properties, etc. (Barzyk et al. 1997; Laine and Lindström 2001; Laine et al. 2000). Several methods available in the literature describe how to introduce acidic groups (i.e. in order to increase the total anionic charge) in cellulose fibres, most of them applying to wood-based cellulose fibres. Widely used methods are as follows (Barzyk et al. 1997; Laine and Lindström 2001; Laine et al. 2000):

(a) Oxidation: The cellulose is oxidised with nitrogen dioxide, ozone, hydrogen peroxide or oxygen.
(b) Addition of sulphonic acid groups to lignin.
(c) Grafting of carboxylic acids via the formation of free radicals by electron beams.
(d) Lacasse treatment.
(e) Carboxymethylation.
(f) Using substantive dyes.

It was reported earlier (Fras Zemljič et al. 2002) that in the use of indirect methods, such as complexometric titration or the methylene blue spectroscopic method for fibre carboxyl groups' determination, the results may be only taken as a trend. For this purpose, a study was performed on the influence of chemical modifications such as selective and nonselective oxidation with different oxidising agents on the carboxyl group content of cellulose cotton fibres.

In previous research, Fras Zemljič et al. (2005) reported on the use of other titrations, such as potentiometric, conductometric direct titrations and polyelectrolyte adsorption, in order to observe changes in the content, accessibility and nature (strength) of acidic groups of selectively oxidised cellulose fibres (cotton). A treatment of potassium periodate (VII) followed by sodium chlorite (III) was used. A predominant acid with a dissociation constant pK of around 3.3 was found in all fibres, nonoxidised as well as oxidised. This pK value corresponds well with the value expected for carboxyl groups in uronic acids (Fras Zemljič et al. 2005) typical for cellulose. There was an additional acid with pK 5.7 in the referenced nonoxidised cotton sample due to the presence of noncellulosic components (low-molecular fractions) on the fibres' surface. It was indicated that selective oxidation caused the elimination of those low-molecular fractions (non-carbohydrate material) that cover the fibres in natural layers, as well as the formation of new carboxyl groups. The combination of polyelectrolyte titration with XPS confirmed that the surface concentration of acidic groups was considerably lower than the bulk concentration of the acidic groups. These results suggest that during oxidation, carboxyl groups on the surface decrease, whilst those in amorphous regions increase (Fras Zemljič et al. 2005). The oxidation reaction of regenerated cellulose fibres mediated by N-hydroxyphthalimide (NHPI) and various co-catalysts at room temperature over different time intervals and various amounts of low-concentration sodium hypochlorite solution was also investigated by Coseri et al. (2009)) in order to convert highly selective C_6 primary hydroxyl groups into carboxylic groups. The amounts of negatively charged groups in oxidised cellulose were determined by means of potentiometric titration showing the introduction of a huge amount of carboxylic groups. Notably, the water retention values for the oxidised fibres increased by 30 % in comparison with the original nonoxidised sample, as a result of the introduction of hydrophilic carboxylate groups (Coseri et al. 2009).

It has become clear that different standard chemical processes such as oxidation procedures usually modify the fibres (structurally and chemically) and are detrimental to the fibres' mechanical properties (Strnad et al. 2008). In addition, these processes are frequently ecologically less desirable. Therefore, over the last decade, the tendency towards cellulose fibre functionalisation has been aimed at using natural, biodegradable and nonaggressive chemicals, such as polysaccharides and their derivates. Using carboxymethyl cellulose (CMC) as an adsorbent for cotton was expected to

(a) Provide higher water absorption.
(b) Improve adhesion capability to moist surfaces.
(c) Exhibit biodegradability and biocompatibility.
(d) Provide antimicrobial activity in cotton fibres.

It has been demonstrated that the obvious advantage of CMC as a method for increasing charge is that the sorption of this macromolecule can be limited to the adsorption on the fibre surfaces, thus yielding a high surface charge without otherwise adversely affecting the fibre properties (Fras Zemljič et al. 2006a, b, 2008b). Any increase in charge was determined indirectly by the phenol-sulphuric acid method and directly by conductometric titration. It was discovered that the total charge for cotton had increased by more than 50 % in the cases of all CMC products used, which increases the adsorption capacity of the fibres for the cationic surfactant (cetylpyridinium chloride N-CPC). It was found that there is a linear relationship between the mass of any of the CMCs adsorbed and the amount of N-CPC adsorbed. The increased charge resulted in a higher fibre capacity for cationic surfactant adsorbance, as well as bettering the mechanical properties of CMC-functionalised fibres (Fras Zemljič et al. 2006a, b, 2008a).

8.3.4.2 Cationisation of Cellulose

The cationic functionalisation of cellulose is of considerable industrial importance and has found applications in the paper industry, cosmetics, textiles, in flotation/flocculation and in drilling fluids (Abbott et al. 2006). Moreover, in the field of developing new medical textiles and devices from natural fibres, consumers and producers have extensively expressed the need to incorporate a biofunctional product within the cellulose fibres in order to increase their safety and obtain bioactive properties. Thus, the use of cationic polysaccharides for cellulose fibre cationisation when developing medical and other advanced materials has recently gained considerable attention, as emphasised by the numerous reviews on the topic (Dumitriu 2002; Liesene 2009; Mesquita et al. 2010; Pecse et al. 2005). It is clear that in all the above-mentioned applications and many others, technical and biomedical, the charged cationic groups of cellulose materials have essential meaning. Even more, the success of applicability is quite often correlated to the density of the cationic groups. Therefore, it is extremely important to understand the protonation behaviour of cationised cellulose (and its derivates) in detail. The protonation of cotton fabric with irreversibly adsorbed chitosan (CO–CT) was measured in an aqueous medium at 0.1 M ionic strength by means of potentiometric titrations and compared with the results obtained for pure cotton (CO) and chitosan (CT) as reported in Čakara et al. (2009). For CO–CT, the charging isotherm exhibits a charge reversal around pH 6.0, which is identified as the point of zero charge (PZC). The pure chitosan and the acid fraction that is present in cotton protonate according to the one-pK model, with $pK_{CT} = 6{:}3$ and $pK_{CO} = 4{:}7$, respectively.

At pH $>$ PZC, the charge of the acid fraction in CO–CT is negative and constant, and the proton binding is attributed purely to the adsorbed chitosan. On the other hand, the cotton-bound acid exhibits a more complex protonation

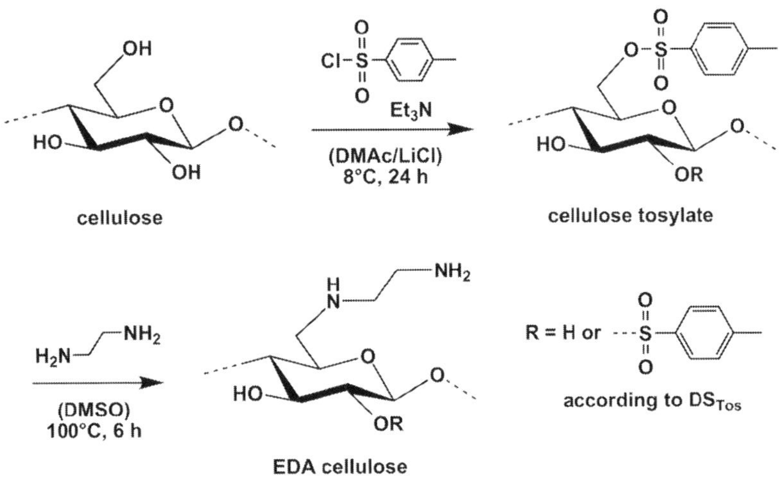

Fig. 8.11 Synthesis of 6-deoxy-6-(2-aminoethyl) amino cellulose (DAEAC) with a degree of substitution of (DS) 0.77 (Fras Zemljič et al. 2011)

mechanism in CO–CT than in the pure fabric, which is evidenced as an excess positive charge at pH < PZC and a deviation from the one-pK behaviour. A comparison of the overall charging isotherm at 0.1 M ionic strength with the sum of the isotherms of the pure components reveals an absence of electrostatic attraction between cotton and chitosan. This is also evidenced from the effective pK value of the composite. It has been shown that the irreversible adsorption of chitosan onto the weakly acidic cotton fabric is, under present condition, predominately driven by a non-electrostatic attraction (Čakara et al. 2009). Unfortunately, the application of chitosan as a cellulose fibre coating is still limited due to its action and solubility under acidic conditions, as well as its difficulties during purification being vital to high-value products and high-level analytical support. Therefore, an appropriate substitute for chitosan is required for specific applications, from among which novel soluble aminocellulose derivatives may be ideal. The synthesis of aminocellulose was investigated systematically and different procedures may be used giving different quantities and qualities (primary, secondary, tertiary, etc.) of amines (Klemm et al. 2001). Fras et al. performed a study where synthesised water-soluble 6-deoxy-6-(2-aminoethyl) amino cellulose (DAEAC), having a well-defined structure and chemical purity degree (see Fig. 8.11), was used for cellulose

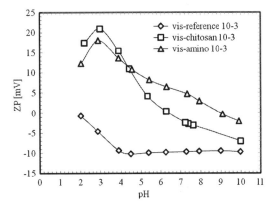

Fig. 8.12 Zeta potential (ζ) as a function of pH ($\zeta = f$ (pH)) for samples treated with chitosan

fibre functionalisation. Before being applied onto fibres, its protonation behaviour was studied by means of potentiometric titrations. The two protonation steps exhibit equal charges of 2.85 mmol/g, which points to the two-pK protonation scheme, where the primary amines protonate around pH ≈ 9, whilst the secondary amines protonate around pH ≈ 4.5. Thus, the ratio between the amounts of primary and secondary amines is 1:1, as is expected from the structure of DAEAC presented in Fig. 8.11.

The cellulose viscose fibres were impregnated with DAEAC solution, according to Liu (Liu et al. 2007). These functionalised fibres were compared to fibres impregnated with chitosan solutions under the same conditions. The amount of dissociable groups (carboxyl groups and amines as present weak acids) was determined by potentiometric titration. It is obvious from the potentiometric titration results that aminocellulose coatings introduce a much higher amount of amino groups to the fibres in comparison with fibres coated with chitosan solution. A lower amount of carboxyl groups was detected in samples impregnated with aminocellulose due to more expansive ionic bond formation between the carboxyls of the fibres and the amino groups of aminocellulose. These results, as well as isoelectric points determined by the potentiometric titrations, were in accordance with the zeta potential of the fibres' surfaces, which was calculated from the measurements of streaming potential as a function of pH (Fig. 8.12).

Pretreated viscose fibre (the reference material) showed a typical ZP = f(pH) function for cellulose, which was negative in practically the whole pH region, with a ZP plateau value -10 mV. After the fibres had been treated with chitosan and aminocellulose, the functions $\zeta = f(pH)$ showed typical amphoteric characteristics due to the introduction of amino groups. The shifting of isoelectric points into a higher pH region reduces the negative ZP plateau values because the cellulose fibre surface acidity is lowered. Negligible negative ZP plateau values were thus determined with the aminocellulose treated viscose fibres, due to the fact that there are almost no accessible carboxyl groups on the fibre surfaces.

8.3.5 Polysaccharide Hybrid Materials

8.3.5.1 Hybrid Materials of Chitosan with Different Organofunctional Silane Compounds

Hybrid materials are able to combine the advantages of organic polymers (e.g. photoluminescence) with those of inorganic compounds (e.g. enhanced stability) (Kickelbick 2006; Gómez-Romero and Sanchez 2004). The most convenient way to synthesise such materials is based on sol–gel chemistry, which can be attributed to the simplicity of influencing the properties of the materials on a molecular level by varying easily accessible experimental parameters such as temperature, pH, etc. (Avnir et al. 2001; Corriu et al. 2001). Although this process is widely used to combine synthetic polymers with inorganic parts, reports on polysaccharide hybrids are much lower in number (Chem et al. 2007; Jun et al. 2010; Rangelova et al. 2008; Samuneva et al. 2008; Shchipunov and Karpenko 2004; Shchipunov et al. 2005a, b; Shirosaki et al. 2009; Smitha et al. 2008; Spirk et al. 2012).

One major problem with performing sol–gel processes with polysaccharides is the compatibility in terms of solubility, pH compatibility, hydrophilicity and charge between the sol–gel system and the polysaccharide of interest. If any of the compatibility parameters is not fulfilled, homogenous materials usually cannot be obtained due to partial phase separation (indeed, in some cases homogeneity is not a prerequisite). Therefore, some polysaccharides can be incorporated into sol–gel processes easier than others. Among all polysaccharides, chitosan, a poly-β-1-4 glucosamine, is one of the few that is soluble in aqueous acidic medium due to protonation of the primary amine functionality, which makes chitosan an interesting candidate for the preparation of hybrid materials. Moreover, chitosan features biocompatibility, biodegradability, low toxicity, low cost, enhancement of wound healing and antibacterial effects, properties which are of great importance in respect to the development of new materials in medical applications.

In principle, the incorporation of inorganic parts embedded into a polymer matrix enhances the mechanical strength of the materials significantly. In a dry state, chitosan has good mechanical properties with reported Young's modulus of about 2,500 MPa, which is in the region of the best performing synthetic polymers like nylon and polystyrene. The addition of inorganic compounds such as alkoxysilanes should therefore result in materials with even higher mechanical strength.

As shown recently, mechanical strength in chitosan–silane hybrid materials is not only dependent on the silane content but also on the incorporation of the silane species into the polysaccharide matrix.

8 Cellulose and Other Polysaccharides Surface Properties and Their Characterisation

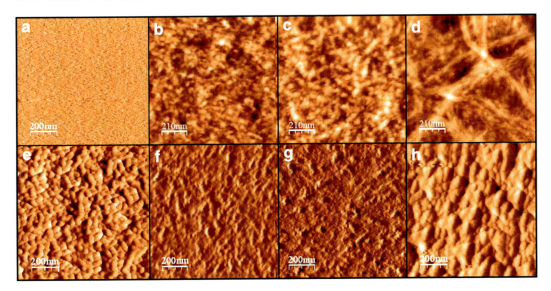

Fig. 8.13 1 μm × 1 μm AFM images of several chitosan–silane hybrids. (**a**) chitosan–TEOS (2:1), (**b**) chitosan–APTES (2:1), (**c**) chitosan–GPTMS (2:1), (**d**) chitosan–TBOS (2:1), (**e**) chitosan–MOMS (4:1), (**f**) chitosan–TVMOS (4:1), (**g**) chitosan–PFTEOS (4:1), (**h**) chitosan–TMEOS (4:1) *TEOS* tetraethoxysilane, *APTES* 3-aminopropyltriethoxysilane, *GPTMS* glycidylpropyltrimethoxysilane, *TBOS* tri-tert-butoxysilanol, *MOMS* methoxymethyltrimethylsilane, *TVMOS* trivinylmethoxysilane, *PFTEOS* (p-trifluoromethyl)tetrafluorophenyltriethoxysilane, *TMEOS* trimethylethoxysilane

Table 8.11 Water contact angle of selected chitosan–silane hybrid membranes at different silane contents

Chitosan/silane ratio	MOMS	ADMEOS	TMEOS	TVMOS	hMODS	PFTEOS
1:3	67 (±3)	No film	102 (±7)	No film	52 (±3)	73 (±5)
1:2	90 (±3)	21 (±2)	96 (±7)	73 (±2)	62 (±3)	81 (±4)
1:1	111 (±2)	31 (±1)	113 (±1)	105 (±3)	83 (±4)	74 (±3)
2:1	105 (±6)	23 (±2)	109 (±3)	107 (±4)	90 (±5)	95 (±4)
3:1	116 (±3)	36 (±4)	105 (±1)	114 (±2)	111 (±2)	112 (±3)
4:1	113 (±3)	75 (±4)	114 (±3)	112 (±4)	110 (±2)	No film

Contact angles are given in (°). Standard deviations are given in brackets. For ADMEOS and TVMOS, homogenous films could not be obtained at a chitosan/silane ratio of 1:3, for PFTEOS not at a ratio of 4:1

From AFM images of different chitosan–silane images, it can be concluded that the surface roughness as well as the size of the formed silica/silane particles plays an important role in the mechanical strength of such materials. Whilst for small, homogenously distributed silica particles in Fig. 8.13 (a), Young's moduli between 4,500 and 5,000 MPa are found, for hybrids where larger particles are present as shown in Fig. 8.13 (b, c, e–h), significantly lower Young's moduli (2,000–2,800 MPa) are observed. For fragile, fibril-like structures like in Fig. 8.13 (d), even lower moduli have been found (1,500 MPa). Of course, the amount of silane plays an important role in the mechanical stability of the synthesised films. Liu et al. investigated the behaviour of chitosan–GPTMS hybrids with varying chitosan content (Liu et al. 2004). For low-silane contents (5 %) in the hybrids, they found Young's moduli of 2,900 MPa, but these values decreased to 1,664 MPa at 50 % silane content. Also the elongation at-break parameter decreased (from 19 % to 1 %) indicating a stiffer, more brittle material at higher silane concentrations.

Besides surface morphology and mechanical strength, the hydrophilicity of the surface is an

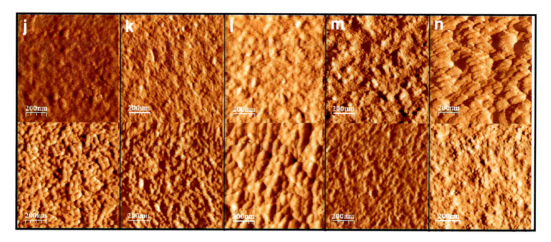

Fig. 8.14 AFM images of chitosan–silane hybrid materials. (**j**) Chitosan–MOMS, (**k**) chitosan–ADMEOS, (**l**) chitosan–TMEOS, (**m**) chitosan–TVMOS, (**n**) chitosan–hMODS. *Upper row:* chitosan/silane ratio of 1:3, for ChiTVMOS and ChiADMEOS, 1:2. Lower row: chitosan/silane ratio of 4:1

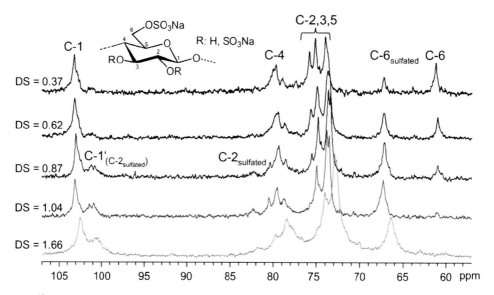

Fig. 8.15 ^{13}C NMR spectra of cellulose sulphates with different degrees of substitution (**a**: DS = 0.89, **b**: DS = 1.04, **c**: DS = 1.66)

important parameter, especially in medical applications. Water contact angle measurements showed that in all known chitosan–silane hybrids, contact angles tend to decrease with increasing silane concentration. For tetraalkoxysilanes like TEOS, where the final products of the sol–gel process consist of hydrophilic silicas with lots of free Si-OH functionalities, hydrophilic behaviour can be expected. Nevertheless, for hydrophobic compounds like PFTEOS or MOMS, hydrophilicity should be reduced in comparison to chitosan; indeed, a tendency to lower water contact angles has been observed. Some selected contact angle data can be found in Table 8.11.

It is obvious that surface morphology is, at least partly, responsible for these effects. When looking at the AFM images of hybrids with high-

and low-silane contents (Fig 8.14), in almost all cases, severe differences can be seen in surface morphology.

In principle, surfaces with a high-chitosan content appear to show higher surface roughness than those with a lower chitosan content. Rod-like structures are present in Fig. 8.14 (j) and (k) at high-chitosan concentrations; however, the morphology changes completely when the chitosan content decreases. Although some inhomogeneities can be seen in the AFM images, surfaces appear smoother in the cases of materials with high-silane contents. In case of Fig. 8.14 (l), a transition from an ordered surface with triangular-shaped particles (high-chitosan content) to a flatter surface with a few inhomogeneities is observed.

Recently, a chitosan–GPTMS hybrid was proven to exhibit high-cell (MG63, osteoblastic cells) compatibility and growth (Amado et al. 2008; Liu et al. 2004; Smitha et al. 2008). A correlation was found between the leaching of Si(IV) species from the material to the surface and growth, as well as coverage of the material by osteoblastic cells. Although high amounts of Si(IV) hinder cell growth; small doses may cause higher growth rates in cells. In addition, a regenerative effect on nerve fibres has been shown when a nerve is enwrapped with chitosan–GPTMS hybrid membranes. Besides regeneration, functional recovery of the nerve fibres was also observed.

8.3.6 Modified Polymer Surfaces with Bioactive Polysaccharides

Besides some well-known derivatisation reactions with OH groups of cellulose, chemical modifications of other polysaccharides, for example, chitosan and galactoglucomannans (GGM), are of growing interest in the scientific community. One interesting field of research is new synthetic procedures resulting in sulphated polysaccharides that can comprise the same antithrombotic features as the well-known anticoagulant heparin. The extent of these new achieved antithrombotic and anticoagulant properties can be tested by two different methods, among others. The first method is activated partial thromboplastin time (APTT) as a blood test, which indirectly measures the formation of thrombin by its action on fibrinogen-forming fibrin (contact activation pathway) (Sterling 2007). The second is the so-called free haemoglobin method, which determines the anticoagulant activities of substances coated onto PET surfaces (Fasl et al. 2010; Doliška et al. 2012). In addition, the common adsorption behaviour of modified polysaccharides onto polyethylene terephthalate (PET) films can be investigated by quartz crystal microbalance technique with a dissipation unit (QCM-D) in combination with an atomic force microscope (AFM) to analyse the surface morphology of these films (Indest et al. 2008, 2009).

Many scientific papers have been published, especially in the last two decades, on the more or less successful introduction of negatively charged sulphate groups in chitosan; Wolfrom and Shen Han (1958) reported the sulphation of chitosan with a mixture of pyridine/chlorosulphonic acid in 1958. Muzzarelli and co-workers (Muzzarelli et al. 1984) investigated the properties of chitosan and sulphated chitosan in the early 1980s. In the 1990s, special interest was directed towards regioselective modification of the OH group at the C-6 position of the glucopyranosyl unit by Baumann et al. (1998, 2000, 2001). This regioselective introduction of sulphate groups at the C-6 position should ensure that the NH_2 group at the C-2 position is not involved in the sulphation process and still remains available for further activities. The sulphation of GGM with chlorosulphonic acid was reported by Pires et al. (2001) and Martinichen-Herrero et al. (2005). Their investigations with ^{13}C-NMR revealed that the preferential substitution takes place at the OH groups on the C-6 position of the galactopyranosyl and mannopyranosyl units of GGM.

The big difference between the sulphation process of chitosan and cellulose on one hand and GGM on the other hand is the solubility of the latter in different organic solvents. Whereas chitosan is soluble only in acidic media, dissolution and subsequent homogeneous derivatisation of cellulose requires special solvent systems. As a consequence, GGM-SO$_4$ can be obtained by homogeneous synthesis, whereas heterogeneous synthesis methods have to be used for the sulphation of chitosan. GGM can be easily dissolved and modified in a mixture of formamide and pyridine according to Martinichen-Herrero et al. (2005). In the case of chitosan, two different synthetic routes were chosen for the introduction of sulphate groups into chitosan: modification with chlorosulphonic acid according to Vongchan et al. (2002) without any intermediate protection of the amino groups at the C-2 position and with the SO$_3$/pyridine complex in combination with intermediate protection of the amino group at the C-2 position according to Baumann and Faust (2001). In both cases, chitosan was suspended in N,N-dimethylformamide (DMF), and this suspension was used for the modification step. These two synthetic routes were tested and both can provide chitosan sulphates (Chito-SO$_4$) with moderate and high degrees of substitutions (DS) (Fasl et al. 2010). For the sulphation of cellulose, different heterogeneous and homogeneous methods have been described, as well as procedures that include the dissolution of the sulphated products during the reaction (Heinze et al. 2010). In the present studies, cellulose sulphates (CS) were prepared according to Gericke et al. (2009), using ionic liquids as reaction media. The completely homogeneous reactions yielded water-soluble CS even at rather low-DS values, which implies a uniform distribution of sulphate groups along the polymer chains. Moreover, only minor chain degradation occurred during the reaction and the DS could be tailored easily.

Important characterisation techniques for sulphated polysaccharides are infrared spectroscopy (IR), nuclear magnetic resonance spectroscopy (NMR) and inductive coupled plasma mass spectroscopy (ICP-MS). IR spectroscopy and ^{13}C-NMR are essential for the detection of sulphate groups in the monomeric building unit of chitosan (Fasl et al. 2010), whereas ICP-MS is one method of determining the sulphur content in the biopolymer (Fasl et al. 2010).

A set of ^{13}C-NMR spectra of cellulose sulphates (CS) is given in Fig. 8.15. The sulphation of cellulose in ionic liquids predominantly occurs

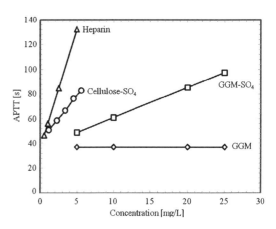

Fig. 8.16 APTT measurements of heparin, cellulose-SO$_4$, GGM and GGM-SO$_4$

at the C-6 position. Above a DS of around 0.9, secondary hydroxyl groups were also sulphated (Zhou et al. 2006).

The adsorption experiments using sulphated polysaccharides onto modified PET films presented in this study were carried out using the QCM-D technique, which recently has become widely used for adsorption studies at solid/liquid interfaces (Doliška et al. 2012; Indest et al. 2008, 2009; Wolfrom and Shen Han 1958). For this purpose, PET foils were dissolved in tetrachloroethane and spin coated onto silica quartz crystals. Unmodified chitosan was used as a spacer (Indest et al. 2008) between the modified PET surface and the sulphated polysaccharides to facilitate their adsorption. Due to structural differences the adsorption behaviour of Chito-SO$_4$ and GGM-SO$_4$ onto modified PET film surfaces is not comparable and leads to quite different results. For Chito-SO$_4$ at pH 7.4, the frequency change during the adsorption process is 10 times higher than for GGM-SO$_4$. It seems that under these conditions, in addition to Chito-SO$_4$ being adsorbed onto a modified PET film surface, precipitation also occurs (Indest et al. 2009). At pH 7, GGM-SO$_4$ forms a stable thin rigid film onto the modified PET film surface after reaching equilibrium conditions, because the adsorption process of GGM-SO$_4$ onto PET is connected with only moderate frequency changes during QCM-D measurements (Doliška et al. 2011).

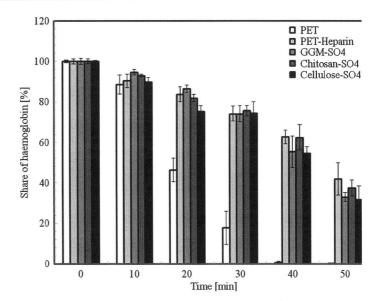

Fig. 8.17 The share of haemoglobin released from surfaces coated with chito-SO$_4$ and GGM-SO$_4$ compared to heparin-coated ones

Comparable behaviour was observed for the adsorption of CS onto PET films that had been previously coated with different amino groups containing polysaccharides (Gericke et al. 2010). The AFM investigations confirmed the results of the QCM-D studies.

The anticoagulant activities of GGM-SO$_4$ in solution were investigated through activated partial thromboplastin time (APTT) measurements and compared to unmodified GGM, as well as heparin as an anticoagulant reference (Doliška et al. 2011). In Fig. 8.16, it can be clearly seen that GGM-SO$_4$ prolonged the APTT significantly in comparison to unmodified GGM, but could not reach the anticoagulant activity of heparin in solution. The same was observed for CS with a DS of 1.04. CS with a rather high DS of 1.66, however, showed an inhibitory effect comparable to the natural anticoagulant, which can be attributed to their higher amount of sulphate groups in the C-2 position (Zuwei et al. 2007). Due to solubility problems, APTT measurements were not suitable for Chito-SO$_4$.

The anticoagulant activities of Chito-SO$_4$ as well as GGM-SO$_4$ and CS (Doliška et al. 2011) coated onto modified PET film surfaces were tested using the free haemoglobin method (Fasl et al. 2010) with heparin as a reference material.

In Fig. 8.17, the percentage of released haemoglobin from a blood drop placed onto the sample surface is described as the release of free haemoglobin after defined times. Despite the differences in their APTT, all semi-synthetic polysaccharide sulphates showed a similar high anticoagulant activity comparable to heparin. The anticoagulant activity of GGM-SO$_4$ is slightly higher than that of Chito-SO$_4$, CS and heparin after 10 and 20 minutes. After 20 min, GGM-SO$_4$ and CS lose activity compared to heparin and chito-SO$_4$. After 50 minutes, nearly 40 % of haemoglobin was still released from the forming clot in all the modified PET surfaces. In contrast, bare PET films showed only modest blood compatibility. The corresponding free haemoglobin release was only around 40 % after 20 min and almost zero after 40 min.

The anticoagulant activity of a PET model surface modified with a GGM-SO$_4$ layer was also investigated using QCM-D in an open QCM-D measurement cell, where the total time needed for clot formation could be determined *in situ*. The coagulation of blood plasma (ORKE 41, Dade Behring) on the surface was triggered with CaCl$_2$ (Fig. 8.18).

Fig. 8.18 Total coagulation time of blood plasma on PET and PET model films modified with a GGM-SO₄ layer determined using QCM-D

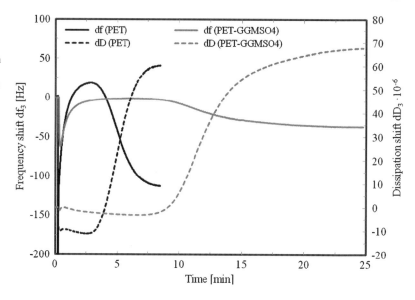

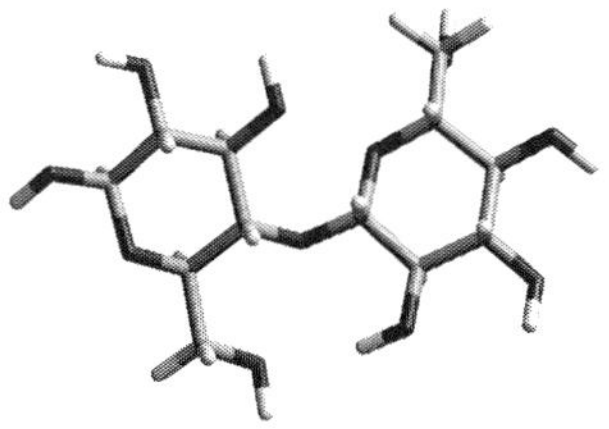

Fig. 8.19 Cellobiose model with $\varphi = 20°$; $\psi = -50°$; O6 conformation is tg and O6' is gg

Cellobiose

φ = H1-C1-**O**-C4 = 20 °
ψ = C1-**O**-C4-H4 = -50 °

O6 : tg (C4-C5-C6-O6 = -60 °)
O6': gg (C4-C5-C6-O6 = 180 °)

A change in frequency and dissipation is caused due to a change in the viscoelastic properties of the blood plasma during the coagulation process. The frequency and dissipation change as a function of time provides information about the onset time, fibrin deposition rate and total coagulation time (Doliška et al. 2012). In Fig. 8.18, it can be seen that the model PET surface modified with a GGM-SO₄ layer prolonged the total coagulation time significantly.

To summarise, the ability to prevent blood from clotting on a surface is remarkably high not only for heparin, but also for sulphated GGM, chitosan, and cellulose over the whole period of time. Thus, these semi-synthetic polysaccharide sulphates are versatile compounds for improving the biocompatibility of artificial polymers.

8.3.7 Computations with Molecular Modelling

The computation of a polysaccharide structure as well as its interaction capability, for example, adsorption, still remains a challenge. For small molecules (number of non-hydrogen atoms <30), the calculation of structural parameters like bond lengths, angles and dihedral angles can be done with high accuracy by using *ab initio* methods where all wave functions are calculated on the basis of Schrödinger's or Dirac's equation, respectively. For polysaccharides with many thousands of atoms per molecule, the situation is much more complicated. Besides the large amount of atoms per molecule, hydrogen bonding and the amount of bound water significantly

Table 8.12 Average and rms fluctuations of φ and ψ calculated from water and vacuum simulations of cellobiose Values are given in degrees (Hardy and Sarko 1993)

Torsion	Solvent Average	rms	Vacuum Average	rms
φ	30.2	13.4	30.3	20.0
ψ	−47.2	13.6	−42.8	15.2

Table 8.13 Average and rms fluctuations in degrees for O6 torsions calculated from water and vacuum simulations of cellobiose (Hardy and Sarko 1993)

Torsion	Solvent Average	rms	Vacuum Average	rms
O6				
tg	291.4	12.4	293.1	11.7
gg	59.9	10.7	55.3	11.5
O6'				
tg	292.2	13.5	288.7	18.1
gg	56.3	7.5	56.9	9.7

influence the structure, making an *ab initio* approach impractical.

8.3.7.1 Molecular Mechanics

Molecular dynamics simulation of polysaccharides and biopolymers is a well-proved technique to study the molecular motions on micro scale. The conformational behaviour of cellobiose in water can be studied by conformational energies and periodic boundary relations. In Fig. 8.19, a model of cellobiose is shown with $\varphi = 20°$ $\psi = -50°$. The O6 conformation is tg and the O6' conformation is gg. The computations were performed on the one hand in vacuum and on the other hand in a periodic cubic box of water with a side length of 20 Å with 195 water molecules. The results of the average and rms (root mean square) fluctuations are shown in Tables 8.12 and 8.13 (Schaub et al. 1993).

8.3.7.2 Semiempirical and Ab Initio

Even for less computer-intensive methods like semiempirical of force field approaches, the sheer size of the polysaccharide is too large. Therefore, smaller oligosaccharides must be used as models in order to get information on structure and reactivity. In the case of cellulose, the smallest possible models consist of cellobiose and cellotriose. Recently, we showed that *ab initio* as well as semiempirical methods are able to give reasonable results on the mono-silylation of cellobiose with various silylation reagents featuring different functional groups (Fig. 8.20) (Spirk et al. 2011).

This model reaction is of great importance for the synthesis of trimethylsilylated cellulose, which is a precursor for the preparation of cellulose model films (Schaub et al. 1993).

In general, the driving force in the silylation reactions of sugars is the formation of strong silicon–oxygen bonds (440 kJ mol^{-1}). Therefore, weaker element–silicon bonds (e.g. Si–N, Si–H, Si–Cl bond energy ca. 300 kJ mol^{-1} for all functionalities) are preferentially replaced by stronger silicon–oxygen bonds. This very simple thermodynamic approach can be easily verified when looking at the calculated reaction energies of the different trihydridosilanes with cellobiose (the use of trihydridosilanes instead of trimethylsilanes accelerates the calculations) using DFT. At the B3LYP 6-311 g* level of theory, exothermic reaction enthalpies ($\Delta G^0 < 0$ kcal/mol) were calculated only for X = H, NH$_2$ and NHSiH$_3$. Zero-point corrections (ΔG^{298}) yielded even more exothermic enthalpies. For X = Cl, endothermic reaction enthalpies were calculated; however, after zero-point correction, ΔG^{298} was close to 0 kcal. This is in agreement with experimental data for O-silylations using chlorosilanes, where the addition of a base (e.g. NEt$_3$) increases the nucleophilicity of the sugar and subsequently the reaction rate of the silylation (Einfeldt et al. 2001).

In contrast, for H$_3$SiF, the reaction enthalpy is strongly positive, which is a consequence of the strong silicon–fluorine bond (465 kJ mol^{-1}, one of the strongest single bonds between two elements). Therefore, the formation of a silicon–oxygen bond is thermodynamically disfavoured. All other calculated silanes show slightly positive enthalpies. Surprisingly, there is no preferred position where the silylation takes place; the largest energy difference between two positions is about 2 kcal/mol, which is within the range of uncertainty of the methods used. This result is a consequence of the relatively simple model used. In cellulose, intermolecular hydrogen bonds decrease the reactivity of the OH functionalities. Therefore, the enthalpy

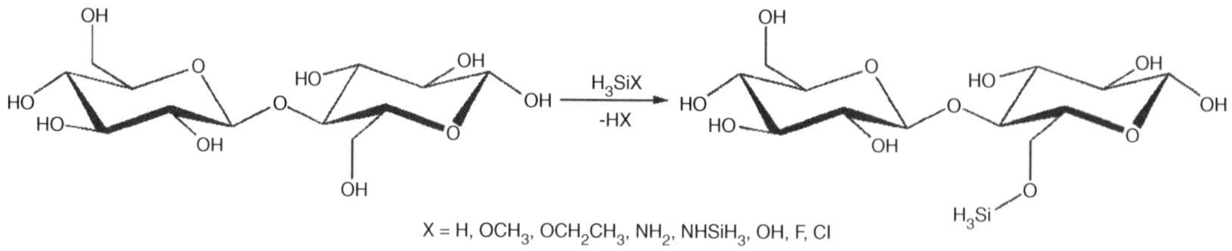

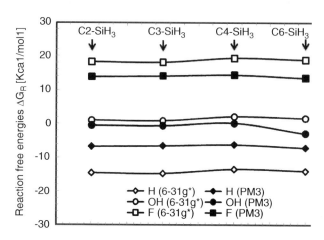

Fig. 8.20 Calculated reactions of cellobiose with various substituted trihydrido silanes

Fig. 8.21 Reaction free energies of different hydroxyl positions of cellobiose; semiempirical PM3 versus DFT 6-31g*

of the silylation reaction is not only dependent on thermodynamics but also on entropy and steric effects, which are both closely related to the accessibility of hydroxyl groups in cellulose. In the case of cellulose, experimental data shows that the C6 position can be trimethylsilylated much more easily than the OH functionalities (nn). In addition, calculations only reflect the total reaction enthalpy, but energy barriers have so far not been considered. Nevertheless, they can help us to guide and design experimental setups and to predict material properties to a certain extent.

8.4 Conclusion and Outlook

The molecular and supermolecular structure and morphology of cellulose-based materials in the dry state is already well described and understood. The properties in the wet state, where most technological processes take place are still barely known. Our knowledge about the interaction of cellulosic materials surfaces with the most frequently process medium water, the swelling processes and the kinetics of penetration is very limited. Even if these processes are the ones which happen at most technological process steps. Even less knowledge is gained about the interaction at the liquid/solid interface. This is the case for the interaction with dissolved molecules of the liquid phase as polymers and surfactants as well as the interaction with inorganic compounds.

This chapter intends to provide an overview about methods and instrumentation applicable to gain knowledge about interface properties and interface processes. This information will allow a better understanding of interface processes and their targeted utilisation in material development and processing. It also demonstrates clearly that it is possible to introduce in a very well-controlled way surface functionalities and surface structures on already existing shaped cellulose materials using physical and chemical processes.

Intensive efforts are ongoing to develop new high-value materials based on cellulose substrates. It is a promising strategy to prevent the excellent properties of this biopolymer and to achieve new features and properties restricting modification and compounding to the surface where they are required. The combination of nanotechnology and well-understood interface processes provides a promising strategic approach to new high-value materials produced with sustainable and environmental friendly production processes.

References

Abbott AP, Bell TJ, Handa S, Stoddart B (2006) Green Chem 8:784–786
Abu-Lail NI, Camesano TA (2003) J Microsc 212:217–238
Ahola S, Salmi J, Johansson LS, Laine J, Österberg M (2008) Biomacromolecules 9:1273–1282
Albrecht W, Reintjes M, Wulfhorst B (1997) Chem Fibres Int 47:289–304
Amado S, Simoes MJ, Armada da Silva PAS, Luıs AL, Shirosaki Y, Lopes MA, Santos JD, Fregnan F, Gambarotta G, Raimondo S, Fornaro M, Veloso AP, Varejao ASP, Maurıcio AC, Geuna S, (2008) Biomaterials 29:4409–4419
Andrade JR, Raphael E, Pawlicka A (2009) Electrochim Acta 54:6479–6483
Araki J, Wada M, Kuga S, Okano T (1998) Colloids Surf A 142(1):75–82
Aulin C, Ahola S, Josefsson P, Nishino T, Hirose Y, Österberg M, Wågberg L (2009) Langmuir 25(13): 7675–7685
Ausserre D, Valignat MP (2006) Nano Lett 6 (7):1384–1388
Avnir D, Klein LC, Levy D, Schubert U, Wojcik AB, Rappoport Z, Apeloig Y (2001) 2 Eds. In Chemistry of Organic Silicon Compounds; Zvi Rappoport and Yitzhak Apeloig, Eds.; Wiley: Chichester, UK
Azizi S, Alloin F, Sanchez JY, Dufresne A (2004) Polymer 45: 4149–4157
Barzyk D, Page DH, Ragauskas AJ (1997) Pulp Paper Sci 23: 520–528
Baumann H, Faust V (2001) Carbohydr Res 331:43–57
Baumann H, Scheen H, Huppertz B, Keller R (1998) Carbohydr Res 308:381–388
Baumann H, Richter A, Klemm D, Faust V (2000) Macromol Chem Phys 201:1950–1962
Beck-Candanedo S, Roman M, Gray DG (2005) Biomacromolecules 6(2):1048–1054

Belu AM, Graham DJ, Castner DG (2003) Biomaterials 24:3635–3653
Berger W (1994) Chemiefasern/Textilindustrie 44/96:747–750
Bergmair M, Bruno G, Cattelan D, Cobet C, de Martino A, Fleischer K, Dohcevic-Mitrovic Z, Esser N, Galliet M, Gajic R, Hemzal D, Hingerl K, Humlicek J, Ossikovski R, Popovic ZV, Saxl O (2009) J Nanoparticle Res 11(7):1521–1554
Berkert U, Allinger NL (1982) American Chemical Society Monograph, vol 177. Washington, DC
Bhattacharya R, Mukherjee P (2008) Adv Drug Deliv Rev 60:1289–1306
Biederman K, Sillen LG (1952) Arkiv Kemi 5:425–440
Biermann CJ (1996) Handbook of Pulping and Papermaking. Academic Press, San Diego
Breitwieser D, Spirk S, Fasl H, Stana-Kleinschek K, Ribitsch V (2012) 243rd American Chemical Society National Meetings and Exposition "Chemistry of Life", San Diego, California, USA
Brüger H (1994) Chemiefasern/Textilindustrie 44/96:701
Čakara D, Fras Zemljič L, Bračič M, Stana-Kleinschek K (2009) Carbohydr Polym 78(1):36–40
Capadona JR, van den Berg O, Capadona LA, Schroeter M, Rohan SJ, Tyler DJ, Weder C (2007) Nat Nanotechnol 2:765–769
Chan CM (1994) Polymer surface modification and characterization. Carl Hanser Verlag, Münich, Germany
Chem JH, Liu QL, Zhan XH, Zhang QG (2007) J Membr Sci 282:125–130
Chen X, Schluesener HJ (2008) Toxicol Lett 176:1–12
Chensong P, Songyun X, Houjiang Z, Yu F, Mingliang Y, Hanfa Z (2007) Anal Bioanal Chem 387:193–204
Choi Y, Simonsen J (2006) J Nanosci Nanotechnol 6:633–639
Ciavatta L (1963) Arkiv Kemi 20:417
Clark T (1985) A handbook of computational chemistry. John Wiley and Sons, New York
Cole DJ (1996) Lenzinger Berichte 75:45–48
Cook G (1984) Handbook of textile fibres, man-made fibres, 5th edn. Merrow, Durham, pp 47–64
Corriu RJP, Boury B, Rappoport Z, Apeloig Y (2001) In: Chemistry of organic silicon compounds; Zvi Rappoport and Yitzhak Apeloig, Eds.; Wiley: Chichester, UK
Coseri S, Nistor G, Fras Zemljič L, Strnad S, Harabagiu V, Simionescu CB (2009) Biomacromolecules 10(8): 2294–2299
Cranston ED, Gray DG (2008) Colloids Surf A – Physiochem Eng Asp 325:44–51
Dewar MJS (1969) The molecular orbital theory of organic chemistry. McGraw-Hill, New York
Doliška A, Willför S, Strnad S, Ribitsch V, Stana Kleinschek K, Eklund P, Xu C (2011) Holzforschung 66(2):149–154
Doliška A, Strnad S, Stana J, Martinelli E, Ribitsch V, Stana Kleinschek K (2012) J Biomater Sci Polym Ed 23:697–714

Donati I, Travan A, Pelillo C, Scarpa T, Coslovi A, Bonifacio A, Sergo V, Paoletti S (2009) Paoletti Biomacromolecules 10:210–213

Dumitriu S (2002) Polymeric Biomater, 2nd ed., Revised and Expanded Marcel Dekker, Inc. New York, pp 1–213

Ebner G, Schiehser S, Potthast A, Rosenau T (2008) Tetrehdron Lett 49:7322–7324

Edgar CD, Gray DG, (2003) Cellulose 10:299–306

Einfeldt L, Petzold K, Günther W, Stein A, Kussler M, Klemm D (2001) Macromol Biosci 1(8):341–347

El-Rafie MH, Mohamed AA, Shaheen TI, Hebeish A (2010) Carbohydr Polym 80:779–782

Eriksson M, Notley SM, Wågberg L (2007) Biomacromolecules 8:912–920

Evanoff DD, Chumanov G (2005) Chem Phys Chem 6:1221–1231

Faibrother F, Mastin HJ (1924) Chem Soc 75:2318–2331

Fairbrother F, Mastin HJ (1924) Chem Soc 75:2318–2331

Fasl H, Stana J, Stropnik D, Strnad S, Stana-Kleinschek K, Ribitsch V (2010) Biomacromolecules 11:377–381

Favier V, Chanzy H, Cavaille JY (1995) Macromolecules 28(18):6365–6367

Ferraria M, Boufi S, Battaglini N, Botelho dR, ReiVilar M (2010) Langmuir 26:1996

Filpponen I, (2009) Ph.D. Thesis, North Carolina State University, Raleigh, NC

Fink HP, Fanter D, Philipp B (1985) Acta Polym 36:1–8

Fowkes FM (1962) J Phys Chem 66:382–382

Fowkes FM (1964) Ind Eng Chem 56:40–52

Fras Zemljič L, Laine J, Stenius P, Stana-Kleinschek K, Ribitsch V, Doleček V (2004) J Appl Polym Sci Ed 92 (5):3186–3195

Fras Zemljič L, Stana-Kleinschek K, Ribitsch V, Sfiligoj-Smole M, Kreže T (2002) Lenzing Ber 81:80–88

Fras Zemljič L, Johansson L, Stenius P, Laine J, Stana-Kleinschek K, Ribitsch V (2005) Colloids Surf A Physicochem Eng Asp [Print ed.] 260(1/3):101–108

Fras Zemljič L, Stenius P, Laine J, Stana-Kleinschek K (2006a) Cellulose (Lond.) 13(6):655–663

Fras Zemljič L, Stenius P, Laine J, Stana-Kleinschek K, Ribitsch V (2006b) Lenzing Ber 85:68–76

Fras Zemljič L, Stenius P, Laine J, Stana-Kleinschek K (2008a) Cellulose (Lond.) 15(2):315–321

Fras Zemljič L, Peršin Z, Stenius P, Stana-Kleinschek K (2008b) Cellulose (Lond.) 15(5):681–690

Fras Zemljič L, Čakara D, Michaelis N, Heinze T, Stana-Kleinschek K (2011) Cellulose (Lond.) 18:33–43

Gericke M, Liebert T, Heinze T (2009) Macromol Biosci 9:343–353

Gericke M, Doliška A, Stana J, Liebert T, Heinze T, Stana-Kleinschek K (2010) Macromol Biosci 11:549–556

Gómez-Romero P, Sanchez C (2004) Functional hybrid materials. Wiley-VCH: Weinheim, Germany

Gran G (1950) Acta Chem Scand A4:97

Gray DG, Weller M, Ulkem N, Lejeune A (2010) Cellulose 17:117–124

Groth T, Wagenknecht W (2001) Biomaterials 22:2719–2729

Gruber M (1998) Ph.D. Dissertation. University of Stuttgart (Germany)

Grunert M, Winter WT (2002) J Polym Environ 10(1): 27–30

Grunnert M, Winter WT (2000) Polym Mater Sci Eng 82:232

Habibi Y, Foulon L, Aguie-Beghin V, Molinari M, Douillard RJ (2007) Colloids Interface Sci 316:388–397

Habibi Y, Lucia LA, Rojas (2010) OJ Chem Rev 110:3479–3500

Hanley SJ, Giasson J, Revol JF, Gray DG (1992) Polymer 33(21):4639–4642

Hardy BJ, Sarko A (1993) J Comput Chem 14(7):848–857

Hehre WJ, Radom L, Schleyer PVR, Pople JA (1986) Ab initio molecular orbital theory. John Wiley and Sons, New York

Heinze T, Daus S, Gericke M, Liebert T (2010) Semi-synthetic sulfated polysaccharides – promising materials for biomedical applications and supramolecular architecture I in "Polysaccharides: development, properties and applications" Nova Science Publishers, Inc., New York, USA

Hermans PH, Weidinger A (1949) J Polym Sci 4:135–144

Herrington TM, Petzold JC (1992) Colloids Surf 64:109–118

Holmberg M, Berg J, Stemme S, Ödberg L, Rasmusson J Claesson PJ (1997) Colloids Interface Sci 186: 369–381

Horvath AT, Horvath AE, Lindström T, Wågberg L (2008) Langmuir 24:10797–10806

Hunter RJ (1981) Zeta potential in colloid science, principles and applications. Academic Press, London, pp 6–9

Hyperchem 7 release (2002) Hypercube Inc.

Ibrahim MM, El-Zawawy WK, Nassar MA (2010) Carbohydr Polym 79(3):694–699

Indest T, Laine J, Ribitsch V, Johansson LS, Stana-Kleinschek K, Strnad S (2008) Biomacromolecules 9:2207–2214

Indest T, Laine J, Johansson LS, Stana-Kleinschek K, Strnad S, Dworczak R, Ribitsch V (2009) Biomacromolecules 10:630–637

Jacobasch HJ, Bauböck GSchurz J (1985) Problems and results of zeta-potential measurements on fibers. Colloids Polym Sci 263(1):3–24

Jianshe C (2007) Surface texture of foods: perception and characterization. Critical Reviews in Food Science and Nutrition 47:583–598

Jun S-H, Lee E-J, Yook S-W, Kim H-E, Kim H-W, Koh Y-H (2010) Acta Biomater 6:302–307

Kaelble DH (1972) Physical chemistry of adhesion. Wiley Interscience, New York, USA

Kälble DH (1969) Adhes J 1:102–123

Kargl R, Mohan T, Köstler S, Spirk S, Doliška A, Stana-Kleinschek, Ribitsch V (2012) Adv Funct Mater DOI:10.1002/adfm.201200607

Katz S, Beatson RP (1984) Scallion AM Svensk. Papperstidning 87:48

Kickelbick G (2006) Hybrid materials. Wiley-VCH, Weinheim, Germany

Kim CW, Kim DS, Kang SY, Marqueze M, Joo YL (2006) Polymer 47:5097–5107

Klemencic D, Simoncic B, Tomsic B, Orel B (2010) Carbohydr Polym 80(2):426–435

Klemm D, Philipp B, Heinze T, Wagenknecht W (1998) Comprehensive cellulose chemistry, vol 1, fundamentals and analytical methods. Wiley-VCH Verlag, Weinheim

Klemm D, Heublein B, Fink H, Bohn A (2005) Angew Chem Int Ed 44:3358–3393

Kocher T, Langenbeck N, Rosin M, Bernhardt O (2002) Methodology of three-dimensional determination of root surface roughness. J Periodont R 37:125–131

Kontturi E, Österberg M, Tammelin T (2006) Chem Soc Rev 35:1287–1304

Kontturi E, Thüne PC, Niemantsverdriet JW (2003) Langmuir 19:5735–5741

Krässig HA (1993) Cellulose, structure, accessibility and reactivity. Gordon and Breach Science Publishers, Y-Parc, Switzerland

Krässig HA (1996) Cellulose – structure, accessibility and reactivity. Gordon and Breach Science Publishers, Amsterdam

Kreže T (2005) Mater Res Innov 9(1):108–129

Kreze T, Malej S, Veder K (2003) Tex Res J 73(8): 675–684

Kreze T, Strnad S, Stana-Kleinschek K, Ribitsch V (2001) Mat Res Innovat 4:107–114

Kreze T, Jeler S, Strnad S (2002) Mat Res Innovat 5:277–283

Krüger R (1994) Lenzinger Ber 74:49–52

Kwok DY, Neumann AW (1999) Adv Colloid Interface Sci 81:167–249

Laine J (1994) Surface properties of unbleached kraft pulp fibers, determined by different methods, Dissertation. Department of Forest Products Technology; Helsinki University of Technology, Espoo, Finland

Laine J, Lindström T (2001) Das Papier 1:40–45

Laine J, Stenius P (1997) Paperi Puu - Paper and Timber 79:257–266

Laine J, Löfgren L, Stenius P, Sjöberg S (1994) Colloid Surf A 88:277–287

Laine J, Buchert J, Viikari L, Stenius P (1996) Holzforschung 50:208–214

Laine J, Lindström T, Glad-Nordnark G, Risinger G (2000) Nordic Pulp Paper Res J 17(1):50–56

Liebert T (2010); Cellulose Solvents - Remarkable History, Bright Future; ACS: Washington DC, USA

Liesene J (2009) Cellulose 17:167–172

Lindgren J, Wiklund L, Öhman LO (2000) Nordic Pulp and Paper Res. J 15:18

Lipkowitz KB, Boyd DB (1990) Eds., Reviews in computational chemistry. VCH Publishers, New York

Liu Y-L, Su Y-H, Lai J-Y (2004) Polymer 45:6831–6837

Liu J, Wang Q, Wang A (2007) Carbohydr Polym 70:166–173

Lu P, Hsieh YL(2010) Carbohydr Polym 82:329–336

MacArtain P, Jacquier JC, Dawson KA (2003) Carbohydr Polym 53:395–400

Martinichen-Herrero JC, Carbonero ER, Sassaki GL, Gorin PAJ, Iacomini M (2005) Int J Biol Macromol 35:97–102

Marx KA (2003) Biomacromolecules 4(5):1102–1120

Mathew AP, Chakraborty A, Oksman K, Sain M (2006) In Cellulose Nanocomposites: Processing, Characterization, and Properties; Oksman, K., Sain, M., Eds.; ACS Symposium Series 938; American Chemical Society: Washington, DC

Mesquita JP, Donnici LC, Pereira FV (2010) Biomacromolecules 11:473–480

Mohan T, Kargl R, Doliška A, Vesel A, Ribitsch V, Stana-Kleinschek K (2011) J Colloid Interf Sci 358 (2):604–610

Mohan T, Spirk S, Kargl R, Doliška A, Vesel A, Salzmann I, Resel R, Ribitsch V, Stana-Kleinschek K (2012a) Soft Matter 8:9807–9815

Mohan T, Spirk S, Kargl R, Doliška A, Ehmann HMA, Köstler S, Ribitsch V, Stana-Kleinschek K (2012b) Colloids and Surfaces A:Physicochem Eng Aspects 400:67–72

Morin A, Dufresne A (2002) Macromolecules 35:2190–2199

Murray RW (2008) Chem. Rev. 108:2688–2720

Muzzarelli RAA, Tanfani F, Emanuelli M (1984) Carbohydr Res 126:225–231

Myers D (1999) Surfaces, interfaces and colloids, John Wiley and Sons, New York

Noorani S, Simonsen J, Atre S (2006) In Cellulose Nanocomposites: Processing, Characterization and Properties; Oksman K, Sain M, Eds.; ACS Symposium Series 938; American Chemical Society:Washington, D.C.

Notley SM, Wågberg L (2005) Biomacromolecules 6:1586–1591

Notley SM, Eriksson M, Wågberg L, Beck S, Gray DG, (2006) Langmuir 22:3154–3160

Owens DK, (1969) J Appl Polym Sci. 13:1741–1747

Park S, Venditti RA, Jameel H, Pawlak JJ Park S et al (2006) Carbohydr Polym 66:97–103

Park W-I, Kang M, Kim H-S, Jin H-J (2007) Macromol Symp 249–250:289–294

Pecse A, Jordane AA, Carluci G, Cintio A (2005) Articles comprising cationic polysaccharides and acidic pH buffering means. US Patent Application Publication: 0124799 A1

Penfold J, Tucker I, Petkov J, Thomas RK (2007) Langmuir 23:8357–8364

Peršin Z Stana-Kleinschek K, Sfiligoj-Smole M, Tatjana K (2004) Tex Res J 74(1):55–62

Pires L, Gorin PAJ, Reicher F, Sierakowski MR (2001) Carbohydr Polym 46:165–169

Rabel W (1977) Physikalische Blätter 33:151–161

Rangelova N, Chernev G, Nenkova S, Samuneva B, Georgieva N, Yotova L, Radev L, Salvado IMM (2008) Nanoscience & Nanotechnology 8:246–249

Räsänen E, Kärkkäinen L, Tervola P, Gullichsen J, Stenius P, Vuorinen T (2001a) Grenoble Workshop on Advanced Methods for Lignocellulosics and Paper Products Characterization, P June 18–19. Grenoble, France

Räsänen E, Stenius P, Tervola P (2001b) Nordic Pulp and Paper Res. J 16:130–139

Reimhulta K, Yoshimatsub K, Risvedenb K, Chena S, Yeb L, Krozer A (2003) Biomacromolecules 4(5): 1103

Reischl M, Stana-Kleinschek K, Ribitsch V (2006) Mater Sci Forum 514–516:1374–1378

Ribitsch V, Stana-Kleinschek K, Tatjana K, Strnad S (2001) Macromol Mater Eng 286(10):648

Roman M (2009) Model cellulosic surface. ACS, Washington, DC, pp 115–136

Rossetti FF, Panagiotou P, Rehfeldt F, Schneck E, Dommach M, Funari SS, Timmann A, Müller-Buschbaum P, Tanaka M (2008) Biointerphases 3:117–127

Ruland W (1961) Acta Crystallogr 14:1180–1195

Saarinen T, Österberg M, Laine J (2008) Colloids Surf A – Physiochem Eng Asp 330: 134–142

Samuneva B, Djambaski P, Kashchieva E, Chernev G, Kabaivanova L, Emanuilova E, Salvado IMM, Fernandes MHV, Wu AJ (2008) Non-Cryst. Solids 354:733–740

Sax AF (2008) Monatsh Chem 139:299–308

Sauerbrey G (1959) Z Phys 155:206–222

Scallan AM, Katz S (1989) Cellulose and wood chemistry and technology. John Wiley, New York, NY

Schaub M, Wenz G, Wegner G, Stein A, Klemm D (1993) Adv Mater 5:919–921

Schenzel K, Fischer S, Brendler E (2005) Cellulose 12:223–231

Schrand M, Braydich-Stolle LK, Schlager JJ, Dai L, Hussain SM (2008) Nanotechnology 19(23):1–13

Schuldt U, Philipp B, Klemm D, Stein A, Jancke H, Nehls I (1994) Papier 48:3–17

Schurz J (1994) Lenzinger Berichte 74:37–40

Sczech R, Riegler HJ (2006) J Colloid Interface Sci 301:376–385

Sfiligoj-Smole M et al (2003) Mater Res Innov 7 (5):275–282

Shchipunov YA, Karpenko TY (2004) Langmuir 20:3882–3887

Shchipunov YA, Karpenko TY, Krekoten AV (2005) Compos Interf 11:587–607

Shchipunov YA, Kojima A (2005) Imaeb T. J Colloid Interf Sci 285:574–580

Shirosaki Y, Tsuru K, Hayakawa S, Osaka A, Lopes MA, Santos JD, Costa MA, Fernandes MH (2009) Acta Biomat 5:346–355

Sjöström E, Enström B (1966) Svensk Papperstidning 69:55

Smitha S, Shajesh P, Mukundan P (2008) Warrier KGK. J Mater Res 23:2053–2060

Smulochowski M (1903) Bull Intern Acad Sci Cracovie 184

Souplet V, Desmet R, Melnyk O (2007) J Peptides Sci 13–7:451–457

Spirk S, Ehmann HMA, Ribitsch V, Stana-Kleinschek K, (2011) In proceedings of the 15th International electronic conference on synthetic organic chemistry, Sciforum electronic conference series

Spirk S, Findenig G, Doliška A, Reischel VE, Swanson NL, Kargl R, Ribitsch V, Stana-Kleinschek K (2012) Carbohydr Polym, DOI:10.1016/j.carbpol.2012.04.030

Spirk S, Ehmann HMA, Kargl R, Hurkes N, Reischl M, Novak J, Resel R, Ming W, Pietschnig R, Ribitsch V (2010) ACS Appl Mater Interfaces 2(10):1412–1424

Stana-Kleinschek K (2002) Mater Res Innov 6(1):13–18

Stana-Kleinschek K et al. (1999) Polym. Eng. Sci. 39 (8):1412–1424

Stana-Kleinschek K, Ribitsch V (1998) Colloids surf., A Physicochem. eng. asp. [Print ed.] 140:127–138

Stana-Kleinschek K, Ribitsch V, Tatjana K, Sfiligoj-Smole M, Peršin Z (2003) Lenzing Ber 82:83–95

Sterling TB (2007) Laboratory Hemostasis – A Practical Guide for Pathologists. Springer Science

Stewart JJP (1990) MOPAC: A Semiempirical Molecular Orbital Program. J Computer-Aided Molecular Design 4:1–105

Strnad S, Šauperl O, Jazbec A, Stana-Kleinschek K (2008) Tex. Res. J. 78(5):390–398

Talapin DV, Lee J-S, Kovalenko MV, Shevchenko EV (2010) Chem. Rev. 110:389–458

Tammelin T, Saarinen T, Österberg M, Laine J (2006) Cellulose 13:519–535

Tanaka M, Wong AP, Rehfeldt F, Tutus M, Kaufmann SJ (2004) Am. Chem. Soc. 126:3257–3260

Twu Y-K, Chen Y-W, Shih C-M (2008) Powder Technology 185. Issue 3:251–257

Usuda M, Suzuki O, Nakano J, Migita N (1967) Kogyo Kagaku Zasshi 70:349–352

Valles C, Drummond C, Saadaoui H, Furtado CA, He M, Roubeau O, Ortolani L, Monthioux M, Penicaud (2008) J Am Chem Soc 130–47:5802–5811

van Oss CJ (1993) Colloids Surf A Phys Chem Eng Asp. 78:1–49

van Oss CJ, Chaudhury MK, Good RJ (1986) J Colloid. Interface Sci 111:378–390

van Oss CJ, Chaudhury MK, Good RJ (1988) Langmuir 4:884–891

Van Riessen A, Winton GH, Ohyi H, Yoshida M (1994) Micron. 25, 6:511–517

Vongchan P, Sajomsang W, Subyen D, Kongtawelert P (2002) Carbohydr Res 337:1239–1242

Wågberg L, Ödberg L, Glad-Nordmark G (1989) Nord Pulp Pap Res J 4(2):71–76

Waltz JE, Taylor GB (1947) Ana.l Chem. 19:448–450

Washburn EW (1921) Phys.Rev. 17(3):273–283

Wegner G, Buchholz V, Ödberg L, Stemme S (1996) Adv Mater 8:399–402

Wolf S, Mobüs BZ (1962) Anal Chemie 86:194

Wolfrom ML, Shen Han TM (1959) J Am Chem Soc 81:1764–1766
Young DC (2001) Computational chemistry: A practical guide for applying techniques to real-world problems. John Wiley and Sons, New York, USA
Zaera F (2012) Chem Rev 112:2920–2986
Zhang Y, Sjögren B, Engstrand P, Htun MJ (1994) Wood Chem Techn. 14:83–102
Zhou Q, Baumann MJ, Brumer H, Teeri T (2006) Carbohydr Polym 63:449–458
Zisman WA (1964) In: Fowkes FM (ed). Contact angle, wettability and adhesion. Washington, DC, USA: American Chemical Society pp 1–51
Zuwei M, Zhengwei M, Changyou G (2007) Colloids Surf., B 60:137–157

Pulp Fibers for Papermaking and Cellulose Dissolution

9

Pedro Fardim, Tim Liebert, and Thomas Heinze

Contents

9.1 General Introduction 254

9.2 Fiber Raw Materials 254
9.2.1 Wood Structure and Chemical Composition .. 255
9.2.2 Fiber Wall Structure 257

9.3 Fiber Separation Methods 259
9.3.1 Mechanical and Chemimechanical Pulping ... 259
9.3.2 Chemical Pulping 261
9.3.3 Recycled Fibers 263

9.4 Fiber Functionalization Methods Used in Pulp and Paper Manufacture 263
9.4.1 Bleaching of Mechanical Pulps 264
9.4.2 Bleaching of Chemical Pulps 265
9.4.3 Low-Consistency Refining 268
9.4.4 Papermaking Chemicals 269

9.5 Fiber Properties and Uses 271
9.5.1 Paper, Tissue, and Packaging Materials 272
9.5.2 Dissolution of Pulp Fibers in Cellulose Solvents 272

9.6 Challenges and Opportunities in Research for Pulp Fibers for Papermaking and Cellulose Dissolution 274

9.7 New Challenges 275
9.7.1 2011–2012: Years of New Insights 275
9.7.2 Year 2030: The Year of the Truth 278
9.7.3 Year 2060: The Year of Counting Black Swans 280

References 281

P. Fardim (✉)
Laboratory of Fibre and Cellulose Technology, Åbo Akademi University, Porthansgatan 3, 20500 Turku/Åbo, Finland
e-mail: pfardim@abo.fi

Abstract

Pulp fibers are extracted from biomass worldwide and extensively utilized for production of paper and as a raw material for products of dissolved cellulose. Different paper grades require diverse fiber properties that are achieved by a proper combination of fiber separation and functionalization methods. Dissolved cellulose can be functionalized to derivatives or regenerated to materials. The research, development, and innovation activities in the utilization of fibers are a constant activity of EPNOE partners, and it is present in several collaborative projects in the network. In this chapter, we describe how pulp fibers are separated from plants and further functionalized to be used in paper products or cellulose dissolution. Challenges and opportunities in research, development, and innovation are also discussed in combination with needs and trends of education of professionals in the topic of bio-based materials. Our main conclusion is that the forest-based industry will need intensive research and innovation activities based on multidisciplinary approach to create applications in different value chains of large commercial interest. Strategic partnership with chemical, energy, and food sectors are crucial for the pulp and paper industry in creating new markets and benefiting of the knowledge-based economy.

9.1 General Introduction

Plant fibers represent the major source of polysaccharides on Earth. Cellulose or pulp fibers are extracted from plant raw materials or recovered paper and further utilized in manufacture of paper products or preparation of dissolving pulp. Pulp fibers are extracted using mechanical, chemimechanical, and chemical methods and further bleaching-treated to remove or to modify lignin. Mechanical and chemimechanical pulps contain cellulose, hemicelluloses, and lignin and are used mainly in papermaking. Chemical pulps contain only cellulose with lignin and hemicelluloses being extensively removed in case of pulps intended for dissolution in cellulose solvents. Chemical pulps which are used in papermaking are mechanically treated in water in a process named low-consistency refining aiming on improvement of mechanical properties. Pulp fibers are treated with several chemical additives in papermaking depending of the type of the paper product. The process to extract pulp fibers from plants and recovery paper are called fiber separation processes, while bleaching, low-consistency refining, and papermaking can be classified as fiber functionalization.

In this chapter, we focus on how plant fibers are separated to be transformed into pulp and subsequently functionalized in bleaching and papermaking. Preparation of pulp fibers for dissolution in cellulose solvents is also mentioned. Challenges and opportunities in research for pulp fibers for papermaking and cellulose dissolution are discussed. In EPNOE, we have three partners dealing with research of fibers for papermaking and three groups focusing on dissolution of pulp fibers to obtain cellulose solutions. Åbo Akademi University and Technical Research Centre of Finland (VTT) have a large project portfolio on fiber technology practically covering all aspects from raw material to paper and cellulose dissolution. The Finnish research environment has been a traditional player in the area of forest products since 1950s, and it is currently shifting to new applications for fibers involving bio-based products and processes. Wood chemistry and pulping has been a strong tradition also in Hamburg, Germany, where the EPNOE partner's Johann Heinrich von Thünen-Institut, Federal Research Institute for Rural Areas, Forestry and Fisheries (vTI)-Institute of Wood Technology and Wood Biology in cooperation with Hamburg University played a key role to European research. Cellulose dissolution is a focus of research for the Centre of Excellence for Polysaccharide Research in Jena, Germany, with unique expertise in polysaccharide functionalization. The Thuringian Institute for Textile and Plastics Research, Rudolstadt, Germany, deals with cellulose dissolution and shaping, and the Mines ParisTech Centre for Material Forming (CEMEF), Sophia Antipolis, France, has expertise in rheology of cellulose solutions and physics of dissolution systems. The University of Natural Resources and Life Sciences, Vienna, has expertise in cellulose chemistry and analytics. Scattered research dealing with aspects of pulping, papermaking, and cellulose dissolution is present research activities in several other EPNOE partners.

9.2 Fiber Raw Materials

Pulp and cellulose fibers are found in wood, non-wood, and recovered paper. Wood is the major source of commercial pulp fibers in Europe while non-wood is used mainly in Asia. The utilization of recovered paper as a fiber raw material is increasing in Europe and taking usual applications of mechanical pulps. Today, the utilization rate in Europe is about 42 %. This is defined as the total weight of recovered paper as a proportion of the total weight of fiber furnish used in the paper and paperboard industry. Wood pulp fibers are extracted or separated from coniferous and deciduous trees growing in commercial forests located all over the planet. Coniferous trees or softwood grow mainly in Northern Hemisphere while deciduous or hardwoods grow mainly in Southern Hemisphere. Strict forest certification practices are available to issue labels where sustainable forest management is monitored and verified.

9.2.1 Wood Structure and Chemical Composition

Tree biomass can be roughly divided into the crown, trunk, and roots. Of these, only stem wood is of interest to the forestry industry. The wood is a natural, complex composite, which is made up of small, short fibers bound together with lignin. What makes the structure complex is that not only the wood itself but the fiber and fiber building blocks, fibrils and microfibrils, are of composite structure. The tree trunk is composed of outer bark, inner bark, cambium, sapwood, heartwood, and pith (Schweingruber 1993). The bark is the outer covering of a tree trunk and protects the tree from mechanical damage and microbial attack and evens out the temperature and humidity variations. Tree radial growth (secondary growth) occurs in cambium just after the inner bark. Cambium consists of a single monolayer of thin-walled living cells (Alén 2000). Sapwood is located near the cambium and transports water and inorganic nutrients from the soil to the tree leaves and pine needles. In most cases, heartwood can be distinguished from sapwood because of its lighter color and higher humidity. The cells near the bone dies and form the so-called heartwood, whose task will be to provide mechanical support to the trunk. The proportion of heartwood in the tree increases as the tree gets older. The heartwood is often darker, drier, harder, and denser than the sapwood. Heartwood proportion is different at different heights of the trunk, but it does not follow a fixed annual ring. Different species of trees begin to form heartwood at different ages, but after the early, ongoing process during the tree lifetime (Saranpää 2003). In the center of the trunk and branches is the pith. It has a diameter of few millimeters and can be seen either as a round dark spot in the stem cross section (Fig. 9.1).

In trees growing in cold or temperate zones, where the period of growth does not continue throughout the year, the annual growth thickness is seen as clear growth rings. The inner part of a ring, which formed in the early growth period, consists of light and porous spring wood (earlywood), because the outer part, formed by the end of growth period, consists of darker and denser summerwood (latewood). The sharp boundary that occurs between the growth rings indicates the tree rest. The rest period is followed by an intensive growth in spring, when the tree vital functions are recovering and the need for new water-transporting cells is large. The cells formed have thin walls and large diameters. Later, when the foliage is completed and linear growth stops, thick-walled latewood fibers are formed. The relation between earlywood and latewood has a significant effect on tree density and affects the fiber properties of resulting pulp.

One tree produces juvenile wood regardless of age. Juvenile wood is formed during the initial phase of radial growth from the pith outward. The transition to mature wood is not sharp, but occurs slowly over several years. The characteristics of juvenile wood vary from annual ring to annual ring, and this is considered one of the main reasons for the variation of wood properties in conifers. Pulps produced from juvenile wood generally have lower strength properties and better optical properties than the pulp produced from sapwood (Alén 2000; Saranpää 2003). Reaction wood is a special cellular tissue which is counteracting any inclination of the strain and restoring it to its original position. The branches form a reaction stung to counter their own weight. In conifers, the formed reaction stung is called compression wood on the lower side and tension wood on the upper side of bent stems. The twigs of a tree grow only from the stem pith. At the annual radial growth of the stem and branches, a twig grows part inside the trunk. These cone-shaped parts inside the trunk are called knots. The wood of a twig is dense and consists largely of reaction wood.

The cells in wood structure are mainly oriented according to the stem direction. In this way, they are best suited to support the tree and to water transport in the stem direction. The pith ray cells (parenchymal cells), which ensures water and nutrient transport in the radial direction, are oriented to the pith of the stem cross direction (Jensen 1977).

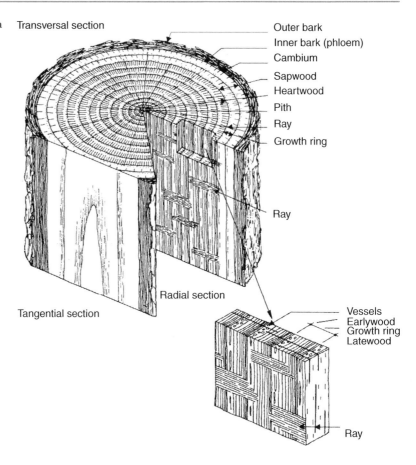

Fig. 9.1 Cross section of a tree trunk (Schweingruber 1993)

Softwoods consist of 90–95 % of tracheids. Tracheids are elongated, tubelike cells, which are sealed at the ends. Tracheids of Scandinavian conifers are 2–4 mm in length and width is ~10–40 μm, depending on the growth season. Earlywood fibers are active in fluid transport and thus have thin walls and large lumens. Earlywood tracheids have hundreds of ring pores that are formed at the interface between two cells which allow fluid transport in the stem and partly in the radial direction. Latewood has a support function with cells with thick walls, small pores, and small lumens, and the cross section has the shape of a rectangle. Tracheids from latewood are slightly longer than tracheids from earlywood (Eero 1981; Jensen 1977). Approximately 5–10 % of the cells in softwood are parenchymal cells. They are both in the pith rays and partly in the so-called resin canals. In the sapwood, where parenchymal cells are alive, there is an accumulation of food reserves. Parenchymal cells are small and brick-shaped. They are only 10–160 μm in length and 20–50 μ in width.

Hardwoods have several cell types with different specialized functions. Libriform cells, tracheids, vessels, and ray parenchymal cells are the major components of hardwoods. Libriform cells are thick-walled, elongated cells with dimensions between 0.8 and 1.6 mm in birchwood. The width of libriform cells is 14–40 μm and cell wall thickness between 3 and 6 μm. Vessels are cells that are placed on the top of each other in the wood structure to form tubes to conduct water and nutrients. They can occupy 25 % of the wood volume in case of birch and aspen. Vessels have a thin wall layer, shorter length (0.3–0.7 mm), and wider width (30–300 μm) in comparison to fibers. Bordered pits are present in vessel walls, and their amount and shape are dependent on hardwood species. Parenchymal cells are the main hardwood rays counting for 5–30 % of the stem volume in aspen wood (Eero 1981).

Wood chemical components can be roughly divided into high molecular weight and low molecular weight compounds. Cell wall components such as cellulose, hemicelluloses, and lignin are the main high molecular mass polymeric components, and organic and inorganic substances such as organic acids and salts are the main low molecular mass components. Cellulose is the most important wood component given the pulp and paper production and dissolution in cellulose solvents. Cellulose is a linear high molecular homopolymer composed only of β-D-glucopyranose units bound together by covalent bonds. The cellulose in the native state consists of ~10,000 units. The chemical pulp-obtained cellulose consists of ~1,000–1,500 units. Cellulose chains form both amorphous and crystalline aggregates. The crystalline structure is held by both intra- and intermolecular hydrogen bonds. The propensity to form hydrogen bonds allows the assembly of fibrils and fibers. Cellulose occurs mainly in the fiber secondary wall. The various structural elements of the fiber wall are illustrated in Fig. 9.2.

Polyoses or hemicelluloses are heterobranched polymers consisting of a number of different monosaccharides. Hemicelluloses are often classified on the basis of main chain components, which consist of about 150–300 units (Eero 1981). Glucuronoxylans or O-acetyl-(4-O-methylglucurono)-xylans are the main hemicelluloses present in industrially used hardwoods, like eucalyptus and birch. The contents of hemicelluloses are between 10 and 30 % on wood weight basis in birch and eucalyptus. Glucomannans (GM) and galactoglucomannans (GGM) are the major hemicellulosic components of the secondary cell walls of softwoods. They are suggested to be present together with xylan and fucogalactoxyloglucan in the primary cell walls of higher plants. Galactoglucomannans are the main hemicelluloses in softwoods, up to about 20 % of the wood. GGM consists of a linear backbone of randomly distributed (1 → 4)-linked β-D-mannopyranosyl and (1 → 4)-linked β-D-glucopyranosyl units, with α-D-galactopyranosyl units as single side units to mannosyl units. More information about molecular and supermolecular structures of polysaccharides can be found in Chap. 2.

Wood mechanical properties are largely determined by the amorphous polymer lignin. In principle, lignin function in wood structure is compared with polyester resin and function of composite plastics, which cellulose fibers are represented by glass fibers. Lignin appears also in the fiber walls, which usually is bound to hemicelluloses. Lignin is composed of aromatic phenols and has a complex three-dimensional structure (Eero 1981). The wood contains a large number of low molecular weight organic compounds commonly known as extractives. In principle, they are all substances other than cellulose, hemicelluloses, and lignin. Extractives can be roughly divided into fatty acids, resin acids, esters of fatty acids (glycerides), and steryl esters. The inorganic components of wood consist mainly of metals in the form of salts enriched in the fiber wall and lumen. The quantity of metals in wood is small, about 1 %. The most common metals are calcium, potassium and magnesium, and smaller amounts of iron and manganese (Fengel and Wegener 1984).

9.2.2 Fiber Wall Structure

The fiber wall is composed of several layers that are different from each other both in terms of structure and chemical composition. Each layer has different amounts of cellulose, hemicelluloses, and lignin. The main structural difference between the layers is due to the fibrillar angle between microfibrils and the fiber longitudinal axis. The middle lamella (ML) does not belong to the actual fiber wall, but its function is to bind fibers to each other. The amorphous middle lamella consists mainly of lignin and has a thickness of 0.2–1.0 μm. The outermost layer of the cell wall is called the primary wall (P). In addition to lignin and hemicelluloses, the primary wall is composed of a sparse network of

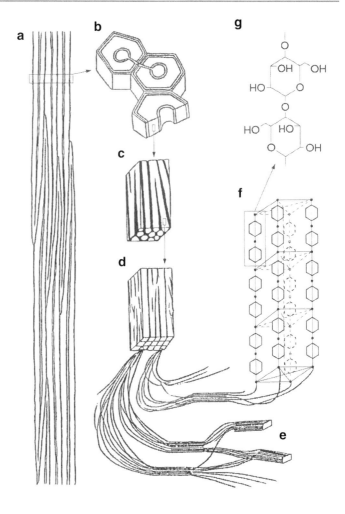

Fig. 9.2 Overall picture of how cellulose is arranged at different fiber layers. Fiber (A), fiber cross section (B), fibrils (C), microfibrils (D), crystals (E), unit cell (F), and cellobiose unit (G) (Fellers and Norman 1998)

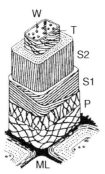

Fig. 9.3 Schematic drawing of the fiber wall and the different fiber wall layers. Primary wall (P), outermost (S1), middle (S2), and innermost (S3) secondary wall layers. Tertiary wall (T) and wart layer (W) (Fengel and Wegener 1984)

microfibrils containing cellulose of low crystallinity. The primary wall has a thickness between 0.1 and 0.2 μm (Fig. 9.3).

The secondary wall of the fiber consists of three layers. The layers are all made up of thin lamellae of cellulose, bound together by hemicelluloses and lignin. They differ mainly on the thickness and microfibrillar angle. The outermost secondary wall layer (S1) consists of 3–4 lamellae, in which microfibrils are alternately oriented to left and right along the fiber axis, creating a net-like structure. The microfibrillar angle of the S1 layer is 50–70° and its thickness between 0.2 and 0.3 μm. The middle layer (S2) makes up about 80 % of the fiber and contains most of the cellulose. The thickness of the S2 layer varies from 1 μm in springwood to 5 μm in summerwood and may thus contain 30–40 or up to more than 150 plates. The lamellae have parallel microfibrils in the same direction along the fiber axis. Technically, the fiber mechanical properties are highly dependent on its microfibrillar angle. A small fibrillar angle shows high stiffness while a large angle results in stretchability and elasticity. The microfibrillar angle of springwood S2 layer is between 20° and 30° and in summerwood between 50° and 10°. The innermost layer of secondary wall (S3), also called tertiary wall, has a microfibrillar network structure similar as in the S1 layer. The layer is thin, i.e., 0.1 μm, and microfibrillar angle can vary between 50° and 90°. This layer has a high resistance to acids, bases, and rotting fungi, suggesting that the chemical composition is likely to be different than in the other layers (Fig. 9.3). The fibers in most conifers and some deciduous trees have a thin film called wart layer (W) as the innermost layer of the lumen (Alén 2000; Eero 1981; Fengel and Wegener 1984).

9.3 Fiber Separation Methods

Wood fibers are embedded in a lignin matrix forming a composite structure in the biomaterial. In order to be processed as fiber-based materials, fibers need to be separated from the wood structure using mechanical, chemimechanical, or chemical methods. The method of separation depends on end use of the fiber-based materials and raw materials available.

9.3.1 Mechanical and Chemimechanical Pulping

Mechanical pulping is a method to separate fibers from wood materials by using mechanical action. Chemimechanical pulping uses chemicals to impregnate the wood prior to mechanical action. Both methods usually require heat to promote the melting of the lignin in the middle lamella between the fibers and that is why they are usually called thermomechanical and chemithermomechanical methods. The wood material can be used in form of logs or cut in form of wood chips with defined dimensions depending on the method used for fiber separation (Lönnberg 1999).

The stone groundwood pulping (SGW) was the first mechanical pulping method developed in Germany in 1840. In this method, the whole log is pressed against a rotating stone and heat is generated by the contact wood-stone surface. The action of heat softens the lignin and hemicelluloses in the middle lamella between the fibers and promotes fiber separation. The physical characteristics of the stone surface play a key role in fiber separation because it affects the compression of wood material and properties of resulting pulp. More recently, a method named pressure groundwood pulping (PGW) was developed. In this method, the grinding is performed under pressure and with addition of water allowing high process temperatures and more energy-efficient fiber separation than in conventional SGW. The PGW process yields a stronger pulp than the SGW. A new variation of PGW process is the Super-PGW (PGW-S) which uses higher pressure (450 kPa instead of 250 kPa) in the grinding zone combined with higher temperature of water (140 °C instead of 105 °C) in comparison with PGW. The PGW-S pulps have more energy efficiency (10 % lower energy consumption) and better strength properties (tear and tensile indexes) in comparison with PGW at the same level of freeness (a property related to water retention in the pulp). Figure 9.4 illustrates a device to produce PGW pulp showing how logs are inserted in the pressure chamber and then pressed against a rotating stone under water showers. PGW pulps are used today in called "wood-containing papers" such as newspapers and catalogues.

Grinding was the main mechanical pulping method until early 1960s when a method based on disc refining was introduced. In this method, the wood was cut in form of chips prior to fiber separation by mechanical actions of refining discs in a technology named refiner mechanical pulp (RMP). The pair of refining discs can be formed by rotating and a stationary element or rotation in opposite directions. The wood chips are compressed between the discs, and fibers are separated through the action of disc grooves. The refining discs can be of conical or circular types. Figure 9.5 illustrates a circular type of disc refiner.

In 1968, a new variation of RMP was developed by treatment of wood chips with heat and performing the refining under pressure. This

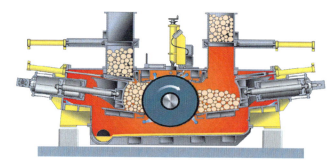

Fig. 9.4 Pressure groundwood pulping (PGW) grinder (Metso Paper undated). Wood logs are pressed against a rotating stone in a pressurized environment

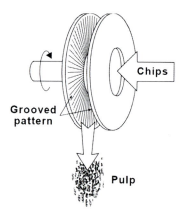

Fig. 9.5 Fiber separation in disc refiner (Lönnberg 1999)

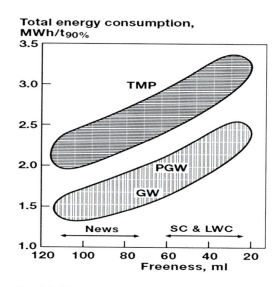

Fig. 9.6 Energy consumption in different process of mechanical pulp production (Lönnberg 1999)

method was called thermomechanical pulping (TMP) and has been used in industrial scale from early 1980s. In TMP, the wood chips are heated with steam before the refining process, which usually occurs in several stages and with a pressurized refiner at least in the first step. However, there are different variations, such as wood chips heating under pressure followed by atmospheric refining and atmospheric wood heating as a result of pressure refining. TMP pulps usually have better strength properties than PGW pulps because of the presence of larger amounts on long fibers. It is important to mention that mechanical and chemimechanical pulps are classified according to the amount of fiber fractions and the freeness level. Pulp properties can only be compared at similar freeness. The advantages of TMP over PGW are still a matter of debate since energy consumption has been a major disadvantage of TMP, while strength properties are drawback to PGW. Figure 9.6 shows the different energy consumption for TMP, PGW, and SGW. In practice, the improvement of PGW pulp properties is dependent on development of new stone surfaces and new strategies for bleaching of pulps.

Chemimechanical pulp (CMP) and chemithermomechanical pulp (CTMP) had their breakthrough after 1970s and during the 1980s with the advent of disc refiners and development of TMP processes. In CMP, the wood chips are impregnated with high dosage of chemicals (usually 10–20 % Na_2SO_3 on dry wood basis) and further refined in atmospheric pressure. The resulting pulping yield is below 90 %. In CTMP, the wood chips are impregnated with low dosages of chemicals (usually 1–3 % Na_2SO_3

on dry wood basis) and further refined under pressurized refining. The typical resulting yield is over 90 %.

The chemical pretreatment in chemimechanical process is dependent of the wood raw material: softwoods are frequently treated with sodium sulfite while hardwoods need two stages, alkali treatment and sulfonation. The impregnation process with sodium sulfite causes sulfonation of the lignin in wood, reducing the glass transition temperature and increasing the hydrophilic character of this polymer and thus promoting the softening of the middle lamella and improving fiber separation. The sulfonation reaction is faster at low pHs and at the same time lignin and carbohydrates can be dissolved. High pH should be avoided due to slow sulfonation, darkening of lignin, and dissolution of carbohydrates. Sulfonation can be monitored by measuring S content in wood or by conductometric titration of pulps.

Recently, some variations of CTMP processes have been developed by different technology suppliers, and other chemicals than sulfonation agents have been suggested. Alkaline peroxide mechanical pulping (APMP) and alkaline peroxide thermomechanical pulping (APTMP) are examples of application of hydrogen peroxide in creating carboxyl groups in lignin to make it hydrophilic and promote fiber separation. Other example is the so-called Bleached CTMP (BCTMP) where sulfonation and alkaline hydrogen peroxide are combined in the process. BCTMP fiber lines can also be combined to kraft pulp mills, using power and steam from recovery system and sharing same effluent treatment station. The end use of chemimechanical pulps is broad. Softwood CMP (freeness of 300–400 ml) is used as reinforcement pulp in newsprint while hardwood CMP and CTMP are used to replace TMP in printing papers. Softwood CTMP is used in tissue paper (freeness of 350–500 ml, added as 20–30 % of fiber furnish), paperboard (freeness of 250–500 ml), and fluff grade pulp (freeness of 650–700 ml). The strength properties and brightness of CTMP pulps are superior than observed for other mechanical pulps (Fig. 9.7).

A schematic representation of the fiber separation regions is present in Fig. 9.8. The fiber separation in RMP is stochastic, and uncontrolled damages to fiber wall cause generation of fines and lower strength of pulp. In case of TMP, the action of heat increase the selectivity of fiber separation to regions close to the primary wall (P) and layer S1. However, higher selectivity in separation can be achieved with the sulfonation of lignin combined to action of heat in CTMP process.

9.3.2 Chemical Pulping

Sulfite and kraft cooking are the main chemical pulping methods currently used. Kraft pulping is by far the most preferred technology used for fiber separation for utilization in papermaking due do its large applicability to different wood raw materials and high strength of pulps in comparison with sulfite cooking. Sulfite cooking is a good alternative for preparation of dissolving pulp because hemicelluloses are extensively removed when the process is operated under acidic conditions (Potthast 2006).

Sulfite pulping processes can be used for preparation of dissolving pulps from both hardwoods and softwoods. This method was a traditional technology to separate fibers from wood to use in papermaking until 1940s when the recovery systems for the kraft pulping process were further developed. Sulfite pulping has several disadvantages in comparison with kraft pulping such as weaker pulps, challenges in recovery of the cooking liquors, and limited application to wood raw materials. The main advantage of sulfite pulping is the capability to obtain fibers rich in cellulose that are easier to bleach than kraft pulps and also the separation of lignosulfonates and other value-added by-products such as vanillin and ethanol. Sulfite pulping is currently used for pulp fibers for specialty papers, tissue, and fine paper and for producing dissolving pulp for cellulose dissolution. The process is based in cooking wood in a liquor containing either sulfites (SO_3^{2-}) or bisulfites (HSO_3^-) linked to different counter ions. Common counter ions are

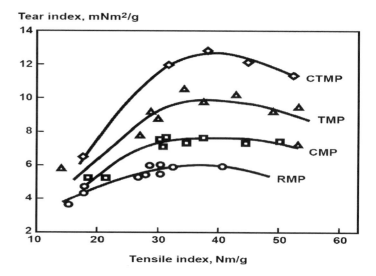

Fig. 9.7 Tear versus tensile indexes of different mechanical pulps (Lönnberg 1999)

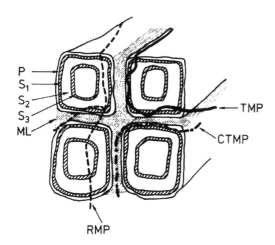

Fig. 9.8 Schematic representation of fiber separation regions in different mechanical pulping processes (Lönnberg 1999)

Ca^{2+}, Mg^{2+}, NH_4^+, K^+, or Na^+ in pH conditions of 2–6. Sulfur dioxide can also be added in the process to regulate the pH. Neutral sulfite (pH 7–8) uses Na^+ as counter ion and Na_2SO_3 as cooking liquor while acidic sulfite (pH < 2) uses a mixture of H_2SO_3 + $XHSO_3$ where X can be any of the counter ions mentioned earlier (Potthast 2006).

Kraft pulping is performed by cooking wood chips with a solution composed of sodium hydroxide and sodium sulfide. This solution is often named "white liquor." The effective alkali load (NaOH) is about 16–20 % of the wood amount. The consumption of effective alkali in a kraft cook is about 150 kg of NaOH per ton of wood. When the wood is treated with heat and alkali, a delignification of the wood occurs. The lignin is dissolved and the cellulose is separated from the most of the lignin (Rydholm 1965).

The main active components during the strong alkaline cooking process are the hydroxide ion and the hydrosulfide ion. The result from the alkaline cooking treatment of wood is strong and flexible but brown fibers. However, the wood chips maintain their wood structure after cooking despite the loss of most of their lignin content. But the structure becomes so weak that they will break down to fibers by slightest mechanical action. The discharge of the digester provides sufficient enough of mechanical action to break the structure of the cooked chips.

After kraft cooking, the pulp is washed in different washers until the cooking liquor, or black liquor, is washed away. The pulp is also removed from knots and screened to remove impurities like shives and bark. The washed and screened brown pulp is subjected to delignification prior to bleaching to reduce lignin content left in the pulp after cooking.

A variation of the conventional kraft pulping process named pre-hydrolysis kraft can be used for preparation of dissolving pulp. Dissolving pulp is a pulp raw material rich in cellulose content for dissolution in cellulose solvents. Softwoods and hardwoods are used as raw material for dissolving pulp grades. The target of the dissolving pulping process is to remove hemicelluloses and lignin using chemical pulping methods such as prehydrolysis kraft and sulfite processes. The prehydrolysis kraft pulping process is usually performed in an impregnation stage using acidic or neutral conditions prior to conventional or modified kraft cooking. The treatment time varies between 0.5 and 3 h and temperature is above 100 °C. In these conditions, hemicelluloses are removed and lignin undergoes condensation reactions; therefore, the process is more suitable for hardwoods than softwoods (Sixta 2006).

9.3.3 Recycled Fibers

Recycled paper is a very good raw material fiber for several applications. The apparent paper collection rate is higher than 70 % in several European countries allowing proper utilization of recycled fibers in manufacture of papers and composites. The recycled pulp contains ink and other residuals such as fillers and usually more fines than fresh pulp. It also has lower retention and dewatering ability and lower bonding capacity than fresh pulp leading to papers with lower bulk, lower strength properties, and higher linting and dusting occurrences. Hence, the deinking process has a key role in separating fibers both from the paper and the coating components and inks used in papermaking and printing. The deinking process has usually four stages (1) mechanical stage where the inked paper is broken into inked fibers, (2) deinking stage where the ink-fiber bond is broken and ink detaches from fibers, (3) dispersing stage where the ink particles are stabilized in solution, and (4) flotation stage where flocculation is used to aggregate ink particles in form of flocks and attach to air bubbles to create a foam that is separated from the fiber furnish. Post-flotation and bleaching can also be used additional cleaning of fibers. The fiber yield varies between 70 and 90 % depending of the process and type of paper used. The deinking process is presented in Fig. 9.9.

The successful detachment of ink particles from fiber surface and consequent removal via flotation can be affected by ink redeposition which gives a low-brightness pulp. The ink-redeposition mechanism can be caused by physicochemical or mechanical redeposition in addition to lumen loading and ink attachment during filtration. Physicochemical redeposition is caused by disturbance of colloidal dispersions and breakage of emulsion. Mechanical redeposition is caused by effects of particle size, turbulence, and trapping of particles into fiber surfaces through collision. Ink redeposition can be solved by using nonionic surfactants which assist the dispersing of ink particles and create a stable emulsion. A multitude of opportunities for developing chemicals for deinking is available, and new processes where fiber properties are preserved or unaffected during recycling have been developed, allowing broader applications in paper products. Currently recycled fibers are used in newsprint, tissue paper, paperboard, and linerboard.

9.4 Fiber Functionalization Methods Used in Pulp and Paper Manufacture

Bleaching, low-consistency refining, and treatment with paper chemicals are the traditional technologies used in the functionalization of fibers to make them suitable for utilization in fiber-based materials. The main goal of bleaching is tailor optical properties by increasing the brightness level of pulps. Low-consistency refining is a method used to treat fibers in water and increase the mechanical properties due to external and internal fibrillation of fibers. The functionalization with paper chemicals include the utilization of polyelectrolytes to build fiber flocks in papermaking, modification of fiber surfaces by sorption of polysaccharides, sizing agents, wet strength agents, and optical brighteners.

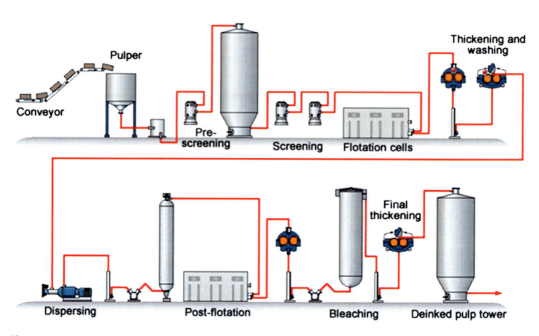

Fig. 9.9 The deinking process used in paper recycling (Metso Paper undated)

9.4.1 Bleaching of Mechanical Pulps

The main objective of bleaching is to increase pulp brightness through chemical treatment of unbleached pulp. There are two strategies used in bleaching (1) removal of lignin such as in the bleaching of chemical pulps and (2) lignin-retaining bleaching such as for mechanical pulp bleaching. Bleaching of mechanical pulps causes increase in brightness, reduces extractive content, and increases pulp strength and fiber-to-fiber bonding. This is caused by elimination of chromophore structures which reduce the light absorption of lignin without removing it from the pulp. As a consequence, the pulp yield is preserved. High brightness requires high dosages of chemicals, and brightness reversion or yellowing usually limits high levels of brightness.

Bleaching chemicals for mechanical pulps can be done using oxidative or reductive reactions. Reductive bleaching is usually performed with sodium dithionite while oxidative bleaching is done with hydrogen peroxide. Removal of transition metals from pulp using acidic treatment or complexation with chelating agents is need both for oxidative and reductive bleaching.

Sodium dithionite (sodium hydrosulfite, $Na_2S_2O_4$) or zinc dithionite (zinc hydrosulfite, ZnS_2O_4) are used in reductive bleaching in dosages of 0.5–1 % $Na_2S_2O_4$ on pulp. pH 4–6 is the optimum pH for ZnS_2O_4 and pH 6–6.5 for $Na_2S_2O_4$. The pH is reduced 0.3–1.0 during bleaching which is done usually in low (3–5 %) or medium consistency (8–12 %) at temperatures of 50–70 °C and retention time of 30–60 min. The reactions of dithionite with the chromophores in mechanical pulps have not been as comprehensively studied as the reactions of peroxide bleaching. It is generally considered, however, that the bleaching effect of dithionite is due to reduction of (1) quinoidic groups into hydroquinones, (2) α-carbonyl, and (3) coniferyl aldehyde groups and probably also reduction of colored Fe^{3+} compounds into less-colored Fe^{2+} compounds (Lönnberg 1999).

Hydrogen peroxide works in two different ways. It is a delignification agent in bleaching of chemical pulps. In bleaching of mechanical

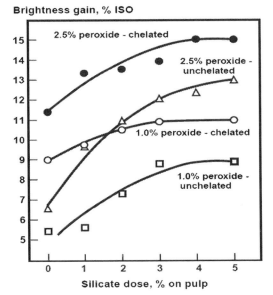

Fig. 9.10 Brightness gain in bleaching of mechanical pulp using sodium silicate as additive (Lönnberg 1999)

pulps, it works as a true bleaching chemical, attacking only colored chromophores. Mechanical pulp bleaching results in a brightness of 80–83 % ISO. This kind of bleaching is lignin retaining, meaning that colored chromophores are removed, while lignin is not degraded. A temperature of 50–70 °C is sufficient for this kind of bleaching. Peroxide bleaching is usually done at alkaline conditions where peroxyl ions act as bleaching agent. The amount of alkali added is dependent upon the hydrogen peroxide dosage in bleaching. The increase in temperature accelerates the reactions in H_2O_2 bleaching (bleaching reaction and decomposition), and at temperatures above 80 °C, brightness developments are fast and maximum brightness is achieved in 30 min. Additives are used in H_2O_2 bleaching such as sodium silicate (Na_2O_{3-4} SiO_2). The exact mechanism is not known. However, it is believed that silicate has stabilizing effect on peroxide and sodium silicate act as a buffer. Silicate can also form metal complexes with magnesium and prevent decomposition of peroxide by metal ions. The effects of additives in peroxide bleaching of mechanical pulp are presented in Fig. 9.10. The increase of peroxide dosage combined with chelation and addition of sodium silicate has clear positive effects in brightness gain.

9.4.2 Bleaching of Chemical Pulps

Bleaching, primarily of clothing, has been taking place since ancient times. Lime, buttermilk or juices of berries, and sunlight were the primary bleaching agents until the 1800. In 1798, Charles Tennant discovered how a suspension of lime could absorb chlorine gas, thus forming hydrogen hypochlorite which became the standard bleaching chemical for the next century. When paper started to be produced from wood instead of cotton and rags, improved bleaching had to be applied to obtain descent brightness. This led to the development of bleaching sequences that better and more effectively bleached the pulp. The first sequence was an H–E–H sequence with two hypochlorite and one extraction stage which severely decreased the chemical consumption compared to using a single stage. Progress led to the use of gaseous chlorine which further improved the bleaching in the C–E–H sequence. The size of the mills increased, and in the 1950s, chlorine dioxide was used in C–E–D–E–D sequences which bleached pulp to a high brightness with a minimum of strength loss (Reeve 1996).

Today, most mills producing pulp from softwood use an oxygen stage before the bleach plant. In 1990, 94 % of the produced bleached pulp was bleached with chlorine. Due to environmental reasons, most mills have replaced chlorine to the less environmentally hazardous chlorine dioxide. In the Nordic countries, there is no elementary chlorine bleaching anymore. Elementary chlorine free (ECF) bleaching is a form of bleaching where no pure form of chlorine gas is used in the bleaching stages. Instead, chlorine dioxide is used which is more environmentally friendly. Totally chlorine free (TCF) bleaching is a bleaching sequence that does not contain any chlorine at all. Oxygen, peroxide, and ozone are used to achieve desired bleaching results. TCF bleaching is more expensive than ECF bleaching and does not reach the same brightness. Also, the environmental impact from ECF bleaching is no longer so huge that the customers are prepared to pay the higher price for TCF. ECF bleaching is today used to a higher extent than TCF bleaching.

It is not possible to bleach from ISO brightness 20–90 in one stage (Reeve 1996). Several different types of stages must be used with washing in between the stages. This way the lignin is attacked in several different ways, and degraded lignin parts are washed away after each stage. It is important not to run the separate stages too long since this will degrade the pulp and make it significantly weaker. Bleaching chemicals have different selectivity, meaning the most economical way is to use the cheapest chemical first and then use the more selective and usually more expensive chemical. Several different bleaching sequences could be used in pulp bleaching. Each mill has their specific sequence with their optimal parameters and sequences to obtain bright pulp. Below are some of the bleaching stages that are used in today's bleaching of pulp.

9.4.2.1 Oxygen Delignification (O_2, O stage)

Kraft mills often use an oxygen stage as a prolongation of the digester cook which is one way to improve bleaching. This means that cooking is aborted earlier and the delignification is continued in the oxygen stage which results in a pulp with a higher brightness at comparable strength for the pulp entering the bleach plant. If the pulp is cooked too long in the digester, the strength will drop dramatically and brightness will only slightly increase. One of the reasons oxygen stages are implemented are environmental reasons since less bleaching chemicals are needed which result in less discharge from the bleach plant. Also, when oxidized white liquor is used as an alkali source, it is possible to use the process water from the oxygen stage in the washers (Pikka et al. 1999). Delignification in a medium consistency (MC) oxygen stage is normally 40–45 % (Reeve 1996). Initial ISO brightness is ~27 after cooking and ~35 after oxygen bleaching. After the oxygen stage, the pulp is washed and sent to the bleach plant where brightness is raised from ISO brightness 35 to its final brightness, usually around ISO brightness 90. If the bleach plant uses chlorine containing bleaching chemicals, the oxygen delignification would result in effluents having less absorbable organic halogens (AOX), less biological oxygen demand (BOD), and less chemical oxygen demand (COD). The oxygen is produced on site and is added directly into the oxygen reactor where it reacts with the pulp. Since oxygen is a weak oxidant, alkali needs to be added to activate the lignin for faster reactions (Reeve 1996; Pikka et al. 1999).

9.4.2.2 Chlorine Dioxide (ClO_2, D stage)

In mills with ECF bleaching, chlorine dioxide has totally replaced elemental chlorine since it is more environmentally friendly. It is an efficient bleaching agent, but the best bleaching result is obtained when chlorine and chlorine dioxide is used together (Singh 1979). It would be possible to bleach kraft pulps to 88–90 ISO units purely with ClO_2 in a D–E–D sequence. Obtaining brightness over 90 ISO units would require five or more stages (Singh 1979).

ClO_2 is explosive in high concentrations so it must be diluted in water (6–12 g/l water) before it is mixed into the pulp. That is the reason why ClO_2 is produced on site. The initial reaction is very fast, 75 % of the ClO_2 reacts within the first 5 min, and the rest of the ClO_2 continues to react during several hours. The temperature should be 70 °C and should not be raised too high to prevent carbohydrate degradation and color reversion.

9.4.2.3 Hydrogen Peroxide (H2O$_2$, P stage)

Peroxide works as a delignifying chemical in bleaching of unbleached chemical pulp and oxygen bleached pulps above 80 °C. Peroxide bleaching gives a higher brightness in comparison to oxygen bleaching, but selectivity is not as good. This results in lower end viscosity for peroxide bleached pulp compared to oxygen bleached pulp. Many aspects of hydrogen peroxide bleaching are similar as in oxygen bleaching since both peroxide and oxygen are formed during oxygen and peroxide bleaching (Gierer 1995). Oxygen is on the other hand more cost effective than peroxide for the same degree of delignification. Hydrogen peroxide requires high pH and metal ions must be removed before bleaching; otherwise the peroxide will decompose without achieving desired brightness. This is usually done by ethylenediaminetetraacetic acid (EDTA) or diethylene triamine pentaacetic acid (DTPA) chelation (Reeve 1996).

9.4.2.4 Alkali Extraction (E Stage)

The role of the alkali extraction stage is to remove the lignin that has been potentially available by the previous bleaching stage. By using alkali, it is also possible to reactivate the pulp for further delignification if the preceding stage has been acidic. Several hypotheses are proposed regarding this subject. One states that the bindings between cellulose and remaining lignin are of a kind that needs an oxidative environment to be broken. One other proposal is that further oxidation is not possible due to deactivation of the remaining lignin and that blocking groups are formed by acidic oxidation that prevents further oxidation. This oxidation is then proposed to continue after alkaline conditions. It was also suggested that alkaline water would increase the solubility of the oxidized material and that the hydrodynamical volume of lignin would decrease, making it easier to remove from the fiber wall (Berry 1996). If the temperature is elevated, there is a risk of hemicellulose losses which in practice imposes the temperature of the extraction stage in industrial facilities to be below 70 °C.

9.4.2.5 Oxidant-Reinforced Alkaline Extraction (EOP Stage)

The extraction stage can be improved by adding an oxidative chemical such as oxygen or peroxide or both to the extraction stage. Using reinforced extraction stages improves delignification without significantly affecting pulp viscosity which results in lower Kappa number and reduces bleaching chemicals in the following stages. EOP stages differ from pure oxygen stages; less retention time and oxygen pressure are needed to obtain significant results. An extra 0.5 % unit of alkali needs to be added compared to a normal extraction stage due to alkali consumption. The first extraction stage is used as a delignifying stage, and the second extraction stage is considered to be used as a reactivation stage of the pulp for further oxidation (Berry 1996).

9.4.2.6 Ozone (Z Stage)

Ozone, O_3, is a powerful oxidizing agent consisting of three oxygen atoms. It is a very toxic gas which rapidly decomposes in room temperature, which means it must be produced on site. Ozone is not as selective as chlorine and chlorine dioxide, but it is being used as a way to lower chlorine emissions. It is sometimes part of the TCF bleaching line and requires a lot of electrical energy to be produced. The kappa number of the pulp entering the ozone stage must be low, and therefore, it is usually preceded with an oxygen stage to reach desirable kappa numbers. The placement of the ozone stage in the bleaching sequence is very important and can dramatically affect the final brightness. The following ISO brightness has been reported using different combinations of stages: O–Q–P–P: 77.9, O–Z–Q–P–P: 83.9, and O–Q–P–Z–P: 89.9 (Lierop et al. 1996). Ozone must be used in low dosages not to degrade pulp, which means that a carrier gas or liquid must be used. Ozone reacts 106 times faster with lignin than with carbohydrates, but by-products from the lignin reaction such as hydroxyl radicals (HO•) and perhydroxyl radicals (HOO•) are formed, and these radicals react readily with carbohydrates. The best delignification and selectivity in the ozone stage is achieved at low temperature (20 °C) due to ozone degradation of pulp in higher temperatures. Since oxygen and peroxide bleaching is performed at elevated temperatures, the pulp must be cooled to achieve optimal ozone bleaching (Owen and Anderson 1996).

9.4.2.7 Peracetic Acid (CH_3COOOH, Paa Stage)

Peracetic acid (Paa) is used as a delignifying and brightening agent in pulp bleaching. It is made from acetic acid and hydrogen peroxide, and a solution of peracetic acid contains both the constituents and the product. Distilled peracetic acid is used in pulp bleaching, and it contains ~40 % peracetic acid. Bleaching is very fast and takes

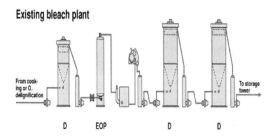

Fig. 9.11 A fiber line showing different bleaching stages (Pikka et al. 1999)

place during the first few minutes. Paa consumption is 1–2 kg/t pulp to increase brightness to 1.5–2.5 ISO units. Higher Paa amounts only slightly improve brightness, but at the same time, brightness reversion and other optical properties are improved. The nonconsumed acid acts as bacterial growth control in the process downstream. Peracetic acid decomposing at room temperature storage should hence be in a controlled temperature between −10 and 6 °C (Jäkärä et al. 2001).

9.4.2.8 Some Typical Sequences Used in Hardwood and Softwood Bleaching

Different stages are combined to create sequences that are optimal for each pulp mill depending on weather a TCF or an ECF sequence is used and type of wood. A typical ECF bleaching sequence is O–D–E–D–E–D, but a typical TCF sequence does not exist. They are all mill specific.

Rauma mill has now an ECF bleaching sequence O–D–EOP–D–P that bleaches softwood to a 90 % ISO brightness. Their former TCF sequence O–Z–EOP–Z–P only produced pulp with 78 % ISO brightness. Fray Bentos in Uruguay which started in 2007 use another ECF sequence to bleach eucalyptus hardwood pulp. Their O–A/D–Eop–D–P sequence produces pulp of brightness 92 ISO units. Östrand mill has been bleaching both softwood and hardwood since 1995 with the Q–OP–Zq–PO sequence to full brightness. Since 2002, they only produce softwood pulp (Fig. 9.11) (Table 9.1).

9.4.3 Low-Consistency Refining

Mechanical treatment such as refining (or beating) is normally used to improve certain physical properties of the final pulp products. Refining of pulp, defined as the repeated passage of pulp through zones of compression and shearing, is carried out to a greater or lesser degree. This mechanical treatment results in structural changes including fiber shortening and both internal and external fibrillation (Clark 1978). In pulp refining, the fibers are tailored mechanically for enhancement of dry paper properties. The treatment affects structural, morphological, and interactive fiber properties and enhances the fiber bonding potential, due to changes in the fiber cell, which increases fiber conformability that result in a consolidation of the paper structure.

Refining of fibers is nowadays done mainly in low-consistency refiners. In low-consistency refining, fibers pass through a narrow gap between rotor and stator bars (Fig. 9.12). The gap between rotor and stator is varied over the range of 10–400 μm. The average gap of 100 μm corresponds to 2–5 swollen fibers or 10–20 collapsed fibers (Clark 1978).

The main aim of refining is to improve the fiber–fiber bonding. It is anticipated that the physical changes of fiber during refining would make the fiber chemical components more accessible and would affect the distribution of these components on the surface as well as within the fiber wall (Fardim and Durán 2003). In other words, fibers become damaged and the inner walls of the fiber partly appear. These new areas on fibers help to create even more hydrogen bonds than just on the outer wall of the fiber. Refined fibers thus have more specific bond area since also fibrillation occurs. Fibrillation helps the fibers to create a web with more fibers linked to each other. An exact detailed explanation of how refining works has not yet been presented since many different theories explains different mechanisms in refining (Everett 1989).

Table 9.1 Different bleaching sequences used in pulp mills

Mill	Hardwood	Softwood	ECF	TCF	ISO brightness
Rauma		O–D–EOP–D–P	X		90
Fray Bentos	O–A/D–EOP–D–P		X		92
Östrand		Q–OP–Zq–PO		X	88–90
Typical	O–D–EOP–D	O–D–E–D–E–D	X		~90

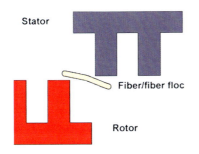

Fig. 9.12 Fiber passage section in refiner (KnowPap 2011)

9.4.4 Papermaking Chemicals

Papermaking chemicals are used to functionalize the fibers in the paper machine, usually at the wet end before the paper is dried. The paper machine wet end is composed by stock preparation, headbox, formation, and press sections (Fig. 9.13). The purpose of the stock preparation is to form macroscopic flocks composed by polymers, colloidal particles, filler, and pulp fibers and also to modify the surface chemistry of the fibers. The purpose of the press section is to remove the excess of water and accelerate interfacial reactions or sorption of chemical during paper drying. Flocks must have enough strength to resist to the turbulent flow in the headbox and also the shear forces in the forming wire. A multitude of interfacial phenomena are involved in the wet end and are directly related with desired paper characteristics.

Flocculation is an aggregation process where the particle identity is maintained and which can be classified as reversible or irreversible. Reversible flocculation is also termed coagulation according to some authors. Irreversible flocculation is a spontaneous process if there is a reduction in the system total free energy. Usually it is assumed that irreversible flocculation is dominated by free energy and the contribution from entropy is not significant (Everett 1989). The addition of electrolytes is not enough for flocculation of pulp fibers and mineral particles in papermaking. The repulsive potential between fibers is then reduced by addition of polymers or polyelectrolytes which hold electric charges with different polarity as those present on the particle and fiber surfaces.

Polymers or polyelectrolytes used to improve flocculation in papermaking are usually termed retention aids. Cationic starch used in stock preparation is also believed to contribute to retention as well as on dry strength. Drainage and retention are important parameters in the paper production and require not only that the flock formed has strength but also that the water on its structure needs to be easily removed. Retention is also affected by mechanical interlacing of flocks in the formation wire. High-molar mass polymers used as retention aids can increase the colloidal attraction forces and induce flocculation by different mechanisms, i.e., bridging, network, depletion, and bridging combined with microparticles. Bridging flocculation involves adsorption of terminal and intermediary regions of the polymeric chain on the particle surface, forming an aggregate fixed by the polymeric chain. High polymer concentrations can hinder the bridge formation between particles due to the tendency to cover their surfaces. It is believed that the optimum polymer concentration to achieve flocculation by this mechanism is half of the amount needed to cover the entire particle surface (Lindström 1989).

Different segments of the polymeric chain bind the particles and an aggregate is formed. Network flocculation can happen if the polymer concentration is very low, being adsorbed on the surface of a few particles and thus transform then onto small electric dipoles with extremes of opposite charges. Under this condition, particles

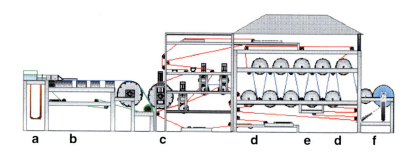

Fig. 9.13 Schematics of a Fourdrinier paper machine. (a) Headbox, (b) forming, (c) press, (d) drying, (e) surface treatment (optional), (f) reel

are oriented in a way that mutual electrostatic attraction is achieved. Polymers used in low concentrations are adsorbed onto the particle surfaces, forming electric dipoles and generating interparticle attraction. Flocculation by depletion mechanism occurs if a weakly or no adsorbed polymer is present when two particles approach each other.

Microparticle systems using either silicates or bentonite combined with high-molar-mass polymers are used in flocculation by bridging microparticle mechanism. The bridging is similar as described previously, but in this case the system highly flocculated after polymer addition is submitted to shear and the macro-flocks are broken into micro-flocks. At this point, microparticles are added and the macro-flock is rebuilt. However, it has higher resistance to turbulence and shear than an ordinary macro-flock formed only using polymer. It is believed that the microparticles interact via electrical charge attraction with the polymer segments attached to the particles and reflocculation is caused by bridging between polymer chains connected to the microparticles (Swerin 1995). Research on flocculation dynamics is traditionally performed in experimental scale using turbidity and dynamic drainage or Britt jar measurements. However, more sophisticated techniques such as scanning tunneling microscopy have been applied, and information about flock size and resistance to shear forces may be obtained (Gerly and Clémençon 1999).

Flock resistance is affected by different types of forces: normal forces between fibers, forces due to fiber flexibility and elasticity, friction forces, and colloidal attractive forces (Kerekes et al. 1985). There are different statements regarding which factor is predominant on flock formation and resistance. Domination of hydrodynamic effects (Beghello 1998) and chemical effects and colloidal attraction domination (Wågberg 1987) have been proposed. There are also some common points between the different hypotheses. The common points are the effects of consistence, fiber length, fiber–fiber contact, turbulence, and shear forces on flocculation. However, the main discordances are related to the effect of the medium, i.e., electrolyte concentration and pH. Fiber surface composition is recognized as an important effect on flocculation by the different hypotheses. Both statements considered that the higher concentration of anionic groups on the surface makes the fiber more sensitive to paper machine medium and flocculation is reduced. Flock size is also affected by surface composition, and small flocks are observed when higher concentration of anionic groups is present on the surface (Beghello 1998). The bleached pulp fiber with lowest total concentration of anionic groups is the pine ECF while the highest concentration is observed for eucalyptus ECF (Fardim et al. 2001). However, effects of formation of a surface layer gel on flocculation and particle retention for pulps with different amounts of anionic groups are presently investigated.

Chemical interference in sizing reactions using alkenyl succinic anhydride (ASA) can be attributed to dicarboxylic acids formed due to the ASA hydrolysis in water. Alkyl ketene dimmer (AKD) can also have hydrolysis in water at high temperatures, and the formation of ketones interferes on the distribution of sizing aids on the fiber surface (Roberts 1992). Flocculation is not the unique phenomenon which involves surface interactions in the wet end (Isogai et al. 1997;

Berg 1993). During stock preparation or ahead of the headbox, components are added to modify the fiber surface free energy. These additives are compounds with hydrophobic character and holding terminal groups which can attach to the fiber surface via electrostatic interaction or chemical reaction. The latter can yield cellulose ester derivatives. Modification of fiber surfaces with the purpose of reducing the water penetration into the paper structure is termed sizing. This process is classified as alkaline, neutral, or acidic sizing, depending on the pH in the reaction medium. Additives used in sizing modify both surface free energy and surface chemical composition (Ozaki and Sawatari 1997; Carlsson 1996). A homogenous sizing aid distribution on the fiber is dependent upon the absence of restrictive effects and availability of superficial groups. Chemical interfering and morphological parameters such as excessive roughness are the main restrictive effects.

Modification of fiber surfaces in papermaking is not limited to sizing; usually other additives such as dry and wet strength aids are also employed and the interaction fiber–fiber is affected. The latter additives are believed to modify the fiber surface composition by insertion of functional groups which hinder the formation of hydrogen bonds between fibers and water molecules with concomitant increase in fiber–fiber bonds. As a consequence, the fibrous network has a higher strength in wet and dry media. Components usually applied for this purpose are anionic and cationic polyacrylamides, polyamines, and polyamides. Imines, urea, and melamine formaldehyde are also employed specifically for wet strength purposes.

9.5 Fiber Properties and Uses

Pulp fibers can be used for a broad range of applications in fiber-based materials and also for dissolution in cellulose solvents. The chemical and surface chemical composition of fibers plays important roles depending on the type of use of fibers. Usually the degree of polymerization and the amount of hemicelluloses and lignin are critical factors for dissolution in cellulose solvents.

In case of papermaking, besides the chemical composition and macromolecular properties, a number of other characteristic properties are taken into consideration. The principal properties for papermaking fibers are: average fiber length, wet compactability, intrinsic fiber strength, sheet strength, cohesiveness, coarseness, latency, drainability or freeness, and optical properties.

Average fiber length is measured using optical devices, and mathematical treatment of distribution data allows calculation of average length as $\Sigma L/N$, where L is the length of the fibers and N is the number of the fibers. However, short fibers affect N but not L, and preferably an average fiber length weighted by weight calculated as $\Sigma WL2/WL$ is used. Wet compactability is the property of fibers that permits them to collapse into a compact mass when made into paper (make a denser sheet). It depends on the thickness and nature of the fiber wall, and it has a direct relation with the water removal from the web in the paper machine. It is measured as the apparent specific volume of a handsheet prepared under standard conditions. Intrinsic fiber strength is usually estimated by measurement of the zero-span tensile strength and determines the rupture of single fibers on an x- and y-plane. It is related to the strength of handsheets and papers when fiber-to-fiber bonds are stronger than the individual fibers. The sheet strength includes tensile strength, tear, burst, and bending strength. Scott bond (tensile strength on z-direction) is also measured for specific papers. Cohesiveness measures the intensity which fibers cohere when they are subjected to tension in the plane of a sheet of paper. It includes the effective contact area between adjacent fibers, and it is increased by wet compactability of fibers. Coarseness measures the weight in milligrams per 100 m of fibers, and it is also known as the titer, usually measured in denier or decitex in the textile industry, being usually measured simultaneously with fiber length and fiber wall width. Latency is a property of lignin-containing fibers and is defined as distortions in fiber shape caused by solidification of lignin and hemicelluloses when the temperature is suddenly reduced after contact with a stone or refiner in mechanical pulping. Cellulose has a high resistance to heat while hemicellulose and lignin soft at 60 °C, and grinding or refining are usually done at 100 °C and at high consistency. Latency is removed by heating the pulp to 85 °C at low consistency during disintegration. It aims at measuring the draining ability of a stock in the paper machine. Freeness and drainability are used extensively by the industry. Canadian Standard Freeness (CSF) is based on the excess of "free" water measured using a standard device and a 3 g/l pulp suspension, and Schopper-Riegler drainability (°SR) is a result of 1,000 minus the excess of "free" water (ml) divided by 10 using a standard device and a 3 g/l pulp suspension. Reflectance, opacity, and brightness are the main optical properties of fibers and papers. When light strikes a paper surface, some light is spectrally reflected and the remainder enters the sheet. The light scatters in all directions with subsequent scattering sequences (multiple scattering), and some light returns vertically from the surface and the remainder in transmitted or absorbed. Reflectance measurements can determine the reflected, transmitted, and absorbed intensity. Standards specify the procedures for reflectance measurements, including the spectral characteristics of the incident light. Reflectance is usually measured using diffuse illumination and normal angle of detection. Opacity and brightness are the most important optical properties of paper. Opacity measures the ability of the paper to conceal text on the other side of the sheet. The "brightness" or reflection factor $R\infty$ is measured by adding sheets to a pile until there is no change to the intensity of reflected light at a wavelength of 457 nm. $R\infty$ is measured relative to a perfectly reflecting diffuser. Opacity is determined by measuring the amount of light reflected by a single sheet, $R0$, when the sheet is backed by a perfect black background. The opacity is then defined as the ratio $R0/R\infty$. Opacity could also be measured directly using a transmission densitometer.

9.5.1 Paper, Tissue, and Packaging Materials

Pulp fibers can be used for manufacture of different paper products and packaging materials. A summary of utilization of fibers used as raw materials for papers is present in Table 9.2.

9.5.2 Dissolution of Pulp Fibers in Cellulose Solvents

Dissolution of biomass, e.g., cellulose, is generally used as a method of forming the natural polymer into industrial products, such as textile fiber, membranes, beads, sponge, etc. Cellulose is first isolated from the other components present in the plant, often wood material due to its high cellulose content and low content of disturbing material. At the moment, this is usually done by kraft cooking. Dissolving grade cellulose with high α-cellulose content is often prepared by sulfite cooking from spruce.

After pulping, cellulose is purified, possibly treated chemically or by other means, and thereafter dissolved in appropriate solvent (Liebert 2010). Mainly there are two types of solvent systems. An example of so-called derivatizing solvent systems is the industrially most common viscose method, where the –OH groups of cellulose are functionalized with –CS2Na in the xanthogenation stage prior to dissolution in aqueous NaOH. Native cellulose has very low solubility to NaOH, possibility due to its crystalline structure. NaOH is not enough powerful solvent to break the intermolecular bonds in this crystalline structure, so that only the amorphous regions dissolve. Cellulose having lower DP than 200 dissolves in aqueous NaOH without any pretreatment. Viscose method's major drawback is the environmental and health risks caused by CS2.

There are also so-called direct, nonderivatizing solvent systems, where cellulose dissolves in the solvent without any chemical modification (see Chaps. 4 and 5). The most important example here is *N*-methylmorpholine-*N*-oxide (NMMO). It is applied in large scale, e.g., for the preparation of textile fibers under different brand names [e.g., Lyocell (Lenzing) or Alceru (TITK Rudolstadt)]. Two problems are still connected with the NMMO process: the instability of the solvent and the tendency of the Lyocell fiber toward fibrillation (Fink et al. 2001). Other systems such

Table 9.2 Pulp raw materials, pulping yield based on wood and use in different types of paper

Pulp fibers	Pulping yield (%)	Used in manufacture of
Groundwood pulp	96–98	Paperboard, newsprint, catalogues
Thermomechanical pulp	96–98	Newsprint, pocketbooks, catalogues
Chemithermomechanical pulp	90–95	Tissue paper, paperboard, liquid packaging printing papers
Unbleached kraft	48–60	Sack paper, wrapping paper, kraftliner, liquid packaging board
Bleached kraft	43–47	Printing paper, the outer layers of cardboard, liquid packaging, fluff
Unbleached sulfite	50–60	Newspaper, magazine
Bleached sulfite	45–47	Printing papers, tissue paper
Recycled fibers	20–40	Newsprint, tissue paper, cardboard, kraftliners

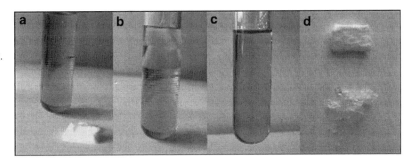

Fig. 9.14 Picture a–c shows dissolution of cellulose (dissolving pulp) in EMIMAc within 15 min. (**d**) shows the starting cellulose and the regenerated material

as N,N'-dimethylacetamide/LiCl (McCormick and Lichatowich 1979) or dimethyl sulfoxide/tetrabutylammonium fluoride (Heinze et al. 2000) utilize often complex solvents and are therefore expensive. The newest solvents having good capability to dissolve cellulose are so-called ionic liquids, which are in fact molten salts (Liebert and Heinze 2008). Several ionic liquids have been studied for their ability to dissolve cellulose, and the results are very promising. The best ILs are liquid at room temperature and can dissolve as much as 20 % w/w cellulose without pretreatment (Fig. 9.14). Cellulose can be precipitated from the solvent by adding water or other nonsolvent. The drawback of the IL systems is still their high price and recyclability (there are a few promising paths for the recycling of the solvent).

Another approach to dissolve cellulose more easily is to increase its solubility by different pretreatments, such as mechanical, chemical, or enzymatical. The idea is to increase the swelling of the native cellulose by opening up its crystalline structure. A method utilizing endoglucanase enzymes has been studied for decades, and the pretreatment with cellulases seems to increase the solubility of the treated fibers, so that they can be dissolved in aqueous NaOH at low temperatures. The so-called Biocelsol method (see Chap. 5) utilizes first mechanical ball mill treatment followed by enzymatic treatment. According to the literature, the DP is not decreasing significantly due to the pretreatments. However, the solutions prepared according to the method possess some uncommon properties that were noticed during the manufacturing of cellulose beads. Even the beads and sponges prepared from this solution had different properties as compared to beads and sponges prepared from other cellulose solutions. The shape of the beads is much more irregular compared to the beads made from viscose. The sponges made from enzyme-treated cellulose are much weaker and harder compared to sponges made from viscose. Even the textile fibers made from this cellulose process are much weaker as compared to Rayon or Lyocell fibers.

9.6 Challenges and Opportunities in Research for Pulp Fibers for Papermaking and Cellulose Dissolution

The pulp and paper industry is used to commodity business and has a traditional practice in low profile in marketing to the general public. The industry is very well prepared for the new demands involving sustainability, environmental concerns, and recycling. A number of efforts were taken by this industry already during the 1990s to fulfill very strict requests involving the emission of pollutants, closing of water circles, reduction of gas emissions, and creation of certification practices for raw materials, processes, and paper products. Those efforts put the pulp and paper industry in the front of many industrial sectors with clear advantages toward a sustainable business. The research and development practices aimed to fulfill requests inside the pulp and paper value chain. Most of the developments were on the basis of trial and error and learn by doing. The developments and inventions were also very much limited to the incremental improvement of previous technologies and focused in their own stage of the value chain. In this practice, forest engineers focused on a productive forest, process engineers on pulp yield and brightness, and paper engineers on increasing speed of their machines, creating a fragmented value chain with several compromises. A few examples to illustrate this practice are given below.

Pulp is produced to achieve a high brightness aiming at good optical properties of paper products. However, whiteness is one of the most important properties for paper because it is the optical property that the consumers can perceive. Brightness measures the relative reflectance of light at well-defined wavelength (457 nm), and the human eye is sensitive to the wavelength range 400–700 nm. A consequence of this practice is that pulp engineers make efforts to gain every single brightness point in the bleaching sequence. When the pulp is refined in stock preparation, the brightness is decreased five points and fluorescent whitening agents (FWA) are added to obtain specified whiteness values. The brightness will increase because of the contribution of FWA. One may ask why does the bleaching is not tailored to fit to the needs of paper machines. The common answer in paper mills is because it has been always like that.

Other example to illustrate the current practices in papermaking is the amount of components in a paper formulation. Usually 14–17 components are used in addition to fibers and fillers. The reason for such a high number of additives is unknown for many papermakers because paper formulations were usually developed using a trial-and-error approach. Functionalization of wood raw materials and pulp fibers to improve papermaking properties are insipient or not existing at the moment.

The examples given above illustrate briefly the effects of incremental research and development practices in product development and performance. There are several opportunities for immediate improvement of pulp and paper products through application of a perspective where the value chain is viewed as whole from forest to recycling. In order to benefit from these opportunities, new approaches involving multidisciplinary aspects are needed. The education of new professionals with multidisciplinary capabilities has a key role to exploit opportunities at the current moment. A number of characteristics of this new professional were selected by a group of professors in one initiative led by the Finnish Pulp and Paper Association. The outcome of this work is summarized in Fig. 9.15.

According to the recommendations of the working group, the education of the new pulp and paper professional needs to include core competences involving basic and applied sciences, biomass chemistry and technology, and plant fiber technology and combine with specialization according to the area of actuation of the professionals. Research and development engineers have to be dynamic and show solid basis in science and engineering. Creativity and flexibility are major talents to identify opportunities and combined with entrepreneurship can create excellent opportunities for innovation.

In addition to creation of new products and processes and satisfaction of customer and

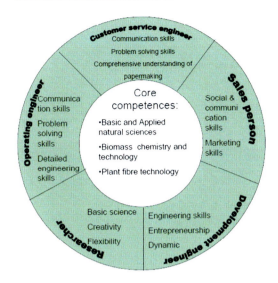

Fig. 9.15 The new engineering knowledge base for pulp and paper professionals as recommended by the working group in pulp and paper education and training of Forest-Based Sector Technology Platform (http://www.forest-platform.org)

environmental needs, the pulp and paper industry is facing new challenges to sustain its business due to competition with electronic media and plastics and other packaging materials. One possible strategic solution is to incorporate and integrate other platforms to the traditional pulp and paper value chain. Platforms of fuels, energy, chemicals, and functional materials are usually seen as the way to improve profits and create new markets to bio-based products. However, the integration of this platforms and interactions with other value chains has several challenges to be met by this industry.

9.7 New Challenges

9.7.1 2011–2012: Years of New Insights

The 2000s marked the increase of pulp production in South America and paper production in Asia. North America remains the largest pulp producer while Asia is the largest paper producer. Market pulp produced in South America is transported overseas to Europe and Asia where the recycled fiber is the major fibrous raw material used in paper production. In Europe, over the decade, there was a reduction of pulp production of nearly 10 % with no decrease in capacity while paper consumption decreased almost 6 %. During this period, two new areas came into attention of the industry: energy and biomass refining. Discussion on how to regulate carbon credits under the Kyoto Protocol in the pulp and paper sector also took the attention of the industry because of issues and political drives related to climate change.

Years 2011–2012 will be very important to define the potential of the pulp and paper industry to adapt to a new working practice where the customer is not only the printing house. Several companies will need to hire or build research and development teams with multidisciplinary skills to be able to address needs of new areas related to other value chains. New insights mean a strategic advance to become the technology leader in 10–20-year period.

9.7.1.1 Pulp and Paper Value Chain

The challenges for the industry are to add value on pulp and paper products and to find new applications for pulp other than paper, particularly for European markets where the paper consumption is already very high. Other alternative to add value is to reduce costs. Product development can fulfill the needs of adding value while process development can partially address the cost reduction. The research and development challenges will be more important for areas where the profits of industry are more critical like Europe and United States. The European industry, particularly the industry operating in the Nordic countries, needs to find ways to utilize the wood in a more comprehensive manner. Research projects to create functional wood are in need of urgent implementation in the Northern Hemisphere. This implies a data bank of chemical and morphological characteristics of wood and capabilities to apply breeding (mainly in Europe) and genetic engineering (e.g., in United States and Asia). The control and tailoring of the morphology and chemistry of the fiber wall is of high priority for the industry to able to design and develop new pulping and papermaking processes and new paper or fiber-based materials. One example could be a hardwood with medium fibers, i.e., uniform fiber length distribution which is longer than short fibers and slightly shorter than long fibers.

The creation of functional wood materials with lower content of lignin or even with tailored characteristics of lignin macromolecule is also of interest and already patented. The engineering of fiber wall chemistry and physical chemistry of polysaccharides such cellulose and hemicelluloses is also of great importance for the industry. The biosynthesis of a xylan free of 4-O-methylglucuronic acid in wood can avoid the formation of hexenuronic acid in cooking and is one example of how to reduce the consumption of alkali in cooking and bleaching chemicals. Other interesting challenge for the engineering of polysaccharide biosynthesis in the fiber wall is the control of cellulose and hemicelluloses assembly in the fiber wall in order to eliminate the need of low-consistency (LC) refining in stock preparation in paper machine. LC refining consumes energy and reduces the bulk and brightness of the pulp at the same time that induces external fibrillation and lumen collapse of fiber wall, increasing fiber-to-fiber bonding and improving the mechanical properties.

In papermaking, the challenges rely in reducing the number of nonfibrous raw materials while creating new products with enhanced properties for printing, functional coatings, and flexible packaging. The number of nonfibrous raw materials can be reduced if a concept of multifunctional chemicals is developed. One example could be nanohybrids of inorganic materials with natural polymers such as starch or chitosan. The natural polymers can be multifunctionalized to hold side groups yielding retention, optical fluorescence, bioactivity, and hydrophobicity and further attached to inorganic substrates before addition to fiber stock. In this concept, the papermaker would need only one supplier, and the consumption of water in the process would be reduced along with reduction of oil-based chemicals. Changes in paper machine technology would also be possible. Creation of a paper product easier to recycle the fibers with no damage to pulp properties and low generation of solid residues is another challenge to be addressed.

Research in functionalization of paper surfaces using surface sizing and coating has also a good potential to add value to paper or create new paper-based materials (e.g., sensor in diagnostics and batteries). Solving of old challenges such as vessel picking, speckles, and double sidedness can also be potentially addressed using this approach.

9.7.1.2 Fuels and Energy

Several companies are currently investigating possibilities to integrate the production of liquid biofuels or utilization of forest residues and other biomass for generation of energy in boilers. Biomass gasification is one of the focuses and allows the utilization of biomass for both energy and chemical production. The production of liquid fuels can be done by removal of hemicelluloses prior to cooking and further fermentation to ethanol. However, losses in pulp yield and mechanical properties of the pulp combined with low prices of ethanol and high prices of wood make this alternative not so attractive to industry. Selling of electricity generated by burning black liquor in recovery boilers of pulp mills is a very well-established practice in many countries and combustion of forest residues is a common practice in the Nordic countries. Advances in biomass gasification and development of fractionation technology of high added-value chemicals prior to combustion are research challenges to be met. The exploitation of energy and fuels requires strategic partnership between the pulp and paper sector and energy companies.

9.7.1.3 Chemicals and Functional Materials

The production of bio-based chemicals can be done through three different routes (1) direct production, (2) biomass refinery, and (3) expression in plants. Direct production uses a combination of biotechnology and chemical technology and has attracted large interest of the chemical companies, the

9 Pulp Fibers for Papermaking and Cellulose Dissolution

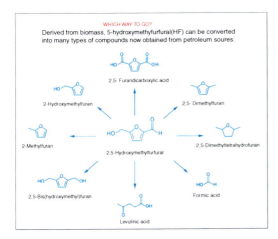 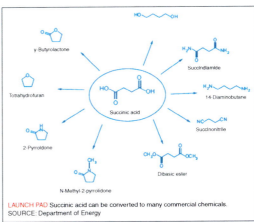

Fig. 9.16 Hydroxymethylfurfural (*left*) and succinic acid (*right*) are products of interest for the chemical industry because they can be used in the manufacture of many commercial chemicals (Mcoy 2010)

potential partners for the pulp and paper industry. Biorefinery can be a great opportunity for the forest companies to integrate platform of chemicals and materials. However, biorefinery is not limited only to forest, and it can be successfully applied to fermentation routes used in direct production of chemicals. Expression in plants by genetic engineering can also be investigated by creation of functional wood, and it is currently in a very preliminary stage for production of chemicals. The integration of platforms of chemicals and functional materials to pulp and paper requires strategic joint ventures with companies of chemical sector. Chemical companies have interest for bio-based hydroxymethyl furfural and succinic acid (Fig. 9.16) for building blocks of chemicals. Currently these companies are investing in patenting of processes based on fermentation while the pulp and sector could easily provide those, using residues or process streams as raw materials. In this area, the pulp and paper industry will have to prove its capabilities to supply the demands and needs of chemical industry in a consistent manner. However, fast developments will only be possible with joint research initiatives involving research and development of both sectors.

Bio-based materials such as polymers have a large market potential. Several companies, including Coca-Cola, Samsung, and Hyundai have announced the utilization of bio-based plastics in their products. The utilization of bio-based plastics is boosted by the potential implementation of the policies regarding carbon credits since those products have a lower carbon footprint as compared to oil-based ones. According to the report "Product overview and market projection of emerging bio-based plastics" (PRO-BIP), the current technical potential for production of bio-based plastics in Western Europe is 85 % of the plastic market, without competition with the fuel or food sectors. This is an excellent opportunity to be evaluated by pulp and paper companies, particularly in utilizing residues or creating new applications for fibers in a market of mature consumption of printing papers. Bioplastics can be fully or partially bio-based and show biodegradability or not (Fig. 9.17).

Late 2000s also brought back the microfibrillar cellulose developed in the 1980s, now with a new name: nanocellulose. Nanocellulose is produced by fluidizing chemical pulp using a commercial device. The pulp is usually pretreated by a mechanical or enzymatic method to reduce the energy consumption in fluidization, and the resulting product has about 98 % of water. This product can be used as substrate for surface modification to improve stability of the suspension or for adding functionality for special applications. Nanocellulose is now in a pilot phase and requires application to other value chains as hydrocolloids, reinforcement to composites, and new nanostructured materials. The challenges for

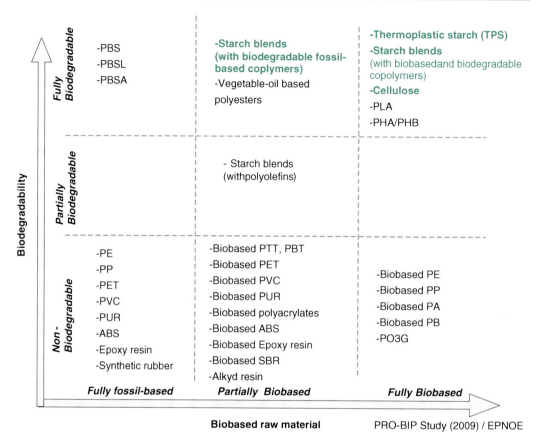

Fig. 9.17 Results of PRO-BIP study indicated that the current potential of bio-based polymers in Western Europe is nearly 42 million ton per year which corresponds to 85 % of the market for polymers (Patel 2009)

those materials involve the reduction of energy consumption in the whole process, the capability to remove water prior to transportation combined with re-dispersion in hydrophilic and hydrophobic media. A balance between adding functionalization and stabilization of suspensions is a difficult research challenge to be addressed.

9.7.1.4 Education

The integration of new multidisciplinary concepts in education programs at the universities is a slow process. The major challenge today is to attract young bright minds to the area of pulp and paper. The pulp and paper industry is generally perceived as old and not interesting by youngsters. This image is a result of a practice of commodity business and low profile in marketing by the industry and very specific education programs at the universities. The new trend in sustainability and green materials can benefit the pulp and paper industry and universities in attracting new students if joint initiatives of offering new alternatives are made available. Furthermore, multidisciplinary programs need to be implemented at the universities to be able to tackle new challenges such as climate change, knowledge bio-based economy, and sustainable business.

9.7.2 Year 2030: The Year of the Truth

Several countries worldwide launched their research strategy and visions for the year 2030. In Europe, a Strategic Research Agenda (SRA) for the year 2030 was prepared considering three shared goals: renewal, improved competitiveness, and sustainable development. The inputs to the SRA had contributions of companies,

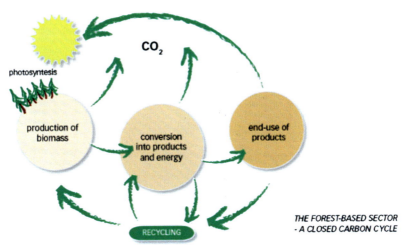

Fig. 9.18 The forest-based sector vision on closed carbon cycle. The sector has excellent business opportunities for a low-carbon economy (http://www.forestplatform.org)

universities, research institutes, associations, national support groups, national experts and authorities at different levels, and European Commission services. According to SRA, the forest-based sector will provide an effective platform for generation of energy in combination with the production of fibers and green chemicals. The biorefinery concepts and extensive use of wood as construction materials have key roles in the SRA as a vision for the year 2030. Other interesting aspect of the SRA is the highlight of the closed carbon cycle as the advantage of the forest-based sector (Fig. 9.18).

The year 2030 will be crucial to confirm or not the prospective visions planed in the 2000s.

9.7.2.1 Pulp and Paper Value Chain

The future of paper in the printing and writing sector is challenged by electronic devices and new forms of media. Utilization of paper as packaging material or in tissue and hygiene has fewer challenges. However, in all cases, functionalization of wood, fibers, and paper will be strongly demanded to fulfill needs of other industrial sectors or value chains that will require new performances. The potential of utilization of fibers and paper in other materials needs to be extensively investigated in a creative manner to occupy markets of composites, plastics, and other synthetic polymers. Fibers and paper as materials have an excellent potential to have good market shares in 2030 if the excellent mechanical and chemical properties of these materials are fully developed for bioplastics and reinforcement applications.

9.7.2.2 Fuels and Energy

Combustion of wood and other raw biomass is not clean energy, and the pulp and paper sector will realize that burning those materials is not a solution for our needs of renewable and sustainable energy. Research on more energy-efficient devices, vehicles, and buildings combined with development of solar, wind, wave, and other currently alternative energy sources will dominate the scenario in 2030. Prospective studies show that renewable energy will account for 20 % of total energy consumption. Biomass fractionation to separate high-added value chemicals from cheap carbon rich fractions will be a mature technology. More energy-efficient processes of black liquor combustion will allow larger amounts of electricity to be transferred to the public networks. The utilization of fibers, paper, cellulose, and hemicelluloses as substrate for solar cells or in light composites for manufacture of devices to be used in alternative energy research is an option to be considered.

9.7.2.3 Chemicals and Functional Materials

The utilization of bio-based chemicals will depend very much on how fuels will be developed for cars and vehicles. In case the use of electricity to cars will be dominating the market,

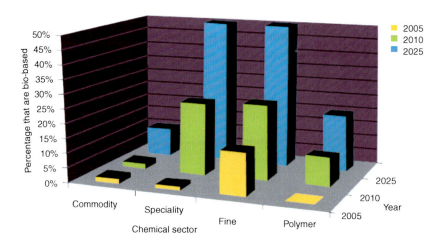

Fig. 9.19 Predicted market penetration of bio-based chemicals in world chemical production (excluding pharmaceuticals) (Schintlmeister 2009)

the production of liquid biofuels and oil-based fuels will be reduced and large quantities of oil will remain for materials. This is a challenge for biorefineries that produce green chemicals and bio-based materials. Another challenge, as already mentioned, is the development of fermentation routes by the chemical sector. The prospective for the increase of market penetration of bio-based chemicals in the chemical sector in the year 2025 shows a larger demand for fine and specialty chemicals (excluding pharmaceuticals) compared with polymers and commodities (Fig. 9.19). Research in the areas of chemical technology, reaction engineering, and organic synthesis needs to be active part of the forest-based sector.

9.7.2.4 Education

The universities will play a key role in helping to find solutions for the population growth and issues related to the climate change and sustainability. Multidisciplinary and cross-disciplinary education programs will need flexibility to be able to adapt fast and interact promptly with quickly emerging needs.

9.7.3 Year 2060: The Year of Counting Black Swans

Events called "black swans" are unexpected developments, coming out of nowhere, for which no one has any kind of contingency plan. They can happen any time. In our view, the year 2060 will be a good time to count them. The prospective vision for research challenges to this year has to consider forecast models based on historical data, and this approach can never predict black swans. However, challenges ahead involve the population growth and impacts of rise of middle classes in countries like China and India in contrast to low birthrates in Europe. Climate change and impacts on water and food supply in different parts of the globe are critical factors to a biomass-based industry. The pulp and paper industry and handling and processing of biomass uses large amounts of water, and this resource will have the key role for continuation of this industry. Processes with closed water cycles will have higher demand. The extinction of oil and other metals and minerals will also create demand for renewable materials. Our guess is that the "water credit" is the next step and will be even more important than the current "carbon credits." The forecast for changes in water supply shows extensive changes for the year 2060 (Fig. 9.20). Desertification is currently increasing in our planet and causing impacts in agriculture and food supply. The pulp and paper industry, perhaps with another name in 2060, will need to adapt to these changes and focus on research to create processes using low amounts of water and solvents and minimize waste.

Fig. 9.20 The predicted streamflow of water in the year 2060 according to average result of 12 climate models suggested by US geological survey. According to this prediction, southern Europe, Amazon region, and west of USA will have considerable reduction in streamflows due to changes in rain pattern (Moyer and Storrs 2010)

The main conclusion is that the future of the pulp and paper industry will be largely dependent on research and development activities which will require multidisciplinary approach and unprecedented success rate in transfer from laboratory and pilot to industrial scale and applications in different value chains of large commercial interest. Strategic partnership with chemical, energy, and food sectors are crucial for the pulp and paper industry in creating new markets and benefiting of the knowledge-based economy. The role of universities, research institutes, and research and development departments in companies will need to be redefined to reduce overlaps and create synergy for a truly innovative research environment required for the new forest-based industry.

References

Alén R (2000) Basic chemistry of wood delignification. In: Stenius P (ed) Forest products chemistry. Fapet Oy, Helsinki

Beghello L (1998) The tendency of fibers to build flocs. Doctoral Thesis, Åbo Akademi University, Turku/Åbo, Finland

Berg JC (1993) The importance of acid base interactions in wetting, coating, adhesion and related phenomena. Nord Pulp Pap Res J 8((1):75

Berry R (1996) (Oxidative) alkaline extraction. In: Dence CW, Reeve DW (eds) Pulping bleaching – principles and practice. TAPPI Press, Atlanta, GA

Carlsson G (1996) Surface composition of wood pulp fibers - relevance to wettability, sorption and adhesion. PhD Thesis, Royal Institute of Technology, Stockholm, Sweden

Clark JA (1978) Pulp technology and treatment for paper. Miller Freeman, San Francisco, CA

Eero S (1981) Wood chemistry, fundamentals and applications. Academic, New York

Everett DH (1989) Basic principles of colloid science. Royal Society of Chemistry, London

Fardim P, Durán N (2003) Modification of fiber surfaces during pulping and refining as analysed by SEM, XPS and ToF-SIMS. Colloids Surf A Physicochem Eng Asp 223:263–276

Fardim P, Holmbom B, Ivaska A, Karhu J, Mortha G, Laine J (2001) Critical comparison and validation of methods for determination of anionic groups in pulp fibres. Nord Pulp Pap Res J 17(3):346

Fellers C, Norman B (1998) Pappersteknik, 3rd edn. TABS, Sweden (Swedish)

Fengel D, Wegener G (1984) Wood: chemistry, ultrastructure, reactions. Walter de Gruyter, Berlin

Fink H-P, Weigel P, Purz HJ, Ganster J (2001) Structure formation of regenerated cellulose materials from NMMO-solutions. Progr Polym Sci 26(9):1473–1524

Gerly A, Clémençon I (1999) The effect of flocculant/microparticles retention programs on floc properties. Nord Pulp Pap Res J 1:23

Gierer J (1995) The chemistry of TCF bleaching. In: Proceedings of the 8th international symposium on wood and pulping chemistry, vol 1, Helsinki, p 285

Heinze T, Dicke R, Koschella A, Kull AH, Klohr EA, Koch W (2000) Effective preparation of cellulose derivatives in a new simple cellulose solvent. Macromol Chem Phys 201:627–631

Isogai A, Kitaoka C, Onabe F (1997) Effects of carboxyl groups in pulp on retention of alkylketene dimer. J Pulp Pap Sci 23(5):215

Jäkärä J, Parén A, Hukkanen P, Autio P, Anderson R (2001) The effect of peracetic acid treatment of bleached kraft pulp in fine paper production. In: Proceedings of the 87th PAPTAC annual meeting, vol B, Montreal, Canada, p B1

Jensen W (1977) "Puukemia", Suomen Paperi- insinöörien Yhdistyksen oppi- ja käsikirja I, Toinen uudistettu painos. Turku, Finnish

Kerekes R S, Soszynski R, Tam Doo PA (1985) In: Punton V (ed) Papermaking raw materials, transactions of the eighth fundamental research symposium. Mechanical Engineering Publications Ltd., London

KnowPap (2011) e-Learning environment for papermaking and automation. http://www.knowpap.com/www_demo/english/paper_technology/general/4_water_stock_systems/frame.htm. Accessed 3 Mar 2011

Liebert T (2010) Cellulose solvents-remarkable history bright future. In: Liebert T, Heinze T, Edgar K (eds) Cellulose solvents: for analysis, shaping and chemical modification. American Chemical Society, Washington DC, ACS Symposium Series 1033, pp 3–54

Liebert T, Heinze T (2008) Interactions of Ionic Liquids with Polysaccharides – 5. Solvents and reaction media for the modification of cellulose. BioResources 3(2): 576–601

Lierop B, Skothos A, Liebergott N (1996) Ozone delignification. In: Dence CW, Reeve DW (eds) Pulping bleaching – principles and practice. TAPPI Press, Atlanta, GA

Lindström T (1989) Fundamentals of papermaking, vol 1. Mechanical Engineering Publishing Ltd., London

Lönnberg B (1999) Mechanical pulping. Fapet Oy, Helsinki

McCormick CL, Lichatowich DK (1979) Homogeneous solution reactions of cellulose, chitin, and other polysaccharides to produce controlled-activity pesticide systems. J Polym Sci Polym Lett Ed 17(8):479–484

Mcoy M (2010) Big plans for succinic acid. Chemical and Engineering News, December, p 23

Metso Paper (undated) (http://www.metso.com)

Moyer M, Storrs C (2010) How Much is Left? The Limits of Earth's Resources, Scientific American, September, On-line version

Owen D, Anderson JR, Homer G (1996) Bleaching Chemicals: Chlorine, Sodium Hydroxide, Hydrogen Peroxide, Peroxy acids, Oxygen and Ozone. In: Dence CW, Reeve DW (eds) Pulping bleaching – principles and practice. TAPPI Press, Atlanta, GA

Ozaki Y, Sawatari A (1997) Surface characterization of a rosin sizing agent in paper by means of EPMA, ESCA and TOF-SIMS. Nord Pulp Pap Res J 12(4):260

Patel M (2009) Product overview and market projection of emerging bio-based plastics. http://www.epnoe.eu

Pikka O, Vesala R, Vilpponen A, Dahlöf H, Germgård Norden S, Bokström M, Steffes G (1999) Bleaching applications. In: Gullichsen J, Fogelholm C-J (eds) Chemical pulping. Fapet Oy, Helsinki

Potthast A (2006) Raw material for pulp. In: Sixta H (ed) Handbook of pulp. Weinheim, Willey-VCH

Reeve DW (1996) Introduction to the principles and practice of pulp bleaching. In: Dence CW, Reeve DW (eds) Pulping bleaching – principles and practice. TAPPI Press, Atlanta, GA

Roberts J (1992) Neutral and alkaline sizing. PIRA Publications, London

Rydholm SA (1965) Pulping processes. Interscience, New York

Saranpää P (2003) Wood density and growth. In: Barnett JR, Jeronimidis G (eds) Wood quality and its biological basis. Blackwell, Oxford

Schintlmeister P (2009) (Chair), Taking bio-based from promise to market: measures to promote the market introduction of innovative bio-based products, European Commission, Enterprise and Industry, November

Schweingruber FH (1993) Trees and wood in dendrochronology – morphological, anatomical, and tree-ring analytical characteristics of trees frequently used in dendrochronology. Springer, Heidelberg

Singh RP (1979) The bleaching of pulp. TAPPI Press, Atlanta, GA

Sixta H, Potthast A, Krostschek, AW (2006) Chemical Pulping Processes. In: Sixta H (ed) Handbook of pulp. Weinheim, Willey-VCH

Swerin A (1995) Flocculation and fiber network strength in papermaking suspensions flocculated by retention aids systems. PhD Thesis, Royal Institute of Technology, Stockholm, Sweden

Wågberg L (1987) Adsorption of polyelectrolytes and polymer-induced flocculation of cellulosic fibres. PhD Thesis, Royal Institute of Technology, Stockholm, Sweden

Cellulose: Chemistry of Cellulose Derivatization

10

Thomas Heinze, Andreas Koschella, Tim Liebert,
Valeria Harabagiu, and Sergio Coseri

Contents

10.1 Introduction 283

10.2 **Novel Synthesis Paths for Cellulose Functionalization** 284
10.2.1 Esterification 284
10.2.2 Regioselective Functionalization 300
10.2.3 Oxidation 306
10.2.4 Brief Discussion of Structure–Property Relationships 312

10.3 **Unconventional Cellulose Chemistry** 315
10.3.1 1,3-Dipolar Cycloaddition Reaction of Azide- and Alkyne-Containing Cellulosics .. 315
10.3.2 Dendronization of Cellulose 315

References 322

Abstract

This chapter gives an overview on various possibilities of cellulose functionalization. Particular attention was paid on homogeneous-phase conversions of cellulose in reaction media based on both aprotic-dipolar solvents in combination with salts and ionic liquids. Paths for cellulose esterification via in situ activated carboxylic acids are discussed. Not only etherification and esterification of hydroxyl groups lead to novel derivatives, but also oxidation reactions, in particular catalytic approaches, are of great importance. Recent developments in the synthesis of regioselectively functionalized cellulose ethers are reviewed as well. It is demonstrated that advanced synthesis methods, in particular the application of silicon-based protecting groups, are important to get the regioselectively functionalized cellulose ethers of high structural uniformity. Moreover, novel derivatives could be prepared thereof in subsequent reactions like oxidation and 1,3-dipolar cycloaddition reaction of azide- and alkyne-containing cellulose derivatives.

10.1 Introduction

One of the main arguments to study cellulose is the fact that it is the most important renewable polymer resource. However, there are other good arguments to consider this fascinating biopolymer

T. Heinze (✉)
Center of Excellence for Polysaccharide Research,
Institute for Organic Chemistry and Macromolecular
Chemistry, Friedrich Schiller University of Jena,
Humboldtstrasse 10, 07743, Jena, Germany
e-mail: Thomas.Heinze@uni-jena.de

as raw material for the future. Cellulose possesses a unique structure combined with reactive groups that allow the design of structure and hence of properties. From the authors' point of view, chemical modification is the most important path to adapt the properties to targeted application and even to develop novel functional polymers and advanced materials based on biomass. The hydroxyl groups show the typical reactions of low molecular weight alcohols, i.e., they can form ethers and esters. It is also possible to oxidize the OH groups to ketones, aldehydes, and carboxylic groups, with and without cleavage of C–C bonds of the repeating units. Moreover, the OH group can be modified with a leaving group enabling nucleophilic displacement reactions and hence formation of deoxypolysaccharide derivatives. Such derivatives possess typical properties of macromolecules, e.g., solubility in various solvents and film-forming and viscosity-regulating ability (Just and Majewicz 1985; El Seoud and Heinze 2005).

Chemical modification of cellulose has been already carried out commercially for many decades, and in particular ethers and esters of cellulose find applications in various fields including cosmetics, food, constructing materials, paper, and oil well drilling as well as in pharmaceutical applications and as filters, membranes, and films. It is interesting to note that commercial chemical modifications have not been limited to some ethers including methyl, ethyl, hydroxyalkyl, and carboxymethyl moieties and some esters including acetate, propionate, butyrate, and phthalic acid half ester but also to products with two or even three of the functional groups mentioned. With respect to the structure of the repeating unit, i.e., the presence of three hydroxyl functions of different reactivity, the resulting polysaccharide derivative is composed of up to eight different repeating units, namely, 6-mono-*O*-, 2-mono-*O*-, 3-mono-*O*-, 2,3-di-*O*-, 2,6-di-*O*-, 3,6-di-*O*-, 2,3,6-tri-*O*-, and unfunctionalized anhydroglucose units. In particular, products with a regioselective functionalization are an indispensable prerequisite to evaluate "real" structure–property relationships.

10.2 Novel Synthesis Paths for Cellulose Functionalization

10.2.1 Esterification

10.2.1.1 Unconventional Esterification of Cellulose

New paths for the esterification of cellulose include the application of homogeneous reaction conditions, i.e., the use of cellulose solvents and the exploitation of new reagents, e.g., for the in situ activation of the carboxylic acids. The homogeneous conditions allow control of the degree of substitution (DS) and a defined distribution of the substituents. The reaction of cellulose with carboxylic acids after in situ activation can significantly enlarge the portfolio of cellulose derivatives accessible.

Solvents for the Homogeneous Esterification of Cellulose

Although a wide variety of solvents for cellulose was developed and investigated in recent years, only a few have shown a potential for a controlled and homogeneous functionalization of the polysaccharide (Table 10.1) (Heinze and Glasser 1998). Limitations of the application of solvents may result from high toxicity, high reactivity of the solvents leading to undesired side reactions, and the loss of solubility during reactions yielding inhomogeneous mixtures by formation of gels and pastes, which can be hardly mixed, and even by formation of deswollen particles of low reactivity, which set down in the reaction medium.

***N,N*-Dimethyl Acetamide (DMAc)/LiCl**

DMAc/LiCl is among the most useful tools for the homogeneous synthesis of complex and tailored cellulose esters, as described below for the reaction after in situ activation of the carboxylic acid or transesterification reactions. Moreover, conversion of cellulose in DMAc/LiCl may lead to a direct access to processable cellulose esters (solvent soluble or melt flowable). This is not only due to the high efficiency of the homogeneous reaction conditions but also due to the fact

Table 10.1 Solvents and reagents exploited for the homogeneous acetylation of cellulose

Solvent[a]	Acetylating reagent	DS$_{max}$[b]	References
N-Ethylpyridinium chloride	Acetic anhydride	Up to 3	Husemann and Siefert (1969)
1-Allyl-3-methylimidazolium chloride	Acetic anhydride	2.7	Wu et al. (2004)
N-Methylmorpholine-N-oxide	Vinyl acetate	0.3	Klohr et al. (2000)
DMAc/LiCl	Acetic anhydride	Up to 3	Ibrahim et al. (1996)
	Acetyl chloride	Up to 3	Heinze et al. (2003)
DMI/LiCl	Acetic anhydride	1.4	Takaragi et al. (1999)
DMSO/TBAF	Vinyl acetate	2.7	Heinze et al. (2000a)
	Acetic anhydride	1.2	Ciacco et al. (2003)

[a] N,N-Dimethyl acetamide (DMA), 1,3-dimethyl-2-imidazolidinone (DMI), dimethyl sulfoxide (DMSO), tetra-n-butylammoniumfluoride trihydrate (TBAF)
[b] Degree of substitution

Table 10.2 Results of the acetylation of different cellulose types in N,N-dimethyl acetamide/LiCl with acetic anhydride [18 h at 60 °C, adapted from El Seoud et al. (2000)]

| Starting materials |||| Molar ratio |||
Cellulose from	M_w[a] (g/mole)	α-Cellulose content (%)	I_c (%)[b]	AGU[c]	Acetic anhydride	DS[d]
Bagasse	116,000	89	67	1	1.5	1.0
Bagasse	116,000	89	67	1	3.0	2.1
Bagasse	116,000	89	67	1	4.5	2.9
Cotton	66,000	92	75	1	1.5	0.9
Sisal	105,000	86	77	1	1.5	1.0

[a] Weight average molecular weight
[b] Crystallinity index
[c] Anhydroglucose unit
[d] Degree of substitution

that acylation without an additional catalyst is possible in this medium. Thus, in recent years, the system cellulose/DMAc/LiCl was studied intensively to develop efficient methods appropriate even for industrial application (McCormick and Callais 1987; El Seoud et al. 2000; Regiani et al. 1999). The thermal cellulose activation under reduced pressure is far superior to the costly and time-consuming activation by solvent exchange. The dissolution procedure and acetylation conditions in DMAc/LiCl allow excellent control of the DS in the range from 1 to 3. A distribution of substituents in the order C-6 > C-2 > C-3 is determined by means of ^{13}C NMR spectroscopy for cellulose acetates prepared with acetic anhydride. In addition to microcrystalline cellulose, cotton-, sisal-, and bagasse-based cellulose may serve as starting material (Table 10.2). There is only a limited influence of the crystallinity of the starting polymer on the homogeneous acetylation.

A comparable efficiency is observed for the conversion of cellulose with carboxylic acid chlorides. In the pioneer work of McCormick and Callais (1987), acetyl chloride was applied in combination with pyridine (Py) to prepare a cellulose acetate with DS of 2.4, which is soluble in acetone. Detailed information on the DS values accessible, concerning solubility of the acetates and distribution of substituents, are given in Table 10.3.

Amazingly, acetylation of cellulose dissolved in DMAc/LiCl with acetyl chloride without an additional base (Table 10.4) succeeds with almost complete conversion of the reagent. In contrast to the application of Py, higher DS values and a more pronounced C-6 selectivity are found. GPC investigations indicate less

Table 10.3 Reaction conditions for and results of the acetylation of cellulose with acetyl chloride in the presence of pyridine in N,N-dimethyl acetamide/LiCl

Reaction conditions			Reaction product					
Molar ratio			Partial DS[a] in position			Solubility[b]		
AGU[a]	Acetyl chloride	Py[d]	6	2,3	Σ	DMSO[e]	Acetone	CHCl$_3$
1	1.0	1.2	0.63	0.37	1.00	+	−	−
1	3.0	3.6	0.94	1.62	2.56	+	−	+
1	5.0	6.0	0.71	2.0	2.71	+	+	+
1	5.0	10.0	0.46	2.0	2.46	+	+	+
1	4.5	−	1.00	1.94	2.94	+	−	−

[a]Degree of substitution
[b]Soluble (+), insoluble (−)
[c]Anhydroglucose unit
[d]Pyridine
[e]Dimethyl sulfoxide

Table 10.4 Reaction conditions for and results of acetylation of cellulose dissolved in N,N-dimethyl acetamide/LiCl with acetyl chloride

Molar ratio	Partial DS$_{Ac}$[a] in position		Σ	Solubility[b]	
Acetyl chloride/AGU[c]	6	2,3		DMSO	CHCl$_3$
1.0	0.77	0.44	1.21[d]	+	−
3.0	0.90	1.95	2.85	+	+
4.5	1.00	1.94	2.94	+	−
5.0	1.00	1.94	2.96	+	+

[a]Degree of substitution of acetyl groups
[b]Dimethyl sulfoxide (DMSO), soluble (+), insoluble (−)
[c]Anhydroglucose unit
[d]This stoichiometrically impossible value may result from fractionation during work-up

pronounced chain degradation during the reaction without a base. In case of Avicel as starting polymer, depolymerization is <2 %. Permethylation, degradation, and HPLC show no hints for a nonstatistic distribution of the substituents along the polymer chain (Heinze et al. 2003).

Conversion of cellulose with acid chlorides in DMAc/LiCl is most suitable for the homogeneous synthesis of freely soluble partially functionalized long-chain aliphatic esters and substituted acetic acid esters (Table 10.5). In contrast to the anhydrides, the fatty acid chlorides are soluble in the reaction mixture and well-soluble polysaccharide esters can be formed with a very high efficiency of the reaction. Even in case of stearoyl chloride, 79 % of the reagent is consumed for the esterification of cellulose (Edgar et al. 1998).

Besides aliphatic esters, a variety of alicyclic, aromatic, and heterocyclic esters are accessible as shown exemplarily in Fig. 10.1.

Well-soluble partially functionalized cellulose acetate (DS 1.4) is obtained by conversion of the polymer with acetic anhydride/Py in 1,3-dimethyl-2-imidazolidinone (DMI)/LiCl (Takaragi et al. 1999). In addition, β-ketoesters with DS up to 2.1 were introduced by reaction of cellulose dissolved in DMI/LiCl with cis-9-octadecenyl ketene dimer (Yoshida et al. 2005).

Dimethyl Sulfoxide (DMSO)/ Tetrabutylammonium Fluoride (TBAF)

The mixture DMSO/TBAF should be considered for polysaccharide modification because it is an easily usable tool for lab scale esterification toward pure and well-soluble products. The acylation of cellulose in DMSO/TBAF using acid chlorides and anhydrides is somewhat limited because the solution contains a certain amount of water due to the use of commercially available TBAF trihydrate and residues of the air-dried polysaccharides. Nevertheless, preparation of

10 Cellulose: Chemistry of Cellulose Derivatization

Table 10.5 Reaction conditions for and results of the preparation of aliphatic esters of cellulose in N,N-dimethyl acetamide/LiCl

Reaction conditions						Reaction product		
Acid chloride	Molar ratio AGU[a]	Agent	Base[b]	Time (h)	Temp. (°C)	DS[c]	Solubility[d]	Ref.
Hexanoyl	1	1.0	Py	0.5	60	0.89	DMSO, Py	Edgar et al. (1998)
Hexanoyl	1	2.0	Py	0.5	60	1.70	Acetone, CHCl$_3$, THF, Py	
Lauroyl	1	2.0	Py	0.5	60	1.83	Py	
Stearoyl	1	1.0	Py	1	105	0.79	Acetone, CHCl$_3$, THF, Py	
Hexanoyl	1	8.0	TEA	8	25	2.8	DMF	McCormick and Callais (1987)
Heptanoyl	1	8.0	TEA	8	25	2.4	Toluene	
Octanoyl	1	8.0	TEA	8	25	2.2	Toluene	
Phenylacetyl	1	15.0	Py	3/1.5	80/120	1.90	CH$_2$Cl$_2$	Pawlowski et al. (1987)
4-Methoxy-phenylacetyl	1	15.0	Py	3/1.5	80/120	1.8	CH$_2$Cl$_2$	
4-Tolylacetyl	1	15.0	Py	3/1.5	80/120	1.8	CH$_2$Cl$_2$	

[a]Anhydroglucose unit
[b]Pyridine (Py), triethylamine (TEA)
[c]Degree of substitution
[d]Dimethyl sulfoxide (DMSO9), pyridine (Py), tetrahydrofuran (THF), N,N-dimethyl formamide (DMF)

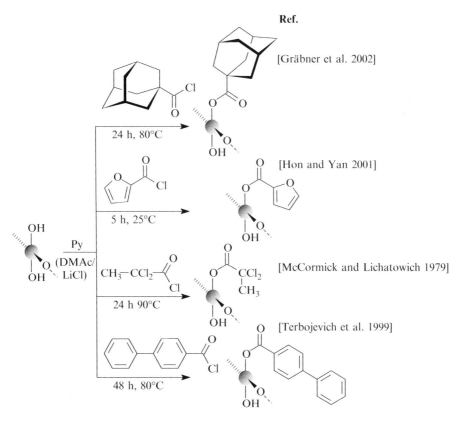

Fig. 10.1 Homogeneous synthesis of adamantoyl, 2-furoyl, 2,2 dichloropropyl, and 4-phenyl-benzoyl cellulose in N,N-dimethyl acetamide (DMAc)/LiCl

Table 10.6 Influence of the amount tetra-*n*-butylammonium fluoride trihydrate (TBAF) on the efficiency of the acetylation of sisal cellulose with acetic anhydride in dimethyl sulfoxide (DMSO)/TBAF

%TBAF in DMSO	Cellulose acetate DS[a]	Solubility[b]
11	0.30	Insoluble
8	0.96	DMSO, Py
7	1.07	DMSO, Py
6	1.29	DMSO, DMF, Py

[a]Degree of substitution
[b]*N*,*N*-Dimethyl formamide (DMF), pyridine (Py)

solutions is very easy and the system showed a remarkable capacity for the esterification of lignocellulosic materials, e.g., sisal cellulose, which contains about 14 % hemicellulose (Ciacco et al. 2003). The DS values of cellulose acetate prepared from sisal with acetic anhydride in mixtures of DMSO/TBAF decrease with increasing TBAF concentration from 6 to 11 % (Table 10.6) due to the increased rate of hydrolysis both of the anhydride and probably of the ester moieties formed as well.

Dewatering of DMSO/TBAF is possible by vacuum distillation. Reactions in the solvent of reduced water content lead to products comparable to reaction of cellulose dissolved in anhydrous DMAc/LiCl. In addition to these basic studies, the conversion of cellulose in DMSO/TBAF with more complex carboxylic acids (e.g., furoyl carboxylic acid) via in situ activation with *N*,*N*'-carbonyldiimidazole and the reaction with acyl-1*H*-benzotriazole (see below) was achieved. A mixture formed with anhydrous TBAF (Sun and DiMagno 2005) represents a cellulose solvent as well and opens up new horizons for the homogeneous functionalization of cellulose in DMSO/TBAF (Heinze and Köhler 2010). Besides DMSO/TBAF, mixtures of DMSO with tetraethylammonium chloride can be exploited for the functionalization of cellulose. To obtain a clear solution, 25 % (w/w) of the salt needs to be added. The cellulose dissolved in this medium is less reactive compared to DMSO/TBAF.

Ionic Liquids

The first ionic liquids with significance for the esterification of cellulose were *N*-alkylpyridinium halides, especially *N*-ethylpyridinium chloride (EPyCl) and *N*-benzylpyridinium chloride (BPyCl) (Graenacher 1934; Husemann and Siefert 1969). The advantage of an easy work-up procedure after modification of polysaccharides in these solvents is ruled out by the fact that they are solid at room temperature and have to be diluted with common organic liquids to give appropriate reaction media. Among the additives were DMF, DMSO, sulfolane, pyridine (Py), and *N*-methyl pyrrolidone, leading to a decreased melting point of 75 °C.

Nevertheless, the most promising ionic liquids for the modification of cellulose are 1-alkyl-3-methylimidazolium salts. In 2002, the use of such ionic liquids as cellulose solvent, in particular for the regeneration of the polysaccharide, was published (Swatloski et al. 2002). The results support the assumption that such ionic liquids can open up new paths for the shaping of polysaccharides. Additionally, they could also lead to a first commercially relevant route toward homogeneous cellulose chemistry, which would significantly broaden the number of tailored cellulose derivatives.

First attempts toward homogeneous acylation of cellulose with carboxylic acid anhydrides or chlorides in mixtures of BPyCl or EPyCl with pyridine were already described by Graenacher (Graenacher 1934; Husemann and Siefert 1969). Thus, these systems were exploited for the synthesis of acetates (**E1**), butyrates (**E2**), benzoates (**E3**), phthalates (**E4**), and anthranilic acid esters (**E5**) of cellulose (Fig. 10.2). Unfortunately, no data concerning the DS were given.

Most of the recent work deals with esterification in imidazolium salts. Here the purity of the reaction medium can have a drastic influence on the reaction. Methyl imidazole and imidazole from the ionic liquid synthesis have to be considered as impurities. They may act as catalysts for the esterification or can initiate side reactions.

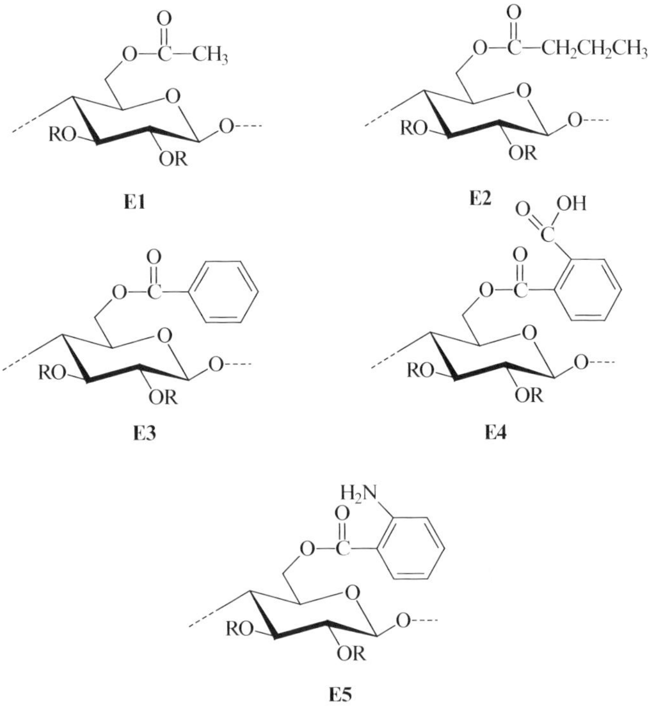

Fig. 10.2 Cellulose esters prepared in the solvent N-benzylpyridinium chloride/pyridine

Table 10.7 Reaction conditions for the preparation of cellulose acetate in N-ethyl-pyridinium chloride

Reaction conditions					
Molar ratios per mol AGU[a]					
Pyridine	Acetic anhydride	Temp. (°C)	Time (min)	Acetyl content (%)	Solubility
16.2	5.4	40	60	12.1	H$_2$O/pyridine 3/1
16.2	5.4	40	295	27.1	CCl$_4$/CH$_3$OH 4/1
32.5	32.5	50	120	37.7	CCl$_4$/CH$_3$OH 4/1
32	32.0	85	55	41.3	CHCl$_3$
32.5	32.5	50	285	42.2	Acetone, CHCl$_3$

[a]Anhydroglucose unit

Aliphatic Acid Esters of Cellulose

Acetylation in EPyCl/pyridine is rather fast (DS 2.65 within 44 min) and can be controlled via the reaction time. Preparation of cellulose triacetate, which is completed within 1 h, has to be carried out at 85 °C (Husemann and Siefert 1969). It was claimed that the acetylation proceeds for cellulose with degree of polymerization (DP) values below 1,000 without degradation, i.e., strictly polymer analogous. Cellulose acetate samples with a defined solubility, e.g., in water, acetone, or chloroform, respectively (Table 10.7), were accessible in one step in contrast to the heterogeneous conversion. A correlation between solubility and distribution of substituents was attempted by means of ^{1}H NMR spectroscopy via this acetylation procedure (Deus et al. 1991). Comparable results can be obtained for solvents containing substituted imidazolium ions. It was stated that the conversion of cellulose with acetic anhydride (Ac$_2$O) in ionic liquids, e.g., 1-allyl-3-methylimidazolium chloride (AMIMCl), succeeds without an additional catalyst (Wu et al. 2004; Zhang et al. 2005; Cao et al. 2007).

Table 10.8 Results of the acetylation of cellulose (DP 650) with acetic anhydride in 1-allyl-3-methylimidazolium chloride [4 wt% cellulose, molar ratio 1:5, temperature 80 °C, adapted from Wu et al. (2004)]

Time [h]	DS[a]	Solubility[b] Acetone	Chloroform
0.25	0.94	−	−
1.0	1.61	−	−
3.0	1.86	+	−
8.0	2.49	+	+
2.30	2.74	+	+

[a]Degree of substitution
[b]All cellulose acetates were soluble in dimethyl sulfoxide, soluble (+), insoluble (−)

Table 10.9 Degree of substitution and solubility of the cellulose acetates homogeneously prepared in 1-butyl-3-methylimidazolium chloride (reaction temperature 80 °C, reaction time 120 min)

Cellulose type	Reagent type	Mol per AGU[a]	DS[b]	Solubility[c] DMSO	CHCl$_3$
Avicel	Acetic anhydride	3.0	1.87	+	−
Avicel	Acetic anhydride	5.0	2.72	+	−
Avicel	Acetic anhydride	3.0[d]	2.56	+	−
Avicel	Acetic anhydride	10.0[d]	3.0	+	+
Avicel	Acetyl chloride	3.0	2.81	+	−
Avicel	Acetyl chloride	5.0	3.0	+	+
Spruce sulfite pulp	Acetyl chloride	3.0	3.0	+	+
Spruce sulfite pulp	Acetyl chloride	5.0	3.0	+	+
Cotton linters	Acetyl chloride	3.0	2.85	+	+
Cotton linters	Acetyl chloride	5.0	3.0	+	+

[a]Anhydroglucose unit
[b]Degree of substitution determined by NMR spectroscopy
[c]Dimethyl sulfoxide, soluble (+), insoluble (−)
[d]2.5 mol pyridine per AGU

Nevertheless, the cellulose acetates obtained started to dissolve in acetone at DS 1.86. The control of the DS by prolongation of the reaction time is nicely displayed in Table 10.8. For a cellulose solution with lower concentrations (2.9 wt%) in the same solvent, a maximum DS of 2.30 is reached. A preferred functionalization of the primary OH group is observed, similar to the acetylation in DMAc/LiCl (El Seoud et al. 2000). The reactions in recycled ionic liquids, which were obtained by removal of volatiles under reduced pressure, gave products of comparable DS.

Still the most promising solvent regarding commercial use is 1-butyl-3-methylimidazolium chloride (BMIMCl), although it is corrosive and shows some toxicity. It was employed for the acetylation of cellulose (Heinze et al. 2005; Barthel and Heinze 2006). Starting materials were microcrystalline cellulose, spruce sulfite pulp, and cotton linters, with DP values of 307, 544, and 814, respectively. Dissolution of the polymer was carried out at 10 °C above the melting point of the ionic liquid. Esterification with acetyl chloride or Ac$_2$O/Py at 80 °C for 2 h yielded pure soluble products of controlled DS (Table 10.9).

Complete functionalization could be achieved with acetyl chloride if a molar ratio reagent/anhydroglucose unit (AGU) of 1/3 was applied, suggesting that complete conversion of the acetylating reagent occurred. For acetylation with acetic anhydride, a higher excess was necessary, and pyridine should be used as catalyst. In addition, the ionic liquids 1-ethyl-3-methylimidazolium chloride (EMIMCl), 1-butyl-2,3-dimethylimidazolium

Table 10.10 Degree of substitution and solubility of the cellulose acetates homogeneously prepared in different ionic liquids with acetic anhydride (3.0 mol per mol AGU, reaction temperature 80 °C, reaction time 120 min)

Ionic liquid	DS[a]	Solubility[b] DMSO	CHCl$_3$
1-Butyl-3-methylimidazolium chloride	1.87	+	−
1-Allyl-3-methylimidazolium bromide	2.67	−	−
1-Butyl-2,3-dimethylimidazolium chloride	2.92	+	−
1-Ethyl-3-methylimidazolium chloride	3.0	+	−

[a]Degree of substitution
[b]Dimethyl sulfoxide (DMSO), soluble (+), insoluble (−)

chloride (BDMIMCl), and 1-allyl-3-methylimidazolium bromide (AMIMBr) were exploited for the acetylation (Barthel and Heinze 2006). Under the same experimental conditions, the following dependence of DS, hence reactivity, on the ionic liquid was observed: EMIMCl > BDMIMCl > AMIMBr > BMIMCl (see Table 10.10). Although BMIMCl showed the lowest reactivity, it seems to be the solvent of choice for the synthesis of cellulose acetates soluble in chloroform. Even the acetylation of high molecular mass cellulose such as bacterial cellulose with a DP of about 6,500 was possible in BMIMCl (Schlufter et al. 2006). The cellulose acetates were found to be soluble in DMSO; the order of reactivity of the OH groups is again C-6 > C-3 > C-2.

Alternative ionic liquids used for the acetylation of cellulose are choline chloride (ChCl)-based compounds, in particular ChCl with ZnCl$_2$ (Abbott et al. 2005). It represents a cheap and readily available alternative to the more commonly employed alkyl imidazolium-aluminum chloride mixtures (Abbott et al. 2004). In an analogous manner to the chloroaluminate systems, it was shown that zinc-based liquids form complex anions such as [ZnCl$_3$]$^-$ and [Zn$_2$Cl$_5$]$^-$. In contrast to the aluminum counterparts, ChCl–ZnCl$_2$ is water insensitive and zinc is supposed to be environmentally more benign. Although it was claimed that the reactivity during an acetylation with acetic acid anhydride in this ionic liquid is comparable to that in AMIMCl, the majority (90 wt%) of the cellulosic product was insoluble in acetone. IR analysis of this insoluble fraction indicated that a significant proportion was acetylated. This uneven acetylation could be due to per-acetylation of low molecular weight cellulose, which was concluded from SEC experiments.

Ionic liquids are very useful media for the preparation of long-chain aliphatic and aromatic esters as well (Dorn et al. 2010a, b; Zhang et al. 2009a). If the reaction is carried out with the acid chlorides, even complete conversion of the reagent was observed as can be seen in Fig. 10.3 for the cellulose benzoyl ester. In case of the long-chain esters, the amount of introduced functions is determined by the chain length.

For reactions with lauroyl chloride, a maximum DS of 1.54 was achieved in BMIMCl (molar ratio acyl chloride/AGU of 3/1). Fully functionalized samples, DS = 3, were unattainable, probably because the system turns heterogeneous. The efficiency of the reaction in dependence on the ionic liquid decreased in the order BMIMCl > EMIMCl > BDMIMCl > AMIMBr. Attempts for the acetylation of cellulose in 3-methyl-N-butylpyridinium chloride and benzyldimethyltetradecylammonium chloride were not successful up to now (Barthel and Heinze 2006).

Nevertheless, cellulose stearates with DS values up to 2.6 were synthesized in imidazolium salts exhibiting good solubility in nonpolar organic solvents. It was indicated by differential scanning calorimetry (DSC) that the melting temperatures of cellulose stearates ranged from 39 to 57 °C (Huang et al. 2011).

Esters of Dicarboxylic Acids and Unsaturated Acids

It was attempted to react cellulose from sugarcane bagasse with succinic anhydride. Experiments were carried out in AMIMCl (Liu et al. 2007b) or in a mixture of BMIMCl/DMSO

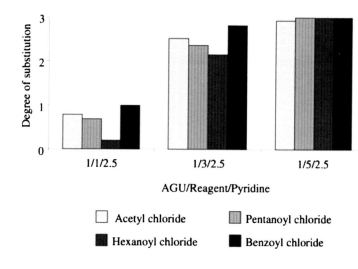

Fig. 10.3 Values of the degree of substitution of cellulose acetate, cellulose pentanoate, cellulose hexanoate, and cellulose benzoate, synthesized in 1-butyl-3-methylimidazolium chloride (2 h, 80 °C)

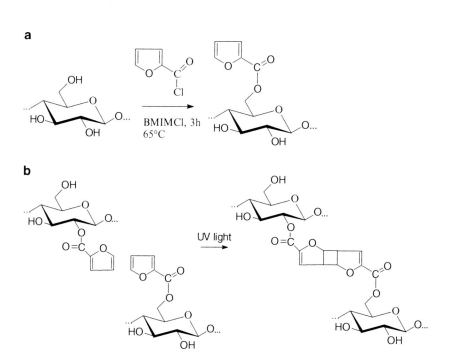

Fig. 10.4 Schematic plot of the reaction of cellulose with 2-furoyl chloride (**a**) yielding photo-cross-linkable material (**b**)

(Liu et al. 2006; Liu et al. 2007b). The DS values obtained were in the range from 0.07 to 0.22 for the conversions in AMIMCl and 0.04 to 0.53 for the BMIMCl/DMSO system. The reason for this inefficient esterification, even under favorable reaction conditions, is unclear. In a comparable manner, phthalate functions were introduced, yielding products with DS values up to 0.73 (Liu et al. 2007a). In addition to the esterification with aliphatic carboxylic acid chlorides, the introduction of unsaturated ester functions leads to membrane-forming photo-cross-linkable materials (Fig. 10.4). It was shown that the reactivity of a 2-furoyl chloride in BMIMCl is comparable to

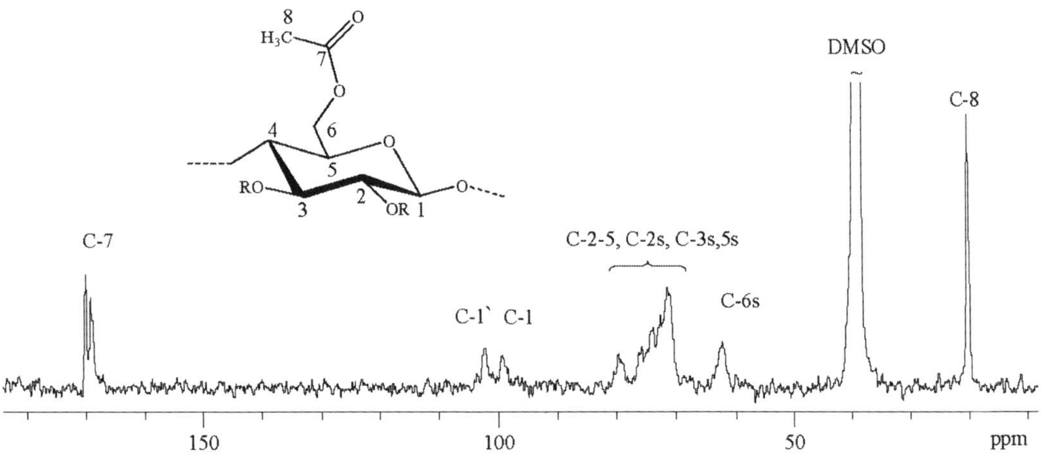

Fig. 10.5 ^{13}C NMR spectrum of a cellulose acetate obtained by conversion of cellulose with furoyl chloride in 1-ethyl-3-methylimidazolium acetate recorded in dimethyl sulfoxide (DMSO)-d_6

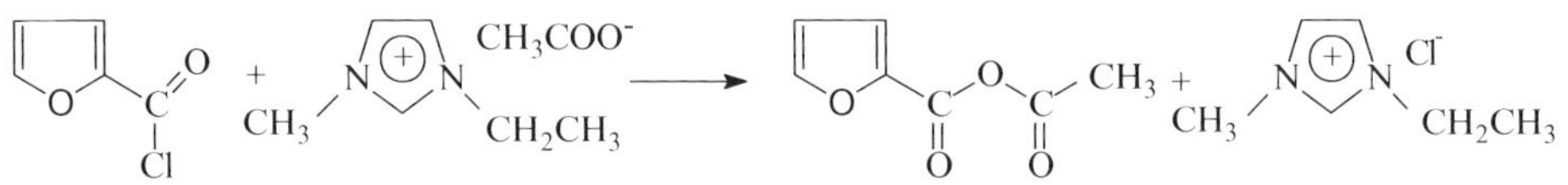

Fig. 10.6 Schematic plot for the formation of the intermediately formed furan carboxylic-acetic acid anhydride

that of acetyl chloride, i.e., with a molar ratio AGU/reagent of 1/5, complete functionalization can be obtained (Köhler and Heinze 2007; Dorn et al. 2010a, b).

Attempts were made to apply the room-temperature ionic liquid 1-ethyl-3-methylimidazolium acetate [EMIMAc, FLUKA, 90 % purity, no Ag(CH$_3$COO)$_2$ detectable] for homogeneous acetylation of cellulose. Thus, 2-furoyl chloride was converted with cellulose in EMIMAc in the presence of pyridine. Surprisingly, the organosoluble cellulose derivative obtained was not the furoate but a pure cellulose acetate, as could be confirmed by ^{13}C NMR spectroscopy (Fig. 10.5) (Köhler et al. 2007). A reasonable explanation for this phenomenon would be the assumption that intermediately the mixed furan carboxylic-acetic acid anhydride is formed (Fig. 10.6). Thus, the first stage of the reaction was studied with NMR spectroscopy. Both ^{1}H and ^{13}C NMR spectra confirmed the formation of this highly reactive intermediate. A comparable experiment with acetyl chloride showed the formation of the more stable and easily detectable symmetric acetic acid anhydride.

Besides the fact that such side reactions have to be considered during the chemical modification of cellulose in ionic liquids with carboxylate as anion, it could also be a tool for the highly efficient synthesis of cellulose derivatives because the ionic liquid can act both as reagent and solvent, as discussed.

New Tools for the Esterification of Cellulose

Homogeneous reaction conditions give the opportunity of esterification with state-of-the-art reagents, e.g., after in situ activation of carboxylic acids, which is characterized by reacting the carboxylic acid with a reagent leading to an intermediately formed, highly reactive carboxylic acid derivative. The carboxylic acid derivative may be formed prior the reaction with the polysaccharide or converted directly in a one-pot reaction. The modification of cellulose with carboxylic acids after in situ activation made a

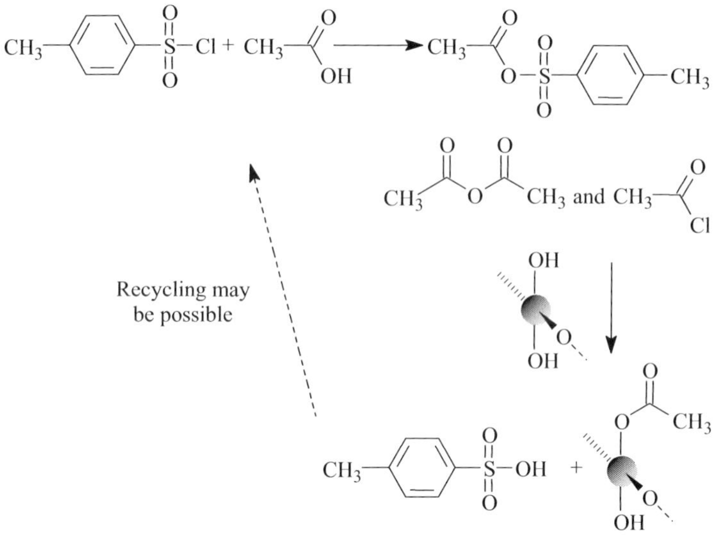

Fig. 10.7 Schematic mechanism for the acetylation of a cellulose with acetic acid using p-toluenesulfonic acid chloride for the in situ activation of the acid

broad variety of new esters accessible because for numerous acids, e.g., unsaturated or hydrolytically instable ones, common reactive derivatives such as anhydrides or chlorides are not accessible. The mild reaction conditions applied for the in situ activation avoid side reactions like pericyclic reactions, hydrolysis, and oxidation.

Typical reagents for in situ activation of carboxylic acids are sulfonic acid chlorides (Shimizu and Hayashi 1988; Shimizu et al. 1991). Various intermediates including the mixed anhydride of p-toluenesulfonic acid (TosOH) and the carboxylic acid and the acid chlorides initiate esterification of the polysaccharide (Fig. 10.7) (Brewster and Ciotti 1955; Sealey et al. 2000).

Cellulose esters, having alkyl substituents in the range from C_{12} (laurylic acid) to C_{20} (eicosanoic acid), could be obtained with almost complete functionalization (Sealey et al. 1996). The method is capable for the introduction of fluorine-containing substituents, e.g., 2,2-difluoroethoxy, 2,2,2-trifluoroethoxy, and 2,2,3,3,4,4,5,5-octafluoropentoxy functions, leading to a stepwise increase of the hydrophobicity of the products and an increased thermal stability (Sealey et al. 2000; Glasser et al. 1995, 2000). Moreover, water-soluble cellulose esters are accessible by derivatization of cellulose with oxocarboxylic acids like 3,6,9-trioxadecanoic acid or 3,6-dioxahexanoic acid (Heinze and Schaller 2000; Dorn et al. 2010a). The cellulosics start to dissolve in water at a DS as low as 0.4. In addition, they are soluble in common organic solvents like acetone or ethanol and resist a thermal treatment of up to 325 °C. Even cellulose derivatives with bulky, fluorescence active anthracene-9-carboxylate moieties are accessible with DS values as high as 1.0 (Koschella et al. 1997). Nevertheless, the reactions are usually connected with pronounced chain degradation. Highly efficient is also the reaction with the alkali or alkaline earth salt of acetic acid in combination with TosCl (Shimizu et al. 1991).

Coupling reagents of the dialkylcarbodiimide type are most frequently utilized for the esterification of polysaccharides with complex carboxylic acids; best-known condensation agent is N,N-dicyclohexylcarbodiimide (DCC) although it is toxic especially via contact with skin (Fig. 10.8) (Haslam 1980). Moreover, oxidation of hydroxyl functions may occur in DMSO due to a Moffatt-type reaction (Fenselau and Moffatt 1966).

Most frequently, DCC is applied both in combination with 4-N,N-dimethylaminopyridine as catalyst and 4-pyrrolidinopyridine, e.g., for the synthesis of aliphatic cellulose esters (Samaranayake and Glasser 1993). Unsaturated esters (e.g., methacrylic, cinnamic, and vinyl acetic acid esters) and esters of aromatic carboxylic acids including (p-N,N-dimethylamino)benzoate

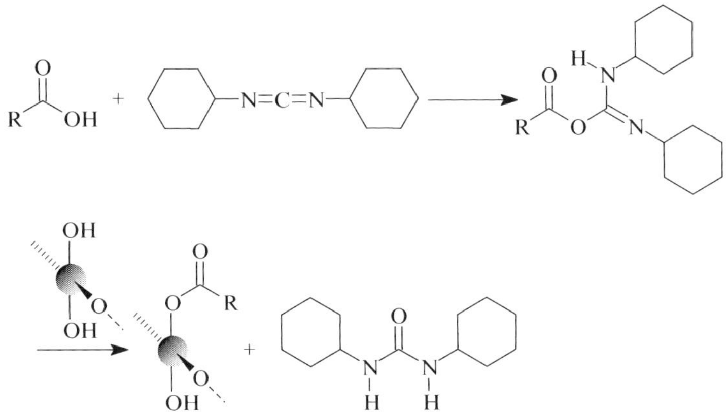

Fig. 10.8 Esterification of cellulose with carboxylic acid in situ activated with dicyclohexylcarbodiimide

of cellulose are accessible (Zhang and McCormick 1997; Williamson and McCormick 1998). The amino group containing ester provides a site for the conversion to a quaternary ammonium derivative, which imparts water solubility.

The homogeneous one-pot reaction after in situ activation of the carboxylic acids with N,N'-carbonyldiimidazole possesses an enormous potential for cellulose modification, which is rather well known from bioorganic chemistry published already in 1962 (Staab 1962). The reactive imidazolide of the acid is generated, and by-products formed are CO_2 and imidazole that are nontoxic (Fig. 10.9). The pH is almost constant during the conversion resulting in negligible chain degradation. In comparison to DCC, the application of N,N'-carbonyldiimidazole is much more efficient, avoids most of the side reactions, and allows the use of DMSO (good solvent for most of the complex carboxylic acids).

The conversion is generally carried out as one-pot reaction; the acid is transformed with N,N'-carbonyldiimidazole to give the imidazolide, which is then converted with solubilized cellulose (see Fig. 10.9, path B). The conversion of cellulose in the first step yields undesired cross-linking via carbonate formation leading to insoluble products (see Fig. 10.9, path A). Esterification of cellulose in DMAc/LiCl or DMSO/TBAF after in situ activation with N,N'-carbonyldiimidazole was achieved for a wide variety of carboxylic acids with chiral [(−)-menthyloxyacetic acid], unsaturated [3-(2-furyl)-acrylcarboxylic acid], heterocyclic [liponic acid], crown ether [4'-carboxybenzo-18-crown-6], and cyclodextrin [carboxymethyl-β-cyclodextrin] containing moieties (Fig. 10.10) (Liebert and Heinze 2005; Hussain et al. 2004a, b).

A similar intermediate is formed during the acylation of cellulose via activation of carboxylic acids with 1H-benzotriazole (see Fig. 10.11). The preparation of the acyl-1H-benzotriazole could be carried out under mild conditions (Katritzky et al. 2003, 2006). The acyl-1H-benzotriazole can be reacted with cellulose leading to cellulose acetate, butyrate, caproate, benzoate, myristate, and stearate with DS values between 1.1 and 1.9. The reaction proceeds completely homogeneously in DMSO/TBAF (Nagel and Heinze 2010).

By conversion of DMF with a variety of chlorinating agents, including phosphoryl chloride, phosphorus trichloride, and most frequently oxalyl chloride, iminium chlorides are generated and subsequent reaction with a carboxylic acids leads to a very reactive intermediate. In the pioneering work of Stadler, it was shown that during the esterification mostly gaseous side products are formed and the solvent is regenerated (Fig. 10.12) (Stadler 1978). Acylation of cellulose with long-chain aliphatic acids (stearic acid and palmitic acid), the aromatic acid 4-nitrobenzoic acid, and adamantane-1-carboxylic acid as bulky alicyclic acid was easily achieved in DMAc/LiCl (Hussain et al. 2004a).

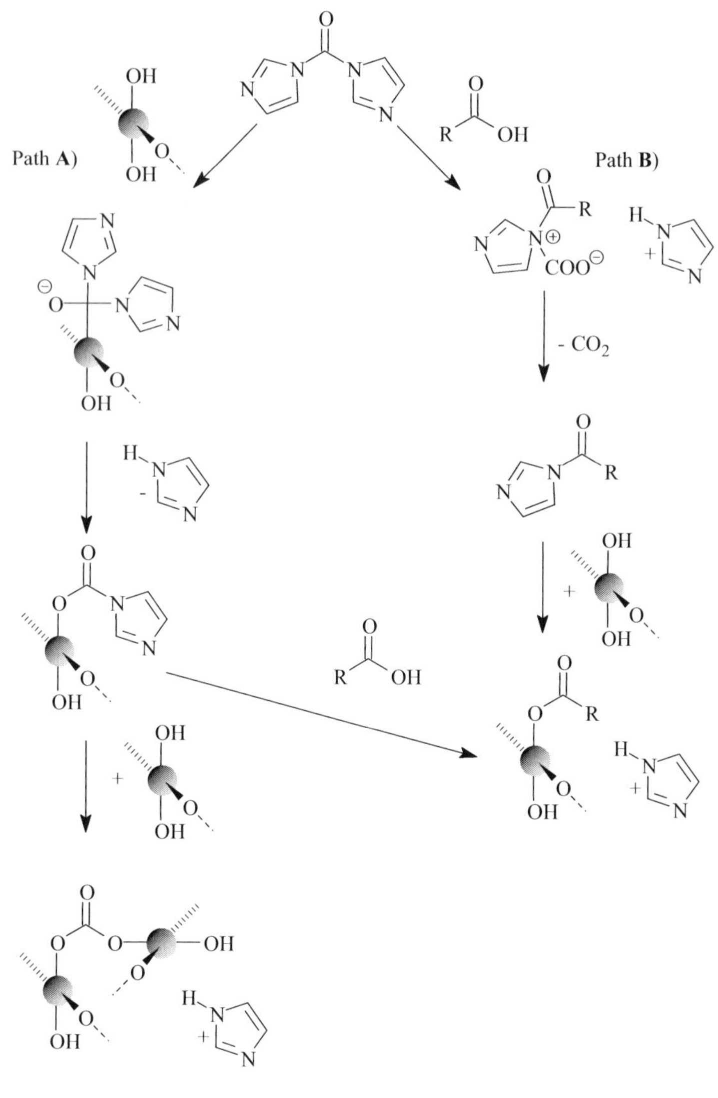

Fig. 10.9 Reaction paths leading exclusively to esterification (*path B*) or cross-linking (*path A*) if the polysaccharide is treated with *N,N'*-carbonyldiimidazole in the first step

An alternative path for cellulose esterification is the transesterification with vinyl acetate using the solvent DMSO/TBAF (Heinze et al. 2000a), which is much more efficient than acetylation with acetic anhydride. An interesting method exploiting a ring-opening reaction in DMAc/LiCl is the conversion with diketene or with a mixture of diketene/carboxylic acid anhydrides giving either pure acetoacetates or mixed acetoacetate carboxylic acid esters of cellulose (Fig. 10.13) (Edgar et al. 1995).

Moreover, ring opening of *N*-methyl-2-pyrrolidone (NMP) can be exploited for the preparation of an ionic ester of cellulose. The glucan is converted homogeneously in NMP/LiCl with an intermediate reagent of a Vilsmeier-Haack type reaction of NMP with TosCl. This reaction procedure yields a cyclic iminium chloride of cellulose. Subsequent hydrolysis of this derivative can follow two possible pathways (Fig. 10.14). One route would regenerate the NMP and cellulose. The other path gives an ester linkage. ^{13}C NMR spectroscopy reveals that the latter hydrolysis is much faster and forms a cellulose 4-(methylamino) butyrate hydrochloride (Brackhagen et al. 2007).

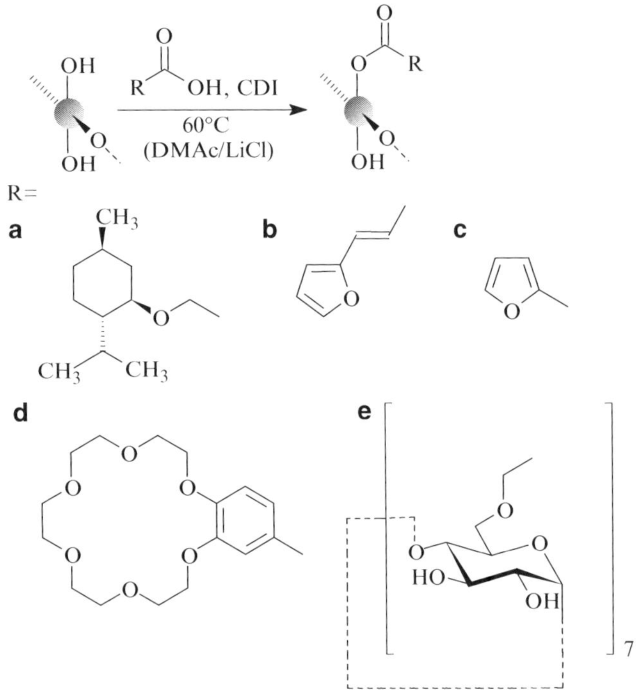

Fig. 10.10 Conversion of cellulose with carboxylic acids applying in situ activation with N,N'-carbonyldiimidazole yielding the esters of (**a**) (−)-menthyloxyacetic acid, (**b**) 3-(2-furyl)-acrylcarboxylic acid, (**c**) furan-2-carboxylic acid, (**d**) 4'-carboxybenzo-18-crown-6, and (**e**) carboxymethyl-β-cyclodextrin

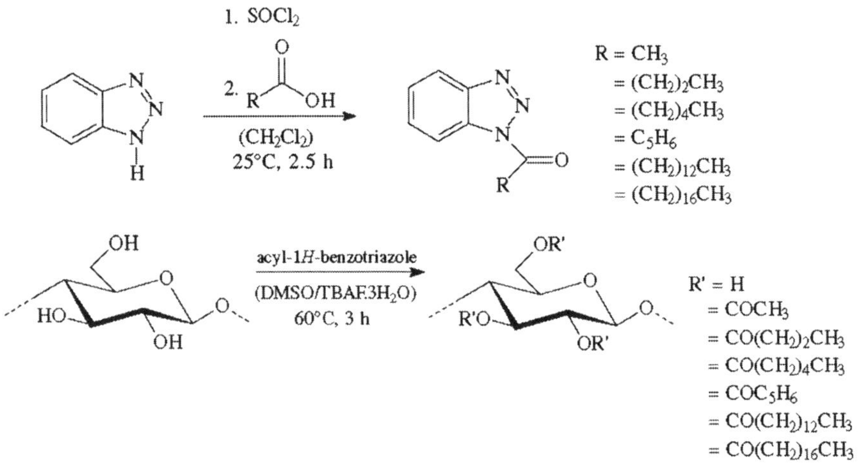

Fig. 10.11 Reaction scheme for the synthesis of acyl-1H-benzotriazole and the subsequent esterification of cellulose with the acylation agent in dimethyl sulfoxide (DMSO)/tetrabutylammonium fluoride trihydrate (TBAF)

10.2.1.2 Sulfation of Cellulose in Ionic Liquids

A reaction of growing interest is the sulfation of cellulose because the products show pronounced bioactivity and can be used for self-assembly systems such as polyelectrolyte complexes. For these applications, defined structures are indispensable. Sulfation of cellulose suspended in DMF with a SO_3 complex starts under heterogeneous conditions and leads to the dissolution of the sulfated product at a certain DS. This method is suitable for the preparation of cellulose sulfate with high DS > 1.5 only. Lower substituted products are sulfated in the swollen amorphous parts

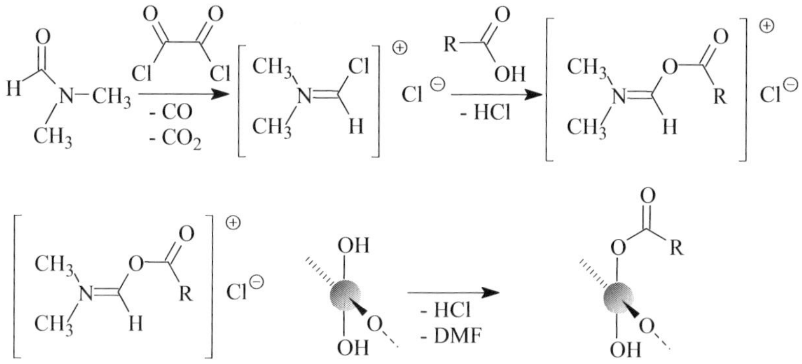

Fig. 10.12 Reaction scheme for the preparation of cellulose esters via in situ activation by iminium chlorides

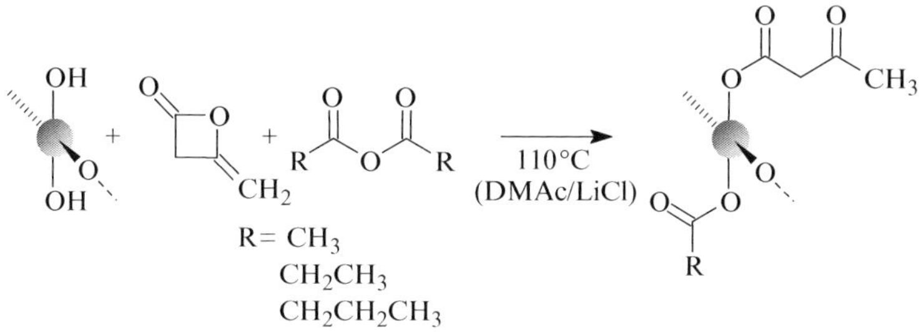

Fig. 10.13 Synthesis of mixed cellulose acetoacetate carboxylic acid esters via conversion with mixture of diketene/ carboxylic acid anhydrides

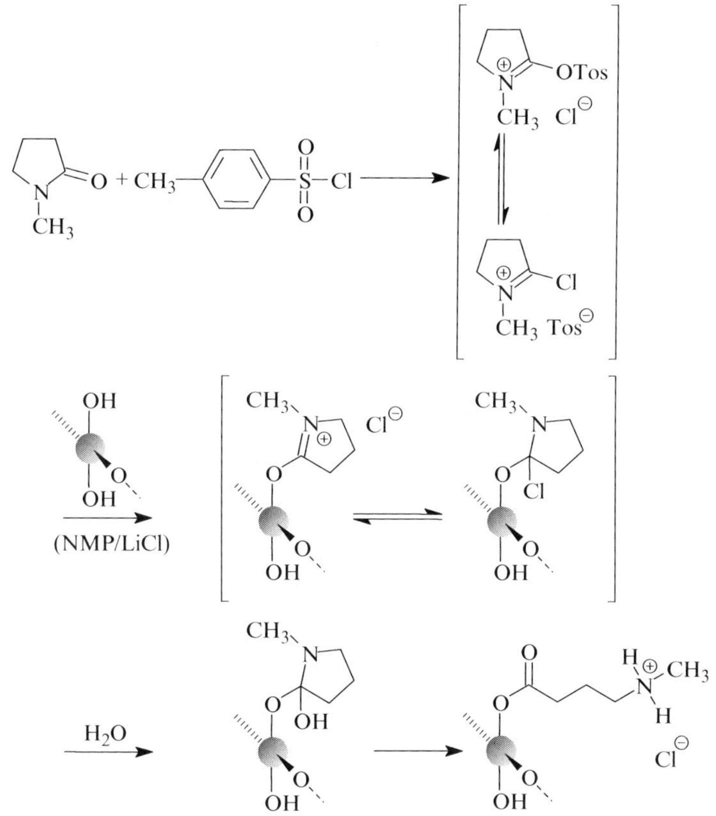

Fig. 10.14 Reaction path for the modification of cellulose with N-methyl-2-pyrrolidone (NMP) in the presence of p-toluenesulfonic acid chloride (TosCl)

of the cellulose while the crystalline parts remain unfunctionalized. Thus, water insolubility and nonuniform sulfation among the cellulose chains result, i.e., different amounts of water-soluble parts (high DS) and water-insoluble parts (low DS) are formed (Schweiger 1972). Nonuniform reaction is additionally favored by the high reactivity of the sulfating reagents and the resulting high reaction velocity.

Homogeneous sulfation of dissolved cellulose can overcome this problem of irregular substituent distribution and yields uniform functionalization patterns. Although widely applied for the esterification and etherification of cellulose, DMAc/LiCl is not the solvent of choice for sulfation, because insoluble products of low DS are obtained due to gel formation upon addition of the sulfating agent (Klemm et al. 1998). Several other cellulose solvents including N-methylmorpholine-N-oxide have also been tested for the homogeneous sulfation of cellulose but showed coagulation of the reaction medium yielding badly soluble cellulose sulfates (Wagenknecht et al. 1985).

A very elegant method for the sulfation offers the application of ionic liquids. Cellulose dissolved in BMIMCl/cosolvent mixtures can be easily converted into cellulose sulfate by using SO_3-Py, SO_3-DMF, or $ClSO_3H$ (Gericke et al. 2009a). Highly substituted cellulose sulfates with DS values up to 3 have been reported for sulfation in BMIMCl at 30 °C (Wang et al. 2009), but it has to be noted that cellulose/ionic liquid solutions slowly turned solid upon cooling to room temperature depending on cellulose and moisture content. Furthermore, they possess rather high solution viscosities making it very difficult to ensure sufficient miscibility and to guarantee even accessibility of the sulfating agent to the cellulose backbone. Consequently, the synthesis of cellulose sulfate with an even distribution of sulfate groups along the polymer chains required a aprotic-dipolar cosolvent that drastically reduces the solution viscosity (Gericke et al. 2009a). The reactivity of the sulfating agent is not significantly influenced by the addition of cosolvent. At a molar ratio of 2 Mol SO_3-DMF per Mol AGU, the sulfation of microcrystalline cellulose in BMIMCl and BMIMCl/DMF mixtures leads to comparable DS values of about 0.86 but different properties. While the cellulose sulfate synthesized without cosolvent is insoluble in water, the other one readily dissolves in water. Temperature increase also yields a considerable decrease of the viscosity of cellulose/ionic liquid solutions, which improves solution miscibility (Gericke et al. 2009b). On the other hand, high temperatures favor the acid-catalyzed chain cleavage leading to rather low solution viscosities of aqueous cellulose sulfate solutions in the range of about 2 mPa s (1 % solution in water).

The homogeneous sulfation in ionic liquid allows tuning of cellulose sulfate properties simply by adjusting the amount of sulfating agent and choosing different types of cellulose (Table 10.11). If conducted at room temperature, the reaction leads only to minor polymer degradation (Gericke et al. 2009a). This makes the procedure very valuable for the preparation of water-soluble cellulose sulfate over a wide DS range. Especially cellulose sulfate with low DS can be prepared efficiently in ionic liquid/cosolvent mixture that are of interest for the bioencapsulation (Gericke et al. 2009a, c).

Sulfation of cellulose in BMIMCl/DMF is regioselective, yielding preferably the 6-sulfated product. A ^{13}C NMR spectrum of cellulose sulfate with DS 0.48 prepared in BMIMCl/DMF and the assignment of the peaks is represented in Fig. 10.15. The signal at 67.3 ppm corresponds to sulfation at position 6. Peaks in the region of 82 ppm, which would correspond to sulfated positions 2 or 3, are missing in the spectrum, and no splitting of the C-1 signal can be observed as would occur for 2-sulfation.

Major drawbacks of ionic liquid are their costs and the high viscosities. This disadvantage is compensated by the ease of recycling due to their negligible vapor pressure. Reusability of ionic liquid for sulfation has already been described for BMIMCl leading to products with similar DS values compared to fresh ionic liquid.

Furthermore, the use of cosolvents and the development of low-viscous task-specific ionic liquid, bearing additional functional groups, can lead to further improvement of the homogeneous

Table 10.11 Degree of substitution values and water solubility of cellulose sulfates obtained by sulfation of spruce sulfite pulp dissolved in 1-butyl-3-methylimidazolium chloride/N,N-dimethyl formamide at different conditions [adapted from Gericke et al. (2009a)]

Reaction conditions				Product	
Sulfating agent					
Type[a]	Mol/mol AGU[b]	Time (h)	Temp. (°C)	DS	Water solubility[c]
SO₃-Py	0.7	2	25	0.14	–
SO₃-Py	0.8	2	25	0.25	+
SO₃-Py	1.1	2	25	0.58	+
SO₃-Py	1.4	2	25	0.81	+
SO₃-Py	1.3	1	80[d]	0.52	+
SO₃-DMF	1.0	2	25	0.34	+
SO₃-DMF	1.5	2	25	0.78	+
SO₃-DMF	1.5	0.5	60[d]	–[e]	–[e]
ClSO₃H	1.0	3	25	0.49	+

[a]Sulfur trioxide-pyridine complex (SO₃-Py), sulfur trioxide-N,N-dimethyl formamide complex (SO₃-DMF)
[b]Anhydroglucose unit
[c]Soluble (+); insoluble (−)
[d]Without cosolvent
[e]Not isolatable

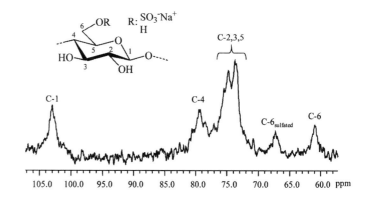

Fig. 10.15 ^{13}C NMR spectrum (in D₂O) of cellulose sulfate (degree of substitution 0.48) obtained by sulfation in 1-butyl-3-methylimidazolium chloride/N,N-dimethyl formamide (DMF) with SO₃-DMF (Gericke et al. 2009a)

sulfation of cellulose. It should be noted that ionic liquid can act as "non-innocent" solvents that participate in the reaction. For instance, sulfation of cellulose in the room-temperature ionic liquid EMIMAc yields cellulose acetate instead of cellulose sulfate (Köhler et al. 2007). Similar side reactions were previously observed for acylation, tritylation, and tosylation of cellulose in EMIMAc (see above).

10.2.2 Regioselective Functionalization

It is well known that the properties of cellulose derivatives are not only determined by the type of substituent and its degree of substitution. An important factor is also the functionalization pattern within the repeating unit and along the polymer chain. Understanding the properties in relation to the functionalization pattern in details requires:

- Synthesis of cellulose derivatives with well-defined structure
- Detailed structure analysis
- Evaluation of properties

These pieces of information are very useful in particular for understanding the action of cellulose ethers in aqueous mixtures. Thus, exact knowledge of such structure–property relationships is an important prerequisite to improve the performance of cellulose-based products.

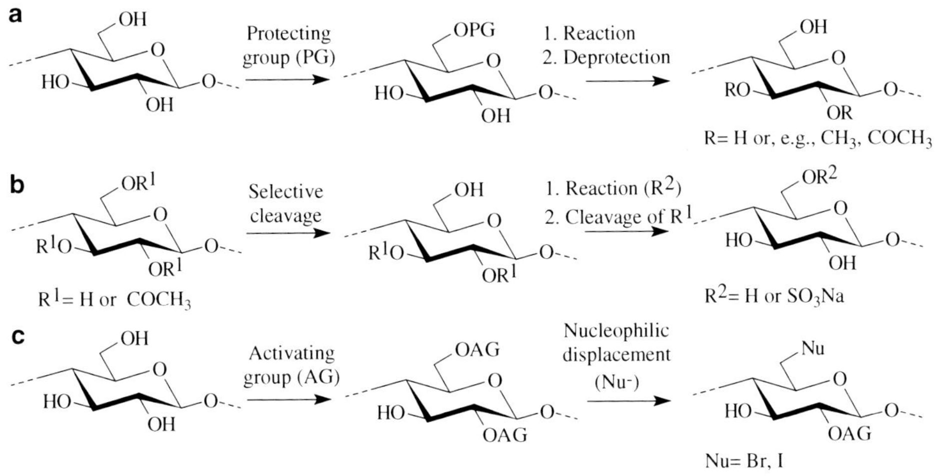

10 Cellulose: Chemistry of Cellulose Derivatization

Fig. 10.16 Pathways for the regioselective functionalization: (**a**) protecting group technique, (**b**) selective cleavage of primary substituents, and (**c**) activating groups

The preparation of regioselectively functionalized cellulose ethers is still a challenge in polymer chemistry. Up to now, the most important approach for the synthesis of cellulose derivatives with controlled functionalization pattern is the application of protecting groups (Fig. 10.16a). Other methods comprising, e.g., selective cleavage of primary substituents play a minor role. Examples are the deacetylation of cellulose acetate under aqueous acidic or alkaline conditions or in the presence of amines (Fig. 10.16b) (Deus et al. 1991; Wagenknecht 1996). In addition, activating groups may also be exploited for selective reactions (Fig. 10.16c). However, sulfonic acid ester moieties cannot be introduced in a selective manner.

In order to achieve selective etherification, blocking-group reagents must consist of at least one bulky alkyl or aryl moiety. In this regard, triphenylmethyl and trialkylsilyl ethers are of special interest (Philipp et al. 1995). A blocking group must meet requirements related to selective introduction, stability during subsequent reactions, and removability without loss of other substituents.

10.2.2.1 Triphenylmethyl Ethers

Triphenylmethyl (trityl) chloride reacts preferably with the primary OH group of cellulose due to its steric demand under heterogeneous (Kondo and Gray 1991) and homogeneous reaction conditions (Camacho Gomez et al. 1996). Due to its stability under alkaline conditions, it is widely used for the preparation of 2,3-O-functionalized cellulose derivatives. Conversion of 6-O-trityl cellulose with acetic acid anhydride or propionic acid anhydride yields the corresponding 2,3-O-functionalized acetates or propionates (Iwata et al. 1992). Detritylation was achieved by treatment with HBr in acetic acid. In a comparable way, the preparation of cellulose sulfuric acid half esters with preferred functionalization of the secondary OH functions was achieved (Heinze et al. 2000b). Sulfur trioxide-pyridine complex was used as reagent followed by detritylation. DS values up to 0.99 were realized. Regioselectively functionalized 2,3-O-methyl and 2,3-O-ethyl celluloses were prepared via 6-O-trityl cellulose (Kondo and Gray 1991).

A fairly difficult synthesis path is the preparation of 6-O-alkyl celluloses shown in Fig. 10.17 (Kondo 1993). This procedure comprises the application of two different protecting groups. First, 6-O-trityl cellulose is converted with allyl chloride in the presence of NaOH yielding allylation of position 2 and 3. After detritylation, isomerization of the 2,3-O-allyl cellulose to 2,3-O-(1-propenyl) cellulose with potassium tert-butoxide is carried out.

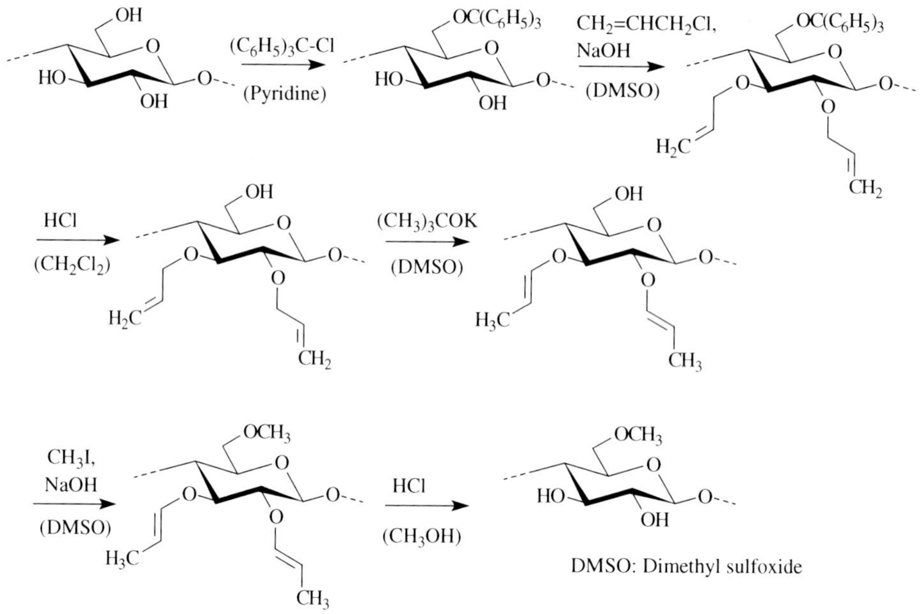

Fig. 10.17 Preparation of 6-O-methyl cellulose (Kondo 1993)

Position 6 is alkylated and the 1-propenyl groups at the secondary positions are cleaved off with HCl in methanol.

Ionic 2,3-O-carboxymethyl cellulose (CMC) was synthesized via 6-O-trityl cellulose applying sodium monochloroacetate as etherifying reagent with solid NaOH as base in DMSO (Heinze et al. 1994, 1999). After a reaction time of 29 h at 70 °C, the product was detritylated with gaseous HCl in dichloromethane for 45 min at 0 °C. Alternatively, the detritylation can be carried out in ethanol slurry with aqueous hydrochloric acid. The 2,3-O-CMC synthesized possesses a DS of up to 1.91. The polymer is water soluble starting with DS 0.3 (Liu et al. 1997).

Very recently, 2,3-O-hydroxyalkyl ethers of cellulose could be prepared starting from 6-O-monomethoxytrityl cellulose (Schaller and Heinze 2005). Homogeneous reaction conditions cannot be applied because the epoxides react with DMSO in the presence of NaOH. Other reaction media like alcohols, which are useful for carboxymethylation reactions, did not work because they do not moisten the hydrophobic monomethoxytrityl cellulose. It has been shown that tensides, in particular mixtures of nonionic and anionic ones, are able to mediate the conversion. The reaction of monomethoxytrityl cellulose with ethylene and propylene oxide in isopropanol/water mixtures containing sodium dodecyl sulfate and polyethylene glycol C_{11}–C_{15} ether (Imbentin AGS-35) afforded 2,3-O-hydroxyalkyl celluloses with a molar substitution (MS) of up to 2.0 after detritylation. Interestingly, the polymers become water soluble starting with a MS 0.25 (hydroxyethyl cellulose) and 0.5 (hydroxypropyl cellulose, HPC), while a conventional HPC is water soluble with MS >4. ^{13}C NMR spectroscopy revealed the etherification of the secondary hydroxyl groups of the AGU. As it is shown in Fig. 10.18, only one signal can be observed for the CH_2 group of position 6. In addition, the peaks of the etherified positions 2 and 3 appear in the range from 80 to 83 ppm.

10.2.2.2 Trialkylsilyl Ethers

Trialkylsilyl ethers like tert-butyl- and thexyldimethylsilyl (TDMS) cellulose are valuable protecting cellulose derivatives. With respect to its performance, the TDMS ether is the most versatile protecting group. It can be used for the preparation of two different types of protected

10 Cellulose: Chemistry of Cellulose Derivatization

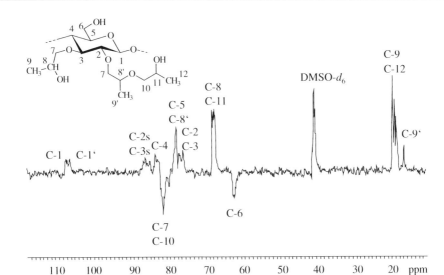

Fig. 10.18 ^{13}C DEPT 135 NMR spectrum of 2,3-O-hydroxypropyl cellulose (molar substitution 0.33) recorded in dimethyl sulfoxide (DMSO)-d_6 at 40 °C (Schaller and Heinze 2005)

Fig. 10.19 Heterogeneous and homogeneous path of silylation leading to cellulose selectively protected at position 6 and position 2 and 6

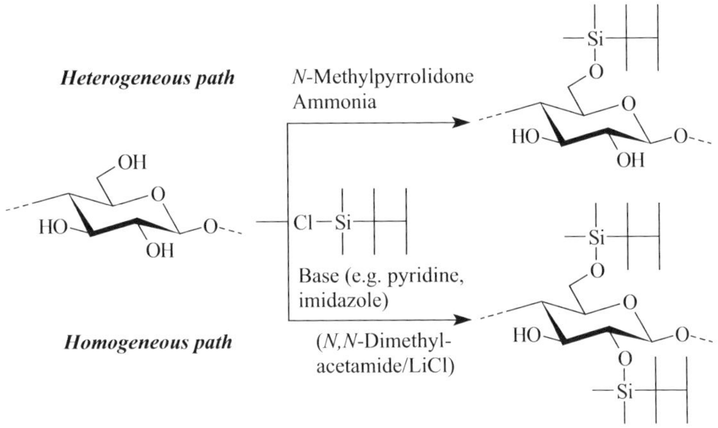

cellulose derivatives because the regioselectivity of the silylation depends on the state of dispersity of cellulose in the reaction mixture, i.e., the selectivity is medium controlled (Fig. 10.19). Starting from cellulose that is swollen in a mixture of ammonia and aprotic-dipolar solvents, in particular NMP at −15 to −25 °C, the conversion with TDMSCl leads to an exclusive silylation of the primary OH group (Klemm and Stein 1995; Koschella and Klemm 1997).

The selectivity toward position 6 is superior compared with the formation of 6-O-triphenyl-methyl ethers of cellulose (Camacho Gomez et al. 1996).

Microcrystalline cellulose is used as staring material in many studies due to their easy handling. However, the degree of polymerization is low, and hence, the products prepared from this starting material are hard to compare with commercially available derivatives. Fenn et al. investigated the 2,6-di-O-silylation of cotton linters with DP 1433 in DMAc/LiCl solution (Fenn et al. 2007). In order to ensure the formation of optically clear cellulose solutions, the cotton

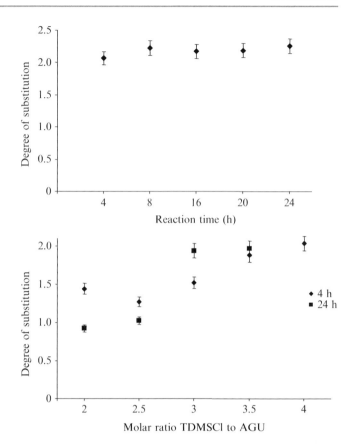

Fig. 10.20 Degree of substitution (DS) obtained by the conversion of cellulose dissolved in N, N-dimethyl acetamide/LiCl with thexyldimethyl chlorosilane (TDMSCl) in the presence of imidazole. Dependency of the DS from the reaction time (*top*, molar ratio anhydroglucose unit, AGU to TDMSCl to imidazole 1:4:4.7) and from the molar ratio AGU to TDMSCl at 4 and 24 h reaction time (*bottom*) [adapted from Fenn et al. (2007)]

linters were mercerized prior to the dissolution process. It was found that the maximum DS was reached already after 4 h reaction time (Fig. 10.20, top). The molar ratio of AGU to TDMSCl may be reduced from 1:4 to 1:3 provided the reaction is conducted for 24 h (Fig. 10.20, bottom). NMR investigations of the polymer after methylation of position 3, desilylation, and acetylation revealed the structural uniformity of the 2,6-di-*O*-TDMS cellulose.

Although the formation of TBDMS cellulose with DS 0.7 was described by Pawlowski et al. (1987), this derivative did not receive much attention in regioselective cellulose chemistry during the past years. Klemm et al. (1990) reported the synthesis of TBDMS cellulose with DS 0.97. In both cases, selective silylation of position 6 was described. Heinze et al. (2008b) investigated the TBDM silylation of cellulose up to DS 2. In contrast to the TDMS cellulose, the synthesis requires a temperature of 100 °C; a reaction already at room temperature yields 2,6-di-*O*-TBDMS cellulose with DS 2. Detailed NMR investigations after methylation, desilylation, and acetylation revealed the structural uniformity of this polymer. Moreover, a conversion at a temperature of 100 °C was found to be inappropriate because side reactions occur, i.e., a change in selectivity; 6-mono-*O*-TBDMS, 3,6-di-*O*-TBDMS, and 2,3,6-tri-*O*-TBDMS moieties were detected.

10.2.2.3 3-*O*-Ethers of Cellulose

The preparation of 3-mono-*O*-functionalized cellulose ethers follows the general scheme of 2,6-di-*O*-protection, 3-*O*-alkylation, and 2,6-di-*O*-deprotection. Alkylating agents are usually alkyl iodides, while alkyl bromides, alkyl chlorides, and sulfonic acid esters are scarcely applied. Due to its high reactivity, sodium hydride is used as base under anhydrous conditions. Sodium hydroxide, which is much easier to handle, cannot be used. Anhydrous conditions

10 Cellulose: Chemistry of Cellulose Derivatization

Table 10.12 3-O-Alkyl celluloses known up to now and their solubility

Alkyl	Solubility[a] Ethanol	DMSO	DMA	H$_2$O (<20 °C)	H$_2$O (room temperature)	References
Methyl	−	−	−	−	−	Koschella et al. (2001)
Ethyl	−	+	+	+	+	Koschella et al. (2006)
Hydroxyethyl	−	+	+	+	+	Fenn and Heinze (2009)
Methoxyethyl	−	+	+	+	+	Heinze and Koschella (2008)
3′-Hydroxypropyl	−	+	+	+	+	Schumann et al. (2009)
Allyl	−	+	+	−	−	Koschella et al. (2001)
Propyl	+	+	+	+	−	Heinze et al. (2011)
Propargyl	−	+	−	−	−	Fenn et al. (2009)
Butyl	+	+	+	−	−	Illy (2006)
n-Pentyl	+	+	+	−	−	Petzold et al. (2004)
Isopentyl	+	+	+	−	−	Petzold et al. (2004)
Dodecyl	−	−	−	−	−	Petzold et al. (2004)
Oligo(ethylene glycol)	−	−	−	−	−	Bar-Nir and Kadla (2009)

[a]Dimethyl sulfoxide (DMSO), N,N-dimethyl acetamide (DMA), soluble (+), insoluble (−)

are hardly to achieve with this reagent, and traces of water induce partial desilylation and, hence, loss of structural uniformity. Finally, the silyl ethers are cleaved off by treatment with tetrabutylammonium fluoride trihydrate. The 3-O-alkyl celluloses known up to now and the typical solubility are summarized in Table 10.12.

The synthesis of 3-mono-O-propargyl cellulose bears some challenges (Fenn et al. 2009). Although the proton of the triple bond is remarkably acidic, there is no hint of chain elongation by formation of C–C bonds, i.e., pure 3-mono-O-propargyl cellulose was obtained. However, hydrogen bonds are present that hinder a detailed structure characterization by means of two-dimensional NMR techniques.

Table 10.12 shows that not only short-chain ethers were prepared. Interestingly, Kadla et al. were able to synthesize 3-O-ethers bearing long-chain methoxypoly(ethylene glycol) ethers with 3–16 ethylene glycol moieties by conversion of 2,6-di-O-TDMS cellulose with methoxypoly(ethylene glycol) tosylates in the presence of imidazole (Kadla et al. 2007). Applying methoxypoly(ethylene glycol) iodide as alkylating agent in the presence of sodium hydride yielded products having a DS as high as 0.8 (Bar-Nir and Kadla 2009).

10.2.2.4 Application of Orthogonal Protecting Groups

The use of protecting group strategies enables the preparation of cellulose ethers, which are not accessible by conventional synthesis procedures. Hydroxyalkyl celluloses are prepared in technical scale by conversion of cellulose with alkylene oxides. As mentioned above, oxyalkylene side chains are formed. In case of hydroxyethyl cellulose, neither ethylene oxide nor 2-bromoethanol will yield a uniform product. This could be avoided by using orthogonal protecting group approach (Fenn and Heinze 2009) (Fig. 10.21).

2-Bromoethanol was allowed to react with 3,4-dihydro-2H-pyran in the presence of p-toluenesulfonic acid as catalyst to yield 2-(2-bromethoxy)tetrahydropyran as the alkylating agent (Arisawa et al. 2000). The order of deprotection of the alkylated 2,6-di-O-TDMS cellulose was found to be very crucial, i.e., the TDMS groups have to be removed first before acid treatment cleaves off the tetrahydropyran moieties (Fenn and Heinze 2009). Obviously, the comparably nonpolar TDMS groups prevent the proton-induced hydrolysis.

The ether moiety regioselectively introduced at position 3 can be used as protecting group itself.

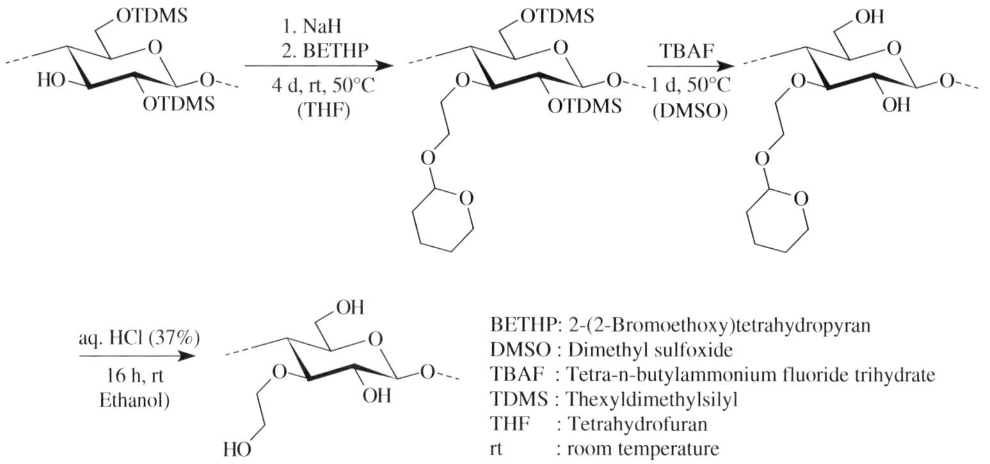

Fig. 10.21 Preparation of 3-mono-O-hydroxyethyl cellulose from 2,6-di-O-thexyldimethylsilyl cellulose via orthogonal protecting groups [adapted from Fenn and Heinze (2009)]

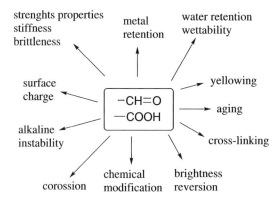

Fig. 10.22 Effects of carbonyl and carboxyl groups on macroscopic properties of cellulose

Despite the stability of the alkyl ethers, allyl groups can be easily removed, e.g., by isomerization to the 1-propenyl ether followed by acidic cleavage (Kondo 1993) or by a palladium-catalyzed reaction (Kamitakahara et al. 2008). This makes the allyl ether to a valuable protecting group. Methylation of 3-mono-O-allyl cellulose and subsequent deallylation affords the corresponding 2,6-di-O-methyl cellulose, which is not accessible by other polymer analogous reactions. This derivative was found to be insoluble in organic media and in water. Moreover, a high crystallinity was detected, which might prevent the polymer from being dissolved.

10.2.3 Oxidation

The oxidation of cellulosic materials is a pivotal reaction in cellulose chemistry, used to impart special properties on the different cellulosic sorts. The cellulose's chemical modification through oxidation leads to added value products, being a prime factor to determine macroscopic properties and chemical behavior of cellulosic materials (Fig. 10.22).

Moreover, the preparation of oxidized cellulose containing carboxyl groups is of special interest because it has several useful medical applications. Oxidized cellulose is completely bioresorbable and easily degradable under physiological conditions, being widely used as absorbable hemostatic scaffolding material (Dias et al. 2003; Galgut 1990) and as a postsurgical adhesion prevention layer (Wiseman et al. 2002). Other applications of oxidized cellulose as carrier material for agricultural, cosmetic, and pharmaceutical applications are reported (Banker and Kumar 1995; Jin and Wu 2005). Oxidized cellulose is also used for the treatment of moderate tubal hemorrhage during laparoscopic sterilization of women and for small uterine perforations (Sharma and Malhotra 2003; Sharma et al. 2003). As blood clotting agents, oxidized cellulose fibers are superior to oxidized regenerated cellulose (Pameijer and Jensen 2007). Remarkably progresses on using oxidized cellulose in osseous regeneration have been reported (Dias et al. 2003), making oxidized cellulose an excellent scaffolding material, able to replace the most widely used material today, i.e., collagen. In the field of separation techniques, oxidized cellulose shows a powerful activity to purify and fractionate proteins, including enzymes, hemoglobins, hormones, and seed proteins. Oxidized cellulose is extensively used as well in the chromatography of peptides, amino acids, alkaloids, nucleic acids, nucleotides, and metallic ions (Guthrie 1971). In the paper industry, it is well known that carbonyl and carboxyl functionalities play a decisive role in the pulping process and therefore in the final paper properties. For example, sheets made from the partially oxidized fibers (below 8 % conversion) experienced higher wet and dry tensile index, presumably due to an increased opportunity of electrostatic interactions between anionic pulp and cationic polyamideamine-epichlorohydrin, which was added as a wet-strength agent (Kitaoka et al. 1999).

10.2.3.1 Cellulose Oxidation

Cellulose, because of its polyhydric alcohol structure, is very sensitive to various oxidizing reagents. The extensive modifications, which occur during oxidation, give rise to products whose physicochemical properties strongly depend upon factors such as the nature of the oxidizing reagent used or the acidity or basicity of the oxidation medium. The chemical structure of cellulose is changed in a way that hydroxyl groups are oxidized into the corresponding carbonyl structure, i.e., an aldehyde at C-6 and a ketone at C-2 and C-3 or carboxyl moiety at C-6. The wide range of potential oxidizing reagents for cellulose can be divided into nonselective [nitrogen oxides (Butrim et al. 2007), alkali metal nitrites and nitrates (Painter 1977), ozone (Johansson and Lind 2005), permanganates (Manhas et al. 2007), and peroxides (Borisov et al. 2004)] and selective, such as periodates (Calvini et al. 2006; Zimnitski et al. 2004; Fras et al. 2005) and nitroxyl radicals (De Nooy et al. 1995; Chang and Robyt 1996; Tahiri and Vignon 2000; Davis and Flitsch 1993; Gomez-Bujedo et al. 2004; Isogai and Kato 1998; Cato et al. 2004; Coseri et al. 2009; Biliuta et al. 2010).

The cellulose oxidation in the presence of nitrogen (IV) oxide requires high temperature (343 K) and high pressure (70 atm) (Zimnitski et al. 2004). The process is rather slow and yields a material with considerable amount of side products containing nitrogen. Recently, the cellulose oxidation by nitrogen dioxide dissolved in pressurized carbon dioxide was reported (Camy et al. 2009). The global reaction is illustrated in Fig. 10.23.

Even if supercritical CO_2 is now growing interest as a reaction medium for many chemical syntheses due to its peculiar physical properties and its specific "green" character in the case of cellulose oxidation, it appears that its role is not as neutral as expected and the degree of oxidation of cellulose depends on the amount of CO_2 introduced. Camy et al. (2009) suspected that CO_2 interacts with NO_2, thus inhibiting the reactivity of this latter toward cellulose.

Selective oxidation of cellulose in the presence of periodates affects bond cleavage between C-2 and C-3 of the anhydroglucose units with concomitant introduction of aldehyde functionalities at those two positions (Fig. 10.24).

The carbonyl groups in oxidized cellulose are indicated by the so-called copper number. However, this method is not very accurate since the copper number is not directly linked with the

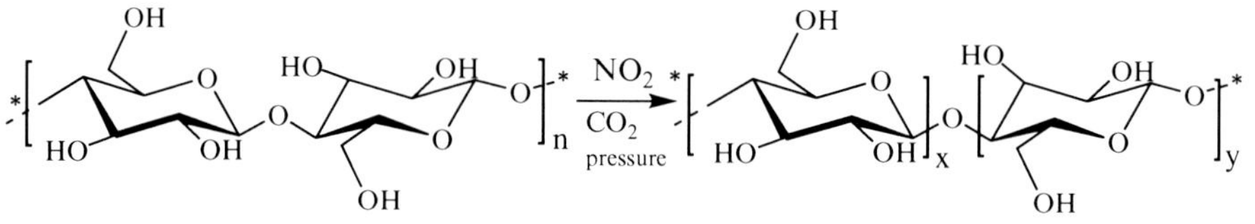

Fig. 10.23 Cellulose oxidation by nitrogen dioxide in pressurized carbon dioxide

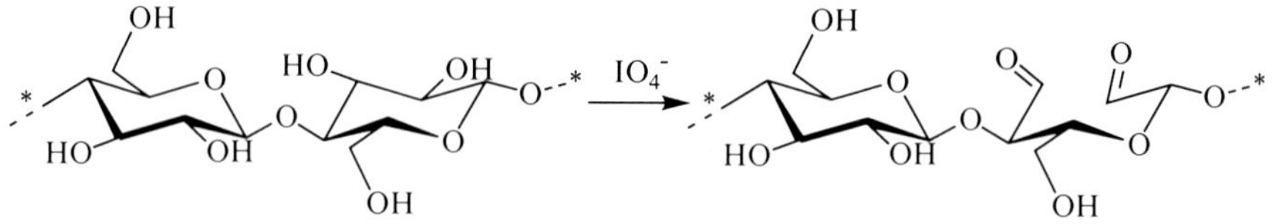

Fig. 10.24 Periodate oxidation of cellulose

quantity of a specific oxidized function. Group-selective fluorescence labeling in combination with multi-detector gel permeation chromatography (multiangle laser light scattering, refractive index, fluorescence) is a valuable and powerful technique developed in the last years for cellulose analytics (Potthast et al. 2003; Kostic et al. 2006).

Since 1994, when De Nooy et al. (1994) reported for the first time the use of 2,2,6,6-tetramethylpiperidine-N-oxyl (TEMPO) for the transformation of primary alcohol groups in polysaccharides to the corresponding polyuronic analogues, this process becomes one of the most studied in the field of selective oxidation of various polysaccharides in general, in particular cellulose. TEMPO-oxidation process is highly effective in the conversion of high molecular weight polysaccharides and ensures high reaction rate and yields high selectivity with a modest degradation of polysaccharide throughout process.

Concerning this new approach, Vignon et al. (Gomez-Bujedo et al. 2004) recently reported: "it appeared that this method was very selective, primary hydroxyl groups being exclusively oxidized, whereas secondary hydroxyl groups remained unaffected."

The actual oxidizing species is the nitrosonium ion (structure C, Fig. 10.25), the oxidized form of TEMPO (structure B, Fig. 10.25). The generation of nitrosonium ion takes place in situ through the reaction of TEMPO with oxidants, such as hypobromide ions, which in turn are being generated from bromide salts and sodium hypochlorite (Fig. 10.25). Upon acting as oxidant species, C is converted into N-hydroxy-2,2,6,6-tetramethylpiperidine (structure A, Fig. 10.25), the reduced form of TEMPO.

When native celluloses were subjected to the TEMPO/NaBr/NaClO oxidation at pH 10, the C6 primary hydroxyl groups of cellulose microfibril surfaces were effectively converted to sodium carboxylate groups, maintaining the original fibrous morphologies, crystallinities, and crystal sizes (Saito and Isogai 2004). The high selectivity of the TEMPO oxidation is substantiated by theoretical calculations (Bailey et al. 2007). According to these calculations, the cellulose oxidation in alkaline medium could begin with a nucleophilic attack of an alkoxide on the nitrogen or oxygen atom of the strongly polarized N=O bond of nitrosonium cation, which results in the formation of either complex I or II (Fig. 10.26). The oxidation via peroxide complex is less probable according to calculations; hence, the reaction path via complex I is much more possible.

The oxidation of cellulose under the action of NaClO was usually carried out in alkaline aqueous media in the presence of TEMPO (0.5–4 mol% with respect the substrate) and NaBr (5–30 mol %) as the catalysts (Bragd et al. 2004). Tahiri and Vignon (2000) investigated the TEMPO oxidation of cellulose, and their results showed the lowest depolymerization of the

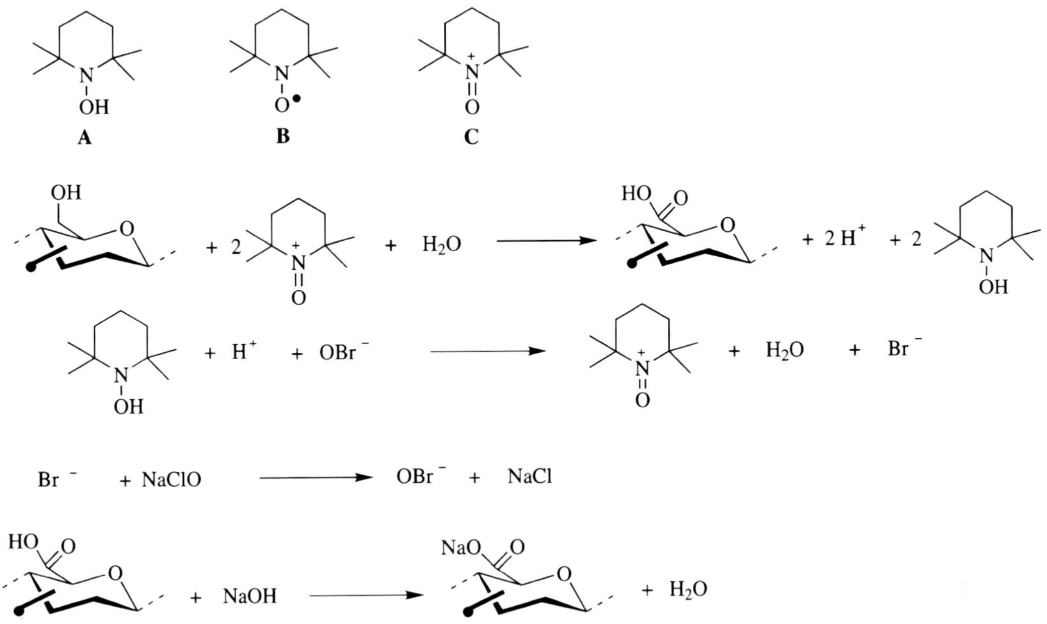

10 Cellulose: Chemistry of Cellulose Derivatization

Fig. 10.25 Oxidation scheme of primary hydroxyl groups of cellulose to carboxyl groups by the 2,2,6,6-tetramethyl-piperidine-N-oxyl (TEMPO)-mediated oxidation

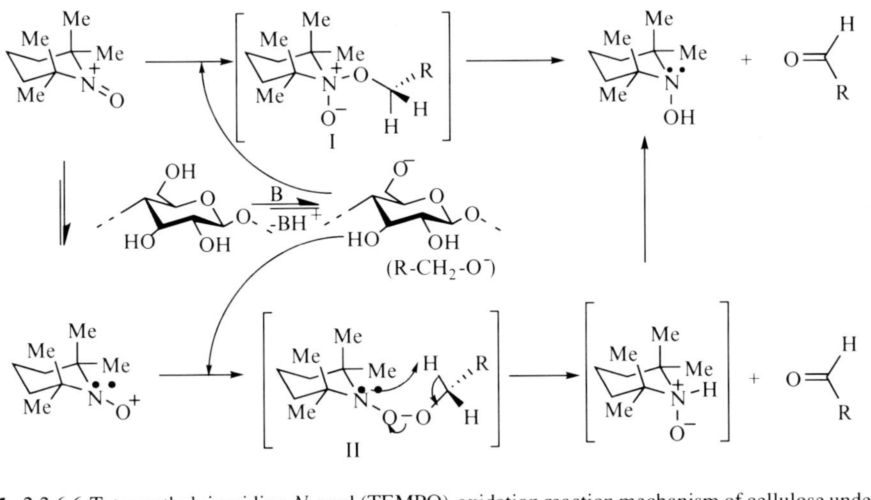

Fig. 10.26 2,2,6,6-Tetramethylpiperidine-N-oxyl (TEMPO)-oxidation reaction mechanism of cellulose under alkaline conditions. B is a base

amorphous cellulose when worked at pH 10 and 4 °C. The oxidation at room temperature using TEMPO–NaBr–NaClO system has been used to increase the carboxyl content of several different cellulosic samples by other authors (Isogai and Kato 1998). They have been found that the key factors controlling the depolymerization of regenerated or mercerized cellulose are the charge of TEMPO, reaction time, and temperature. The depolymerization of cellulose during TEMPO-mediated oxidation was attributed to sodium hypochlorite present in the system, which causes 2,3-scissions of glucose unit, forming dialdehyde and dicarboxylic groups (De Nooy et al. 1994). The formation of carbonyl groups at C-2, C-3 in glucose unit, facilitates depolymerization of

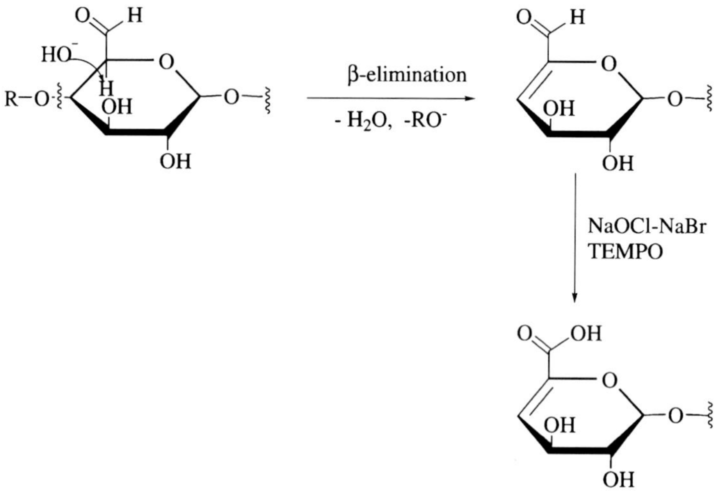

Fig. 10.27 Depolymerization during 2,2,6,6-tetramethylpiperidine-*N*-oxyl (TEMPO)-oxidation via β-elimination reaction. R is a polysaccharide residue

celluloses via β-alkoxy fragmentation in alkaline medium (Calvini et al. 2004). Depolymerization during TEMPO oxidation can occur also via β-elimination reaction (Fig. 10.27).

Several attempts to overcome the depolymerization issues during TEMPO/NaClO/KBr oxidation of cellulose have been reported. Thus, working with TEMPO derivates, such as 4-acetamide-TEMPO/NaClO/NaClO$_2$ system, at 60 °C and pH 4.8–6.8 for 1–5 days (Hirota et al. 2009a, b), high yields of water-soluble cellouronic acids with higher degree of polymerization were obtained by oxidation of regenerated celluloses. In this case, NaClO$_2$ acts as a primary oxidant, whereas NaClO added in catalytic amounts starts the oxidation cycle (Fig. 10.28). NaClO oxidizes 4-acetamide-TEMPO to the corresponding *N*-oxoammonium ion, which then rapidly oxidizes the primary hydroxyl to aldehyde under acidic conditions, forming the hydroxylamine. The aldehyde is oxidized to carboxyl by the primary oxidant NaClO$_2$, forming NaClO, and the hydroxylamine is oxidized to the *N*-oxoammonium ion again by the NaClO thus generated (Fig. 10.28).

However, it can be noted that relatively higher oxidation temperatures around 40–60 °C and longer oxidation times of 1–5 days are needed. From an environmental point of view, it may be better not to use chlorine-containing chemicals such as NaClO or NaClO$_2$. Moreover, side reactions caused by NaClO or NaClO$_2$ cannot be avoided as long as the above TEMPO-mediated oxidation systems are adopted. In the last years, development of suitable electrochemical methods for oxidation of organic compounds (i.e., electro-organic oxidation) has attracted worldwide attention as a green chemistry way (Zhang et al. 2010; Danaee et al. 2010). The main future of the electro-organic oxidation of cellulose in the presence of TEMPO is the absence of depolymerization. Recently, the TEMPO-electromediated oxidation of some polysaccharides including regenerated cellulose fiber has been reported (Isogai et al. 2010). The reaction takes place in a 0.1 M phosphate buffer at pH 6.8 and at room temperature for 45 h. 4-Acetamide-TEMPO catalyst was found to be quite specific, differing from other TEMPO-mediated oxidations. Significant amounts of C6-carboxylate and C6-aldehyde groups (1.1 and 0.6 mmol/g, respectively) were formed. Interestingly, the original fibrous morphology as well as the fine surface structures was maintained.

Among the different kinds of catalysts reported in the literature, *N*-hydroxyphthalimide (NHPI) is another catalyst having excellent

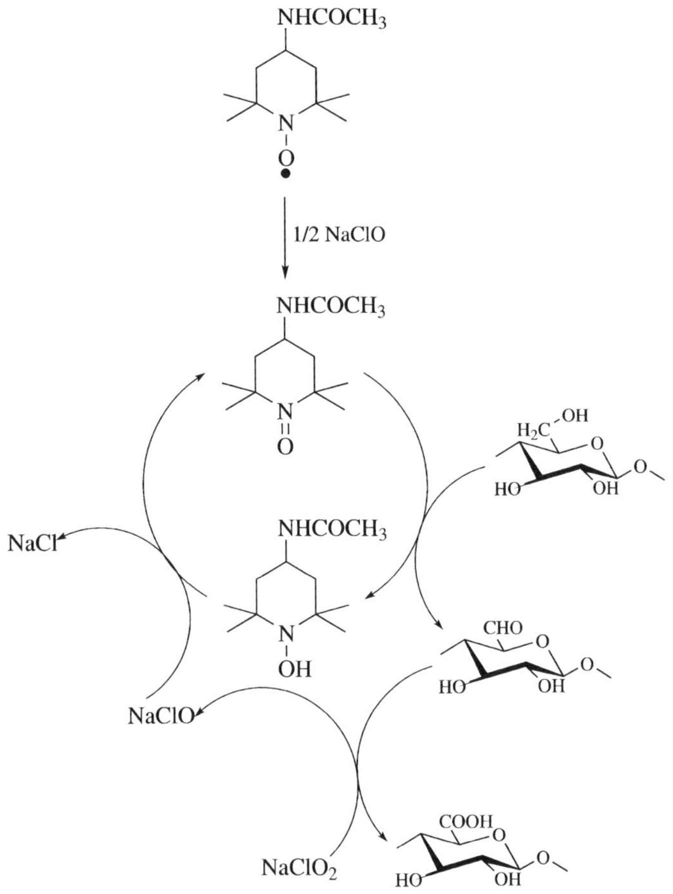

Fig. 10.28 Catalytic cycle of the 4-acetamide-2,2,6,6-tetramethylpiperidine-*N*-oxyl (TEMPO)/NaClO/NaClO$_2$ system for oxidation of cellulose

performance in oxidation reactions (Fig. 10.29) (Coseri 2009; Coseri et al. 2005). These oxidative reactions occur via the intermediary phthalimide *N*-oxyl (PINO) radical.

The mild and selective oxidation of viscose fibers in the presence of PINO radical (in situ generated from NHPI and various cocatalysts, i.e., lead tetraacetate, cerium (IV) ammonium nitrate, anthraquinone) with diluted solutions of sodium hypochlorite (2 %) and sodium bromide was very recently reported (Coseri et al. 2009). Soon after, several other nonpersistent free radicals, generated in situ from their OH correspondents, *N*-hydroxybenzotriazole (HBT), violuric acid (VA), and *N*-hydroxy-3,4,5,6-tetraphenylphthalimide (NHTPPI), have shown their efficiency for the cellulose fibers oxidation at pH 10 and room temperature in the presence of dilute solutions of NaClO and NaBr (Fig. 10.29) (Biliuta et al. 2010). The reaction mechanism involves in the first step the activation of the parent hydroxyl precursors through their transformation into corresponding free radical compounds. In the presence of hypobromide ions, these nitroxyl radicals are oxidized to *N*-oxoammonium ions, which are actually the oxidizing species. At this point, the primary –OH groups of cellulose are converted into carboxylic groups, process which proceeds via aldehyde stage. However, the aldehyde groups are undetectable in the final products, being rapidly oxidized to carboxylic groups, in the presence of hypobromide ions. Due to the presence of NaOH, the carboxylic groups are obtained in their sodium salt form (Fig. 10.30).

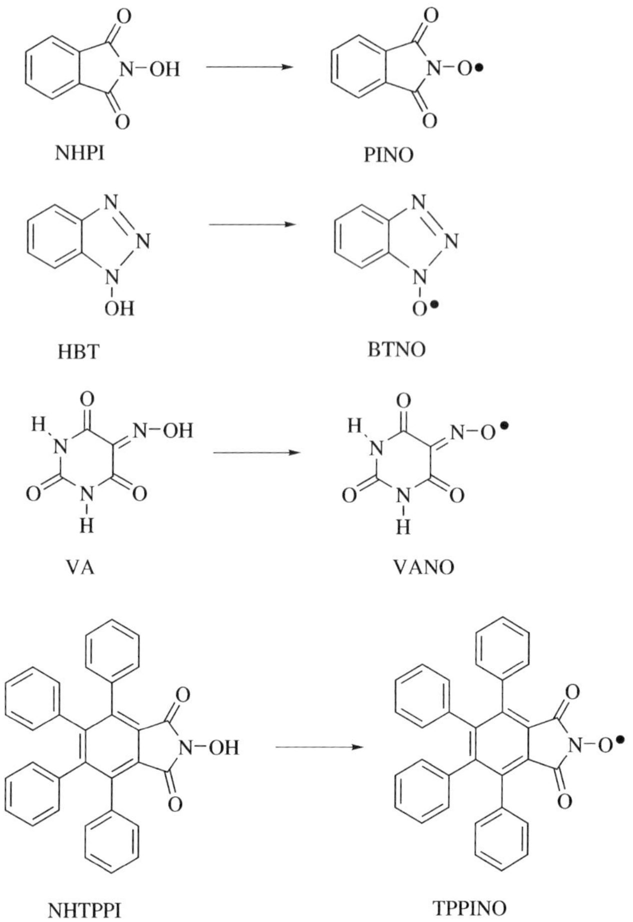

Fig. 10.29 The –OH-bearing precursors for the nonpersistent free radicals: N-hydroxyphthalimide (NHPI), phthalimide N-oxyl (PINO), N-hydroxybenzotriazole (HBT), violuric acid (VA), and N-hydroxy-3,4,5,6-tetraphenylphthalimide (NHTPPI)

10.2.4 Brief Discussion of Structure–Property Relationships

The 6-O- and 2,3-O-functionalized alkyl cellulose samples were subjected to a number of investigations, e.g., study of gelation effects (Itagaki et al. 1997), behavior in blends with synthetic polymers (Kondo et al. 1994), hydrogen bond system (Kondo 1997), and enzymatic degradation (Nojiri and Kondo 1996).

Cellulose ethers may exhibit the phenomenon of thermoreversible gelation that strongly depends on the functionalization pattern as shown for selectively methylated cellulose (Hirrien et al. 1998). It was found that slight differences in the polymer composition result in remarkable changes of the thermal behavior.

A set of 2,3-O-methyl celluloses was prepared starting from 6-O-trityl and 6-O-monomethoxytrityl cellulose (Kern et al. 2000). Due to varied reaction conditions, the total DS and composition of the repeating units are slightly different. Typical samples are summarized in Fig. 10.31, which are water soluble and studied as aqueous solutions by DSC measurements. It can be clearly seen that thermal events are in strong correlation with the polymer composition regarding differently functionalized repeating units. If the polymer contains tri-O-methylated glucose units in combination with monomethylated ones, a distinct thermal behavior is observed, i.e., the methyl cellulose exhibits thermoreversible gelation (Fig. 10.31a, b). In contrast, a uniform 2,3-O-methyl cellulose shows no thermal gelation. It becomes obvious that 2,3-O-methyl glucose

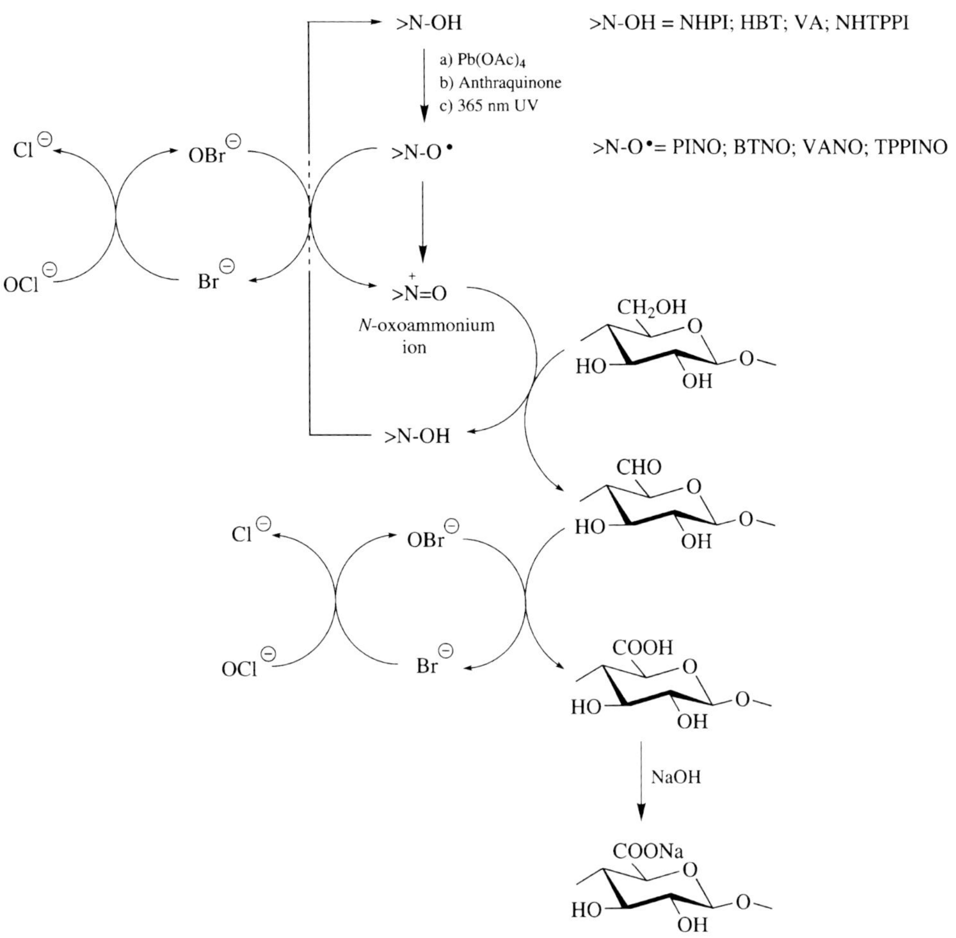

10 Cellulose: Chemistry of Cellulose Derivatization

Fig. 10.30 Oxidation scheme of cellulose fibers performed by *N*-hydroxyphthalimide (NHPI), *N*-hydroxybenzotriazole (HBT), violuric acid (VA), and *N*-hydroxy-3,4,5,6-tetraphenylphthalimide (NHTPPI)

units do not cause significant intermolecular interactions that are necessary for the gelation (Fig. 10.31c).

3-*O*-Alkyl celluloses exhibit an interesting solubility behavior. For instance, 3-mono-*O*-methyl cellulose was found to be insoluble in common organic solvents and water (Koschella et al. 2001). However, swelling in aprotic-dipolar solvents was observed. Addition of LiCl to the swollen polymer in aprotic-dipolar solvents gives polymer solutions, which may be explained with the breakage of hydrogen bonds. FTIR spectroscopy in combination with a line-fitting method exhibited that instead of intramolecular hydrogen bonds between the hydroxyl group of position 3 and the ring oxygen of the adjacent anhydroglucose unit known for cellulose, another intramolecular hydrogen bond between the hydroxyl groups at position 2 and 6 may exist in 3-mono-*O*-methyl cellulose. The strong interchain bonds indicated by the absorption at $3,340 \text{ cm}^{-1}$ were also formed, causing crystallization with relatively high degree of crystallinity (Kondo et al. 2008).

Extension of the alkyl chain by one methylene moiety changes the solubility remarkably. 3-mono-*O*-ethyl cellulose is soluble in aprotic-dipolar organic media without addition of LiCl. Moreover, this compound is soluble in water (Koschella et al. 2006). The impact of functionalization pattern on thermal behavior became obvious by comparing the properties of 3-*O*-ethyl cellulose and randomly functionalized

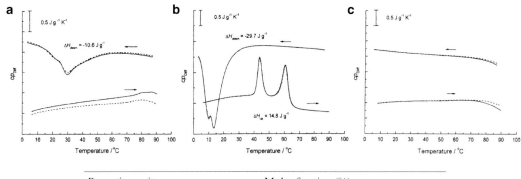

Repeating unit	Molar fraction (%)		
	Sample a	Sample b	Sample c
Glucose	5.79	3.04	3.09
mono-O-methyl	26.08	2.14	7.17
di-O-methyl	67.26	84.05	87.94
tri-O-methyl	2.78	10.77	1.81
Total DS	1.69	2.03	1.89

Fig. 10.31 Thermal behavior of 2,3-O-functionalized methyl celluloses dependent on the polymer composition (differential heat capacity, cp$_{Diff}$, degree of substitution, DS) [reprinted from Kern et al. (2000)]

Fig. 10.32 Micro-DSC results for aqueous solutions (4 %, w/w) of ethyl cellulose with random functionalization pattern (**a**) and 3-mono-O-ethyl cellulose (**b**)

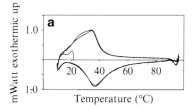 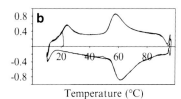

ethyl cellulose with similar DS. While the randomly functionalized ethyl cellulose became insoluble at ~30 °C, the cloud point temperature of the regioselectively functionalized product is 60 °C. This could be assessed by eye and measured using the techniques of ^{1}H NMR spectroscopy, oscillatory rheology, and microdifferential scanning calorimetry (Fig. 10.32) (Sun et al. 2009).

3-mono-O-propyl cellulose dissolves also in DMF and DMSO. Solubility tests in water at room temperature revealed an insoluble polymer (Heinze et al. 2011). However, an almost clear solution was formed upon storage in a refrigerator. Temperature-dependent turbidity measurements indicated cloud point temperatures between 15.2 °C (DS 1.02) and 23.5 °C (DS 0.71) that depend on the DS.

3-mono-O-hydroxyethyl (Fenn and Heinze 2009), 3-mono-O-methoxyethyl (Heinze and Koschella 2008), and 3-mono-O-(3′-hydroxypropyl) celluloses (Schumann et al. 2009) were found to be water soluble as well, but they do not exhibit gelation or flocculation at elevated temperatures.

Starting with 3-mono-O-allyl cellulose, the higher 3-O-alkyl celluloses lose their water solubility. However, the pentyl, isopentyl, and dodecyl ethers become soluble in less polar solvents like methanol and ethanol and even tetrahydrofuran (dodecyl ether) (Petzold et al. 2004). Light scattering measurements revealed a distinct aggregation behavior that depends on the type of substituent. The highest aggregation number of 182 was found for the 3-*O*-pentyl ether, while the 3-*O*-dodecyl ether does not show aggregation. Once the OH group at position 3 is blocked, the remaining OH groups will randomly form hydrogen bonds among different chains, until a particle with a surface of mainly alkyl chains is obtained. The monomolecular solubility of the 3-*O*-dodecyl ether demonstrates that such aggregation can be prevented if the side chain is long enough.

10.3 Unconventional Cellulose Chemistry

10.3.1 1,3-Dipolar Cycloaddition Reaction of Azide- and Alkyne-Containing Cellulosics

The copper-catalyzed 1,3-dipolar cycloaddition reaction of alkynes and azides (Huisgen reaction, click reaction) (Kolb et al. 2001) is a valuable synthetic tool, which can also be applied in cellulose chemistry. Cellulose *p*-toluenesulfonic acid ester (tosyl cellulose) is an easily accessible and valuable starting material for the preparation of derivatives capable of undergoing a click reaction. Conversion of tosyl cellulose with sodium azide yields 6-deoxy-6-azido cellulose (Heinze et al. 2006). The versatility of the click reaction was demonstrated with 6-deoxy-6-azido cellulose dissolved in aprotic-dipolar media that was allowed to react with various alkynes yielding unconventional cellulose derivatives (Fig. 10.33) (Liebert et al. 2006).

Activated alkynes do react without a catalyst. Click reactions can also be used for the introduction of anionic moieties that are derived from acetylenedicarboxylic acid (Fig. 10.34) (Koschella et al. 2010). Acetylenedicarboxylic acid ester readily reacted with 6-deoxy-6-azido cellulose, and the methyl ester was saponified with aqueous sodium hydroxide as monitored by NMR spectroscopy (Fig. 10.35).

Tosyl cellulose can also be converted with propargyl amine to the corresponding 6-deoxy-6-aminopropargyl cellulose (Pohl and Heinze 2008). Alternatively, the reactive alkyne moiety can be attached to cellulose via an ether linkage. This was demonstrated by conversion of 2,6-di-*O*-protected cellulose with propargyl bromide in the presence of a sodium hydride followed by deprotection. The resulting 3-mono-*O*-propargyl cellulose was starting material for click reactions (Fenn et al. 2009).

Kato et al. immobilized oligosaccharides on the surface of filter paper by a click-chemistry approach. Thus, cellulose was treated with bromoacetic acid in the presence of aqueous sodium hydroxide to introduce carboxymethyl moieties. The carboxyl groups were subsequently aminated with propargyl amine. Finally, azide-modified oligosaccharides were attached in the presence of copper(I) (Kato et al. 2009).

In a similar way, Ringot et al. attached porphyrin moieties to the surface of cotton fabrics. Cellulose fabrics were allowed to react with triphenylphosphine, sodium azide, and carbon tetrabromide to introduce the deoxyazido moieties. Afterward, an alkyne-modified zinc porphyrin complex was bound via a triazolo moiety to obtain a photobactericidal material (Ringot et al. 2009).

Zhang et al. (2009b) prepared thermosensitive hydrogels by cross-linking 6-deoxy-6-azido cellulose with an alkyne-modified poly(*N*-isopropyl acrylamide). Click reactions are not only limited to the reaction of alkynes with azides. Zhao et al. prepared cellulose derivatives bearing alkene moieties by conversion of cellulose with unsaturated carboxylic acids or with unsaturated epoxides (Fig. 10.36). These derivatives were able to react with thiol compounds upon irradiation with UV light (Zhao et al. 2010) (Fig. 10.37).

10.3.2 Dendronization of Cellulose

Dendrons are molecules that possess a treelike branched structure. They contain a core (focal moiety) that can be linked to other molecules. Regular structures are attached to the core. They are branched at their end, which forms the next generation of the dendron.

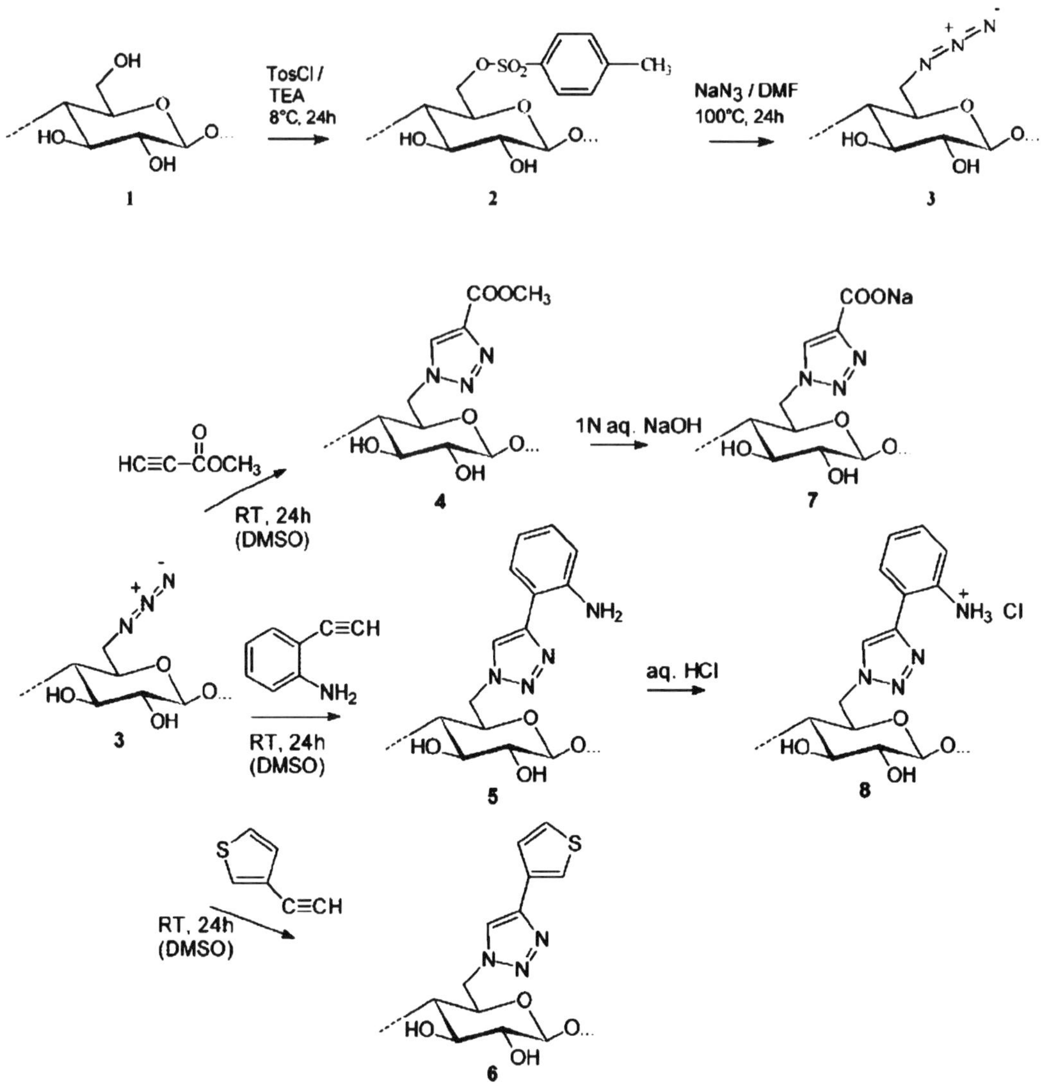

Fig. 10.33 Reaction scheme for preparation of 6-azido-6-deoxy cellulose (3) and subsequent copper(I)-catalyzed Huisgen reaction of 1,4-disubstituted 1,2,3-triazoles used as linker for the modification of cellulose with methylcarboxylate (4), 2-aniline (5), and 3-thiophene (6) moieties (Liebert et al. 2006)

First studies described the transformation of hydroxypropyl cellulose with 2,2-bis(hydroxymethyl)propionic acid anhydride generating a first generation of dendritic cellulose derivative, which was characterized by ATR-FTIR spectroscopy (Östmark et al. 2007). Deprotection of the peripheric acetal groups and the conversion of the hydroxyl groups formed with 2,2-bis(hydroxymethyl)propionic acid anhydride resulted in higher generations of dendronized cellulose derivatives. It was assumed that the polymers formed nanoscaled, rod-shaped objects.

Hassan et al. (2004, 2005) used isocyanate-functionalized dendrons, which were linked to the cellulose backbone via a carbamate function. A regioselective functionalization of position 6 (primary OH group) was claimed.

Further studies illustrated the preparation of a poly(propylene imine) and poly(amidoamine) cellulose derivative based on cyanoethyl

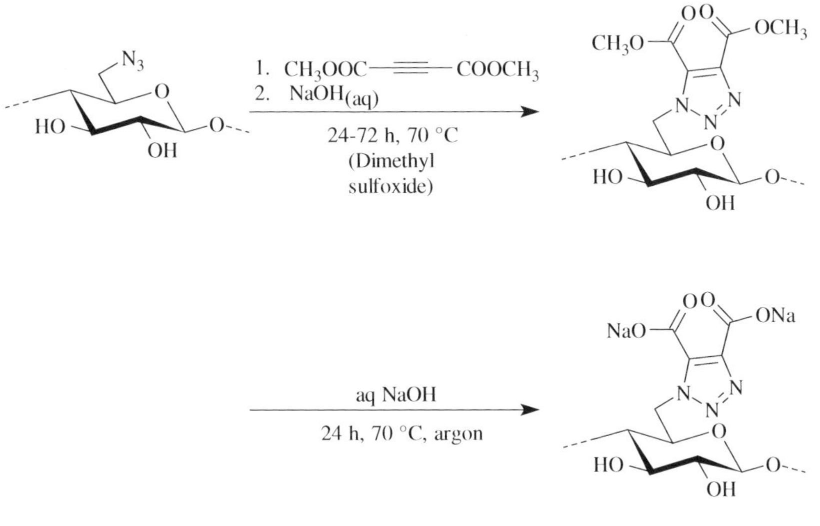

10 Cellulose: Chemistry of Cellulose Derivatization

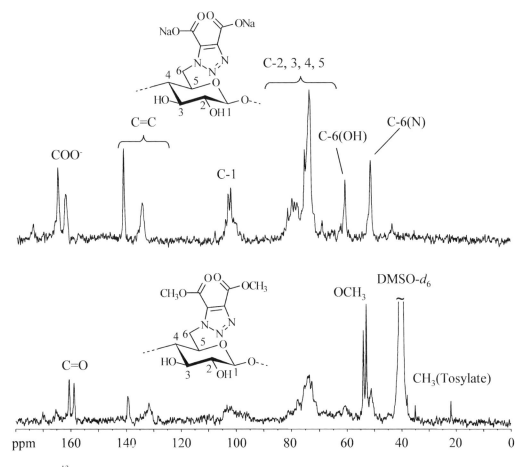

Fig. 10.34 1,3-Dipolar cycloaddition reaction of 6-deoxy-6-azido cellulose with acetylenedicarboxylic acid ester and subsequent saponification (Koschella et al. 2010)

Fig. 10.35 ^{13}C NMR spectra of 6-deoxy-6-(1-triazolo-4,5-dicarboxylic acid dimethyl ester) cellulose (*bottom*) recorded in dimethyl sulfoxide-d_6 (DMSO-d_6) and 6-deoxy-6-(1-triazolo-4,5-disodium dicarboxylate) cellulose (*top*) recorded in D$_2$O (Koschella et al. 2010)

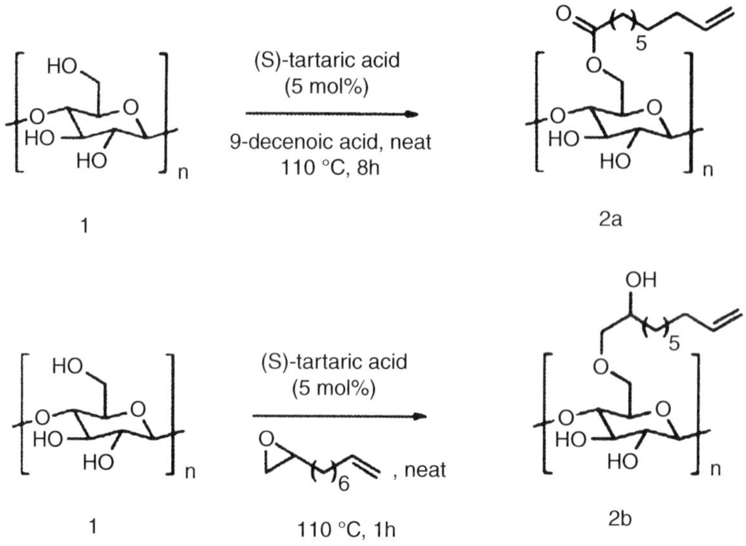

Fig. 10.36 Organocatalytic syntheses of ene-functionalized cellulose (Zhao et al. 2010)

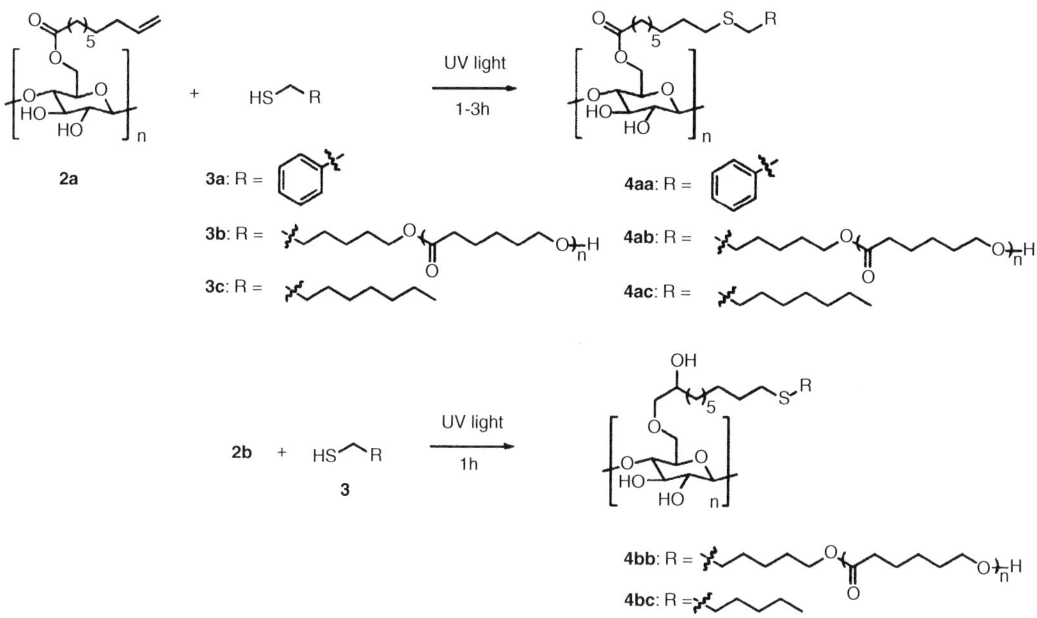

Fig. 10.37 Thiol-ene functionalization of cellulose (Zhao et al. 2010)

cellulose with a DS of 2.9 (Hassan 2006). Only the first generation could be obtained without a high degree of imperfection in the structure of the introduced hyperbranched moieties. Zhang et al. synthesized two different types of dendronized cellulose applying a stepwise attachment of Behera's amine (aminotriester-based dendrons) to carboxymethyl cellulose (DS 0.7). Subsequent conversion of the tert-butyl ester peripheric groups with N,N-dimethyl-1,3-propanediamine forms the water-soluble derivative with second-generation polyamide dendrons as the

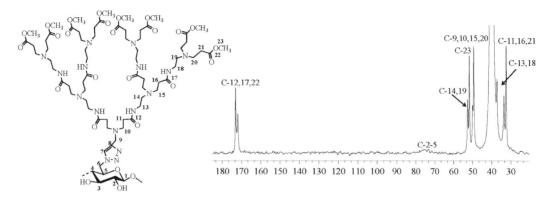

Fig. 10.38 ^{13}C NMR spectrum of the dendritic polyamidoamine-triazolo cellulose of third generation (Pohl et al. 2008a)

interior and amidoamine substitutes as hydrophilic tips (Zhang and Daly 2005; Zhang et al. 2006).

Pohl and Heinze investigated the conversion of cellulose with carboxylic acid dendrons of different generation. Thus, cellulose was converted with 3,5-bis(benzyloxy)benzoyl chloride as well as with 3,5-bis(benzoyloxy)benzoic acid in situ activated with N,N'-carbonyldiimidazole under homogeneous reaction conditions (Heinze et al. 2007; Pohl et al. 2008b). A DS of 0.21 was obtained using the quite bulky 3,5-bis(3,5-bis[3,5-bis(benzoyloxy)benzoyloxy)benzoyloxy)benzoic acid as reagent. Both primary and secondary hydroxyl groups were esterified.

The copper-catalyzed 1,3-dipolar cycloaddition reaction of azides and alkynes is a valuable tool to introduce the dendrons in a regioselective manner. Pohl et al. used 6-deoxy-6-azido cellulose with a DS of 0.75 as starting material. Propargyl-functionalized polyamidoamine dendrons of the first to third generation were attached to the cellulose backbone via triazolo linkages. Almost all azide moieties were converted using the first-generation dendrons (maximum DS obtained 0.67). Even a DS of 0.31 was obtained with the third-generation dendrons (Pohl et al. 2008a). Structure characterization by means of FTIR and NMR spectroscopy was carried out. However, the ^{13}C NMR spectrum is dominated by the peaks of the dendrons (Fig. 10.38).

6-Deoxy-6-azido cellulose was found to be soluble in the ionic liquid EMIMAc. Obviously, the reactivity in this solvent is higher, because a DS of up to 0.60 using the third-generation propargyl-polyamidoamine (PAMAM) dendrons could be realized (Fig. 10.39) (Heinze et al. 2008a).

6-Deoxy-6-aminopropargyl cellulose is prepared from tosyl cellulose, which is then converted with azidopropyl-polyamidoamine dendrons of first to second generation. In case of the second-generation dendrons, about 50 % of the available alkyne moieties reacted, probably due to steric reasons (Pohl and Heinze 2008).

An interesting approach is the application of propargyl-PAMAM dendrons of the 2.5th generation, which contains amino groups at the end of the branches (Pohl et al. 2009b). This compound can be used to decorate cellulose derivatives with a large number of amino groups. A mixture with cellulose acetate can be coated on a solid support. Subsequent coupling of enzymes (e.g., glucose oxidase) affords bioactive surfaces (Fig. 10.40).

Both 6-deoxy-6-azido cellulose and 6-deoxy-6-aminopropargyl cellulose are water insoluble. It could be demonstrated that carboxymethylation renders 6-deoxy-6-azido cellulose water soluble (Pohl et al. 2009a). The click reaction with propargyl-PAMAM dendrons of the first and second generation could be accomplished in aqueous solution. Viscometric techniques and analytical ultracentrifugation were employed to

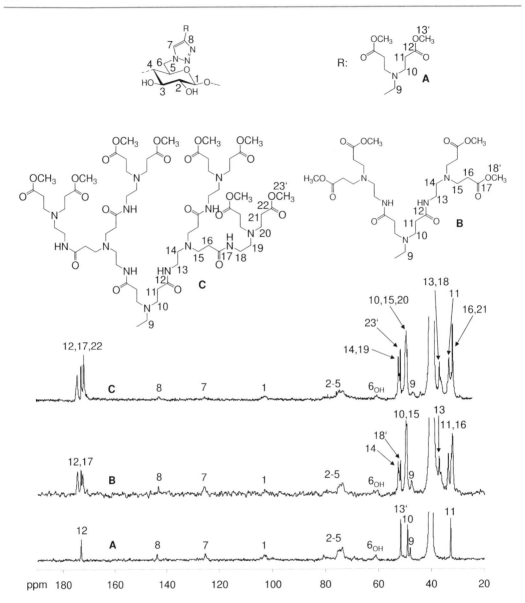

Fig. 10.39 ^{13}C NMR spectra A of first (degree of substitution, DS 0.60), B second (DS 0.48), and C third (DS 0.28) generation of polyamidoamine-triazolo-cellulose recorded in dimethyl sulfoxide-d_6 at 60 °C (Heinze et al. 2008a)

characterize the macromolecular properties of these derivatives. An important finding is that dendronization leads to an increase in sedimentation coefficient (with increasing dendritic generation), however, a decrease in intrinsic viscosity. It could be shown from sedimentation conformation zoning that the carboxymethyl deoxyazido cellulose and the carboxymethyl 6-deoxy-(1-N-[1,2,3-triazolo]-4-PAMAM) cellulose derivatives possess a semiflexible coil conformation. The values of persistence length obtained from combined analysis for these samples also indicate a semiflexible coil conformation in solution. Thus, dendronization of carboxymethylated deoxyazido cellulose has no or only a rather small influence on the chain stiffness and the conformation in solution.

Bulky dendrons can be attached to position 3 of the repeating unit by conversion of 3-mono-*O*-propargyl cellulose with azidopropyl-PAMAM

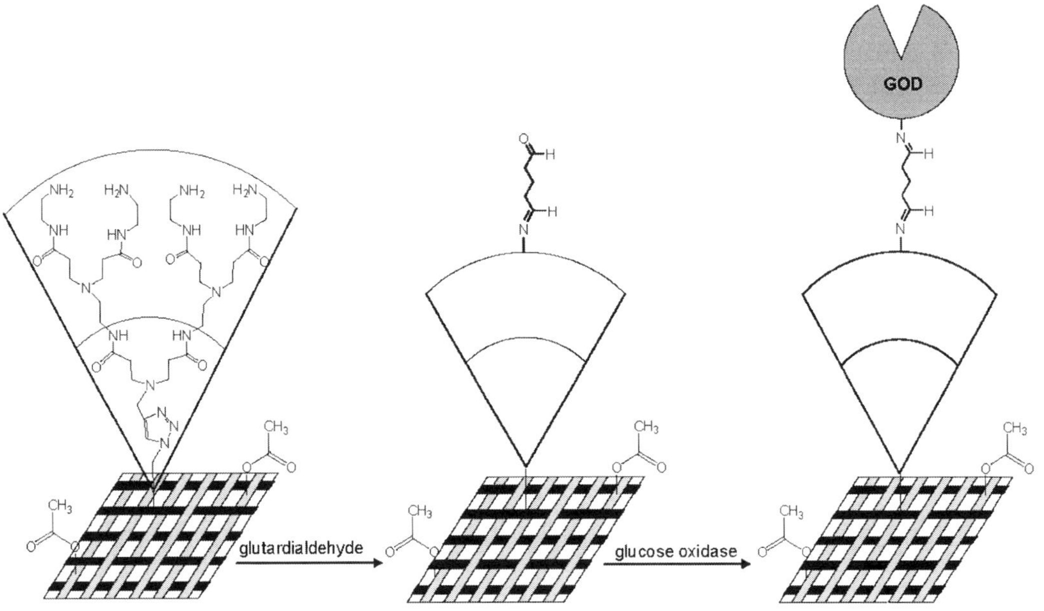

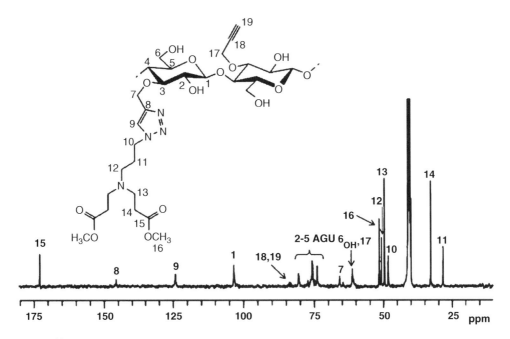

Fig. 10.40 Schematic preparation of a blend of 6-deoxy-6-(1,2,3-triazolo)-4-polyamidoamine cellulose (degree of substitution, DS 0.25) and cellulose acetate (DS 2.5) and subsequent surface activation with glutardialdehyde for covalent immobilization of glucose oxidase (GOD; *light gray bars* = cellulose acetate, *black bars* = 6-deoxy-6-(1,2,3-triazolo)-4-PAMAM cellulose) (Pohl et al. 2009b)

Fig. 10.41 ^{13}C NMR spectrum of 3-*O*-(4-methyl-1-*N*-propyl-polyamidoamine-(1,2,3-triazolo)) cellulose of first generation, recorded in dimethyl sulfoxide-d_6 at 60 °C (Fenn et al. 2009)

dendrons of the first and second generation (Fig. 10.41) (Fenn et al. 2009). However, the reactivity is comparably small, i.e., 25 % of the alkyne moieties were converted in case of the first generation and 13 % in case of the second generation.

Conclusions

The preparation of cellulose derivatives possessing sophisticated and well-defined structures is still a very challenging task to design the structure and hence the properties of products based on this unique biopolymer. Novel routes of the chemical modification includes (1) activation of the polymer by dissolution in suitable solvents, (2) conversion of the dissolved cellulose with reactive compounds, and (3) control of the functionalization pattern by using protecting groups. Derivatives resulting from such advanced synthesis strategies, e.g., homogeneous esterification, click reactions, oxidation or regioselective functionalization via selectively protected cellulose derivatives, and subsequent conversions, are valuable materials for the establishment of structure–property relationships and for applications in the medical and pharmaceutical field and in biotechnology or as high-value additives in different materials such as paper or tensides. There is still a lot of space to prepare novel cellulose derivatives both highly engineered ones and products of bulk applications using heterogeneous reaction paths that were not discussed in this chapter. From the authors' point of view, the use of cellulose as starting material contributes to a sustainable development that is one prerequisite for the well-being of humans in the future.

Acknowledgments We are indebted to Marcel Meiland, Marc Kostag, Jana Wotschadlo, and Torsten Jordan for their technical assistance. One of the authors (S. Coseri) acknowledges the financial support of European Social Fund—"Cristofor I. Simionescu" Postdoctoral Fellowship Programme (ID POSDRU/89/1.5/S/55216), Sectoral Operational Programme Human Resources Development 2007–2013.

References

Abbott AP, Capper G, Davies DL, Rasheed RK (2004) Ionic liquids based upon metal halide/substituted quaternary ammonium salt. Inorg Chem 43: 3447–3452

Abbott AP, Bell TJ, Handa S, Stoddart B (2005) O-acetylation of cellulose and monosaccharides using a zinc based ionic liquid. Green Chem 7: 705–707

Arisawa M, Kato C, Kaneko H, Nishida A, Nakagawa M (2000) Concise synthesis of azacycloundecenes using ring-closing metathesis (RCM). J Chem Soc Perkin Trans 1 12:1873–1876

Bailey WF, Bobbitt JM, Wiberg KB (2007) Mechanism of the oxidation of alcohols by oxoammonium cations. J Org Chem 72:4504–4509

Banker GS, Kumar V (1995) Microfibrillated oxycellulose. US Patent 5405953

Bar-Nir BB, Kadla JF (2009) Synthesis and structural characterization of 3-O-ethylene glycol functionalized cellulose derivatives. Carbohydr Polym 76:60–67

Barthel S, Heinze T (2006) Acylation and carbanilation of cellulose in ionic liquids. Green Chem 8:301–306

Biliuta G, Fras L, Strnad S, Harabagiu V, Coseri S (2010) Oxidation of cellulose fibers mediated by nonpersistent nitroxyl radicals. J Polym Sci A Polym Chem 48:4790–4799

Borisov IM, Shirokova EN, Mudarisova RK, Muslukhov RR, Zimin YS, Medvedeva SA, Tolstikov GA, Monakov YB (2004) Kinetics of oxidation of an arabinogalactan from larch (Larix sibirica L.) in an aqueous medium in the presence of hydrogen peroxide. Russ Chem Bull 53:318–324

Brackhagen M, Heinze T, Dorn S, Koschella A (2007) Verfahren zur Herstellung aminogruppenhaltiger Cellulosederivate in ionischer Flüssigkeit. EP Patent 2072530A1

Bragd PL, Van Bekkum H, Besemer AC (2004) TEMPO-mediated oxidation of polysaccharides: survey of methods and applications. Top Catal 27:49–66

Brewster JH, Ciotti CJ Jr (1955) Dehydrations with aromatic sulfonyl halides in pyridine. A convenient method for the preparation of esters. J Am Chem Soc 77:6214–6215

Butrim SM, Bil'dyukevich TD, Butrim NS, Yurkshtovich TL (2007) Structural modification of potato starch by solutions of nitrogen (IV) oxide in CCl4. Chem Nat Compd 43:302–305

Calvini P, Conio G, Lorenzoni M, Pedemonte E (2004) Viscometric determination of dialdehyde content in periodate oxycellulose. Part I. Methodology. Cellulose 11:99–107

Calvini P, Gorassini A, Luciano G, Franceschi E (2006) FTIR and WAXS analysis of periodate oxycellulose: evidence for a cluster mechanism of oxidation. Vib Spectrosc 40:177–183

Camacho Gomez JA, Erler UW, Klemm D (1996) O,4-methoxy substituted trityl groups in 6-O protection of cellulose: homogeneous synthesis, characterization, detritylation. Macromol Chem Phys 197:953–964

Camy S, Montanari S, Rattaz A, Vignon M, Condoret JS (2009) Oxidation of cellulose in pressurized carbon dioxide. J Supercrit Fluids 51:188–196

Cao Y, Wu J, Meng T, Zhang J, He JS, Li HQ, Zhang Y (2007) Acetone-soluble cellulose acetates prepared by one-step homogeneous acetylation of cornhusk cellulose in an ionic liquid 1-allyl-3-methylimidazolium chloride (AmimCl). Carbohydr Polym 69:665–672

Cato Y, Kaminaga J, Matsuo R, Isogai A (2004) TEMPO-mediated oxidation of chitin, regenerated chitin and N-acetylated chitosan. Carbohydr Polym 58:421–426

Chang PS, Robyt JF (1996) Oxidation of primary alcohol groups of naturally occurring polysaccharides with 2,2,6,6-tetramethyl-1-piperidine oxoammonium ion. J Carbohydr Chem 15:819–830

Ciacco GT, Liebert TF, Frollini E, Heinze TJ (2003) Application of the solvent dimethyl sulfoxide/tetrabutyl-ammonium fluoride trihydrate as reaction medium for the homogeneous acylation of Sisal cellulose. Cellulose 10:125–132

Coseri S (2009) Phthalimide-N-oxyl (PINO) radical, a powerful catalytic agent: its generation and versatility towards various organic substrates. Cat Rev 51:218–292

Coseri S, Mendenhall GD, Ingold KU (2005) Mechanisms of reactions of aminoxyl (nitroxide), iminoxyl, and imidoxyl radicals with alkenes and evidence that in the presence of lead tetraacetate, N-hydroxyphthalimide reacts with alkenes by both radical and nonradical mechanisms. J Org Chem 70:4629–4636

Coseri S, Nistor G, Fras L, Strnad S, Harabagiu V, Simionescu BC (2009) Mild and selective oxidation of cellulose fibers in the presence of N-hydroxyphthalimide. Biomacromolecules 10:2294–2299

Danaee I, Jafarian M, Mirzapoor A, Gobal F, Mahjani MG (2010) Electrooxidation of methanol on NiMn alloy modified graphite electrode. Electrochim Acta 55:2093–2100

Davis NJ, Flitsch SL (1993) Selective oxidation of monosaccharide derivatives to uronic acids. Tetrahedron Lett 34:1181–1184

De Nooy AEJ, Besemer AC, Van Bekkum H (1994) Highly selective TEMPO mediated oxidation of primary alcohol groups in polysaccharides. Recl Trav Chim Pays Bas 113:165–166

De Nooy AEJ, Besemer AC, Van Beckum H (1995) Highly selective nitroxyl radical-mediated oxidation of primary alcohol groups in water-soluble glucans. Carbohydr Res 269:89–98

Deus C, Friebolin H, Siefert E (1991) Partiell acetylierte Cellulose-Synthese und Bestimmung der Substituentenverteilung mit Hilfe der ^{1}H NMR Spektroskopie. Makromol Chem 192:75–83

Dias GJ, Peplow PV, Teixeira F (2003) Osseous regeneration in the presence of oxidized cellulose and collagen. J Mater Sci Mater Med 14:739–745

Dorn S, Schöbitz M, Schlufter K, Heinze T (2010a) Novel cellulose products prepared by homogeneous functionalization of cellulose in ionic liquids. ACS Symp Ser 1033:275–285

Dorn S, Pfeifer A, Schlufter K, Heinze T (2010b) Synthesis of water-soluble cellulose esters applying carboxylic acid imidazolides. Polym Bull 64:845–854

Edgar KJ, Arnold KM, Blount WW, Lawniczak JE, Lowman DW (1995) Synthesis and properties of cellulose acetoacetates. Macromolecules 28:4122–4128

Edgar KJ, Pecorini TJ, Glasser WG (1998) Long-chain cellulose esters: preparation, properties, and perspective. In: Heinze T, Glasser WG (eds) Cellulose derivatives; modification, characterization and nanostructures. ACS Symposium Series 688, American Chemical Society, Washington, DC, p 38

El Seoud OA, Heinze T (2005) Organic esters of cellulose: New perspectives for old polymers. In: Heinze T (ed) Advances in polymer science: polysaccharides I, structure, characterization and use, vol 186. Springer, Heidelberg, p 103ff

El Seoud OA, Marson GA, Ciacco GT, Frollini E (2000) An efficient, one-pot acylation of cellulose under homogeneous reaction conditions. Macromol Chem Phys 201:882–889

Fenn D, Heinze T (2009) Novel 3-mono-O-hydroxyethyl cellulose: synthesis and structure characterization. Cellulose 16:853–861

Fenn D, Pfeifer A, Heinze T (2007) Studies on the synthesis of 2,6-di-O-thexyldimethylsilyl cellulose. Cell Chem Technol 41:87–91

Fenn D, Pohl M, Heinze M (2009) Novel 3-O-propargyl cellulose as a precursor for regioselective functionalization of cellulose. React Funct Polym 69:347–352

Fenselau AH, Moffatt JG (1966) Sulfoxide-carbodiimide reactions. III. Mechanism of the oxidation reaction. J Am Chem Soc 88:1762–1765

Fras L, Johansson LS, Stenius P, Laine J, Stana-Kleinschek K, Ribitsch V (2005) Analysis of the oxidation of cellulose fibres by titration and XPS. Colloid Surf A 260:101–108

Galgut PN (1990) Oxidized cellulose mesh: I. Biodegradable membrane in periodontal surgery. Biomaterials 11:561–564

Gericke M, Liebert T, Heinze T (2009a) Interaction of ionic liquids with polysaccharides – 8. Synthesis of cellulose sulfate suitable for symplex formation. Macromol Biosci 9:343–353

Gericke M, Schlufter K, Liebert T, Heinze T, Budtova T (2009b) Rheological properties of cellulose/ionic liquid solutions: from dilute to concentrated states. Biomacromolecules 10:1188–1194

Gericke M, Liebert T, Heinze T (2009c) Polyelectrolyte complex formation in ionic liquids. J Am Chem Soc 131:13220–13221

Glasser WG, Samaranayake G, Dumay M, Dave VJ (1995) Novel cellulose derivatives. III. Thermal analysis of mixed esters with butyric and hexanoic acid. J Polym Sci B Polym Phys 33:2045–2054

Glasser WG, Becker U, Todd JG (2000) Novel cellulose derivatives. VI. Preparation and thermal analysis of two novel cellulose esters with fluorine-containing substituents. Carbohydr Polym 42:393–400

Gomez-Bujedo S, Fleury E, Vignon MR (2004) Preparation of cellouronic acids and partially acetylated cellouronic acids by TEMPO/NaClO oxidation of water-soluble cellulose acetate. Biomacromolecules 5:565–571

Gräbner D, Liebert T, Heinze T (2002) Synthesis of novel adamantoyl cellulose using differently activated carbonic acid derivatives. Cellulose 9:193–201

Graenacher C (1934) Cellulose solution. US Patent 1943176

Guthrie JD (1971) Ion-exchange celluloses. In: Bikales NM, Segal L (eds) Cellulose and cellulose derivatives. Wiley-Interscience, New York

Haslam E (1980) Recent developments in methods for the esterification and protection of the carboxyl group. Tetrahedron 36:2409–2433

Hassan ML (2006) Preparation and thermal stability of new cellulose-based poly(propylene imine) and poly (amido amine) hyperbranched derivatives. J Appl Polym Sci 101:2079–2087

Hassan ML, Moorefield CN, Newkome GR (2004) Regioselective dendritic functionalization of cellulose. Macromol Rapid Commun 25:1999–2002

Hassan ML, Moorefield CN, Kotta KK, Newkome GR (2005) Regioselective combinatorial-type synthesis, characterization, and physical properties of dendronized cellulose. Polymer 46:8947–8955

Heinze T, Glasser WG (1998) The role of novel solvents and solution complexes for the preparation of highly engineered cellulose derivatives. In: Heinze T, Glasser WG (eds) Cellulose derivatives, modification, characterisation and nanostructures. ACS Symposium series (688), p 2

Heinze T, Köhler S (2010) Dimethyl sulfoxide and ammonium fluorides – novel cellulose solvents. ACS Symp Ser 1033:103–118

Heinze T, Koschella A (2008) Water-soluble 3-O-methoxyethyl cellulose: synthesis and characterization. Carbohydr Res 343:668–673

Heinze T, Schaller J (2000) New water soluble cellulose esters synthesized by an effective acylation procedure. Macromol Chem Phys 201:1214–1218

Heinze T, Röttig K, Nehls I (1994) Synthesis of 2,3-O-carboxymethylcellulose. Macromol Rapid Commun 15:311–317

Heinze U, Heinze T, Klemm D (1999) Synthesis and structure characterization of 2,3-O-carboxymethylcellulose. Macromol Chem Phys 200:896–902

Heinze T, Dicke R, Koschella A, Kull AH, Klohr EA, Koch W (2000a) Effective preparation of cellulose derivatives in a new simple cellulose solvent. Macromol Chem Phys 201:627–631

Heinze T, Vieira M, Heinze U (2000b) New polymers based on cellulose. Lenzinger Ber 79:39–44

Heinze T, Liebert T, Pfeiffer K, Hussain MA (2003) Unconventional cellulose esters: synthesis, characterization, and structure property relations. Cellulose 10:283–296

Heinze T, Schwikal K, Barthel S (2005) Ionic liquids as reaction medium in cellulose functionalization. Macromol Biosci 5:520–525

Heinze T, Koschella A, Brackhagen M, Engelhardt J, Nachtkamp K (2006) Studies on non-natural deoxyammonium cellulose. Macromol Symp 244:74–82

Heinze T, Pohl M, Schaller M, Meister F (2007) Novel bulky esters of cellulose. Macromol Biosci 7:1225–1231

Heinze T, Schöbitz M, Pohl M, Meister F (2008a) Interactions of ionic liquids with polysaccharides: IV. Dendronization of 6-azido-6-deoxy cellulose. J Polym Sci A Polym Chem 46:3853–3859

Heinze T, Pfeifer A, Petzold K (2008b) Regioselective reaction of cellulose with tert-butyldimethylsilyl chloride in N,N-dimethyl acetamide/LiCl. BioResources 3:79–90

Heinze T, Pfeifer A, Sarbova V, Koschella A (2011) 3-O-Propyl cellulose: cellulose ether with exceptionally low flocculation temperature. Polym Bull 66:1219–1229

Hirota M, Tamura N, Saito T, Isogai A (2009a) Oxidation of regenerated cellulose with $NaClO_2$ catalyzed by TEMPO and NaClO under acid-neutral conditions. Carbohydr Polym 78:330–335

Hirota M, Tamura N, Saito T, Isogai A (2009b) Preparation of polyuronic acid from cellulose by TEMPO-mediated oxidation. Cellulose 16:841–851

Hirrien M, Chevillard C, Desbrieres J, Axelos MAV, Rinaudo M (1998) Thermogelation of methylcelluloses: new evidence for understanding the gelation mechanism. Polymer 39:6251–6259

Hon DN, Yan HJ (2001) Cellulose furoate. I. Synthesis in homogeneous and heterogeneous systems. J Appl Polym Sci 81:2649–2655

Huang K, Xia J, Li M, Lian J, Yang X, Lin G (2011) Homogeneous synthesis of cellulose stearates with different degrees of substitution in ionic liquid 1-butyl-3-methylimidazolium chloride. Carbohydr Polym 83:1631–1635

Husemann E, Siefert E (1969) N-äthyl-pyridiniumchlorid als Lösungsmittel und Reaktionsmedium für Cellulose. Makromol Chem 128:288–291

Hussain MA, Liebert T, Heinze T (2004a) Acylation of cellulose with N,N'-carbonyldiimidazole-activated acids in the novel solvent dimethyl sulfoxide/tetrabutylammonium fluoride. Macromol Rapid Commun 25:916–920

Hussain MA, Liebert T, Heinze T (2004b) First report on a new esterification method for cellulose. Polym News 29:14–17

Ibrahim AA, Nada AMA, Hagemann U, El Seoud OA (1996) Preparation of dissolving pulp from sugar cane bagasse, and its acetylation under homogeneous solution condition. Holzforschung 50:221–225

Illy N (2006) Diploma Thesis, Friedrich Schiller University of Jena

Isogai A, Kato Y (1998) Preparation of polyuronic acid from cellulose by TEMPO-mediated oxidation. Cellulose 5:153–164

Isogai T, Saito T, Isogai A (2010) TEMPO electromediated of some polysaccharides including regenerated cellulose fiber. Biomacromolecules 11:1593–1599

Itagaki H, Tokai M, Kondo T (1997) Physical gelation process for cellulose whose hydroxyl groups are regioselectively substituted by fluorescent groups. Polymer 38:4201–4205

Iwata T, Azuma J, Okamura K, Muramoto M, Chun B (1992) Preparation and n.m.r. assignments of cellulose mixed esters regioselectively substituted by acetyl and propanoyl groups. Carbohydr Res 224:277–283

Jin B, Wu W (2005) Compositions for veterinary and medical applications. WO Patent 2005020997

Johansson EE, Lind J (2005) Free radical mediated cellulose degradation during high consistency ozonation. J Wood Chem Technol 25:171–186

Just EK, Majewicz TG (1985) Cellulose ethers. In: Mark HF, Bikales NM, Overberger CG, Menges G, Kroschwitz JI (eds) Encyclopedia of polymer science and engineering (2nd ed). Wiley, New York, p III/226ff

Kadla JF, Asfour FH, Bar-Nir B (2007) Micropatterned thin film honeycomb materials from regiospecifically modified cellulose. Biomacromolecules 8:161–165

Kamitakahara H, Koschella A, Mikawa Y, Nakatsubo F, Heinze T, Klemm D (2008) Syntheses and comparison of 2,6-Di-O-methyl celluloses from natural and synthetic celluloses. Macromol Biosci 8:690–700

Kato T, Miyagawa A, Carmelita M, Kasuya Z, Hatanaka K (2009) Development of membrane filter with oligosaccharide immobilized by click chemistry for influenza virus adsorption. Open Chem Biomed Methods J 2:13–17

Katritzky AR, Zhang Y, Singh SK (2003) Efficient conversion of carboxylic acids into N-acylbenzotriazoles. Synthesis 18:2795–2798

Katritzky AR, Cai C, Singh SK (2006) Efficient microwave access to polysubstituted amidines from imidoylbenzotriazoles. J Org Chem 71:3375–3380

Kern H, Choi S, Wenz G, Heinrich J, Erhardt L, Mischnik P, Garidel P, Blume A (2000) Synthesis, control of substitution pattern and phase transitions of 2,3-di-O-methylcellulose. Carbohydr Res 326:67–79

Kitaoka T, Isogai A, Onabe F (1999) Chemical modification of pulp fibers by TEMPO-mediated oxidation. Nord Pulp Pap Res J 14:279–284

Klemm D, Stein A (1995) Silylated cellulose materials in design of supramolecular structures of ultrathin cellulose films. J Macromol Sci A Pure Appl Chem A32:899–904

Klemm D, Schnabelrauch M, Stein A, Philipp B, Wagenknecht W, Nehls I (1990) Recent results from homogeneous esterification of cellulose using soluble intermediate compounds. Das Papier 44:624–632

Klemm D, Philip B, Heinze T, Heinze U, Wagenknecht W (1998) Cellulose sulfates. In: Comprehensive cellulose chemistry, vol 2, Wiley VCH, New York, pp 115–133

Klohr EA, Koch W, Klemm D, Dicke R (2000) Regioselektiv substituierte Ester von Oligo- und Polysacchariden und Verfahren zu ihrer Herstellung. DE Patent 19951734, CAN Patent 133:224521

Köhler S, Heinze T (2007) Efficient synthesis of cellulose furoates in 1-N-butyl-3-methylimidazolium chloride. Cellulose 14:489–495

Köhler S, Liebert T, Schöbitz M, Schaller J, Meister F, Günther W, Heinze T (2007) Interactions of ionic liquids with polysaccharides - 1: unexpected acetylation of cellulose with 1-ethyl-3-methylimidazolium acetate. Macromol Rapid Commun 28:2311–2317

Kolb HC, Finn MG, Sharpless KB (2001) Click-Chemie: diverse chemische Funktionalität mit einer Handvoll guter Reaktionen. Angew Chem 113:2056–2075

Kondo T (1993) Preparation of 6-O-alkylcelluloses. Carbohydr Res 238:231–240

Kondo T (1997) The relationship between intramolecular hydrogen bonds and certain physical properties of regioselectively substituted cellulose derivatives. J Polym Sci B Polym Phys 35:717–723

Kondo T, Gray DG (1991) The preparation of O-methyl- and O-ethyl-celluloses having controlled distribution of substituents. Carbohydr Res 220:173–183

Kondo T, Sawatari C, Manley R, St J, Gray DG (1994) Characterization of hydrogen bonding in cellulose-synthetic polymer blend systems with regioselectively substituted methylcellulose. Macromolecules 27:210–215

Kondo T, Koschella A, Heublein B, Klemm D, Heinze T (2008) Hydrogen bond formation in regioselectively functionalized 3-mono-O-methyl cellulose. Carbohydr Res 343:2600–2604

Koschella A, Klemm D (1997) Silylation of cellulose regiocontrolled by bulky reagents and dispersity in the reaction media. Macromol Symp 120:115–125

Koschella A, Haucke G, Heinze T (1997) New fluorescence active cellulosics prepared by a convenient acylation reaction. Polym Bull 39:597–604

Koschella A, Heinze T, Klemm D (2001) First synthesis of 3-O-functionalized cellulose ethers via 2,6-di-O-protected silyl cellulose. Macromol Biosci 1:49–54

Koschella A, Fenn D, Heinze T (2006) Water soluble 3-mono-O-ethyl cellulose: synthesis and characterization. Polym Bull 57:33–41

Koschella A, Richter M, Heinze T (2010) Novel cellulose-based polyelectrolytes synthesized via click reaction. Carbohydr Res 345:1028–1033

Kostic M, Potthast A, Rosenau T, Kosma P, Sixta H (2006) A novel approach to determination of carbonyl groups in DMAc/LiCl-insoluble pulps by fluorescence labeling. Cellulose 13:429–435

Liebert T, Heinze T (2005) Tailored cellulose esters: synthesis and structure determination. Biomacromolecules 6:333–340

Liebert T, Hänsch C, Heinze T (2006) Click chemistry with polysaccharides. Macromol Rapid Commun 27:208–213

Liu H-Q, Zhang L-N, Takaragi A, Miyamoto T (1997) Water solubility of regioselectively 2,3-O-substituted carboxymethylcellulose. Macromol Rapid Commun 18:921–925

Liu CF, Sun RC, Zhang AP, Ren JL, Wang XA, Geng ZC (2006) Structural and thermal characterization of sugarcane bagasse cellulose succinates prepared in ionic liquid. Polym Degrad Stabil 91:3040–3047

Liu CF, Sun RC, Zhang AP, Qin M-H, Ren J-L, Wang X-A (2007a) Preparation and characterization of phthalated cellulose derivatives in room-temperature ionic liquid without catalysts. Agricult Food Chem 55:2399–2406

Liu CF, Sun RC, Zhang AP, Ren JL, Wang XA, Qin MH, Chaod ZN, Luod W (2007b) Homogeneous modification of sugarcane bagasse cellulose with succinic anhydride using a ionic liquid as reaction medium. Carbohydr Res 342:919–926

Manhas MS, Mohammed F, Khan Z (2007) A kinetic study of oxidation of β-cyclodextrin by permanganate in aqueous media. Colloid Surf 295:165–171

McCormick CL, Callais PA (1987) Derivatization of cellulose in lithium chloride and N-N-dimethylacetamide solutions. Polymer 28:2317–2323

McCormick CL, Lichatowich DK (1979) Homogeneous solution reactions of cellulose, chitin, and other polysaccharides to produce controlled-activity pesticide systems. J Polym Sci C Polym Lett 17:479–484

Nagel MCV, Heinze T (2010) Esterification of cellulose with acyl-1 H-benzotriazole. Polym Bull 65:873–881

Nojiri M, Kondo T (1996) Application of regioselectively substituted methylcelluloses to characterize the reaction mechanism of cellulase. Macromolecules 29:2392–2395

Östmark E, Lindqvist J, Nyström D, Malmström E (2007) Dendronized hydroxypropyl cellulose – synthesis and characterization of biobased nanoobjects. Biomacromolecules 8:3815–3822

Painter TJ (1977) Preparation and periodate oxidation of C-6-oxycellulose: conformational interpretation of hemiacetal stability. Carbohydr Res 55:95–103

Pameijer CH, Jensen S (2007) Agents and devices for providing blood clotting functions to wounds. US Patent 20070190110

Pawlowski WP, Sankar SS, Gilbert RD, Fornes RD (1987) Synthesis and solid state ^{13}C-NMR studies of some cellulose derivatives. J Polym Sci A Polym Chem 25:3355–3362

Petzold K, Klemm D, Heublein B, Burchard W, Savin G (2004) Investigations on structure of regioselectively functionalized celluloses in solution exemplified by using 3-O-alkyl ethers and light scattering. Cellulose 11:177–193

Philipp B, Wagenknecht W, Wagenknecht M, Nehls I, Klemm D, Stein A, Heinze T, Heinze U, Helbig K (1995) Regioselective esterification and etherification of cellulose and cellulose derivatives. 1. Problems and description of the reaction systems. Das Papier 49:3–7

Pohl M, Heinze T (2008) Novel biopolymer structures synthesized by dendronization of 6-deoxy-6-aminopropargyl cellulose. Macromol Rapid Commun 29:1739–1745

Pohl M, Schaller J, Meister F, Heinze T (2008a) Novel bulky esters of biopolymers: dendritic cellulose. Macromol Symp 262:119–128

Pohl M, Schaller J, Meister F, Heinze T (2008b) Selectively dendronized cellulose: synthesis and characterization. Macromol Rapid Commun 29:142–148

Pohl M, Morris GA, Harding SE, Heinze T (2009a) Studies on the molecular flexibility of novel dendronized carboxymethyl cellulose derivatives. Eur Polym J 45:1098–1110

Pohl M, Michaelis N, Meister F, Heinze T (2009b) Biofunctional surfaces based on dendronized cellulose. Biomacromolecules 10:382–389

Potthast A, Rohrling J, Rosenau T, Borgards A, Sixta H, Kosma P (2003) A novel method for the determination of carbonyl groups in cellulosics by fluorescence labeling. 3 Monitoring oxidative processes. Biomacromolecules 4:743–749

Regiani AM, Frollini E, Marson GA, Arantes GM, El Seoud OA (1999) Some aspects of acylation of cellulose under homogeneous solution conditions. J Polym Sci A Polym Chem 37:1357–1363

Ringot C, Sol V, Granet R, Krausz P (2009) Porphyrin-grafted cellulose fabric: new photobactericidal material obtained by "Click-Chemistry" reaction. Mater Lett 63:1889–1891

Saito T, Isogai A (2004) TEMPO-mediated oxidation of native cellulose. The effect of oxidation conditions on chemical and crystal structures of the water-insoluble fractions. Biomacromolecules 5:1983–1989

Samaranayake G, Glasser WG (1993) Cellulose derivatives with low DS. I. A novel acylation system. Carbohydr Polym 22:1–7

Schaller J, Heinze T (2005) Studies on the synthesis of 2,3-O-hydroxyalkyl ethers of cellulose. Macromol Biosci 5:58–63

Schlufter K, Schmauder HP, Dorn S, Heinze T (2006) Efficient homogeneous chemical modification of bacterial cellulose in the ionic liquid 1-N-butyl-3-methylimidazolium chloride. Macromol Rapid Commun 27:1670–1676

Schumann K, Pfeifer A, Heinze T (2009) Novel cellulose ethers: synthesis and structure characterization of 3-mono-o-(3'-hydroxypropyl) cellulose. Macromol Symp 280:86–94

Schweiger RG (1972) Polysaccharide sulfates. I. Cellulose sulfate with a high degree of substitution. Carbohydr Res 21:219–228

Sealey JE, Samaranayake G, Todd JG, Glasser WG (1996) Novel cellulose derivatives. IV. Preparation and thermal analysis of waxy esters of cellulose. J Polym Sci B Polym Phys 34:1613–1620

Sealey JE, Frazier CE, Samaranayake G, Glasser WG (2000) Novel cellulose derivatives. V. Synthesis and thermal properties of esters with trifluoroethoxy acetic acid. J Polym Sci B Polym Phys 38:486–494

Sharma JB, Malhotra M (2003) Topical oxidized cellulose for tubal hemorrhage hemostasis during laparoscopic sterilization. Int J Gynecol Obstet 82:221–222

Sharma JB, Malhotra M, Pundir P (2003) Laparoscopic oxidized cellulose (Surgicel) application for small uterine perforations. Int J Gynecol Obstet 83:271–275

Shimizu Y, Hayashi J (1988) A new method for cellulose acetylation with acetic acid. Sen-i Gakkaishi 44:451–456

Shimizu Y, Nakayama A, Hayashi J (1991) Acetylation of cellulose with carboxylate salts. Cell Chem Technol 25:275–281

Staab HA (1962) Neuere Methoden der präperativen organischen Chemie IV Synthesen mit heterocyclischen Amiden (Azoliden). Angew Chem 74:407–423

Stadler PA (1978) Eine einfache Veresterungsmethode im Eintopf-Verfahren. Helv Chim Acta 61:1675–1681

Sun H, DiMagno SG (2005) Anhydrous tetrabutylammonium fluoride. J Am Chem Soc 127:2050–2051

Sun S, Foster TJ, MacNaughtan W, Mitchell JR, Fenn D, Koschella A, Heinze T (2009) Self-association of cellulose ethers with random and regioselective distribution of substitution. J Polym Sci B Polym Phys 47:1743–1752

Swatloski RP, Spear SK, Holbrey JD, Rogers RD (2002) Dissolution of cellulose with ionic liquids. J Am Chem Soc 124:4974–4975

Tahiri C, Vignon MR (2000) TEMPO-oxidation of cellulose: synthesis and characterisation of polyglucuronans. Cellulose 7:177–188

Takaragi A, Minoda M, Miyamoto T, Liu HQ, Zhang LN (1999) Reaction characteristics of cellulose in the LiCl/1,3-dimethyl-2-imidazolidinone solvent system. Cellulose 6:93–102

Terbojevich M, Cosani A, Focher B, Gastaldi G, Wu W, Marsano E, Conio G (1999) Solution properties and mesophase formation of 4-phenyl-benzoylcellulose. Cellulose 6:71–87

Wagenknecht W (1996) Regioselectively substituted cellulose derivatives by modification of commercial cellulose acetates. Das Papier 50:712–720

Wagenknecht W, Phillip B, Keck M (1985) Zur Acylierung von Cellulose nach Auflösung in O-basischen Lösemittelsystemen. Acta Polym 36:697–698

Wang Z-M, Li L, Xiao K-J, Wu J-Y (2009) Homogeneous sulfation of bagasse cellulose in an ionic liquid and anticoagulation activity. Bioresour Technol 4:1687–1690

Williamson SL, McCormick CL (1998) Cellulose derivatives synthesized via isocyanate and activated ester pathways in homogeneous solutions of lithium chloride N,N-dimethylacetamide. J Macromol Sci A Pure Appl Chem A35:1915–1927

Wiseman DM, Saferstein L, Wolf S (2002) Bioresorbable oxidized cellulose composite material for prevention of postsurgical adhesions. US Patent 6500777

Wu J, Zhang J, Zhang H, He J, Ren Q, Guo M (2004) Homogeneous acetylation of cellulose in a new ionic liquid. Biomacromolecules 5:266–268

Yoshida Y, Yanagisawa M, Isogai A, Suguri N, Sumikawa N (2005) Preparation of polymer brush-type cellulose β-ketoesters using LiCl/1,3-dimethyl-2-imidazolidinone as a solvent. Polymer 46:2548–2557

Zhang C, Daly WH (2005) Synthesis and characterization of a trifunctional amidoamine cellulose derivative. Polymer Prepr 46:707–710

Zhang ZB, McCormick CL (1997) Structopendant unsaturated cellulose esters via acylation in homogeneous lithium chloride/N,N-dimethylacetamide solutions. J Appl Polym Sci 66:293–305

Zhang H, Wu J, Zhang J, He JS (2005) 1-Allyl-3-methylimidazolium chloride room temperature ionic liquid: a new and powerful nonderivatizing solvent for cellulose. Macromolecules 38:8272–8277

Zhang C, Price LM, Daly HD (2006) Synthesis and Characterization of a trifunctional amidoamine cellulose derivative. Biomacromolecules 7:139–145

Zhang J, Wu J, Cao Y, Sang S, Zhang J, He J (2009a) Synthesis of cellulose benzoates under homogeneous conditions in an ionic liquid. Cellulose 16:299–308

Zhang Y, Zhang L, Shuang S, Feng F, Qiao J, Guo Y, Choi MMF, Dong C (2010) Electro-oxidation of methane on roughened palladium electrode in acidic electrolytes at ambient temperatures. Anal Lett 43:1055–1065

Zhao GL, Hafren J, Deiana L, Cordova A (2010) Heterogeneous "Organoclick" Derivatization of Polysaccharides: Photochemical Thiol-ene Click Modification of Solid Cellulose. Macromol Rapid Commun 31:740–744

Zimnitski DS, Yurkshtovich TL, Bychkovsky PM (2004) Synthesis and characterization of oxidized cellulose. J Polym Sci A Polym Chem 42:4785–4791

Chitin and Chitosan as Functional Biopolymers for Industrial Applications

Iwona Kardas, Marcin Henryk Struszczyk, Magdalena Kucharska, Lambertus A.M. van den Broek, Jan E.G. van Dam, and Danuta Ciechańska

Contents

11.1 Chitin and Chitosan 329
11.2 Chitin/Chitosan Processing, Isolation, and Purification 330
11.2.1 General Procedure for the Isolation of Chitin 330
11.2.2 General Procedure for the Isolation of Chitosan 333
11.2.3 Alternative Methods to Isolate Chitin/Chitosan 334
11.2.4 Microorganisms and Their Application in the Isolation of Chitin/Chitosan 336
11.2.5 Isolation of Chitin/Chitosan from Other Sources as Crustacean Waste 337
11.2.6 Future 338

11.3 Applicable Forms of Chitin and Chitosan 338
11.3.1 Chitosan Derivatives 339
11.3.2 Applicable Forms 340

11.4 Application of Chitin and Chitosan 343
11.4.1 Food and Feed 343
11.4.2 Medical and Pharmaceutical Application 347
11.4.3 Bionanotechnology 358
11.4.4 Agriculture 358
11.4.5 Cosmetics 362

11.5 Future Trends in Chitin and Its Derivatives in Medicine 363

11.6 Research on Chitin/Chitosan Within EPNOE 364

References 364

D. Ciechańska (✉)
Institute of Biopolymers and Chemical Fibres, Marie Sklodowskiej-Curie 19/27, 90-570 Łódź, Poland
e-mail: dciechan@ibwch.lodz.pl

Abstract

Chitin research and development seems to be under intensive progress during the last years. Attractive properties of chitin and its derivative—chitosan, for example, biological behavior, and development of their applications caused increased interest of scientists and companies. More and more practical development has been applied on the market, mainly in medicine and food. Firstly, this chapter reviews processing, isolation, and purification of chitin/chitosan. Then it discusses applicable forms of chitin and chitosan and chitosan derivatives. The main part of this chapter contains applications of chitin and chitosan in food, medicine, agriculture, bionanotechnology, and cosmetics. Future trends in chitin and its derivatives in medicine are presented and the end of this book includes research conducted within the EPNOE.

11.1 Chitin and Chitosan

Chitin is a polymer of $[\beta\text{-}(1\rightarrow 4)\text{-}2\text{-acetamido-}2\text{-deoxy-}D\text{-glucopyranose}]_n$ residues having up to 1,000–3,000 repeat units. It shows structural similarity to cellulose, except that the C(2)-hydroxyl group is exchanged by an acetamide group. Chitin is one of the most abundant biological materials in the world and it is after cellulose, the second most biosynthesized polymer (Li et al.

1992; Struszczyk 2002a, b, c). The occurrence of chitin in nature is shown in Fig. 11.1. Crustacean shells production has been estimated by Teng et al. (2001) to be approximately 1.2×10^9 kg and this is the main source for chitin.

Chitin, like cellulose, is characterized by a highly ordered structure with an excess of crystalline regions. It is present in three polymorphic forms: α, β, and γ-chitin which differ in their arrangement of the polymer chains into crystalline regions. The polymer chains are antiparallel in α-chitin, whereas in β-chitin they are parallel. In γ-chitin, two of the three chains are parallel and the third is antiparallel. The chains associate with each other by very strong hydrogen bonding between the amide groups and the carbonyl groups of the nearby chains. These intra- and intermolecular hydrogen bonds are responsible for the high insolubility of chitin. Compared with the most common α-chitin, β-chitin is distinguished by its loose packing of molecules because of their arrangement in a parallel fashion (Roberts 1992a, b, c, d; Kurita et al. 1997). Therefore, β-chitin shows a higher susceptibility to enzymatic degradation as well as chemical and enzymatic N-deacetylation.

Chitin can be converted to chitosan by deacetylation. When the degree of deacetylation is below 50 %, it becomes soluble in aqueous media and it is called chitosan. In general, the degree of N-acetylation of chitosan is between 0.05 and 0.30 according to Mima et al. (1983).

Chitin and chitosan are presently mainly manufactured from crustacean wastes in the USA, Japan, India, and, to a lesser extent, Malaysia, Canada, China, South Korea, Russia, Norway, and Kenya and in EU countries like France and Germany, according to Struszczyk (2002a, b, c). The production has some aspects to be taken into account such as:

- Chitinous organisms, mainly crabs, shrimps, prawns, lobsters, and krills, are very abundant all over the world, and only a small amount of natural resources are exploited by the marine food industry.
- Manufacturing of chitin is usually carried out as a secondary activity related to the marine food industry.
- The chitinous waste from the marine industry constitute a high natural environmental load.

The preparation of chitin and chitosan from insects or fungi does not involve the above-described limitations. However, the accessibility of sufficient amounts of raw material for industrial application is still a problem. Economic restrictions make it necessary to connect this production with other applications. In this way, insect chitin and chitosan can be manufactured, similar to crustacean biopolymers, as coproducts (Struszczyk 2002a, b, c; InfoFish Shrimp Wastes Utilisation Technical Handbook 1997).

11.2 Chitin/Chitosan Processing, Isolation, and Purification

Chitin, especially originating from crustacean food product wastes like crab and shrimp shells, is tightly bound in complexes with other substances and harsh conditions are needed for their isolation. In general, chitin is prepared using acid hydrolysis also known as acidic or chemical treatment. Subsequently, an alkaline treatment is performed for removing proteins and a decolorization step is included to remove pigments and lipids. The order of the different steps may be reversed depending on the application of chitin. The obtained chitin can be converted into chitosan by another alkaline treatment which results in the (partial) deacetylation of chitin (Fig. 11.2). In some cases, an enzymatic treatment is used instead of an acidic and/or alkaline treatment (Shahidi et al. 2005; Kurita 2006; Synowiecki and Al-Khateeb 2003; Muffler and Ulber 2005).

11.2.1 General Procedure for the Isolation of Chitin

Despite the widespread occurrence of chitin (e.g., shells of crustaceans; shells and skeletons of mollusks, krill, and insects; and the cell walls of fungi), the main commercial sources up to

11 Chitin and Chitosan as Functional Biopolymers for Industrial Applications

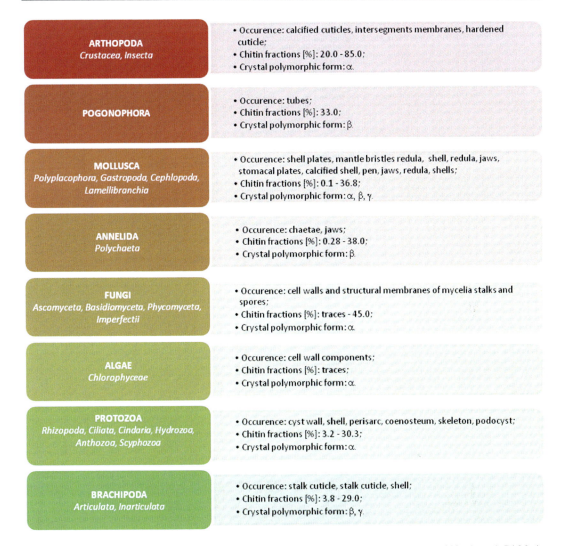

Fig. 11.1 Occurrence of chitin in nature (Struszczyk 2002a, b, c; Teng et al. 2001; Roberts 1992a, b, c, d; Di Mario et al. 2008)

now have been crustacean wastes. Struszczyk (2002a, b, c) reported that the common source for crustacean chitin is krill. The total amount of this crustacean specimen in nature is annually estimated up to 800 million tons. Taking into the account the economical aspect of the production, the most important species is Antarctic krill (*Euphausia superba*). Another important source for chitin manufacture is the deep-water shrimp (*Pandalus borealis*) being caught in large quantities (~700,000 kg annually), crab (*Scylla serrata*), lobster (*Panulirus ornatus*), or prawn (*Penaeus indicus*) (Stenberg and Wachter 1996; Oduor-Odote et al. 2005; Struszczyk et al. 1997a, b).

In general, the raw material is first washed in (boiling) water to remove soluble organics and adherent proteins. After drying the material, it is generally ground to 2–5 mm particle size, although smaller particle size will improve the extraction process, as described by Cosio et al. (1982). The decalcification or demineralization step is performed to dissolve the calcium and magnesium carbonates and phosphates present in the samples. Commonly used conditions are 0.25–2M HCl at 0–100 °C for 1–48 h (Acosta

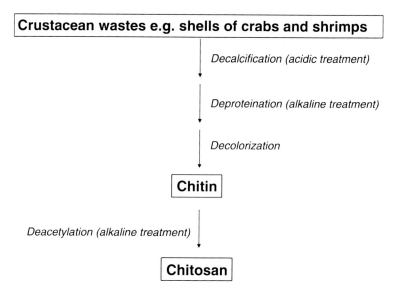

Fig. 11.2 Schematic overview of the general procedure for isolation of chitin and chitosan from crustacean wastes

et al. 1993; Tolaimate et al. 2003; Naznin 2005). In some procedures, repeated acid treatments are applied. The solid-to-solvent ratio is in the range of 1:10 till 1:15 (w/v) (Acosta et al. 1993; Tolaimate et al. 2003; Shahidi and Synowiecki 1991; No et al. 1989). Percot et al. (2003a, b) reported that the demineralization of shrimp shells was completed within 15 min using 0.25 M HCl at ambient temperature with a solid-to-solvent ratio of 1:40. The material is neutralized by washing with water. The extraction has also been reported (No et al. 1989; Oduor-Odote et al. 2005; Rødde et al. 2008; Peniston and Johnson 1980) using other inorganic and organic acids solutions, such as formic acid, nitric acid, or sulfuric acid, as demineralization agents with vigorous agitation at temperatures 0–100 °C for 1–4 h. It was shown by Mahmoud et al. (2007) that demineralization of deproteinated Northern pink shrimp (*P. borealis*) was similar by using lactic acid or acetic acid instead of HCl. The amount of chitin extracted from deproteinated black tiger shrimp (*Penaeus monodon*) was equal using 0.25 M formic acid/0.25 M citric acid in comparison with demineralization with 0.25 N HCl, according to Charoenvuttitham et al. (2006). To prevent the risk of chitin depolymerization, ambient temperature is favored (Oduor-Odote et al. 2005; Rødde et al. 2008). An alternative approach is the use of ethylenediaminetetraacetic acid (EDTA) that can be used for removal of minerals salts. The total amount of residual inorganic substances is usually defined as ash content. It is the most important indication of the purity of chitin and its derivatives.

In general, the deproteination is performed in 0.125–2.5M NaOH at 65–100 °C for 1–72 h, and in some cases, the treatment is repeated. The solid-to-solvent ratio varies from 1:10 till 1:20. Chemicals like potassium hydroxide, sodium carbonate, potassium carbonate, calcium hydroxide, or sodium sulfate at various concentrations in aqueous solutions have also been applied (Shahidi and Synowiecki 1991; Roberts 1992b). After extraction, the material is neutralized by washing with water (Tolaimate et al. 2003; Naznin 2005; Shahidi and Synowiecki 1991; No et al. 1989; Percot et al. 2003a, b).

Carotenoids and pigments recovery can be important for the economical feasibility of chitin isolation. In this case, the deproteination step is performed before the acid treatment to obtain intact pigments like astaxanthin. For this purpose, solvent extraction can be done with, for example, acetone, chloroform, ethyl acetate, and ethanol, to recover pigments, melanins, and carotenoids. If decolorization of the material is

Table 11.1 Chitin yield from crustacean shell wastes

Source	Type of chitin[a]	Yield (%)[b]	References
American crayfish (*Astacus cambarus*)	n.d.	14	Kurita et al. (1997)
(Spanish) crayfish (*Astacus fluviatilis*)	–, α	15, 36	Kurita et al. (1997) and Mima et al. (1983)
Snow crab (*Chionoecetes opilio*)	n.d.	27	InfoFish Shrimp Wastes Utilisation Technical Handbook (1997)
Shrimp (*Crangon crangon*)	n.d.	18	Struszczyk and Peter (1998)
Krill (*Euphausia superba*)	n.d.	24	Cosio et al. (1982)
Marbled crab (*Grapsus marmoratus*)	α	10	Mima et al. (1983)
Lobster (*Homarus vulgaris*)	α	17	Mima et al. (1983)
Barnacle (*Lepas anatifera*)	α	7	Mima et al. (1983)
Squid (*Loligo vulgaris*)	β	40, 42	Acosta et al. (1993) and Mima et al. (1983)
Spider crab (*Maia squinado*)	α	16	Mima et al. (1983)
Shrimp (*Metapenaeus monoceros*)	n.d.	23	Di Mario et al. (2008)
Norway lobster (*Nephrops norvegicus*)	n.d.	24	Kurita et al. (1997)
Shrimp (*Pandalus borealis*)	n.d.	17	InfoFish Shrimp Wastes Utilisation Technical Handbook (1997)
Shrimp (*Palaemon fabricius*)	α	22	Mima et al. (1983)
Lobster (*Palinurus vulgaris*)	–, α	14	Kurita et al. (1997) and Mima et al. (1983)
Prawn (*Penaeus caramote*)	n.d.	23	Kurita et al. (1997)
Crawfish (*Procambarus clarkii*)	n.d.	24	Shahidi and Synowiecki (1991)
Red crab (*Portunus puber*)	α	10	Mima et al. (1983)
Cuttlefish (*Sepia officinalis*)	β	20	Mima et al. (1983)
Squilla (*Squilla mantis*)	–, α	24, 24	Kurita et al. (1997) and Mima et al. (1983)

[a] n.d. is not determined
[b] Dry weight

needed, it can be oxidized with $KMnO_4$ with or without H_2O_2, NaClO/HCl, or H_2O_2/HCl which will result in bleaching (Muffler and Ulber 2005; Acosta et al. 1993; No et al. 1989).

In Table 11.1, some examples of isolated chitin from crustacean shell wastes are presented. The yield varies from 7 to 42 % dry weight, depending on peeling conditions during processing, as well as the part of the organism; state of development and nutrition of the organism; and stage or reproductive cycle and may vary from batch to batch as described by Synowiecki and Al-Khateeb (2003). In addition, the procedure of the isolation process plays also a very important role in the quality and yield of chitin. However, for shrimp shells from *P. borealis*, hardly any seasonal variations in their chemical composition were found and also no differences in molecular weight of the isolated chitin polymers were observed by Rødde et al. (2008).

11.2.2 General Procedure for the Isolation of Chitosan

The chitin obtained after demineralization, deproteination, and decolorization can be used for the production of chitosan. The main process for chitosan production is *N*-deacetylation directly from chitin using vigorous conditions because of the difficulty to break amide bonds under standard conditions. Acetamide groups adjacent to *trans* hydroxyl groups are much more resistant to the *N*-deacetylation than *cis*-related analogues; thus chitin having 2,3-*trans* arrangements is remarkably stable to most reagents, according to Muzzarelli (1977). Many studies (Methacanona et al. 2003; Castelli et al. 1997) have shown that the reaction of *N*-deacetylation of α-chitin and β-chitin follows the pseudo-first-order kinetics during an initial period. However, under various concentration of alkali in heterogeneous *N*-deacetylation of shrimp chitin, the reaction appeared to be more complicated and cannot be described by pseudo-first-order kinetics. Moreover, Gagne and Simpson (1993) demonstrated that it might be controlled by a higher-order reaction and a diffusion controlled reaction.

Chitin is treated with base so that partially deacetylation will occur resulting in chitosan. Depending on the chitin source 30–60 % (w/v) aqueous NaOH or KOH is used at 60–140 °C. To prevent polymer degradation, NaBH$_4$ and thiophenol can be added and the extraction is preferably done under nitrogen (Synowiecki and Al-Khateeb 2003; Tolaimate et al. 2003; Naznin 2005). It is also possible to obtain chitosan by treating chitin with 50 % (w/v) aqueous NaOH at 15 psi/121 °C for 15 min (Tajik et al. 2008; Youn et al. 2009). Another approach, presented by Tolaimate et al. (2003), is the addition of chitin in a mixture of 50 % (w/v) solid KOH in 96 % ethanol 25 % (w/v) and monoethlyene glycol 25 % (w/v). Some authors, like Yaghobi and Hormozi (2010) and Batista and Roberts (1990), proposed to add an organic solvent to the alkaline medium, such as isopropanol, 2-methyl-2-propanol, or acetone. However, Struszczyk (2002a, b, c) demonstrated that the yield of N-deacetylation and molecular weight is lower than obtained with unmodified aqueous sodium hydroxide alone. In industrial facilities, sodium hydroxide is mainly used as N-deacetylation agent with concentrations from 30 to 50 % (w/v) at temperatures of 120–150 °C. The effect of the various factors of alkaline N-deacetylation of chitin on the final properties of chitosan is described in Table 11.2.

As described in Table 11.2, the alkaline N-deacetylation of chitin proceeds rapidly until the polymer is N-deacetylated for more than 75 %. Prolonged incubation time has only a limited effect on the extent of deacetylation, but has significant influence on the crystallinity and molecular weight of chitosan (Struszczyk et al. 1997a, b, 1998). Furthermore, it can be concluded that to obtain a highly N-deacetylated product, it is necessary to provide N-deacetylation at temperature not higher than 100 °C for 1 h during multiple treatments. This can be more effective than a single treatment in a similar procedure for the total time; however, the final effect is strongly dependent on the initial parameters (e.g., degree of N-acetylation) of chitin as well as its origin (Roberts 1992b; Struszczyk et al. 1997a, b).

Tsigos et al. (2000) demonstrated that bioconversion of chitin to chitosan is also possible using chitin deacetylase (EC 3.5.1.41), although most chitins are hardly accessible resulting only in a low degree of deacetylation. The mode of action of chitin deacetylase is shown in Fig. 11.3. Chitin deacetylase catalyzes the reaction of N-deacetylation of chitin or fully or partially N-acetylated chitooligosaccharides by participating of water molecules resulting in chitosan or glucosamine oligomers and acetate. N-Deacetylation can be carried out using chitin deacetylase isolated from *Zygomycetes* (*Mucor rouxii*, *Aspergillus nidulans*) and *Deuteromycetes* (*Colletotrichum lindemuthianum*) (Jaworska and Konieczna-Moras 2009; Araki and Ito 1975; Tokuyasu et al. 2000; Stevens et al. 1998; Christodoulidou et al. 1996, 1998; Martinou et al. 1998; Gauthier et al. 2008). The extracellular chitin deacetylase from *C. lindemuthianum* is characterized by an endotype pattern of action. Other enzymes like acetyl xylan esterase catalyze the hydrolysis of N-acetyl groups in chitinous materials of variable degrees of polymerization and acetylation. Morley et al. (2006) presented that it is possible to N-deacetylate chitin and chitosan to a similar extent as chitin deacetylase; however, it has higher effectiveness toward chitooligosaccharides.

11.2.3 Alternative Methods to Isolate Chitin/Chitosan

It is clear from the above-mentioned isolation processes that harsh chemical conditions are used. For example, the use of strong acids harms chitin, resulting in lower molecular weight and degree of acetylation. In addition, the waste liquid leads to environmental pollution and in increasing processing costs.

Instead of strong alkali (NaOH and KOH), enzymes can be used to degrade proteins. The

Table 11.2 Factors of alkaline N-deacetylation affecting the properties of chitosan

Factor	Effect
Alkali concentration	A reduction in the alkali concentration increases the time required to obtain soluble product, whereas the increase in deacetylation agent concentration reduces the N-deacetylation time and results in the reduction in molecular weight of chitosan (Alimuniar and Zainuddin 1992; Mima et al. 1983; Roberts 1992a; Yen and Mau 2007)
Incubation time	The degree of N-deacetylation increases dynamically during the initial phase (first 1–2 h), while prolongation of the process causes a significant reduction of improvement in the degree of deacetylation. Moreover, a significant depolymerization of the polymer during this first phase is observed. In summary, the prolongation of N-deacetylation increases the deacetylation degree with a reduction in the average molecular weight of chitosan (Benjakul and Wisitwuttikul 1994; Muzzarelli 1982; Roberts 1992a; Struszczyk and Peter 1997; Wu and Bough 1978; Yen and Mau (2007)
Ratio of chitin to alkali	A low ratio of chitin to alkali increases the solubility of the obtained chitosan with a reduction in N-deacetylation time. This affects the homogeneity of the deacetylation media (Moorjani et al. 1978; Yen and Mau 2007)
Temperature	A relatively high temperature and high alkali concentration yields an improvement of polymer solubility at shorter reaction time (Lusena and Rose 1953; Shahidi and Synowiecki 1991; Yaghobi and Hormozi 2010)
Atmosphere	Free access of oxygen during N-deacetylation has a substantial effect on chitosan degradation, whereas N-deacetylation in an atmosphere of nitrogen yields chitosan with a much higher average molecular weight than prepared under standard conditions. However, differences in nitrogen and ash compositions were found (Bough et al. 1978; Moorjani et al. 1978; Roberts 1992a; Yen and Mau 2007). The degradation effect of air becomes more pronounced by reduction of the N-deacetylation time (Yen and Mau 2007). Application of reducing agents, such as sodium borohydride, resulted in a lower degree of chitosan degradation, even when high temperature for N-deacetylation has been applied
Type of the source (chitin origin), polymorph type	Chitin originating from crustaceans is more difficult to N-deacetylate than from other sources, such as chitin from insects or fungal chitin (No et al. 1989; Struszczyk et al. 1997b; Struszczyk and Peter 1997, 1998, Wojtasz-Pajak 1997, 1998; Wojtasz-Pajak and Brzeski 1998). Above phenomenon can be a result of the difference in molecular structure of chitins of various origins as well as accompanying compounds such as minerals, that is, β-chitin (parallel chain alignment) is more reactive and susceptible for N-deacetylation than its α-polymorph type (antiparallel chain alignment) (Roberts 1992a). β-Chitin indicates higher susceptibility for dissolution and swelling because it contains significantly weaker intermolecular hydrogen bonding (Chang et al. 1997; Shepherd et al. 1997)
Particle size of chitin	N-deacetylation of chitin with a lower particle size distribution results in chitosan with a lower average molecular weight. The deacetylation yield depends on the extent of swelling of chitin particles. Chitin with smaller particles requires a shorter swelling time, resulting in a higher N-deacetylation rate (Bough et al. 1978; Lusena and Rose 1953; Struszczyk and Peter 1997; Yen and Mau 2007)
Heterogeneous or homogeneous N-deacetylation	N-deacetylation can be carried out under two conditions: Conventional-heterogeneous, distinguished by block copolymers structure of polymers and an increase in crystallinity for high N-deacetylated chitosan

(continued)

Table 11.2 (continued)

Factor	Effect
	Homogeneous using alkali chitin dispersion resulting in an amorphous product with an increase in random position of N-acetylglucosamine and glucosamine units (Kurita et al. 1977)
Single or multiple process	The multiple process of N-deacetylation of chitin is favored to obtain a high N-deacetylated chitosan with a low effect on its average molecular weight and crystallinity. This process involves water washing between successive N-deacetylation phases or sometimes dissolution/reprecipitation of chitosan between successive deacetylation treatments. The gradual N-deacetylation can be repeated several times (Roberts 1992a). In literature (Roberts 1992c; Yen and Mau 2007), two explanations of above-mentioned phenomenon have been proposed. The first explanation is based on the effect of sodium hydroxide concentration observed on the swelling of cellulose, suggesting that during washing the concentration of sodium hydroxide in the chitin/chitosan particles gradually increases to the maximum swelling concentration. This increase in swelling facilitates diffusion of alkali into the crystalline regions (Roberts 1992a; Yaghobi and Hormozi 2010)
	The second explanation suggests that chitin forms a complex with the alkaline medium. The chitin-alkali complex shows a less constant rate of the N-deacetylation step as compared with chitin that forms no complex. Washing and drying are considered to destroy the formed complex, thereby converting the remaining N-acetylated glucosamine residues back. It yields much more reactive forms for the subsequent N-deacetylation (Roberts 1998)

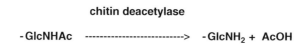

Fig. 11.3 Enzymatic N-deacetylation of chitin or chitooligosaccharides catalyzed by chitin deacetylase [EC 3.5.1.41] (Jaworska and Konieczna-Moras 2009)

advantage of enzymes is that they can be used under mild conditions, because prolonged chemical alkaline treatment can cause depolymerization and deacetylation of chitin together with denaturation of proteins. A drawback of the enzymatic treatment could be a longer incubation time than needed for the chemical treatment. Proteases from *Pseudomonas aeruginosa*, trypsin, chymotrypsin, and papain are examples which have been investigated (Wang and Chio 1998; Gagne and Simpson 1993; Giles et al. 1958; Broussignac 1968; Taked and Aiba 1962; Shimahara et al. 1982). A commercial protease preparation (Alcalase) from *Bacillus licheniformis* can also be used, according to Synowiecki and Al-Khateeb (2000). Bustos and Healy (1994) demonstrated that the enzymatic treatments do not affect the changes in chitin properties. However, the use of enzymes does not always result in complete removal of protein.

11.2.4 Microorganisms and Their Application in the Isolation of Chitin/Chitosan

Apart from the addition of enzymes, microbial deproteination and/or demineralization is of highly interest. The deproteination step for Northern pink shrimp (*P. borealis*) was performed by Mahmoud et al. (2007) in a bioreactor containing *Aspergillus niger*. A fermentation for 15 days with *B. subtilis* resulted in 84 % deproteination and 72 % demineralization of shrimp

(*Metapenaeopsis dobsoni*), as reported by Sini et al. (2007). A fermentation of a mixed culture of lactic acid bacteria (*Lactobacillus plantarum*, *L. acidophilus*, and *L. lactis*) demineralized shrimp waste (*P. jordani*) within 72 h at 30 °C. Chitosan was produced from the remaining material and showed lower protein and ash content as when the shrimp waste was demineralized with 1 N HCl, according to Phuvasate and Su (2010). In addition, Prameela et al. (2010a, b) demonstrated the fermentation of shrimp (*P. monodon*) within 72 h using only *L. plantarum*, which resulted in 89–97 % deproteination and 73–76 % demineralization. Red crab shell waste (*Chionoecetes japonicus*) was fermented by Jung et al. (2007) in two steps by *L. paracasei* followed by *Serratia marcescens*. The chitin obtained was 94 % deproteinized and 69 % demineralized. A two-stage fermentation process was developed by Xu et al. (2008) for deproteination and demineralization of shrimp waste (*Penaeus monodon* and *Crangon crangon*). In the first step, a mixed culture was used for deproteination, and in the second step, lactobacilli were used for demineralization. The fermentation of crustacean shells with lactic bacteria (such as *L. pentosus*) at the pH of the reaction medium to approximately pH 4 facilitated the hydrolysis of crustacean proteins and did not affect the structural changes of the obtained intact chitin (Bautista et al. 2001; Hall and Silva 1994). Recently a strain of *B. licheniformis* was genetic improved for deproteination of shrimp shells by Hoffmann et al. (2010).

11.2.5 Isolation of Chitin/Chitosan from Other Sources as Crustacean Waste

Although at this moment crustacean waste is the main commercial source of chitin, it can also be isolated from other sources. Due to the absence of minerals in cuticles of insects, chitin extraction from insect species requires the implementation of the deproteinization and/or decoloration steps only without the removal of inorganic compounds. However, addition of ascorbic acid or phenylthiourea inhibits tyrosinase activity and prevents darkening of the resulting chitin (Dwivedi and Om 1995; Struszczyk et al. 1998). Currently, silk worm (*Bombyx mori*) pupa exuvia chitin is the most important industrially source of insect chitin, according to Synowiecki and Al-Khateeb (2003). It can be extracted by treatment with 1 N HCl at 100 °C for 20 min followed by an extraction of 1 N NaOH for 24–36 h at 80 °C and refluxed with 0.4 % (w/v) Na_2CO_3 for 20 h and the chitin yield was 15–20 %, as demonstrated by Zhang et al. (2000).

Fungi contain chitin and chitosan and there is an increasing interest to obtain these polymers from this source. Thirty-three fungal strains were screened by Hua et al. (2004) as potential bioproducers of fungal chitosan as industrial source. Among them, *Absidia glauca* was the most promising one. Di Mario et al. (2008) presented that chitosan is also naturally found in *Mucorales* such as *Mucor rouxii*, *Absidia glauca*, *A. niger*, *Gongronella butleri*, *P. sajor-caju*, *Rhizopus oryzae*, *Lentinula edodes*, and *Trichoderma reesei*. Chitosan is present because fungi may contain the enzyme chitin deacetylase, according to Tsigos et al. (2000). The advantage of chitin/chitosan from fungi is that it needs less harsh isolation conditions as described above for chitin/chitosan from crustacean waste. Mycelia of lower fungi, like *A. niger*, *M. rouxii*, *Absidia coerulea*, *R. oryzae*, can be used for chitin and chitosan isolation. Due to enhanced use of fungi in the industry, their availability is increasing (Synowiecki and Al-Khateeb 2003; Cai et al. 2006). Chitosan was obtained from *A. niger* mycelium by using lysozyme, snailase, neutral protease, and chitin deacetylase instead of the acid and alkali treatments. The obtained chitosan had a molecular weight which was three times higher than chitosan purified by the acid/alkali treatment, and the other characteristics were the same, according to Cai et al. (2006). The direct extraction of chitosan from the fungus *A. coerulea* without thermal or chemical depolymerization has been proposed by Niederhofer and Muller (2004). The procedure consisted of deproteinization of fungal biomass with 2 % (w/v) NaOH for 3 h and dissolution of alkali-insoluble material in 4 % (v/v) acetic acid for ~12 h. The centrifuged, clear solution was treated with sodium hydroxide to precipitate chitosan. Chitin

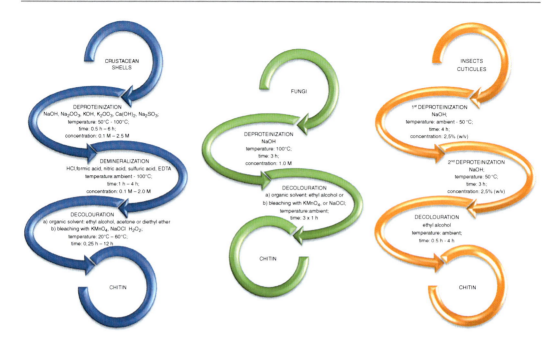

Fig. 11.4 Schematic overview of parameters used for conventional crustacean, fungal, and insects chitin isolation (Struszczyk et al. 1997a, b, 2002a, b, c; Roberts 1992a, b, c, d; Yen and Mau 2007; Oduor-Odote et al. 2005)

from higher fungi can be isolated from the cell walls in the form of a chitin-glucan complex. An optimized extraction process for the honey mushroom (*Armillariella mellea*) and yellow morel (*Morchella esculenta*), described by Ivshina et al. (2009), consisted of a deproteination step (2 % (w/v) NaOH and 0.1 % (w/v) sodium stearate for 2 h at 83–85 °C), a demineralization step (1 % (v/v) HCl for 2 h at 55–60 °C), a decoloring step (5 % (v/v) H_2O_2 in ammonia for 4 h at 30–35 °C), and a deglucanization step (2 % (w/v) NaOH for 2 h at 83–85 °C). Kitozyme (Belgium) is the first company selling chitosan from fungal origin, namely, from *A. niger* and the edible mushroom *Agaricus bisporus*.

An overview of the different extractions procedures of chitin is shown in Fig. 11.4.

11.2.6 Future

The industry is continuously seeking for improved chitin/chitosan isolation techniques to decrease processing costs. To make chitin more accessible is one of the key research areas. Most research in the past was dedicated to the optimization of acid and base treatments by varying the parameters like time, temperature, and kind of acids/base, etc. A gradual shift is observed nowadays toward research focusing on application of enzymatic/microbial deproteination and demineralization. Another interesting novel approach is ionic liquids to purify chitin. Recently it was reported by Qin et al. (2010) that it was possible to extract 94 % chitin from shrimps in just one single step using [C_2mim][OAc], which is considered to be a green solvent. However, the application of ionic liquids is at this moment not cost-effective. Next to this improvement, the use of novel extraction technologies is of interest for making chitin more accessible, such as microwave and freeze pump-out thaw cycles.

11.3 Applicable Forms of Chitin and Chitosan

The most important factor indicating the potential of chitin applications is its solubility, which is not dependent on its molecular weight but is related to the degree of *N*-acetylation as well as its polymorph type. Roberts (1992a, b, c, d) reported that α-chitin is soluble in concentrated inorganic acids such as hydrochloric acid or phosphoric acid, whereas β-chitin dissolves in concentrated formic acid. The dissolution in formic acid results in significant reduction of the average molecular weight. Other organic acids, such as carboxylic acids (di- or trichloroacetic acid), are also applied for degenerative dissolution of chitin. Chitin is also soluble in hexafluoroacetone, chloroalcohols, and hexafluoroisopropanol in conjugation with aqueous solutions of mineral acids, according to Kumar (2000).

Chemical modification of chitin carried out by protecting the glycosidic and acetamido bonds leads to chitin derivatives characterized by higher solubility and in some cases higher biodegradability, bioactivity, and/or reactivity. The most well-known derivative of chitin is chitosan obtained by chemical N-deacetylation as described above. Fujii et al. (1980) demonstrated that acylation of the hydroxyl groups of chitin in alkali results in water-soluble O-acylchitin that may differ according to the acylation agent used in chain length. O-Carboxyalkylation of chitin prepared in alkali yields mainly in substitution of the hydroxyl groups. The chemical treatment of chitin at heterogeneous conditions with 2-chloroethanol resulted in 2-hydroxyethylchitin (glycochitin), according to the literature (Muzzarelli et al. 1997). Chilarski et al. (2007) found that chitin can undergo also sulfation (in the presence of pyridine) and cyanoethylation. A useful, biodegradable diester of chitin is dibutyrylochitin (Szosland and Struszczyk 1997). It is used for the design of microspheres, transparent films, nonwovens, fibers, yarns, and as a source for electrospinning (Szosland 1998; Chilarski et al. 2004a, b, 2007; Krucińska et al. 2006). Chitin fibers were also manufactured by Struszczyk et al. (1999) by alkaline hydrolysis at mild conditions using dibutyrylochitin filament as a precursor. The liquid system of chitin-dimethylacetamide-lithium chloride was used for the design of fibers (Agboh 1986), semitransparent films, or as a homogenous system for synthesis of other chitin derivatives (Quong et al. 1998).

11.3.1 Chitosan Derivatives

Chitosan is insoluble in aqueous solutions of alkali and in most organic solvents. In laboratory or industrial practice, it is usually dissolved in aqueous solutions of organic (such as formic acid, acetic acid, lactic acid, citric acid, etc.) or inorganic acids (such as hydrochloric acid, nitric acid, perchloric acid, phosphoric acid, etc.). The most important factor affecting chitosan solubility is pH. When the pH is lower than ~6.3, chitosan shows complete dissolution, whereas the increase in pH results in precipitation of chitosan in the form of gel-like flocks, as described by Burkinshaw and Karim (1991). Moreover, chitosan manufactured using alkali chitin by homogenous N-deacetylation is soluble in water at neutral pH, according to Sannan et al. (1975).

Nowadays, different derivatives of chitosan are produced to improve solubility and other properties. A well-known water-soluble chitosan derivative is a carboxyalkyl derivative of chitosan known as carboxymethylchitosan that is obtained in forms of N- and O-, N-, or O-carboxymethylchitosan. Kumar (2000) reported that the place and the degree of substitution depend on the reaction conditions. Controlled N-acylation with acetic anhydride resulted in water-soluble, partially N-acetylated chitosan, whereas reactions carried out with other carboxylic acid anhydrides produces N-acyl derivatives (formyl, acetyl, propionyl, butyryl, hexanoyl, octanoyl, decanoyl, dodecanoyl, tetradecanoyl, lauroyl, myristoyl, palmitoyl, stearoyl, benzoyl, monochloroacetoyl, dichloroacetyl, trifluoroacetyl, carbamoyl, succinyl, acetoxybenzoyl) (Kumar 2000; Hirano 1998). According to the literature, sulfation of chitosan is carried out using various conditions such as sulfur trioxide and pyridine (Gamzazade et al. 1997), sulfur trioxide and trimethylamine (Je et al. 2005; Holme and Perlin 1997), sulfur trioxide in the presence of sulfur dioxide and chlorosulfonic acid–sulfuric acid (Naggi et al. 1986), or concentrated sulfuric acid (Nagasawa et al.1971) and chlorosulfonic acid (Zhang et al.

2003) at homogeneous or heterogeneous conditions and with various temperatures with or without additional physical factors, such as microwave irradiation (Mourya and Inamdar Nazma 2008; Xing et al. 2004). Other types of chitosan derivatives like imines are created based on Schiff reaction with aldehydes (such as glutaraldehyde, oxygenated starch, phenylacetaldehyde) or ketones (diphenyl ketone) due to the presence of free amine groups. The resulted Shiff's base is converted to *N*-alkyl derivatives by a hydrogenation process using, that is, sodium tetrahydroborate or sodium cyanoborohydride (more applicable but produces toxic compounds). Epichlorohydrin, as a cross-linking agent, is used for an increased chelating ability of carboxymethylchitosan to insoluble and amorphous derivatives, as reported by Roberts (1992a, b, c, d). Kumar (2000) demonstrated that semisynthetic resins of chitosan consist in various forms of chitosan copolymers with methyl methacrylate polyureaurethane, poly(amide ester), and acrylamidemaleic anhydride.

Chitosan and some of its derivatives and degradation products have a wide range of biological activities, such as bacteriostatic, antibacterial, antifungal, and/or antiviral properties (Struszczyk et al. 1997a, b, 1999; Pospieszny and Giebel 1997). The presence of free, protonated amino groups in chitosan makes it possible to form complexes with negatively charged derivatives, such as anionic synthetic polymers, polysaccharides, proteins, dyes, and lipids, and also with cholesterol, fat, enzymes, tumor cells, bacteria cell wall proteins or DNA and RNA, according to Naggi et al. (1986). Chitosan has the ability to chelate with various metal ions, due to the neutral or negatively charged hydroxyl groups of D-glucosamine (GlcN) residues and amino groups. The chelating properties of chitosan, due to the presence of a significantly higher number of free amine groups, are much more effective than the *N*-acetyl groups present in chitin. The susceptibility for metal chelatin could be improved by several adjustments such as change in ionic strength and/or pH, cross-linking with low-molecular-weight derivatives (Kurita et al. 1986), controlled *N*-acetylation or *N*-deacetylation (Roberts 1992a, b, c, d; Naggi et al. 1986), as well as chemical binding with other natural or synthetic polymers (Muzzarelli et al. 1980).

Commercial products of various companies differ in physical and/or microbiological purity, minerals and protein content, granulation, color, degree of deacetylation, average molecular weight, and solubility. According to the literature, commercial chitosan is available as flakes, powder, and processed products, such as films (Muzzarelli 1998), beads or microcapsules (Muzzarelli 1998; Quong et al. 1998), fibers (Urbanczyk et al. 1997; Struszczyk 1998a, b), and microcrystalline chitosan (MCCh) (Struszczyk and Kivekas 1990, Polish Patent No. 125995). The potential commercial applications of the different chitosan forms are shown in Table 11.3.

11.3.2 Applicable Forms

The production of chitosan beads or microcapsules is performed by dropping aqueous chitosan salt solutions in a low-concentrated alkali solution with slow or vigorous agitation. The size of the beads is determined by the diameters of the drop; viscosity of the chitosan solution, solution stream, concentration of alkali, and velocity of agitation (Kawamura et al. 1997, Fin patent No. 93-4616 931019). According to the literature, the properties of the chitosan beads, including stability and mechanical resistance, can be improved by several cross-linking agents: mono-, di-, or polyaldehyde; terephthaloyl chloride; and hexamethylene diisocyanate (Pospieszny and Giebel 1997; Pellegrino et al. 1990; Baba et al. 1996; Hayashi et al. 1992; Griethuysen-Diblber et al. 1988).

Chitosan fibers are manufactured according to many conventional methods, although most frequently an organic acid solution of chitosan for the spinning process is used (Japan Patent No. 60159123, East 1993; Struszczyk 1994). The process consists in spinning and regeneration of formed chitosan salt fibers in low-concentrated alkali solution. Additionally, the mechanical resistance of fibers can be improved using cross-linking agents, that is, bi- or polyfunctional aldehydes. Nevertheless, the application of cross-linking agents reduces the fiber elasticity.

Table 11.3 Potential of chitosan applications (268)

Application	Applicable Form of Chitosan OR Chitosan Derivative	Examples of potential product
Medical devices, pharmaceutical, parapharmaceutical applications	• Salts, microcrystalline chitosan, fibers, amorphous gel, hydrogel, lyophilized sponges, microfibrides, fibrides • Fibrides, powder, lyophilized sponges • Salts, microcrystalline chitosan • Salts, solution • Membranes, films • Nanofibers and nanoporous fibers • Powders, salts, microcrystalline chitosan • Microcrystalline chitosan, salts, gel-like dispersions, films • Salts • Microcapsules, microcrystalline chitosan, gel-like dispersions • Microcapsules • Film, fibers, microcrystalline chitosan, lyophilized sponge, bacterial cellulose containing glucosamine or chitosan • Bacterial cellulose containing glucosamine or chitosan, microcrystalline chitosan • O-carboxymethylchitosan, film • Microcrystalline chitosan, chitosan lactate differing in pH, cross-linked gels • Monofilaments, multifilaments • Hydrogels, beads, microcapsules, chitosan-alginate complex, nanoparticles, nanospheres • Powder, beads, membrane, chitosan precipitate, films • Hydrogels, hybrids composite, modified chitosan • Nonfibrous modified membrane	• Wound dressings (Kucharska et al. 1997; Oungbho and Muller 1997, Prodas-Drozd and Gwiezdinski) • Topical hemostatic agents (Kucharska et al. 2011a, b; Struszczyk and Struszczyk 2007) • Blood cholesterol control (Veroni et al. 1996; Yihua and Binglin 1997) • Tumor inhibition (Guminska et al. 1997; Jameela et al. 1996) • Membranes (Kim et al. 1995; Kumar 2000) • Filtration and ultrathin implants (Li and Hsieh 2006) • Dental/plague inhibition (Kochanska 1997) • Skin burns/artificial skin (Kochanska 1997; Shahabeddin et al. 1991) • Contact lens (Yu et al. 1991) • Control release drugs (Akbuga and Bergisadi 1996; Bodek 1997) • Bone disease treatment (Japan Patent No. 92-257696 (920928)) • Hernia treatments (Niekraszewicz et al. 2008; Niekraszewicz et al. 2007a, b; Polish Patent No 380861 • Sealed vascular prostheses (Ciechańska et al. 2010b) • Nerve regeneration (Crompton et al. 2007; Lu et al. 2007) • Orthopedic applications (Niekraszewicz et al. 2009; Ratajska et al. 2008) • Surgical suture (Gamzazade et al. 1997; Hirano 1998) • Drug, active substances delivery systems or vector for the gene transfection (Issa et al. 2005; Kumar 2000; Lee et al. 2000; Liu and Yao 2002) • Diagnosis and treatment of diseases (Pellegrino et al. 1990) • Tissue engineering (Liu and Yao 2002; Twu et al. 2005) • Biosensors (Issa et al. 2005)
Pulp and papermaking	• Solution of salts, microcrystalline chitosan • Solution of salts forming films • Solution of salts forming films • Solution of salts forming films	• Surface treatment (Allan and McConnell 1975; Struszczyk et al. 2000, Struszczyk et al. 2001a) • Photographic paper (Japan Patent No 63189859 2999) • Carbonless copy paper (U.S. patent No. 2712507) • Flexibilizer (Struszczyk et al. 2007, World Patent No. WO 9723390 A1 970703)
Technical textiles	• Fibers, threats, solution, gel-like dispersions, microcrystalline chitosan • Solution, gel-like dispersions, microcrystalline chitosan, threats, fibers, yarns • Nanofibers, nonwoven	• Sanitary fibrous materials or disposable hygienic products (Gamzazade et al. 1997; Hirano 1998; Struszczyk and Brzoza-Malczewska 2007) • Textile material (Baba et al. 1996; Gajdziecki et al. 1995; Gamzazade et al. 1997; Hirano 1998; Kumar 2000; Pellegrino et al. 1990; Yoshida et al. 1993) • Filtration (Desai et al. 2009)

(continued)

Table 11.3 (continued)

Application	Applicable Form of Chitosan OR Chitosan Derivative	Examples of potential product
Water treatment	• Beads, membranes, powders, microcrystalline chitosan • Beads, powders, microcrystalline chitosan, fibers • Beads, salt, films • Membranes	• Removal of metal ion (Baba et al. 1996; Pellegrino et al. 1990) • Flocculent/coagulant of: - Dyes (Liu and Yao 2002; Pellegrino et al. 1990) - Amino acids (Pellegrino et al. 1990) • Filtration (Lee et al. 1997)
Cosmetics	• Powder, microcrystalline chitosan • Gel-like dispersions • Gel-like dispersions, solution, powder, gel-like dispersion • Gel-like dispersions, salts • Gel-like dispersions, salts • Gel-like dispersions • Gel-like dispersions	• Make-up powder (Japan Patent No. 63-161001) • Nail polish (U.S patent No. 94-349661 (941205)) • Moisturizers and fixtures (Lu et al. 2007) • Bath lotion (Lu et al. 2007; Muzzarelli 1989) • Face, hand, and body creams (Lu et al. 2007; Muzzarelli 1989) • Toothpaste (Crompton et al 2007) • Foam enhancing (Crompton et al 2007)
Biotechnology	• Membranes, microcapsules, beads, powder • Membranes, beads, films • Beads • Beads • Beads, membranes, films • Membranes, films	• Enzyme immobilization (Galas et al. 1996; Itozawa and Kise 1995; Krajewska 2004; Muzzarelli et al. 1980; Niekraszewicz et al. 2008) • Protein separation (Senstad and Mattiasson 1989) • Chromatography (Hirano 1998; Muzzarelli et al. 1990) • Cell recovery (Roberts 1992d; Senstad and Mattiasson 1989) • Cell immobilization (Freeman and Dror 1994; Hisamatsu 1998) • Electrodes and sensors (Kurauchi and Ohga 1998; Ohashi and Karube 1995)
Agriculture	• Salts, microcrystalline chitosan • Salts, microcrystalline chitosan • Salts, microcrystalline chitosan • Microcapsules, microcrystalline chitosan, oligosaccharides	• Seed coating (Hadwiger et al. 1984) • Leaf coating (Agboh 1986) • Hydroponic/fertilizer (Polish Patent No 141381 2999) • Controlled agrochemical release (Struszczyk et al. 2001b; Teixeira et al. 1990)
Food chemistry	• Salts, beads, films • Salts forming films • Salts • Gel • Gel, powder, microcrystalline chitosan, oligomers of chitin or chitosan • Acid-soluble fungal chitosan • Film, coating, biocomposite films, and multilayers system	• Removal of dyes, solid, acids (Hsien and Rorrer 1995; Seo et al. 1988) • Preservatives (Japan Patent No 63169975 (2999)) • Color stabilization (Japan Patent No. 62297365 (2999)) • Increasing viscosity aids (Kumar 2000) • Diet supplement (Gades and Stern 2005; Harish Prashanth and Tharanathan 2007; Morley et al. 2006) • Clarifying agent (Rungsardthong et al. 2006) • Antimicrobial preservation and bio-packaging (Dutta et al. 2009)
Membranes for various applications	• Membranes, films • Membranes, films	• Permeability control (King et al. 1989) • Solvent separation (Burkinshaw and Karim 1991; Uragami 1989)

Microcrystalline chitosan (MCCh), in the form of a gel-like water dispersion or, after drying, as a powder, is prepared by precipitation of an acidic aqueous chitosan solution in an alkali solution (most often a NaOH solution) (Je et al. 2005; Holme and Perlin 1997; Struszczyk 1987). The process of MCCh manufacture consists of many stages, such as a neutralization, coagulation, and aggregation, of polymer chains. The applied process of MCCh results in its specific properties in relation to molecular and supermolecular structures that are different from the initial chitosan, that is, microcrystalline form of chitosan has a more developed internal surface, reduced crystallite size, and lower average molecular weight. Moreover, Wieczorek and Mucha (1997) reported that the final properties of MCCh depend on the processing conditions. According to the literature (Hirano 1998; Gamzazade et al. 1997; Je et al. 2005; Bodek 1997, Japan Patent No 92-257696 (920928)), MCCh has unique and specific useful properties, that is, high crystallinity connected with the ability to form powerful hydrogen bonds, high water absorption, high adhesiveness, enhanced chelating and sorption properties, direct film-forming behavior, long-term stability in aqueous dispersion, controlled biodegradability, higher susceptibility to hydrolytic degradation, high ability for solvent exchange, high chemical reactivity, and bioactivity, including bacteriostatic properties and biocompatibility. Due to its direct film-forming properties, MCCh has improved practical application, as reported by Hirano (1998). The improvement in the properties of MCCh films is achieved by incorporation of humectants interfering with the hydrogen bonds between the chitosan chains. Briston (1974) reported that evaporation of water during MCCh formation results in the ordering of polymer chains and results in an increase in intermolecular hydrogen bonds mediated by humectants.

Most aqueous solutions of chitosan salts are able to form semi- or transparent films. They could be regenerated to chitosan films by treatment with diluted alkali solution, either aqueous or containing low amount of evaporating organic solvent, such as ethanol. The chitosan salt films are characterized by dissolution in water, with an increase in solubility when the pH is lowered. Moreover, the regenerated chitosan films are characterized by high water absorbancy, selective permeability to various gases, controlled release of incorporated agents, high reactivity, and electrochemical properties (Chilarski et al. 2007; Naggi et al. 1986). The insolubility can be improved by reaction with low-molecular-weight cross-linking agents. The cross-linking modification introduces changes in the permeability as well as in mechanical properties or polyionic properties (Kittur et al. 1998; Andrady and Xu 1997).

Hetero-cross-linking, by using various types of mediator polymers, is also applied. For example, chitosan films containing glutamate and alginate cross-linked by calcium chloride (Remunan et al. 1997) or chitosan films containing proteins (collagen or gelatin) enzymatically bound together using tyrosinase (Muzzarelli et al. 1980). Other types of modification are possible, that is, formation of chitosan films by physical bonding of globular or fibrous proteins (casein, keratin, etc.), resulting in significant change in mechanical properties, dissolution, and absorption behavior. The resulting mixtures showing direct film-forming properties are applicable in nonwoven and papermaking industries (Rawls 1984, Japan Patent No. 92-257696 (920928), Niekraszewicz et al. 2007b). Depending on their origin (i.e., crustaceans, insects, fungal) or species (i.e., shrimp, crabs, krill, etc.), the microstructure of films formed from chitosan can differ significantly. The existence of more-ordered areas and amorphous regions plays an important role. The films with a low degree of crystallinity formed from high-molecular polymers possess better mechanical endurance.

11.4 Application of Chitin and Chitosan

Due to its physical and chemical properties, chitin and chitosan are being used in a vast array of widely different products and applications, ranging from pharmaceutical and cosmetic products to water treatment and plant protection (Fig. 11.5). In different applications, different properties of chitin or chitosan are required.

11.4.1 Food and Feed

The use of chitosan in the food industry is well known because it is not toxic for warm-blooded animals. Microcrystalline chitin (MCC) shows

Fig. 11.5 Applications of chitin and chitosan

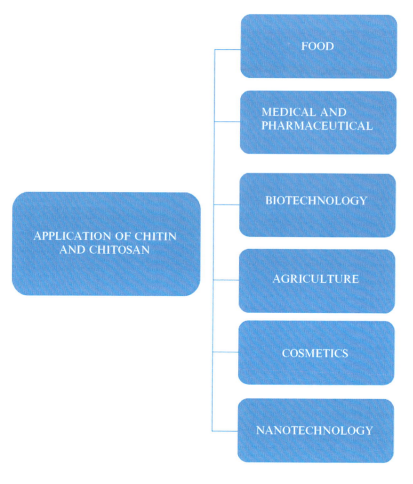

good emulsifying properties and is a superior thickening and gelling agent for stabilizing foods. It is also used as a dietary fiber in baked foods. The use of MCC solved some of the problems introduced by other sources of fiber with flavor, color, and shelf-life of food products. It could be of special importance for manufacturing protein-enriched bread, even without such ingredients as emulsifiers and shortenings. Chitin has been used in immobilization of enzymes. It can be used as a nonabsorbable carrier for highly concentrated food ingredients, for example, food dyes and nutrients. In India, incorporation of chitin in poultry feed at a level of 0.5 % decreased the food consumption ratio and increased body weight by 12 % in comparison with birds fed a chitin-free diet. Similarly, nutritional studies in the USA have shown that chicks fed on a diet containing dried whey and chitin utilized whey more efficiently and gained more weight than those fed similar but with a chitin-free diet. Trials also showed that small amounts of chitin added to the diets of chicks and calves enabled the animals to digest milk lactose through increased growth of specific intestinal bacteria. These bacteria impede the growth of Rother types of organisms and generate the enzyme required for lactose digestion. This property may be of immense importance, since certain groups of human and many animals have lactose intolerance. There is no complete study on the metabolism of chitin and chitosan in the human body; therefore, the use of these polymers in food processing industries still needs to be further explored (http://www.dawn.com).

To increase the application of chitin and chitosan, conversion of processing discards into valuable by-products and alternative specialty materials has been identified as a timely challenge for food research and development associated with numerous applications of chitinous polymers. In that sense, these biopolymers offer a wide range of unique applications including as an additive in bioconversion processing for the production of value-added food products (Shahidi and Synowiecki 1991; Carroad and Tom 1978; Revah-Moiseev and Carroad 1981), preservation of foods from microbial deterioration (Papineau et al. 1991; El Ghaouth et al. 1992; Sudarshan et al. 1992; Chen et al. 1998), formation of biodegradable films (Tajik et al. 2008; Youn et al. 2009; Synowiecki and Al-Khateeb 2000), recovery of waste material from food processing discards (Senstad and Mattiasson 1989; Hwang and Damodaran 1995; Sun and Payne 1996; Pinotti et al. 1997), purification of water (Jeuniaux 1986; Micera et al. 1986; Muzzarelli et al. 1989; Deans and Dixon 1992), and clarification and deacidification of fruit juices (Imeri and Knorr 1988; Soto-Perlata et al. 1989; Chen and Li 1996; Spagna et al. 1996) (Table 11.4).

11.4.1.1 Cholesterol-Lowering Effects

Fibers with a range of abilities to perturb cholesterol homeostasis were used to investigate how the serum cholesterol-lowering effects of insoluble dietary fibers are related to parameters of intestinal cholesterol absorption and hepatic cholesterol homeostasis in mice. Cholestyramine, chitosan, and cellulose were used as examples of components with high, intermediate, and low bile acid-binding capacities, respectively. The serum cholesterol levels in a control group of mice fed a high-fat/high-cholesterol (HFHC) diet for 3 weeks increased about two-fold to 4.3 mM and inclusion of any of these fibers at 7.5 % of the diet prevented this increase from occurring. In addition, the amount of cholesterol accumulated in hepatic stores due to the HFHC diet was reduced by these fiber treatments. The three kinds of fibers showed similar hypocholesterolemic activity; however, cholesterol depletion of liver tissue was greatest with cholestyramine. The mechanisms underlying the cholesterol-lowering effect of cholestyramine were (1) decreased cholesterol (food) intake, (2) decreased cholesterol absorption efficiency, and (3) increased fecal bile acid and cholesterol excretion. The latter effects can be attributed to the high bile acid-binding capacity of cholestyramine. In contrast, incorporation of chitosan or cellulose in the diet reduced cholesterol (food) intake, but did not affect either intestinal cholesterol absorption or fecal sterol output. The study of van Bennekum et al. (2007) provides strong evidence that above all, satiation and satiety effects underlie the cholesterol lowering.

11.4.1.2 Food Additives

Chitosan as dietary food additive or as a dietary supplement does not need FDA approval. Although, chitosan is not approved as a food additive in many countries, chitin, chitosan, and their derivatives have a range of potential uses as food additives, packaging agents and aids to beverage processing. Both chitin and chitosan provides fiber in diets. The addition appears entirely safe. In laboratory studies, animals have grown normally while consuming up to 10 % of their diet in the form of chitin. Chitin imposes stability on food emulsions. For example, whipped dessert topping containing chitin can be frozen and thawed without breaking up. As in the case of health care, introduction to the market will require the time and expense of obtaining regulatory approval. The ability of chitosan molecules to scavenge fat and cholesterol in the digestive system, plucking it from the stomach and excreting it in the duodenum, has significant implications for its use as a beneficial food additive. Foods containing chitosan, or chitosan complexes with fatty acids, could be designed to reduce obesity, cholesterol levels, and the incidence of colon cancer.

When broken down into small polymers, known as microcrystalline chitin (MCC), it can be used as an additive to enhance the flavor and taste of foods. Unlike conventional chitin, this form of the biopolymer distributes itself evenly throughout aqueous solutions, as tiny particles. Heated to normal cooking temperatures, MCC forms pyrazines, which are responsible for the roasted taste and aroma of several foods. Chitosan, meanwhile, has potential use in masking

Table 11.4 Food applications of chitin, chitosan, and their derivatives in the food industry

Area of application	Examples
Antimicrobial agent	Bactericidal
	Fungicidal
	Measure of mold contamination in agricultural commodities
Edible film industry	Controlled moisture transfer between food and surrounding environment
	Controlled release of antimicrobial substances
	Controlled release of antioxidants
	Controlled release of nutrients, flavors, and drugs
	Reduction of oxygen partial pressure
	Controlled rate of respiration
	Temperature control
	Controlled enzymatic browning in fruits
	Reverse osmosis membranes
Additive	Clarification and deacidification of fruits and beverages
	Natural flavor extender
	Texture controlling agent
	Emulsifying agent
	Food mimetic
	Thickening and stabilizing agent
	Color stabilization
Nutritional quality	Dietary fiber
	Hypocholesterolemic effect
	Livestock and fish feed additive
	Reduction of lipid absorption
	Production of single-cell protein
	Antigastritis agent
	Infant feed ingredient
Recovery of solid materials from food processing wasters	Affinity flocculation
	Fractionation of agar
Purification of water	Recovery of metal ions, pesticides, phenols, and PCBs
	Removal of dyes

flavor components from seafood processing streams. Free amino acids present in the streams, such as arginine, alanine, glutamic acid, serine, and glycerin, contribute significantly to the taste of seafood. They can be removed from the streams using ligand-exchange chromatography; the amino acids form complexes with copper ions, which are fixed on a chitosan support.

11.4.1.3 Films

Films made from chitosan have two characteristics highly desirable to the food industry: they are biodegradable and they have low permeability to oxygen. At present, those beneficial characteristics of chitosan films come at the expense of other technical properties such as tensile strength. However, the application of genetic engineering techniques can potentially change the distribution of molecular weights in chitosan, particularly that derived from fungi. That would permit scientists to change such characteristics of films as tensile strength, flexibility, gas permeability, and rate of degradation in the environment.

Cross-linking chitosan films with epichlorohydrin in alkaline conditions improves the film tensile strength by factors up to 100, bringing it close to that of synthetics such as polyethylene and polypropylene that are used as package film. In addition, cross-linked films have better wet strength than non-cross-linked films. Studies indicate that films of higher molecular weight forms of chitosan are more brittle, even when plasticizer is added. An alternative application in food packaging involves sprinkling chitosan powder on synthetic packaging film or spraying a chitosan solution on the film. If viscosity increasing agents are added, the material can be deposited on the film by letterpress. In these cases, the chitosan is used as an antibiotic or antimold treatment.

Chitosan also has potential as a coating to preserve fruits. *N*-Carboxymethyl chitosan, a derivative made when chitosan reacts with monochloroacetic acid, forms a strong film that is selectively permeable to such gases as oxygen and carbon dioxide. The film is made by spraying an aqueous solution of *N*-carboxymethyl chitosan on the fruit or by dipping the fruit into the solution. Apples coated with the material and left in cold storage retain their freshness for more than 6 months and keep their titratable acids for about 250 days. The film can be removed by washing with water before consumption of the fruits.

11.4.1.4 Preservation

Chemical food preservatives can be replaced with chitin-based formulations. The advantages are twofold. First, chitinous materials are safer, and second, with their antimicrobial activity, they can protect food products against microbial invasion. Chitosan-based films have been developed for this purpose by Begin and van Calsteren (1999). It has also been applied by Ouattara et al. (2000) to improve the preservation of vacuum-packed processed meat and it could delay the growth of *Enterobacteriaceae*, which are indigenous bacteria in food products. It seems this antibacterial activity is due to the positive charge of the C2 amino group of glucosamine. Dutta et al. (2009) described that this positive charge interacts with negatively charged microbial cell membranes and leads to leakage of the intracellular constituents of the microorganisms.

11.4.2 Medical and Pharmaceutical Application

11.4.2.1 Regulation for Medical Devices and the Use of Chitin and Its Derivatives

In Europe, three regulations describe the requirements for medical devices:
- Active Implantable Medical Devices (AIMDD)—Directive 90/385/EEC
- Medical Devices Directive (MDD)—Directive 93/42/EEC
- In Vitro Diagnostic Directive (IVDD)—Directive 98/79/EC

In 2007, the new approach for the regulation was implemented on the base of European Directive 2007/47/EC amending the above-mentioned Council Directives for new requirements including guidelines for the medical devices utilizing animal tissue-originated substances, such as chitin and chitosan. The harmonized EN-ISO 22442-1/2/3 Standards present the most important factors that should be taken into account for the estimation of chitosan and chitin safety, when used in medical devices (EN-ISO 22442-1:2007; EN-ISO 22442-2:2007; EN-ISO 22442-3:2007). The use of chitin and its derivatives ought to be evaluated for possible risks related to:
- Parasites and other unclassified pathogenic entities
- Bacteria, mold, and yeast
- Viruses (acc. EN-ISO 22442-2:2007 and EN-ISO 22442-3:2007)
- Undesired pyrogenic reaction (acc. relevant part of EN-ISO 10993 Standard)
- Undesired immunological reaction (acc. relevant part of EN-ISO 10993 Standard)
- Undesired toxicological reaction (acc. relevant part of EN-ISO 10993 Standard)

There are several guidelines for the chemical characterization of chitin and its derivatives (mostly chitosan) to be used in medical devices. European Pharmacopeia as well as EN-ISO 10993-18 and ISO/DIS 10993-19 make the guidelines that are used for chemical characterization of chitin and its derivatives. The compressed flowchart for primary chemical characterization and identification of chitosan utilizing in medical devices is shown in Fig. 11.6. However, the chitosan in medical devices must not have undesired immunological and toxicological reactions. Above-mentioned aspects should by verified in series of biocompatibility studies according to EN-ISO 10993-1 guidelines. This standard gives the optimal indications to design biocompatibility studies, taking into account the anatomical position of clinical use and time of use (short, intermediate, and permanent).

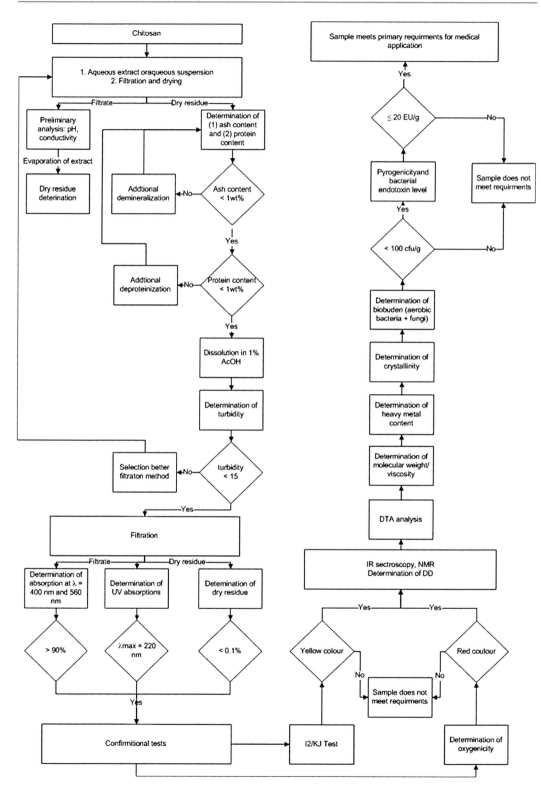

Fig. 11.6 Chemical characterization and identification of chitosan purity for application in medical devices (Struszczyk and Struszczyk 2007; Struszczyk 2006; Hein et al. 2001)

11.4.2.2 Medical Application of Chitin and Its Derivatives

Chitin and chitosan have been applied in the design of medical devices (biomaterials). The most important medical applications of chitin and its derivatives are shown in Fig. 11.7.

Wound Dressings

Wound dressings, especially topical agents, accounts for the most important part of chitin and chitosan utilization in medical devices. The range of possible applications covers various types of wound dressings and is shown in Fig. 11.8.

Primary and Secondary Wound Dressings

Primary wound dressings should assure a suitable environment for wound healing. According to the literature (Struszczyk 2002a, b, c; Muzzarelli 2009; Kucharska et al. 2007, 2011a, b; Niekraszewicz et al. 2007a, b, c), there is an increasing demand for bioactive materials that synergistically protect the wound and have strictly and clearly defined bioactive properties, such as:

- Acceleration of wound healing
- Antimicrobial (antibacterial, bacteriostatic, antifungal) properties
- Controlling the connective tissue regeneration (due to prevention of scar formation)
- Local hemostatic behavior
- Enhance the function of polymorphonuclear leukocytes, macrophages, and fibroblast
- Inhibition of the matrix metalloproteinases

Chitin derivatives possess several advantages connected to the passive preventing of infection and assisting in healing of wounds, such as:

- Formation of gel preventing over-adhesion to wound bed
- Pain reduction
- Physical entrapping the bacteria to gel structures
- Absorption of the exudate excess
- Remaining non-adherent to the wound and easy to remove after use
- Formation of transparent or semitransparent gels or films
- Controlled wet environment during wounds healing

However, Muzzarelli (2009) reported that chitosan, especially with a higher degree of deacetylation (DD > 70 %), is degraded by several enzymes found in body fluids, such as lysozyme.

Histological studies of chitosan wound dressings revealed that the regenerated collagen fibers were more mature when compared with competitive controls. The arrangements of collagen fibers were similar to that found in natural non-defected skin. The tensile strength of regenerated connective tissue was significantly superior compared to controls and showed complete re-epithelialization and granulation.

The action of chitin and its derivatives primary wound dressings is connected with the synergistic mode of their physical properties (moisture control, formation of absorbing exudate gels, etc.) and bioactivity.

Secondary wound dressings containing chitin and its derivatives have mainly two important properties: antibacterial guarding (bioactivity) as well as the possibility to form semipermeable or absorbing exudate usable forms of biopolymers.

Various types of secondary wound dressings were designed between 2005 and 2010 but not yet implemented on the market (Table 11.5). The wound dressing researches were selected on base of properly provided preclinical date (according to harmonized standards including biocompatibility studies). However, till now there is no implemented CE-certified or FDA-agreed primary or secondary wound dressing, except the hemostatic topical agents.

Composite wound dressings in form of needle-punched nonwovens and sponges containing antibacterial fibers and/or various functional forms of chitosan were developed by Niekraszewicz et al. (2007a, b, c). They may be used over the initial healing period, often on wounds accompanied by inflammations. The antibacterial agents, such as bioactive staple polypropylene fibers containing triclosan, Irgaguard B7000, or aluminosilicate with Ag^+ and Zn^{2+}, would suppress the infection, whereas chitosan would facilitate the wound healing and shorten the recovery period. Modified staple polypropylene fibers characterized by

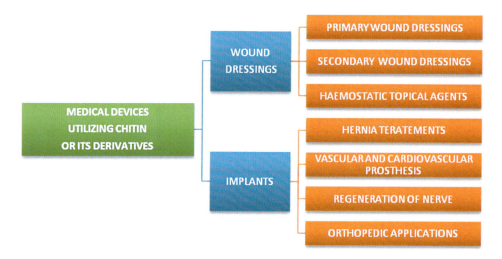

Fig. 11.7 The most important usage of chitin and its derivatives for the design of medical devices (Struszczyk 2002a, b, c)

antibacterial activity and selected chitosan forms, such as microcrystalline chitosan (MCCh) or chitosan fibers, were investigated by Kucharska et al. (2007). The designed dressings were characterized for there bacteriostatic properties against *E. coli* and *Staphylococcus aureus* (Niekraszewicz et al. 2007d; Kucharska et al. 2007).

Preliminary studies of Wiśniewska-Wrona et al. (2010) related to the development of composite films containing chitosan, sodium alginate, and anti-inflammatory agents suggested a potential opportunity to use this material for the treatment of bedsores in the initial phase of wound healing.

The method of UV-photocrosslinking of a solution of chitosan containing azide groups and lactose moieties with incorporated paclitaxel is described by Masayuki Ishihara et al. (2006). The resulting hydrogel showed potential application in the design of hydrogel wound dressings containing integrated fibroblast growth factors or paclitaxel for tumor treatments. The bioactive substances are gradually released during in vivo studies. Additionally, the modified chitosan hydrogels effectively stop bleeding from the carotid artery and air leakage from the lung of tested animals. The sealing strength of hydrogels is higher than fibrin glue.

Validated studies of the nonwoven wound dressings made of dibutyrylchitin (soluble product of chitin substitutions by butyric anhydride) have been reported (Chilarski et al. 2004a, b, 2007; Struszczyk et al. 2004; Krucińska et al. 2006). These studies contained the estimation of the performance (sorption, mechanical properties, etc.) as well as safety (biocompatibility according EN-ISO 10993-1:2003 and clinical studies). Dibutyrylchitin particles were also examined by infrared spectrometry, size exclusion chromatography, and viscometry. Satisfactory results during clinical study of wound healing were achieved in most cases, especially in cases of burn wounds and postoperative/post-traumatic wounds and various other conditions causing skin/epidermis loss (Chilarski et al. 2007).

In vivo studies of acceleration of wound healing by chitosan topical gels with different molecular weight (M_w) and with different degrees of deacetylation (DD) have been examined by Alsarra (2009). The treated wounds were found to contract at the highest rate with high M_w and high DD of chitosan-treated animals (rats) as compared to untreated animals. Wounds treated by chitosan having high M_w showed significantly high epithelialization as well as quicker wounds closure. Histological

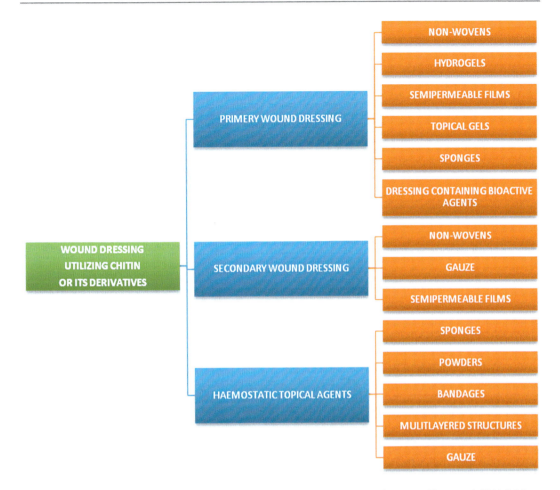

Fig. 11.8 Types of wound dressings utilizing chitin or its derivatives (Struszczyk 2002a, b, c, 2004; Muzzarelli 2009; Kucharska et al. 2011a, b, 2007; Niekraszewicz et al. 2007a, b, c; Wiśniewska-Wrona et al. 2010; Ishihara et al. 2006; Chilarski et al. 2004a, b, 2007)

study and collagenase activity estimation revealed advanced granulation tissue formation and epithelialization in wounds treated by chitosan with high M_w. Topical gel made of chitosan having high M_w and high DD showed potential for use as a treatment system for dermal burns; however, the mode of action may be synergistic via bioactivity of chitosan oligomers as well as suitable moisture remaining for wound healing by chitosan gel with high fluid handling capacity.

The research carried by Madhumathi et al. (2010) described novel α-chitin/nanosilver composite materials for wound healing applications. These α-chitin/nanosilver composite materials indicated excellent antibacterial activity against E. coli and S. aureus combined with good blood clotting ability.

According to Sudheesh Kumar et al. (2010), composite of β-chitin hydrogel containing silver nanoparticles are characterized by their antibacterial properties (E. coli and S aureus), whole blood clotting, swelling, and absence of cytotoxicity. Moreover, β-chitin/nanosilver hydrogels were evaluated for their cell adhesion properties using epithelial cells (Vero cells).

The performance of the designed wound dressing has been described in EN 13726 Standards. The research conducted by Kucharska et al. (2011a, b) evaluates the comprehensive performance of several types of chitosan prototypes of innovative wound dressings (such as

Table 11.5 Primary and secondary wound dressings utilizing chitin and its derivatives designed in 2005–2010

Type of chitin derivative/usable form	Properties of chitin derivatives utilized in wound dressing	Additional substances participating in design of wound dressing	Form of wound dressing	Potential scope of the applications	References
Chitosan	M_w = 400 kDa DD = 80 %	Bioactive PP[a] fibers containing Triclosan, Irgaguard B7000, or aluminosilicate with Ag^+ and Zn^{2+}	Nonwoven	Initial healing period, often on wounds accompanied by inflammations or infection (primary or secondary). Antibacterial properties against: • *E. coli* • *S. aureus*	Niekraszewicz et al. (2007c)
Microcrystalline chitosan	M_w = 445 kDa DD = 82 %	Bioactive PP[a] fibers containing Triclosan, Irgaguard B7000, or aluminosilicate with $Ag+$	Nonwoven obtained by wet forming method containing bioactive fibers	Initial healing period, often on wounds accompanied by inflammations or infection (primary or secondary). Antibacterial properties against: • *E. coli* • *S. aureus*	Kucharska et al. (2007)
Chitosan microfibers	M_w = 330 kDa DD = 80 %		Chitosan sponges containing bioactive fibers or active agents obtained using lyophilization method		
Chitosan lactate	M_w = 319.7 kDa DD = 77.5 %	Sodium alginate; $ZnSO_4$	Semipermeable and semitransparent film	Bedsores treatment Antibacterial properties against *E. coli*	Wiśniewska-Wrona et al. (2010)
Photo-cross-linked chitosan	M_w = 800–1,000 kDa DD = 80 %	Lactobionic acid	Topical gel	Bioactive hydrogel incorporating fibroblast growth factor (FGF-2)	Masayuki Ishihara et al. (2006)
Dibutyrylchitin (DBC)	M_w = 132 kDa	–	Nonwoven	Burns treatments	Chilarski et al. (2004a, b, 2007), Krucińska et al. (2006), Struszczyk et al. (2004)
Chitosan	M_w = 2,000 kDa DD = 92 %	Methyl paraben sodium salt—as a preservative	Topical gel (1 wt% chitosan acetate)	Burns treatments	Alsarra (2009)
Chitin/chitosan blend	Chitosan DD = 93.2 % Chitin DD = approx. 0 %	Alginate, fucoidan Ethylene glycol diglycidyl ether—as cross-linking agent	Hydrogel blends	Wet treatment of chronic wounds	Murakami et al. (2010)
Chitosan	DD = 90 %	Silver sulfadiazine (1 wt%)	Films	Chronic and infected wounds	Shuangyun et al. (2008)

α-Chitin	–	Nanosilver	Composite	Chronic and infected wounds and burns	Madhumathi et al. (2010)
α-Chitin	–	Nanosilver	Hydrogels	Chronic and infected wounds and burns	Sudheesh Kumar et al. (2010)
Microcrystalline chitosan	–	–	Sponges (lyophilization)	Chronic wounds	Kucharska et al. (2011a, b)
Chitosan fibers Chitosan/alginate fibers	–	Alginate	Sponges	Primary wounds treatment	Kucharska et al. (2011a)
Chitosan/carboxymethyl cellulose	–	Carboxymethyl cellulose	Sponges	Wounds treatments at first phase of healing	Kucharska et al. (2011a)
Chitosan	–	–	Sponges (lyophilization) made of chitosan gel	Artificial skin replacement	Kucharska et al. (2011a)
Chitosan	–	–	Hydrogel	Topical amorphous wound dressing for wet treatments	Kucharska et al. (2011a)
Chitosan	–	–	Films (semitransparent; semipermeable)	Primary wound dressings	Kucharska et al. (2011a)

[a]PP is polypropylene

sponges, hydrogels, films) made of various usable form of chitosan (microcrystalline chitosan, chitosan fibers, chitosan/alginate fibers, and chitosan/carboxymethylcellulose mixture).

The evaluation of the above-mentioned prototypes of wound dressings was carried out using the guidelines from series of EN 13726 Standards in scope of:
- Free swell absorptive capacity
- Fluid handling capacity
- Gelling properties
- Dispersion characteristics
- Moisture vapor transmission rate (MVTR) of the wound dressing when in contact with water vapor or liquid
- Waterproofness
- Extensibility
- Permanent set

Commercially applied chitin or its derivative-based wound dressings for dermis regeneration are listed in Table 11.6 (Muzzarelli 2009).

Hemostatic Topical Agents

The hemostatic properties of chitosan and its oligomers have been commercially applied in several topical hemostatic dressings which differ in their usable form. Many researches showed the ability of chitosans, which differ in molecular weight and degree of deacetylation to local activation of platelets and turnover of the intrinsic blood coagulation cascade (Muzzarelli 2009; Kucharska et al. 2011a, b). Moreover, the gelling properties of chitosans promote the formation of a physical barrier against massive bleeding (Kucharska et al. 2011a, b; Struszczyk and Struszczyk 2007). The mode of hemostatic action of chitosan strongly depends on its chemical as well as physical nature and structure. The commercially approved topical hemostatic agents are listed in Table 11.7.

Implantable Application of Chitin and Its Derivatives

Hernia Treatments. Hernias are a very serious health problem and statistically appearing in 2 % of the human population. The procedure of hernia treatments is among the main surgical interventions in general surgery. The non-resorbable implant, so-called surgical mesh or hernia mesh, is typically used. An essential disadvantage of above methods is the patients' discomfort listed as a long-term complication of hernioplasty. This phenomenon is connected with the stiffening of the implant related with the formation of non-regular and massive structures of connective tissue around the implant. Other complication of hernioplasty is an inadequate blood supply to the organs lying in the implant's surroundings, as well as the fistula formation or adhesion type in the case of an implant's direct contact with viscera (most frequently appearing in several types of laparoscopic hernia treatments) (Amid 1997; Niekraszewicz et al. 2007a, b, c). The implantation of partially resorbable surgical meshes reduces significantly most of the conventional disadvantages.

The aim of the research of Niekraszewicz et al. (2007a, b, c) was to develop new/novel ideas of partially resorbable surgical meshes or partially resorbable composite implants dedicated to less-invasive procedures of hernia treatment. The following variants of implants were designed:
- Variant I was manufactured by knitting technique, where two types of yarns were used: a resorbable multifilament chitosan thread and a non-resorbable polypropylene monofilament.
- Variant II consisted of a complex technique: both or single surface of non-resorbable and low-weight knits was covered by a micro porous resorbable chitosan coating.

Both variants of implants were tested after sterilization by ethylene oxide (EO) for the estimation of their biomechanical features as well as their chemical and biological purity. Moreover, the in vitro tests of the susceptibility to enzymatic degradation of surgical implants, as well as carrying out biomedical investigations into estimating the cytotoxicity, irritation, allergic effects, and implantation tests (according to harmonized standards with European Directive for medical devices) with experimental animals were performed according to harmonized standards by Niekraszewicz et al. (2007a, b, c).

Table 11.6 Commercial wound dressings for dermis regeneration utilizing chitin or its derivatives

Trade name	Description
ChorioChit®	Biological dressing manufactured by the lyophilization of human placenta blended with microcrystalline chitosan (Polish Patent 320491 (1997)) ChorioChit® was characterized by good handling, good wound isolation, and the ability to the reduction of pathogens growth The application range was chronic wounds, mainly skin ulceration. Due to the strictly defined composition and form of the wound dressing, the material exerts stimulation and bacteriostatic action as well as absorbs wound exudates (Niekraszewicz 2005) The wound dressing was not CE-certificated[a] and not approved by FDA[b]. Available in Poland till 2004
Beschitin® Unitika	The wound dressing in form of nonwoven manufactured using chitin. It accelerates granulation phase and affects no scar formation The application range is connected with the surgical intervention (external connective tissue) (Muzzarelli 2009) The wound dressing was not CE-certificated[a] and not approved by FDA[b]. Available only in Japan
ChitiPack® S Eisai Co.	The wound dressing made by the freeze-drying of squid pen chitin dispersion. The usable form of wound dressing is a sponge-like product. The wound dressing favors the acceleration of wound healing and gives no scar formation (Muzzarelli 2009) The wound dressing was not CE-certificated[a] and not approved by FDA[b]. Available only in Japan
ChitiPack® P Eisai Co.	The wound dressing is designed by drifting the chitin suspension onto poly-(ethylene terephthalate] nonwoven fabric. The recommended range of application is large skin defects, especially with difficulty to suture (Muzzarelli 2009) The wound dressing was not CE-certificated[a] and not approved by FDA[b]. Available only in Japan
Chitopack®C Eisai Co.	The fibrous wound dressing made by spinning of chitosan acetate solution coagulated in a bath containing mixture of ethylene glycol (humectant), cold water, and sodium or potassium hydroxide. The application range considers the regeneration and reconstruction of body tissue, subcutaneous tissue, as well as skin (Muzzarelli 2009) The wound dressing was not CE-certificated[a] and not approved by FDA[b]. Available only in Japan
Chitodine® IMS	Powdered chitosan containing elementary iodine. Applicable for disinfection and cleaning of wounded skin and as a primary wound dressing (Muzzarelli 2009) Wound dressing has obtained CE-mark
Vulnsorb® Tesla	The composition of collagen and chitosan in form of freeze-dried sponges (Muzzarelli 2009) Wound dressing has obtained CE-mark since 1996 (Muzzarelli 2009)

[a]CE certification—conformity assessment with the requirements of EU Directive made by producer (medical devices class I or with participation of notified body (higher classes of medical devices)
[b]FDA—Federal Drug Agency (USA)

Textile implants (variants II) coated by chitosan on only one side were selected for further research as they promised better commercial and clinical application possibilities (Niekraszewicz et al. 2007a, b, c, 2008, Niekraszewicz et al. PL Patent Appl. No. 380861, 2006).

A patent application by Niekraszewicz et al. (PL Patent Appl. No. 380861, 2006) describes the method of manufacturing a multilayer hernia mesh made of monofilaments, polypropylene knit as a non-resorbable layer, and modified bacterial cellulose synthesized biotechnologically by *Acetobacter* strains. Biotechnologically modified microbiological cellulose has been carried out using chitosan (having M_w from 50 to 200 kDa and DD from 75 to 95 %) or its oligomers during its biosynthesis. Application of bacterial cellulose modified by chitosan as a layer of implant significantly decreased the inflammatory reaction against implant, reduced the complications connected to stiffed implant due to the massive scar formation around the whole of implant, and

Table 11.7 Commercially available hemostatic topical agents based on chitin or its derivatives

Trade name/company	Description
HemCon® Bandage ChitoFleX® Hemostatic Dressing HemCon Medical Technologies	HemCon products are made of positively charged chitosan (chitosan acetate) by lyophilization. Positively charged chitosan salt has strong affinity to bind with red blood cells, activates the platelets, and forms a clot that stops massive blood bleeding. Moreover, it forms an antibacterial barrier resulting in the reduction of the potentiality of the secondary infection. It adheres to tissue surfaces when in contact with blood or moisture forming a strong, flexible barrier that seals and stabilizes the wound. Extremely robust, the bandage tolerates highly compressive forces while retaining adequate flexibility to conform to irregular wound surfaces (Muzzarelli 2009; Struszczyk and Struszczyk 2007) HemCon® Bandage was approved by FDA[a] in 2002, whereas ChitoFleX® Hemostatic Dressing was approved in 2007 (Struszczyk and Struszczyk 2007)
Syvatek® Patch Marine Polymer Technologies	Syvatek® Patch has been implemented for the control of bleeding at vascular access sites in interventional cardiology and radiology procedures. It consists of poly-N-acetyl glucosamine (pGlcNAc) isolated in a unique fiber crystalline structural form. It is able to significantly reduce the fibrin clot formation time and has the ability to cause aggregation of red blood cells form (Struszczyk (2002a)) Approved by FDA[a] and CE-certificated[b] (Struszczyk and Struszczyk 2007)
Clo-Sur® PAD Scion Cardio-Vascular	Clo-Sur® PAD is made as a nonwoven sealed by a soluble form of chitosan Approved by FDA[a] and CE-certificated[b] (Struszczyk and Struszczyk 2007)
ChitoSeal® Abbott Vascular Devices	ChitoSeal® is made of soluble chitosan salt. It is intended for external temporary use to control moderate to severe bleeding Approved by FDA[a] and CE-certificated[b] (Struszczyk and Struszczyk 2007)
Traumastat® Ore-Medix	Traumastat® is made on the nonwoven substrate comprised of porous polyethylene fibers highly filled with precipitated silica. This substrate is coated with chitosan (ChitoClear™). It is intended for external temporary use to control moderate to severe bleeding Approved by FDA[a] and CE-certificated[b]
ExcelArrest® Hemostasis LLC Co.	ExcelArrest® is comprised of modified chitin particles and polysaccharide binders. Hemostat is manufactured in form of foam by lyophilization process from the chitin and polysaccharides suspension. The addition of sodium carboxymethyl cellulose and hydroxyethyl cellulose binders has an affinity to hold hydrophilic fluids. The lyophilized foam quickly dehydrates blood cells, and indirectly causing rapid hemoconcentration of platelets, serum proteins, and fibrinogen. Above phenomenon leads to clotting that limits and controls bleeding in moderate to severe lacerations. Approved by FDA[a] in 2007
Tromboguard® TRICOMED SA	Multifunctional Tromboguard® consist in three layers: 1. External: made of semipermeable, microporous film selectively passing gases and protecting against external factors including bacteria 2. Middle: made of hydrophilic polyurethane foam with innovative structure "pore-in-pore"—responsible for adsorption of amorphous parts of blood and local concentration of natural blood coagulation factors, such as platelets, serum proteins, and fibrinogen 3. Internal (contacted with wound or trauma): made of a mixture of chitosan salt, sodium/calcium alginate and silver salt—responsible for direct blood coagulation activation, adhesion, and resistance against primary or secondary infection (Kucharska et al. 20062011b) It shows multifunctional behavior: local hemostasis, antibacterial, and acceleration of wound healing During CE-certification[b] (Kucharska et al. 2011b)

[a]FDA—Federal Drug Agency (USA)
[b]CE certification—conformity assessment with requirements of EU Directive

reduced the risk of the adhesion of implant to viscera during the laparoscopic hernia treatment. Ciechańska et al. (PL Patent Appl. No. P 390253, 2010a) demonstrated that the presence of chitosan resulted in the antibacterial behavior of the final product.

Vascular Prosthesis. The research by Ciechańska et al. (2010a, b) presents new ideas on application of bacterial cellulose modified by chitosan during its biosynthesis for design of vascular prostheses for microsurgery. The biosynthesis was carried out under both static (using a silicone tube reactor) and dynamic conditions (in a reactor equipped with a rotating shaft partly immersed in the medium) using selected *Acetobacter* strains. The resulted tubes were tested and assessed with respect to its structural, biomechanical (acc. ISO 7198:1998 Standard), and biological properties as well as its chemical purity. The study showed a high biocompatibility of the modified bacterial cellulose that is essential for the design material for vascular prostheses. It was found that the mechanical properties of the prosthesis, such as the peripheral and longitudinal tenacity as well as suture retention strength, can be improved by applying a special designed polyester plaited carrier in the course of the dynamic synthesis of the modified cellulose. The biocompatibility studies conducted by Ciechańska et al. (2010a, b) did not show cytotoxicity, whereas in vivo studies in range of irritation, allergic reaction as well as local effect after implantation indicate designed vascular prostheses as a promising material for the vascular microsurgery. Introduction of blocks of glucosamine units in chains of bacterial cellulose yielded new properties, such as change in surface charge (important for the resistance against aggregation and activation of platelets as well as blood coagulation) and antibacterial properties.

Nerve Regeneration. Peripheral nerve injuries and/or defects are among the serious causes of human disability. Kardas et al. (2010) demonstrated that regeneration of peripheral nerves is still unsatisfactory even through evolution of the microsurgical procedures that improve their efficiency. Yamaguchi et al. (2003a, b) presented methods of crystalline chitin tubes production by direct removal of the calcium phosphate and proteins from crab tendons. The obtained tubes have a tubular structure and a suitable size for induction of peripheral nerve reconstruction. However, its mechanical strength is still too low resulting in the swelling of the tube wall and causing stenosis. The improvement of the mechanical behavior has been attained after application of a heat-press method as well as alternative soaking method for modification of the tube surface (Yamaguchi et al. 2003a, b; Itoh et al. 2003a, b; Taguchi et al. 1998). Itoh et al. (2003a, b) described the evaluation of nerve conduit of the hydroxyapatite-coated and artificial laminin-peptides (accelerated nerve regeneration) adsorbed tubes in the range of their mechanical properties, biocompatibility, and efficacy. Also, semi-artificial bonding of laminin was obtained by Lu et al. (2007) using *O*-carboxymethyl chitosan and 1-ethyl-3(3-dimethylaminopropyl) carbodiimide as a cross-linker. Poly-D-lysine was linked with chitosan using azidoaniline photocoupling to improve cell adhesion and neurite regeneration (Muzzarelli 2009; Crompton et al. 2007). The above-discussed literature review unfortunately does not cover any compliant with the EU (harmonized standards) or USA (FDA Guidelines) studies. The reliability of commercial implementation of nerve prosthesis is still under discussion.

Orthopedic Applications. The future trends in orthopedic surgery and neurosurgery are to find solutions for easy and quickly regeneration or repair of bones. Ratajska et al. (2008) have elaborated a novel group of products which could be used as an adhesive or implants in orthopedic surgery consisting of various chitosan forms (microcrystalline chitosan, chitosan lactate differing in pH) and hydroxyapatite extracted from the cortical part of long pig bones. All prepared mixtures of chitosan forms and hydroxyapatite showed affinity to form a three-dimensional structure—sponges regardless of the composition and types of chitosan form used.

Osteosynthesis ability, mechanical properties susceptibility to biodegradation, and biocompatibility of the prepared hydroxyapatite/chitosan glue composites were estimated by Niekraszewicz et al. (2009). The most important factor in the design of bone glue or paste is the proper selection of a usable form of chitosan and its physical and chemical properties, such as pH, polymer content, degree of deacetylation, and viscosity. The composition of bone glue was consistent in chitosan (DD—83 % and M_w—334 kDa) and hydroxyapatite (decacalcium dihydroxy hexaphosphate) when cross-linked by glutaraldehyde and glycerol glue was added as a humectant. Glue, film, and lyophilized sponges were resulting products of the research. The glue joints in form of a film underwent enzymatic degradation in the presence of lysozyme in vitro. After 28 days of the above-mentioned enzymatic treatment, the mass loss of the glue was ~21 %. The elaborated chitosan-hydroxyapatite glue is characterized by optimal adhesion to the bone surface tested in vitro. A maximum of the glue joint is attained after maximal 3 days, which is acceptable in bone surgery both for reconstruction of porous and smooth bones. If glutaraldehyde was used as a cross-linker, the joint glue is characterized by tenacity from 40 to 130 %, regardless of the kind of the bones and joining methods. The compressed chitosan/hydroxyapatite sponges present the potential use as defect fillers in maxillofacial surgery.

11.4.3 Bionanotechnology

Bionanotechnology, a marriage between biology and nanotechnology, is an emerging field. Through biomimetic approaches and strategies, many micro-/nano-systems can be produced. Chitosan has been used by Koev et al. (2006) to immobilize and pattern biomolecules on microfabricated surfaces. A photolithographic method has been applied by Cheng and Pisano (2008) to integrate chitosan to micro- and nano-structures, which is an important step toward the fabrication of bioinspired micro-electromechanical systems. Jeong (2007) used nanoimprinting lithography to micro- and nanopattern chitosan. This micro-/nanopatterning enables researchers to use chitosan for bionanotechnology applications such as nanobiodevices. Chitosan has recently been used by Wang et al. (2010) in the preparation of graphitic carbon nanocapsules, tungsten carbide, and tungsten carbides/graphitic carbon composites. In this system, after preparation of the precursors of chitosan and metal ions, they were carbonized. The system can be regulated by changing the type and/or ratio of the metal. Chitin whiskers have been proposed by Gopalan and Dufresne (2003) to reinforce nanocomposites. It seems this ability mainly depends on chitin whiskers being able to form 3D networks. Any modification by which this network is disrupted results in lowering or loss of this ability.

11.4.4 Agriculture

Both chitin and chitosan have demonstrated antiviral, antibacterial, and antifungal properties and have been explored for many agricultural uses. They have been utilized to control disease or reduce their spread; to chelate nutrient and minerals, preventing pathogens from accessing them; or to enhance plant innate defenses. Chitosan exhibits a variety of antimicrobial activities (Pośpieszny et al. 1991; Rabea et al. 2003; Kulikov et al. 2006), which depend on the type of chitosan (native or modified), its degree of polymerization, the host, the chemical and/or nutrient composition of the substrates, and environmental conditions. For example, a commercial chitosan preparation (Biochikol 020 PC) showed antiviral, antibacterial, and antifungal effects and belongs to the inducers of plants resistance (Benhamou et al. 1994; Patkowska and Pięta 2004; Pięta et al. 2006; Pośpieszny 1995). In some studies, for example, conducted by Rabea et al. (2003), oligomeric chitosans (pentamers and heptamers) have been reported to exhibit a better antifungal activity than larger units. In others, the antimicrobial activity increased with the increase in chitosan molecular weight (Kulikov et al. 2006) and seems to be faster on fungi and algae than on bacteria (Savard et al. 2002).

11.4.4.1 Viruses

Chitosan has shown to inhibit the systemic propagation of viruses and viroids throughout the plant and to enhance the host's hypersensitive response to infection (Pośpieszny et al. 1991; Faoro et al. 2001; Chirkov 2002). According to Kulikov et al. (2006), the level of suppression of viral infections varied according to chitosan molecular weight. Similar observations were reported with the potato virus X, tobacco mosaic and necrosis viruses, alfalfa mosaic virus, peanut stunt virus, and cucumber mosaic virus (Struszczyk 2002c; Pośpieszny et al. 1991; Chirkov 2002; Pośpieszny et al. 1996; Pośpieszny 1997).

11.4.4.2 Bacteria

Muzzarelli et al. (1990) reported that chitosan inhibits the growth of a wide range of bacteria. The minimal growth-inhibiting concentrations vary among species from 10 to 1,000 ppm. Quaternary ammonium salts of chitosan, such as *N,N,N*-trimethylchitosan, *N*-propyl-*N,N*-dimethylchitosan, and *N*-furfuryl-*N,N*-dimethylchitosan, were shown by Jia et al. (2001) to be effective in inhibiting the growth and development of *E. coli*, especially in acidic media. Similarly, according to Kim et al. (1997), several derivatives of chitin and chitosan were shown to inhibit *E. coli*, *S. aureus*, some *Bacillus* species, and several bacteria infecting fish.

11.4.4.3 Fungi and Oomycetes

Fungicidal activity of chitosan has been documented against various species of fungi and oomycetes (Muzzarelli et al. 1990, Vasyukova et al. 2005). The minimal growth-inhibiting concentrations varied between 10 and 5,000 ppm (Stössel and Leuba 1984; Sudarshan et al. 1992; Wang 1992; Tsai and Su 1999; Rhoades and Roller 2000). Badawy et al. (2005) reported on the fungicidal activity of 24 new derivatives of chitosan (i.e., *N*-alkyl, *N*-benzylchitosans) and showed that all derivatives have a higher fungicidal action than the native chitosan. Recently, Palma-Guerrero et al. (2008) demonstrated that chitosan is able to permeabilize the plasma membrane of *Neurospora crassa* and kills the cells in an energy-dependent manner. In general, chitosan, applied at a rate of 1 mg mL^{-1}, is able to reduce the in vitro growth of a number of fungi and oomycetes except *Zygomycetes*, which have chitosan as a component of their cell walls, as demonstrated by Allan and Hadwiger (1979). Another category of fungi that seems to be resistant to the antifungal effect of chitosan is the nemato-/entomopathogenic fungi that possess extracellular chitosanolytic activity.

11.4.4.4 Insects

As more and more derivatives of chitosan (i.e., *N*-alkyl-, *N*-benzylchitosans) are made available through chemical synthesis, their insecticidal activities are being reported using an oral larvae feeding bioassay (Palma-Guerrero et al. 2008; Muzzarelli et al. 2001). Twenty-four new derivatives were shown to have significant insecticidal activity when administered at a rate of 5 g kg^{-1} in an artificial diet. The most active derivative, *N*-(2-chloro-6-fluorobenzyl)chitosan, caused 100 % mortality of larvae and its LC$_{50}$ was estimated at 0.32 g kg^{-1}. All synthesized derivatives highly inhibited larvae growth as compared to chitosan by 7 % and the most active derivative was the *O*-(decanoyl)chitosan, with 64 % growth inhibition after 5 days of feeding on the treated artificial diet.

11.4.4.5 Applications of Chitosan in Plant Disease Control

The literature provides information about the possibilities of using chitosan in the protection of plants against different diseases (Lafontaine and Benhamou 1996; Orlikowski and Skrzypczak 2003a, b; Wojdyła 2001). Chitosan used to control plant pathogens has been extensively explored with more or less success depending on the pathosystem, the used derivatives, concentration, degree of deacylation, viscosity, and the applied formulation (i.e., soil amendment, foliar application; chitosan alone or in association with other treatments). For example, Muzzarelli et al. (2001) tested the effectiveness of five chemically modified chitosan derivatives in restricting the growth of *Saprolegnia parasitica*. Results indicated that methylpyrolidinonechitosan, *N*-phosphonomethylchitosan, and *N*-carboxymethylchitosan, as opposed to *N*-dicarboxymethylchitosan, did not allow the fungus to grow normally. Substratum amendment with chitosan was reported to enhance plant growth and suppress some of the notorious soilborne diseases. For example, in soilless tomato, root rot caused by *Fusarium oxysporum* f. sp. *radicis-lycopersici* was suppressed using chitosan amendments by Lafontaine and Benhamou (1996). Similarly, according to Benhamou (2004), in order to control post-harvest diseases, addition of chitosan stimulated microbial degradation of pathogens in a way resembling the application of a hyper-parasite. This area of application is important because it suggests alternatives to the use of pesticides on fresh products in storage. Recent investigations on coating tomatoes with chitosan have shown that it delayed ripening by modifying the internal atmosphere, which reduced decays due to pathogens (El Ghaouth et al. 1992, 2000). Various methods of application of chitosan and chitin are practiced to control or prevent the development of plant diseases or trigger plant innate defenses against pathogens. Many studies have shown that chitosan as the biopreparation can be used for the control or retardation of fungi pathogens on tomato and onion (Borkowski et al. 1998, 2001; 2002, 2003, 2004), runner bean (Pięta et al. 2002), rose (Wojdyła 2001), and several ornamental plants (Orlikowski et al. 2001; Wojdyła et al. 1996, 2001; Orlikowski and Skrzypczak 2003a, b). Borkowski et al. (2007) found that spraying the tomato plants with chitosan may increase vitality of those plants after some months. This phenomenon was probably connected with larger resistance of tomato roots to fungal pathogens. During the following years the possibility was checked if Biochikol 020 PC (chitosan) was able to control or retard the development of *Orobanche ramosa* L. on tomato roots or stem.

11.4.4.6 Applied as Seed Coating Agents

Guan et al. (2009) examined the use of chitosan to prime maize seeds. Although chitosan had no significant effect on germination at low temperatures, it enhanced germination index, reduced the mean germination time, and increased shoot height, root length, and shoot and root dry weights in two tested maize lines. In both tested lines, chitosan induced a decline in malonyldialdehyde content, altered the relative permeability of the plasma membrane, and increased the concentrations of soluble sugars and proline and of peroxidase and catalase activities. In other studies, conducted by Shao et al. (2005), seed priming with chitosan improved the vitality of maize seedlings. It was also reported by Reddy et al. (1999) to increase wheat seed resistance to certain diseases and improve their quality and/or their ability to germinate. Similarly, peanut seeds soaked in chitosan were reported to exhibit an increased rate of germination and energy, lipase activity, and gibberellic acid and indole acetic acid levels (Zhou et al. 2002). Ruan and Xue (2002) demonstrated that rice seed coating with chitosan may accelerate their germination and improve their tolerance to stress conditions. In carrot, seed coating help to restrain further development of *Sclerotinia* rot, according to Cheah and Page (1997). Chitosan has also been extensively utilized as a seed treatment to control *F. oxysporum* in many host species (Rabea et al. 2003).

11.4.4.7 Applied as Foliar Treatment Agents

Foliar application of chitosan has been reported in many systems and for several purposes. For instance, Khan et al. (2002) reported that foliar application of a chitosan pentamer affected the net photosynthetic rate of soybean and maize 1 day after application Khan et al. (2002). This correlated with increase in stomatal conductance and transpiration rate. Chitosan foliar application did not have any effect on the intercellular CO_2 concentration. The observed effect on the net photosynthetic rate is common in maize and soybean after foliar application of high molecular weight chitosan. Foliar applications of these oligomers did not, on the other hand, affect maize or soybean height, root length, leaf area, or total dry mass. Bittelli et al. (2001) suggested that chitosan might be an effective anti-transpiring to preserve water resources use in agriculture. In their investigation, they examined the potential of foliar applications of chitosan on pepper plants transpiration in the growth room and in the field. In both experiments, the authors monitored plant water use directly and indirectly. The plant biomass and yield were determined to calculate biomass-to-water ratios, and the differences in canopy resistance between control and chitosan-treated plants were analyzed. Using scanning electron microscopy and histochemical analyses, stomata were shown to close in response to treatment with chitosan, resulting in a decrease in transpiration. Reduced water use of pepper plants upon treatment with chitosan was estimated at 26–43 %, while there was no change in biomass production or yield (El Ghaouth et al. 2000). Iriti et al. (2009) unveiled some of the aspects through which chitosan was able to reduce transpiration in bean plants after being used as a foliar spray. The authors showed that this activity was likely occurring thanks to the increase in abscisic acid (ABA) content in the treated leaves. Using scanning electron microscopy and other histocytochemistry techniques, the authors showed that upon treatment and increase in ABA content, a partial stomatal closure occurred and led, among others, to a decrease in conductance for water vapor and in the overall transpiration rate. Interestingly, the authors revealed a new chitosan antitranspirant mechanism in bean plants that was not described by their commercial supplier Vapor Gard® and in which a formation of a thin antitranspirant film at the surface of the leaves was much more efficient than stomatal closure. This difference in mechanisms also suggested an important consideration for the environmental conditions under which chitosan is applied as shown by the authors but may also depend on the intrinsic properties of the tested plant species.

Chitosan has also been extensively utilized as a foliar treatment to control the growth, spread, and development of many diseases involving viruses, bacteria, fungi, and pests, according to Rabea et al. (2003). It has also been used by Kowalski et al. (2006) to increase yield and tuber quality of micropropagated greenhouse-grown potatoes. Similarly, Faoro et al. (2008) showed that the use of chitosan applied as a foliar spray on barley reduced locally and systemically the infection by powdery mildew pathogen *Blumeria graminis* f. sp. *hordei*.

11.4.4.8 Applied as Soil Amendment

Chitosan utilized as a soil amendment was shown to control *Fusarium* wilts in many plant species (Rabea et al. 2003). Applied at an optimal concentration, this biomaterial is able to induce a delay in disease development, leading to a reduced plant wilting, as reported by Benhamou et al. (1994). Similar results were reported by Laflamme et al. (1999) in forest nurseries suffering from *F. acuminatum* and *Cylindrocladium floridanum* infections. These infections were dramatically reduced upon the use of chitosan as soil amendment. *A. flavus* was also completely inhibited in field-grown corn and peanut after soil treatment with chitosan, according to El Ghaouth et al. (1992). Part of the effect observed by chitosan on the reduction of soil-borne pathogens comes from the fact that it enhances plant defense responses. The other part is linked to the fact that this biopolymer is composed of polysaccharides that stimulate the activity of beneficial microorganisms in the soil such as *Bacillus, Pseudomonas*, actinomycetes, mycorrhiza, and rhizobacteria (Bell et al. 1998; Murphy et al. 2000). This alters the microbial equilibrium in the rhizosphere disadvantaging plant pathogens. Beneficial organisms, on the other hand, are able to outcompete them through mechanisms such as parasitism, antibiosis, and induced resistance (Daayf et al. 2003; El Hassni et al. 2004; Uppal et al. 2008). Vruggink (1970) reported on the effect of chitin amendment on actinomycetes in soil and on the infection of potato from susceptible cultivar 'Bintje' by *S. scabies*, the causal agent of tuber scab. The percentage scab on tubers from the control and the soil amended with antagonist was about 22 % while only 4 % of the tubers from the soil amended with chitin and chitin with antagonist had scab at harvest. After planting these tubers, for a second time, the scab was 21 % on tubers from untreated soil and 9.5 % from soil amended with chitin. Investigation of the effect of chitin amendment on the actinomycete population in the soil, a few months after chitin amendment, revealed that chitin had a greater increase in total actinomycete population (24–30 times as compared to the untreated control). The study also showed that some actinomycetes (i.e., *Micromonospora*) had disappeared, while others including *S. scabies* were isolated less frequently.

11.4.5 Cosmetics

Usually organic acids are used as solvents for cosmetic applications. The natural aminopolysaccharide, chitosan, can be classified as hydrocolloid. However, unlike most other hydrocolloids which are polyanions, chitosan is the only natural cationic gum that becomes viscous on being neutralized with acid. It facilitates its interaction with common integuments (skin covers) and hair. Chitin and chitosan are fungicidal and fungistatic in nature. Chitosan is compatible with many biologically active components incorporated in cosmetic products. Chitosan or chitosan-alginate composites in the range of 1–10 μ, as well as microcapsules including various hydrophobic substances find a wide application in cosmetics. It may be noted that substances absorbing the harmful UV radiation or different dyes can be easily covalently linked to chitosan amino groups. Compositions based on chitosan and other hydrocolloids containing antioxidants, antiallergic, and anti-inflammatory substances of vegetable origin, new types of depilatory creams, and hair tonics for curling are being elaborated by Sonat Co and Kitozyme. Chitin, chitosan, and their derivatives can be used in three areas of cosmetics: hair care, skin care, and oral care (Dutta et al. 2004; Thanou and Junginger 2005; Ravi Kumar et al. 2006). Products have already reached the market in all three areas.

11.4.5.1 Hair Care

Chitosan and hair are complementary because they carry opposite electrical charges—chitosan positive and hair negative. A clear solution that contains chitosan forms a clear, elastic film on hair, increasing its softness and mechanical strength. The material can also form a gel when added to mixtures of alcohol and water. Chitosan shows an antistatic activity, which enables hair protection from waving and bleaching agents. It helps to retain moisture in low humidity and to maintain hair's style in high humidity. The material can be used in shampoos, rinses, permanent wave agents, hair colorants, styling lotions, hair sprays, and hair tonics. Several derivatives of chitosan and chitin have potential applications in hair care. They include glyceryl chitosan, an adduct of an oligomer of hydrolyzed chitosan, n-hydroxypropyl chitosan, quaternary hydroxypropyl-substituted chitosan, polyoxyaikylene chitosan, chitosan oligosaccharides, chitin sulfate, and carboxymethyl chitin. Some derivatives of chitosan can form foams and act as emulsifying agents. Chitin powder can be used directly in shampoo (Dutta et al. 2004; Pandya 2007).

11.4.5.2 Skin Care

Chitosan and its derivatives have two advantages that make it suitable for skin care: their positive electrical charge and the fact that the molecular weights of most chitosan products are so high that they cannot penetrate the skin. Thus, for example, chitosan can function as a moisturizer for skin. Because of its lower costs, it might compete with hyaluronic acid in this application. Both chitosan and chitin are already found in creams, packaging material, lotions, nail enamel, foundation, eye shadow, lipstick, cleansing materials, and bath agents. In many cases, cosmetics companies use the same derivatives and formulations for skin care that they apply in products for hair care. For skin care, also chitosan acylated with an organic diacid anhydride and fine particles of chitin or chitosan (MCC) are used (Dutta et al. 2004; Pandya 2007).

11.4.5.3 Oral Care

Both chitin and chitosan can be used in toothpaste, mouthwashes, and chewing gum; they freshen the breath and prevent the formation of plaque and tooth decay. Salts of chitosan added to toothpaste mask the unpleasant taste of silicon oxide and bind powders so that they maintain their granular shapes. Chitin can also be applied as a dental filler material. Both chitin and chitosan absorb *Candida*, a fungus that sticks to teeth, making them candidates to clean false teeth (Dutta et al. 2004; Pandya 2007; Tozaki et al. 2002).

11.5 Future Trends in Chitin and Its Derivatives in Medicine

According to the literature (Struszczyk and Struszczyk 2007; Struszczyk 2006), the most crucial limitations that are encountered in the implementation and use of chitin and its derivatives for medical devices are:

- Collection and supply of the appropriate quality raw materials.
- Difficulty to manufacture reproducible product batches from various sources of raw material and different collection periods.
- Production cost is still too high; however, the efficiency is expected to be improved in future.
- Lack of knowledge on the exact physiological mechanism of chitosan sources required for advanced applications in medicine.
- Absence of the standardization for chitinous raw materials with respect to their use in medical applications.
- Absence of validated processes of biopolymer manufacture.
- Unavailability of a reliable quality assessment system for chitinous derivatives manufacturing (such as ISO 9001:2008 Quality Management Systems: Requirements) or manufacturing of medical devices (such as ISO 13485:2003 Medical devices—Quality management systems: Requirements for regulatory purposes).
- Absence of standardization of product quality and product assay methods for chitin and its derivatives in scope of medical application.
- High cost of biocompatibility studies as well as clinical assessment (including clinical studies)

A prerequisite for the reproducible processing of chitinous raw products is understanding and control of the important parameters of sources and final products. It needs standardized test methods for biological, physical, and chemical characterization.

Overcoming the above-described difficulties as encountered in ASTM Standards for Tissue-Engineered Medical Product (TEMP), by ASTM F04 Division IV, includes the following issues:

- Patient safety
- Relation of function
- Reproducible results
- Independent assessment(s)

These issues improve manufacturing efficiency, reduce costs of development and manufacture processes significantly, enhance the clinical effectiveness, as well as reduce the regulatory hurdles and timelines (Struszczyk and Struszczyk 2007; Struszczyk 2006).

The existing standards describing the requirements for chitosan, if applicable in medicine, are as follow:

- F2103 Standard—*Standard Guide for Characterization and Testing of Chitosan Salts as Starting Materials Intended for Use in Biomedical and Tissue-Engineered Medical Product Applications.* The guide covers the evaluation of chitosan salts suitable for use in biomedical or pharmaceutical applications, or both, but not limited to tissue-engineered medical products (TEMPs).
- F2260-03 Standard—*Test Method for the Determination of the Degree of Deacetylation of Chitosan Salts by Proton Nuclear Magnetic Resonance (^1H NMR) Spectroscopy.* The test method covers the determination of DD of chitosan and chitosan salts intended for use in biomedical and pharmaceutical applications as well as in Tissue-Engineered Medical Products (TEMPs) by high-resolution proton NMR (^1H NMR).
- WK965 Standard—*Test Method for the Determination of the Molecular Weight of CHITOSAN and CHITOSAN Salts by Size Exclusion Chromatography with Multi-Angle Light Scattering Detection (SEC-MALS).* The test method covers the determination of the molecular weight of chitosan and chitosan salts by size exclusion chromatography with multi-angle light scattering detection (SEC-MALS).

The future application of chitin and its wide scope of derivatives in medical fields significantly depend on the proper description of the base requirements in range of chemical and physical characterizations and chemical purity determination as well as biological conformity assessments.

11.6 Research on Chitin/Chitosan Within EPNOE

Within the European Polysaccharide Network of Excellence (EPNOE), the common research project in Fundamental Theme (FT) entitled "Polysaccharides in human technologies" was studied by the Institute of Biopolymers and Chemical Fibres (Poland) and Wageningen UR Food & Biobased Research (The Netherlands) (Persin et al. 2011). The FT "Polysaccharides in human technologies" was developed around the concept of the enhancement of the utilization of polysaccharides in pharmaceutical, medical, nutraceutical, and cosmetic applications. The research work focused on the investigation of new routes for tailoring the properties of native polysaccharides to the requirement of the applications by chemical and/or enzymatic modifications, for understanding the relation between the structural features of selected polysaccharides and their functional properties, and the development of new products derived from polysaccharides for biomedical applications. For this FT, the specific topic entitled "Biocompatible polysaccharide-based nano-carriers for drug delivery systems" was defined. The aim of this project was to develop new concepts for polysaccharide-based nano-carriers for target delivery and controlled release of bioactive compounds and drugs.

References

Acosta N et al (1993) Extraction and characterization of chitin from crustaceans. Biomass Bioenergy 5(2):145–153

Agboh OB (1986) The production of fibres from chitin, PhD thesis, University of Leeds

Akbuga J, Bergisadi N (1996) J Microencapsul 13:161–168

Alimuniar A, Zainuddin R (1992) In: Brine CJ, Sandford PA, Zikakis JP (eds) Advances in chitin and chitosan, vol 1. Elsevier Applied Science, London, p 627

Allan C, Hadwiger LA (1979) The fungicidal effect of chitosan on fungi of varying cell wall composition. Exp Mycol 3:285–287

Allan GG, McConnell WJ (1975) C. Ong, symposium papers, third technical symposium, Nonwoven Products Technology, International Nonwovens and Disposable Association, New York, pp 109–121

Alsarra IA (2009) Chitosan topical gel formulation in the management of burn wounds. Int J Biol Macromol 45:16–21

Amid PK (1997) Classification of biomaterials and their related complications in abdominal wall hernia surgery. Hernia 1:15–21

Andrady AL, Xu P (1997) J Polym Sci Part B Polym Phys 35:517–521

Araki Y, Ito E (1975) A pathway of chitosan formation in Mucor rouxii. Eur J Biochem 55:71–78

Baba Y, Kawano Y, Hirakawa H (1996) Bull Chem Soc Jpn 69:1255–1260

Badawy MEI, Rabea EI, Rogge TM, Stevens CV, Steurbaut W, Höfte M, Smagghe G (2005) Fungicidal and insecticidal activity of O-acyl chitosan derivatives. Polym Bull 54:279–289

Batista I, Roberts GAF (1990) Macromol Chem 191:429–434

Bautista J, Jover M, Gutierrez JF, Corpas R, Cremades O, Fontiveros E, Iglesias F, Vega J (2001) Process Biochem 37:229–234

Begin A, Van Calsteren MR (1999) Antimicrobial films produced from chitosan. Int J Biol Macromol 26:63–67

Bell AA, Hubbard JC, Liu L, Davis RM, Subbarao KV (1998) Effects of chitin and chitosan on the incidence and severity of Fusarium yellows in celery. Plant Dis 82:322–328

Benhamou N (2004) Potential of the mycoparasite, Verticillium lecanii, to protect citrus fruit against Penicillium digitatum, the causal agent of green mold: a comparison with the effect of chitosan. Phytopathology 94:693–705

Benhamou N, Lafontaine PJ, Nicole M (1994) Induction of systemic resistance to Fusarium crown and root rot in tomato plants by seed treatment with chitosan. Phytopathology 84(12):1432–1444

Benjakul S, Wisitwuttikul P (1994) ASEAN Food J 9:136

Bittelli M, Flury M, Campbell GS, Nichols EJ (2001) Reduction of transpiration through foliar application of chitosan. Agric Forest Meteorol 107:167–175

Bodek KH (1997) In: Struszczyk H (ed) Progress on chemistry and application of chitin and its derivatives, vol 3. Polish Chitin Society, Lodz, p 109

Borkowski J, Dyki B, Felczyńska A, Kowalczyk W (2007) Effect of BIOCHIKOL 020 PC (chitosan) on the plant growth, fruit yield and healthiness of tomato plant roots and stems. In: Progress on chemistry and application of chitin and its derivatives. Monograph XII, pp 217–223

Borkowski J, Dyki B, Niekraszewicz A, Struszczyk H (2004) Effect of preparations Biochikol 020 PC, Tytanit, Biosept 33 SL and others on the healthiness of tomato plants and their fruiting in glasshouse. In: Struszczyk H (ed) Progress on chemistry and application of chitin and its derivatives. Monograph X, pp 167–173

Borkowski J, Kotlińska T, Niekraszewicz A, Struszczyk H (2003) Comparison on the effect of chitosan and tytanit on the growth and healthiness of top onion (Allium proliferum) and onion (Allium cepa) in field conditions. In: Struszczyk H (ed) Progress on chemistry and application of chitin and its derivatives. Monograph IX, pp 107–112

Borkowski J, Kowalczyk W, Struszczyk H (1998) Effect of spraying tomato plants with chitosan and other treatments on the growth of plants, their healthiness and fruit yield. In: Struszczyk H (ed) Progress on chemistry and application of chitin and its derivatives. Monograph IV, Lodz, pp 149–155

Borkowski J, Nowosielski O, Kotlińska T, Niekraszewicz A, Struszczyk H (2001) Influence of chitosan and tytanit on the growth and healthiness of the lettuce, top onion and the glasshouse tomato. In: Struszczyk H (ed) Progress on the chemistry and application of chitin and its derivatives. Monograph VII, pp 159–168

Borkowski J, Struszczyk H, Niekraszewicz A (2002) Effect of chitosan and other preparations on the infection of powdery mildew on tomato, plant growth and red spider population. In: Struszczyk H (ed) Progress on chemistry and application of chitin and its derivatives. Monograph VIII, pp 141–148

Bough WA, Salter WL, Wu ACM, Perkins BE (1978) Biotech Bioeng 20:1931–1943

Briston JH (1974) Plastic films. The Plastic Institute, ILIFFE Books, London, pp 63–82

Broussignac P (1968) Chim Ind Genie Chim 99:1241

Burkinshaw SM, Karim MF (1991) J Soc Leath Technol Chem 75:203–208

Bustos RO, Healy MG (1994) Second international symposium on environmental biotechnology, biotechnology, pp 15–25

Cai J et al (2006) Enzymatic preparation of chitosan from the waste Aspergillus niger mycelium of citric acid production plant. Carbohydr Polym 64(2): 151–157

Carroad PA, Tom RA (1978) Bioconversion of shellfish chitin wastes: process conception and selection of microorganisms. J Food Sci 43:1158–1161

Castelli A, Bergamasco L, Beltrame PL, Focher B (1997) In: Domard A, Jeuniaux C, Muzzarelli R, Roberts G (eds) Advances in chitin science, vol 1. Jacques André Publ, Lyon, pp 198–203

Chang KLB, Tsai G, Lee J, Fu W-R (1997) Carbohydr Res 303:327–332

Charoenvuttitham P, Shi J, Mittal GS (2006) Chitin extraction from black tiger shrimp (Penaeus monodon) waste using organic acids. Sep Sci Technol 41 (6):1135–1153

Cheah LH, Page BBC (1997) Trichoderma spp. for potential biocontrol of clubroot of vegetable brassicas. Crop Food Res., pp 150–153, Proc. 50th New Zealand Plant Protection Conf. 1997

Chen C, Liau W, Tsai G (1998) Antibacterial effects of N-sulfonated and N-sulfobenzoyl chitosan and application to oyster preservation. J Food Prot 61:1124–1128

Chen Y, Li C (1996) Studies on the application of chitosan to clarification of grapefruit juice. Food Sci 23:617–628

Cheng JC, Pisano AP (2008) Photolithographic process for integration of the biopolymer chitosan into micro/nanostructures. J Microelectromech Syst 17:402–409, Park I, Cheng J, Pisano AP, Lee E-S

Chilarski A, Szosland L, Krucinska I, Błasinska A, Cisło R (2004) The application of chitin derivatives as biological dressing in treatment of thermal and mechanical skin injuries, The Annual of Pediatric Traumatic Surgery, The Division of Pediatric Traumatic Surgery 8 (XXXII), pp 58–61

Chilarski A, Szosland L, Krucińska I, Błasińska A, Cisło R (2004) Non-wovens made from dibutyrylchitin as novel dressing materials accelerating wound healing. In: Proceedings of 6th international conference of the European Chitin Society, EUCHIS'04, Poznan-Poland

Chilarski A, Szosland L, Krucińska I, Kiekens P, Błasińska A, Schoukens G, Cisło R, Szumilewicz J (2007) Novel dressing materials accelerating wound healing made from dibutyrylchitin. Fibres Text East Eur 15(4 (63)):77–81

Chirkov SN (2002) The antiviral activity of chitosan (review). Appl Biochem Microbiol 38:1–8

Christodoulidou A, Bouriotis V, Thireos GJ (1996) Biol Chem 271:31420–31425

Christodoulidou A, Tsigos I, Martinou A, Tsandaskalaki M, Kafetzopoulos D, Bouriotis V (1998) Adv Chitin Sci 2:188–193

Ciechańska D, Struszczyk MH, Kucharska M et al (2010) Multilayer surgical mesh and method of multilayer surgical mesh manufacture, PL Patent Appl. No. P 390253

Ciechańska D, Wietecha J, Kaźmierczak D, Kazimierczak J (2010b) Fibres Text East Eur 18(5 (82)):98–104

Cosio IG, Fisher RA, Carroad PA (1982) J Food Sci 42:901

Crompton KE, Goud JD, Bellamkonda RV, Gengenbach TR, Finkelstein DI, Horne MK et al (2007) Polylysine-functionalised thermoresponsive chitosan hydrogel for neural tissue engineering. Biomaterials 28:441–449

Daayf F, El Bellaj M, El Hassni M, J'aiti F, El Hadrami I (2003) Elicitation of soluble phenolics in date palm (Phoenix dactylifera L.) callus by Fusarium oxysporum f. sp. albedinis culture medium. Environ Exp Bot 49:41–47

Deans JR, Dixon BG (1992) Bioabsorbents for wastewater treatment. In: Brine CJ, Sandford PA, Zikakis JP (eds) Advances in chitin and chitosan. Elsevier Applied Science, Oxford, pp 648–656

Desai K, Kit K, Li J, Davidson PM, Zivanovic S, Meyer H (2009) Polymer 50:3661–3669

Dutta PK, Tripathi S, Mehrotra GK, Dutta J (2009) Perspectives for chitosan based antimicrobial films in food applications. Food Chem 114:1173–1182

Dwivedi J, Om P (1995) Agrawal. Physiol Entomol 20:318–322

East GC (1993) J Appl Polym Sci 50:1773

El Ghaouth A, Arul J, Asselin A, Benhamou N (1992) Antifungal activity of chitosan on post-harvest pathogens: induction of morphological and cytological alterations an Rhizopus stolonifer. Mycol Res 96:769–779

El Ghaouth A, Smilanick JL, Wilson CL (2000) Enhancement of the performance of Candida saitoana by the addition of glycolchitosan for the control of postharvest decay of apple and citrus fruit. Postharvest Biol Technol 19:103–110

El Hassni M, El Hadrami A, Daayf F, Chérif M, Ait Barka E, El Hadrami I (2004) Chitosan, antifungal product against Fusarium oxysporum f. sp. albedinis and elicitor of defence reactions in date palm roots. Phytopathol Mediterr 43:195–204

EN 13726-1:2002 Test methods for primary wound dressings – Part 1: Aspects of absorbency

EN 13726-2:2002 Test methods for primary wound dressings – Part 2: Moisture vapour transmission rate of permeable film dressings

EN 13726-3:2002 Non-active medical devices – Test methods for primary wound dressings – Part 3: Waterproofness

EN 13726-4:2002 Non-active medical devices – Test methods for primary wound dressings – Part 4: Conformability

EN-ISO 10993-1:2008 Biological evaluation of medical devices – Part 1: Evaluation and testing

EN-ISO 22442-1:2007 Medical devices utilizing animal tissues and their derivatives – Part 1: Application of risk management

EN-ISO 22442-2:2007 Medical devices utilizing animal tissues and their derivatives – Part 2: Controls on sourcing, collection and handling

EN-ISO 22442-3:2007 Medical devices utilizing animal tissues and their derivatives – Part 3: Validation of the elimination and/or inactivation of viruses and transmissible spongiform encephalopathy (TSE) agents

Faoro F, Maffi D, Cantu D, Iriti M (2008) Chemical-induced resistance against powdery mildew in barley: the effects of chitosan and benzothiadiazole. BioControl 53:387–401

Faoro F, Sant S, Iriti M, Appiano A (2001) Chitosan-elicited resistance to plant viruses: a histochemical and cytochemical study. In: Muzzarelli RAA (ed) Chitin enzymology. Atec, Grottammare, Italy, pp 57–62

Fin. patent No. 93–4616 931019

Freeman A, Dror Y (1994) Biotechnol Bioeng 44:1083–1108

Fujii S, Kumagai H, Noda M (1980) Carbohydr Res 83:389–393

Gades M, Stern JS (2005) J Am Diet Assoc 105:72–77

Gagne N, Simpson BK (1993) Use of proteolytic enzymes to facilitate the recovery of chitin from shrimp wastes. Food Biotechnol 7(3):253–263

Gajdziecki B, Mozyszek Z, Struszczyk H, Oczkowski M, Rybicki E (1995) In: Progress on Chemistry and Application of Chitin and its Derivatives, vol 1, pp 106–117

Galas E, Kubik C, Struszczyk MH (1996) In: Struszczyk H (ed) Progress on chemistry and application of chitin and its derivatives, vol 2. Polish Chitin Society, Lodz, pp 55–73

Gamzazade A, Sklyar A, Nasibov S, Sushkov I, Shashkov A, Knirel Y (1997) Carbohydr Polym 34:113

Gauthier C, Clerisse F, Dommes J, Jaspar-Versali M-F, Protein Expression and Purification 59 (2008), pp 127–137

Giles CH, Hassan ASA, Laidlaw M, Subramanian RVR (1958) J Soc Dyers Colou 74:645

Gopalan NK, Dufresne A (2003) Crab shell chitin whisker reinforced natural rubber nanocomposites. 1. Processing and swelling behavior. Biomacromolecules 4:657–666

Griethuysen-Diblber E, Flaschel E, Renken A (1988) Process Biochem 4:55

Guan YJ, Hu J, Wang XJ, Shao CX (2009) Seed priming with chitosan improves maize germination and seedling growth in relation to physiological changes under low temperature stress. J Zhejiang Univ Sci B 10:427–433

Guminska M, Ignacak J, Kedryna T, Struszczyk H (1997) In: Struszczyk H (ed) Progress on chemistry and application of chitin and its derivatives, vol 3. Polish Chitin Society, Lodz, pp 124–132

Hadwiger LA, Fristensky B, Riggleman RC (1984) In: Zikakis JP (ed) Chitin, chitosan and related enzymes. Academic, Orlando, FL, pp 291–302

Hall GM, Silva SD (1994) Biotechnology, pp 633–638

Harish Prashanth KV, Tharanathan RN (2007) Trends Food Sci Technol 18:117–131

Hayashi T, Ikada Y, Itoyama K, Tanibe H (1992) Proceedings from 4th world biomaterial congress, Berlin, p 369

Hein S, Ng C.H., Chandrkrachang S, Stevens WF (2001) "A Systematic Approach to Quality Assessment System of Chitosan" in Chitin and Chitosan: Chitin and Chitosan in Life Science, Yamaguchi, pp 332–335 ed. by Urgami T., Kurita K., Fukamizo T., Kodansha Scientific, Tokyo

Hirano S (1998) In: Muzzarelli RAA, Peter MG (eds) Chitin handbook. Atec Edizioni, Italy, pp 71–83

Hisamatsu M (1998) In: Muzzarelli RAA, Peter MG (eds) Chitin handbook. Atec Edizioni, Italy, p 411

Hoffmann K et al (2010) Genetic improvement of bacillus licheniformis strains for efficient deproteinization of shrimp shells and production of high-molecular-mass chitin and chitosan. Appl Environ Microbiol 76 (24):8211–8221

Holme KR, Perlin AS (1997) Carbohydr Res 302:7

Hsien T-Y, Rorrer GL (1995) Sep Sci Technol 30:2455–2475

Hua K-J, Hua J-L, Hob K-P, Yeung K-W (2004) Carbohydr Polym 58:45–52

Hwang D, Damodaran S (1995) Selective precipitation and removal of lipids from cheese whey using chitosan. J Agric Food Chem 43:33–37

Imeri AG, Knorr D (1988) Effect of Chitosan on yield and compositional data of carrot and apple juice. J Food Sci 53:1707–1709

InfoFish, Shrimp Wastes Utilisation (1997) Technical Handbook

Iriti M, Picchi V, Rossoni M, Gomarasca S, Ludwig N, Garganoand M, Faoro F (2009) Chitosan antitranspirant activity is due to abscisic acid-dependent stomatal closure. Environ Exp Bot 66:493–500

ISO 10993–18:2005 Biological evaluation of medical devices – Part 18: Chemical characterization of material

ISO 13485:2003 Medical devices – Quality management systems – Requirements for regulatory purposes

ISO 7198:1998 – Cardiovascular implants – Tubular vascular prostheses

ISO 9001:2008 Quality Management Systems Requirements

ISO/DIS 10993-19 Biological evaluation of medical devices – Part 19: Physico-chemical, mechanical and morphological characterization

Issa MM, Koping-Hoggard M, Artursson P (2005) Drug Discov Today Technol 2(1):1–6

Itoh S, Yamaguchi I et al (2003a) Brain Res 993: 111–112

Itoh S, Yamaguchi I, Shinomiya K, Tanaka J (2003b) Sci Technol Adv Mater 4:261–268

Itoyama K (1996) Gekkan Fudo Kemikaru 12:19–23

Itozawa T, Kise H (1995) J Ferment Bioeng 80:30–34

Ivshina TN et al (2009) Isolation of the chitin-glucan complex from the fruiting bodies of mycothallus. Appl Biochem Microbiol 45(3):313–318

Jameela SR, Latha PG, Subramoniam A, Jayakrishnan A (1996) J Pharm Pharmacol 48:685–688

Japan Patent No 63169975 (88169975)

Japan Patent No. 60159123

Japan Patent No. 62297365 (877297365)

Japan Patent No. 63–161001

Japan Patent No. 63189859

Japan Patent No. 92-257696 (920928)

Jaworska MM, Konieczna-Moras E (2009) Inhibition of chitin deacetylase by acetic acid preliminary investigation. In: Jaworska MM (ed) Progress on chemistry and application of chitin and its derivatives, vol XIV. Polish Chitin Society, Lodz, pp 83–88

Je JY, Park PJ, Kim SK (2005) Carbohydr Polym 60:553

Jeong J-H (2007) Low temperature, low pressure nanoimprinting of chitosan as a biomaterial for bionanotechnology applications. Appl Phys Lett 90:093902–093903

Jeuniaux A (1986) Chitosan as a tool for the purification of waters. In: Muzzarelli RAA, Jeuniaux C, Gooday GW (eds) Chitin in nature and technology. Plenum Press, New York, NY, pp 551–570

Jia Z, Shen D, Xu W (2001) Synthesis and antibacterial activities of quaternary ammonium salt of chitosan. Carbohydr Res 333:1–6

Jung WJ et al (2007) Production of chitin from red crab shell waste by successive fermentation with Lactobacillus paracasei KCTC-3074 and Serratia marcescens FS-3. Carbohydr Polym 68(4):746–750

Kardas I, Marcol W, Niekraszewicz A, Kucharska M, Ciechańska D, Wawro D, Lewin-Kowalik J, Właszczuk A (2010) Utilization of biodegradable polymers for peripheral nerve reconstruction. In: Jaworska M (ed) Progress on chemistry and application of chitin and derivatives, vol 15. Polish Chitin Society, Lodz, pp 159–167

Kawamura Y, Yoshida H, Asai S, Kurahashi I, Tanibe H (1997) Sep Sci Technol 32:1959–1974

Khan W, Prithiviraj B, Smith DL (2002) Effect of foliar application of chitin and chitosan oligosaccharides on photosynthesis of maize and soybean. Photosynth Res 40:621–624

Kim JH, Shin JH, Lee HJ, Chung IS, Lee HJ (1997) Effect of chitosan on indirubin production from suspension culture of Polygonum tinctorium. J Ferm Bioeng 83:206–208

Kim SS, Lee YM, Cho CS (1995) Polymer 36:4497–4501

King GA, Daugulis AJ, Faulkner P, Bayly D, Goosen MFA (1989) Biotech Bioeng 34:1085

Kittur FS, Kumar KR, Tharanathan RN (1998) Zeitschrift Fur Lebensmittel Untersuchung Und Forschung A. Food Res Technol 206:44–47

Kochanska B (1997) In: Struszczyk H (ed) Progress on chemistry and application of chitin and its derivatives, vol 3. Polish Chitin Society, Lodz, pp 103–108

Koev ST, Powers MA, Park JJ, Yi H, Wu L, Bentley WE, Payne GF, Rubloff GW, Ghodssi R (2006) Chitosan as a functional interface between biology and microsystems. In: Bio Micro and Nanosystems Conference, BMN '06, 15–18 Jan 2006, pp 82–82

Kowalski B, Jimenez Terry F, Herrera L, Agramonte Peñalver D (2006) Application of soluble chitosan in vitro and in the greenhouse to increase yield and seed quality of potato minitubers. Potato Res 49:167–176

Krajewska B (2004) Enzym Microb Technol 35:126–139

Krucińska I, Komisarczyk A, Paluch D, Szumilewicz J (2006) Biological estimation of dibutyrylchitin nonwovens manufactured by the spraying of polymer solution technique. In: Progress in the chemistry and application of chitin and its derivatives, Polish Chitin Society, Monograph XI, pp 129–135

Kucharska M, Niekraszewicz A, Lebioda J, Malczewska-Brzoza K, Wesołowska E (2007) In: Jaworska M (ed) Progress on chemistry and application of chitin and its derivatives. Polish Chitin Society, Lodz, pp 131–138

Kucharska M, Niekraszewicz A, Struszczyk H, Bursig H (1997) In: Struszczyk H (ed) Progress on chemistry and application of chitin and its derivatives, vol 3. Polish Chitin Society, Lodz, pp 141–151

Kucharska M, Struszczyk MH et al. (2006) Haemostatic multilayer wound dressing and the method of haemostatic multilayer wound dressing manufacture, PL Patent Appl. No. P 390253

Kucharska M, Struszczyk MH, Cichecka M, Brzoza K (2011) Preliminary studies on the usable properties of innovative wound dressings. In: Progress on chemistry and application of chitin and its derivatives, vol 16, pp 131–137

Kucharska M, Struszczyk MH, Cichecka M, Brzoza K (2011) Preliminary Studies on the Usable Properties of Innovative Wound Dressings, Progress on Chemistry and Application of Chitin and Its Derivatives, Monograph of Polishchitin Society, vol 16, pp 131–137

Kulikov SN, Chirkov SN, Il'ina AV, Lopatin SA, Varlamov VP (2006) Effect of the molecular weight of chitosan on its antiviral activity in plants. Prik Biokhim Mikrobiol 42(2):224–228

Kumar MNVR (2000) React Funct Polym 46:1–27

Kurauchi Y, Ohga K (1998) In: Muzzarelli RAA, Peter MG (eds) Chitin Handbook. Atec Edizioni, Italy, p 431

Kurita K (2006) Chitin and chitosan: functional biopolymers from marine crustaceans. Mar Biotechnol 8 (3):203–226

Kurita K, Akao H, Kaji Y, Kojima T, Hirakawa M, Kato M, Ishii S, Mori T, Nishijama Y (1997) In: Muzzarelli RAA (ed) Chitin enzymology, vol 2. Atec Edizioni, Italy, pp 483–490

Kurita K, Koyama Y, Taniguchi A (1986) J Appl Polym Sci 21:1169

Kurita K, Sannan T, Iwakura Y (1977) Macromol Chem 178:3197–3202

Laflamme P, Benhamou N, Bussiéres G, Dessureault M (1999) Differential effect of chitosan on root rot fungal pathogens in forest nurseries. Can J Bot 77:1460–1468

Lafontaine JP, Benhamou N (1996) Chitosan treatment: an emerging strategy for enhancing resistance of greenhouse tomato plants to infection by Fusarium oxysporum f. sp. radicis-lycopersici. Biocontrol Sci Technol 6:111–124

Lee KY, Kim JH, Kwon LC, Jeong SY (2000) Colloid Polym Sci 278:1216–1219

Lee YM, Nam SY, Woo DJ (1997) J Membr Sci 133:103–110

Li L, Hsieh Y-L (2006) Carbohydr Res 341:374–381

Li Q, Dunn E, Grandmaison EW, Goosen MFA (1992) J Bioact Comp Polym 7:370–397

Liu WG, Yao H (2002) J Control Release 83:1–11

Lu GY, Kong LJ, Sheng BY, Wang G, Gong YD, Zhang XF (2007) Degradation of covalently cross-linked carboxymethyl chitosan and its potential application for peripheral nerve regeneration. Eur Polym J 43:3807–3818

Lusena CV, Rose RC (1953) J Fish Res Board Can 10:521

Madhumathi K, Sudheesh Kumar PT, Abilash S, Sreeja V, Tamura H, Manzoor K et al (2010) Development of novel chitin/nanosilver composite scaffolds for wound dressing applications. J Mater Sci Mater Med 21:807–813

Mahmoud NS, Ghaly AE, Arab F (2007) Unconventional approach for demineralization of deproteinized crustacean shells for chitin production. Am J Biochem Biotechnol 3(1):1–9

Di Mario F, Rapana P, Tomati U, Galli E (2008) Int J Biol Macromol 43:8–12

Martinou A, Tsigos I, Bouriotis V (1998) In: Muzzarelli RAA, Peter MG (eds) Chitin handbook. Atec Edizioni, Italy, p 501

Ishihara M et al (2006) Chitosan hydrogel as a drug delivery carrier to control angiogenesis. J Artif Organs 9:8–16

Methacanona P, Prasitsilpa M, Pothsreea T, Pattaraarchachaib J (2003) Carbohydr Polym 52: 119–123

Micera G, Deiana S, Dessi A, Decock P, Dubois B, Kozlowski H (1986) Copper and vanadium complexes of chitosan. In: Muzzarelli RAA, Jeuniaux C, Gooday GW (eds) Chitin in nature and technology. Plenum, New York, NY, pp 565–567

Mima S, Miiya M, Iwamoto R, Yoshikawa S (1983) Polym Sci 28:1909–1917

Moorjani MN, Khasim DI, Rajalakshmi S, Puttarajappa P, Amla BL (1978) In: Muzzarelli RAA, Pariser ER (eds) Proceedings of 1st international conference on chitin and chitosan, MIT Sea Grant Program, p 210

Morley KL, Chauve G, Kazlauskas R, Dupont C, Shareck F, Marchessault RH (2006) Carbohydr Polym 63:310–315

Mourya VK, Inamdar Nazma N (2008) React Funct Polym 68: 1013–1051

Muffler K, Ulber R (2005) "Downstream processing in marine biotechnology" in: Y. Le-Gal, R. Ulber (Eds.); Marine Biotechnology II; Advances in Biochemical Engineering/Biotechnology; Springer Verlag, Berlin, 63–103

Murakami K et al (2010) Hydrogel blends of chitin/chitosan, fucoidan and alginate as healing-impaired wound dressings. Biomaterials 31:83–90

Murphy JG, Rafferty SM, Cassells AC (2000) Stimulation of wild strawberry (Fragaria vesca) arbuscular mycorrhizas by addition of shellfish waste to the growth substrate: interaction between mycorrhization, substrate amendment and susceptibility to red core (Phytophthora fragariae). Appl Soil Ecol 15: 153–158

Muzzarelli RAA (1977) Chitin. Pergamon of Canada Ltd., Toronto

Muzzarelli RAA (1982) Proceedings of the 2nd international conference on chitin and chitosan. The Japanese Society of Chitin and Chitosan, Sapporo, Japan, p 25

Muzzarelli RAA (1989) In: Skjak-Braek G, Anthonsen T, Standford P (eds) Chitin and chitosan. Elsevier Applied Science, New York, pp 87–99

Muzzarelli RAA (1998) In: Muzzarelli RAA, Peter MG (eds) Chitin handbook. Atec Edizioni, Italy, p 47

Muzzarelli RAA (2009) Chitins and chitosans for the repair of wounded skin, nerve, cartilage and bone. Carbohydr Polym 76:167–182

Muzzarelli RAA, Muzzarelli C, Tarsi R, Miliani M, Gabbanelli F, Cartolari M (2001) Fungistatic activity of modified chitosans against Saprolegnia parasitica. Biomacromology 2:165–169

Muzzarelli RAA, Muzzarelli C, Terbojevich M (1997) Carbohydr Europe 19:10–18

Muzzarelli RAA, Tafani F, Scarpini G (1980) Biotech Bioeng 22:885–896

Muzzarelli RAA, Tarsi R, Filippini O, Giovanetti E, Biagini G, Varaldo PE (1990) Antimicrobial properties of N-carboxybutyl chitosan. Antimicrob Agents Chemother 34: 2019–2023

Muzzarelli RAA, Weckx M, Fillipini O (1989) Removal of trace metal ions from industrial waters, unclear effluents and drinking water, with the aid of cross-linked N-carboxymethyl chitosan. Carbohydr Polym 11:293–296

Nagasawa K, Tohira Y, Inoue Y, Tanoura N (1971) Carbohydr Res 18: 95

Naggi AM, Torri G, Compagnoni T, Casu B (1986) Chitin in nature and technology. Plenum, New York, pp 371–377

Naznin R (2005) Extraction of chitin and chitosan from shrimp (Metapenaeus monocerus) shell by chemical method. Pak J Biol Sci 8(7):1051–1054

Niederhofer A, Muller BW (2004) "A method for direct preparation of chitosan with low molecular weight from fungi", European Journal of Pharmaceutics and Biopharmaceutics, 57, pp. 101–105

Niekraszewicz A (2005) Chitosan medical dressings. Fibres Text East Eur 13(6 (54)):16–18

Niekraszewicz A, Kucharska M, Struszczyk MH, Gruchała B, Brzoza K (2006) Composite surgical mesh and method of composite mesh manufacture, PL Patent Appl. No. 380861

Niekraszewicz A, Kucharska M, Struszczyk MH, Rogaczewska A, Struszczyk K (2008) Investigation into biological, composite surgical meshes. Fibres Text East Eur 16(6 (71)):117–121

Niekraszewicz A, Kucharska M, Wawro D, Struszczyk MH, Kopias K, Rogaczewska A (2007a) Development of a manufacturing method for surgical meshes modified by chitosan. Fibres Text East Eur 15(3 (62)): 105–109

Niekraszewicz A, Kucharska M, Wawro D, Struszczyk MH, Rogaczewska A (2007) Partially resorbable hernia meshes. In: Jaworska M (ed) Progress on chemistry and application of chitin and its derivatives, vol 12. Polish Chitin Society, pp 109–114

Niekraszewicz A, Kucharska M, Wiśniewska-Wrona M, Ciechańska D, Ratajska M, Haberko K (2009) Surgical biocomposites with chitosan. In: Jaworska M (ed) Progress on chemistry and application of chitin and its derivatives, vol 14. Polish Chitin Society, pp 167–178

Niekraszewicz A, Lebioda J, Kucharska M, Wesołowska E (2007) Research into developing antibacterial dressing materials. Fibres Text East Eur 15 1 (60): 99–103

No HK, Meyers SP, Lee KS (1989) Isolation and characterization of chitin from crawfish shell waste. J Agric Food Chem 37(3):575–579

Oduor-Odote PM, Struszczyk MH, Peter MG (2005) Western Indian ocean. J Mar Sci 4(1):99–107

Ohashi E, Karube I (1995) Fish Sci 61:856–859

Orlikowski LB, Skrzypczak C (2003a) Chitosan induces some plant resistance to formae sp. Fusarium oxysporum. In: Progress on chemistry and application of chitin and its derivatives. Monograph IX, pp 101–106

Orlikowski LB, Skrzypczak C, Niekraszewicz A, Struszczyk H (2001) Influence of chitosan on the development of Fusarium wilt of carnation. In: Struszczyk H (ed) Progress on chemistry and application of chitin and its derivatives, Monograph VII, pp 155–158

Orlikowski LB, Skrzypczak CZ (2003b) Biocides in the control of soil-borne and leaf pathogens. Hortic Veget Grow 22(3):426–433

Ouattara B, Simard RE, Piette G, Begin A, Holley RA (2000) Inhibition of surface spoilage bacteria in processed meats by application of antimicrobial films prepared with chitosan. Int J Food Microbiol 62:139–148

Oungbho K, Muller BW (1997) Int J Pharm 156: 229–237

Palma-Guerrero J, Jansson HB, Salinas J, Lopez-Llorca LV (2008) Effect of chitosan on hyphal growth and spore germination of plant pathogenic and biocontrol fungi. J Appl Microbiol 104:541–553

Papineau AM, Hoover DG, Knorr D, Farkas DF (1991) Antimicrobial effect of water-soluble chitosans with high hydrostatic pressure. Food Biotechnol 5:45–57

Patkowska E, Pięta D (2004) Introductory studies on the use of biopreparations and organic compounds for seed dressing of runner bean (Phaseolus coccineus L.). Folia Univ Agric Stetin Agricultura 239(95):295–300

Pellegrino JJ, Geer S, Maegley K, Rivera R, Steward T, Ko M (1990) Ann N Y Acad Sci Biochem Eng 589:229–244

Peniston QP, Johnson EL (1980) US patent 4 195 175

Percot A, Viton C, Domard A (2003a) Characterization of shrimp shell deproteinization. Biomacromolecules 4 (5):1380–1385

Percot A, Viton C, Domard A (2003b) Optimization of chitin extraction from shrimp shells. Biomacromolecules 4(1):12–18

Persin Z, Stana-Kleinschek K, Foster T, van Dam JEG, Boeriu CG, Navard P (2011) Challenges and opportunities in polysaccharides research and technology: The EPNOE views for the next decade in the areas of materials, food and health care. Carbohydr Polym 84:22–32

Phuvasate S, Su YC (2010) Comparison of lactic acid bacteria fermentation with acid treatments for chitosan production from shrimp waste. J Aquat Food Prod Technol 19(3):170–179

Pięta D, Pastucha A, Struszczyk H, Wójcik W (2002) The effect of Chitosan and runner bean (Phaseolus coccineus L.) cultivation on the formation of microorganisms communities in the soil. In: Struszczyk H (ed) Progress of chemistry and application of chitin and its derivatives. Monograph VIII, Lodz, pp 133–140

Pięta D, Patkowska E, Pastucha A (2006) Influence of Biochikol 020 PC used as seed dressing of bean on healthiness and yield of plants. In: Progress on chemistry and application of chitin and its derivatives, vol 11, pp 159–170

Pinotti A, Bevilacqua A, Zaritzky N (1997) Optimization of the flocculation stage in a model system of a food emulsion waste using chitosan as polyelectrolyte. J Food Eng 32:69–81

Polish Patent 320491 (1997) A biological dressing and method to its manufacture

Polish Patent No. 125995

Polish Patent No. 141381

Polish Patent No. 380861

Pośpieszny H (1995) Inhibition of tobacco mosaic virus (TMV) infection by chitosan. Phytopath Polonica 10 (XXII):69–74

Pośpieszny H (1997) Antiviroid activity of chitosan. Crop Prot 16:105–106

Pośpieszny H, Chirkov S, Atabekov J (1991) Induction of antiviral resistance in plants by chitosan. Plant Sci 79:63–68

Pospieszny H, Giebel J (1997) In: Muzzarelli RAA (ed) Chitin enzymology, vol 2. Atec Edizioni, Italy, pp 379–383

Pośpieszny H, Struszczyk H, Cajza M (1996) In: Muzzarelli RAA (ed) Chitin enzymology, vol 2. Atec, Grottammare, Italy, pp 385–389

Dutta PK, Dutta J, Tripathi VS, Tripathi VS (2004) Chitin and chitosan: chemistry, properties and applications. J Sci Ind Res 63:20–31

Prameela K, Mohan CM, Hemalatha KPJ (2010a) Extraction of pharmaceutically important chitin and carotenoids from shrimp biowaste by microbial fermentation method. J Pharm Res 3(10):2393–2395

Prameela K, Mohan CM, Hemalatha KPJ (2010b) Optimization of fermentation of shrimp biowaste under different carbon sources for recovery of chitin and carotenoids by using Lactic acid bacteria. J Pharm Res 3(12):2888–2889

Prodas-Drozd F, Gwiezdinski Z (1997) In: Struszczyk H (ed) Progress on chemistry and application of chitin and its derivatives, vol 3. Polish Chitin Society, Lodz, p 99

Qin Y et al (2010) Dissolution or extraction of crustacean shells using ionic liquids to obtain high molecular weight purified chitin and direct production of chitin films and fibres. Green Chem 12(6):968–971

Quong D, Groboillot A, Darling GD, Poncelet D, Neufeld RJ (1998) In: Muzzarelli RAA, Peter MG (eds) Chitin handbook. Atec Edizioni, Italy, p 405

Rabea EI, El Badawy MT, Stevens CV, Smagghe G, Steurbaut W (2003) Chitosan as antimicrobial agent: applications and mode of action. Biomacromolecules 4:1457–1465

Ratajska M, Haberko K, Ciechańska D, Niekraszewicz A, Kucharska M (2008) Hydroxyapatite – chitosan biocomposites. In: Jaworska M (ed) Progress on chemistry and application of chitin and its derivatives, vol 13. Polish Chitin Society, Lodz, pp 89–94

Ravi Kumar MNV, Muzzarelli RAA, Muzzarelli C, Sashiwa H, Domb AJ (2004) Chitosan chemistry and pharmaceutical perspectives. Chem Rev 104:6017–6084

Ravi Kumar MNV, Muzzarelli RAA, Muzzarelli C, Sashiwa H, Domb AJ (2006) Marguerite Rinaudo – chitin and chitosan: properties and applications. Prog Polym Sci 31:603–632

Rawls RL (1984) Technology 14(5):42–45

Reddy MV, Arul J, Angers P, Couture L (1999) Chitosan treatment of wheat seeds induces resistance to Fusarium graminearum and improves seed quality. J Agric Food Chem 47:1208–1216

Remunan C, Lopez C, Bodmeier R (1997) J Control Release 44:215–225

Revah-Moiseev S, Carroad A (1981) Conversion of the enzymatic hydrolysate of shellfish waste chitin to single-cell protein. Biotechnol Bioeng 23:1067–1078

Rhoades J, Roller S (2000) Antimicrobial actions of degraded and native chitosan against spoilage organisms in laboratory media and foods. Appl Environ Microbiol 66:80–86

Roberts GAF (1992a) Chitin chemistry. MacMillan Press Ltd., Houndmills, pp 1–5

Roberts GAF (1992b) Chitin chemistry. MacMillan Press Ltd., Houndmills, pp 54–58

Roberts GAF (1992c) Chitin chemistry. MacMillan Press Ltd., Houndmills, pp 64–74

Roberts GAF (1992d) Chitin chemistry. MacMillan Press Ltd., Houndmills

Roberts GAF (1998) In: Domard A, Roberts GAF, Varum KM (eds) Advances in chitin and chitosan, vol 2. Jacques Andre Publisher, Lyon, p 22

Rødde RH et al (2008) Carbohydr Polym 71:388–393

Ruan SL, Xue QZ (2002) Effects of chitosan coating on seed germination and salt-tolerance of seedlings in hybrid rice (Oryza sativa L.). Acta Agron Sinica 28:803–808

Rungsardthong V, Wongvuttanakul N, Kongpien N, Chotiwaranon P (2006) Process Biochem 41:589–593

Sannan T, Kurita K, Iwakura Y (1975) Macromol Chem 176:1191–1195

Savard T, Beaulieu C, Boucher I, Champagne CP (2002) Antimicrobial action of hydrolyzed chitosan against spoilage yeasts and lactic acid bacteria of fermented vegetables. J Food Prot 65:828–833

Senstad C, Mattiasson B (1989) Purification of wheat germ agglutinin using affinity flocculation with chitosan and a subsequent centrifugation or floatation step. Biotechnol Bioeng 34:387–393

Seo T, Kantabara T, Iijima T (1988) J Appl Polym Sci 36:1443–1451

Shahabeddin L, Damour O, Berthod F, Rousselle P, Saintigny G, Collombel C (1991) J Mater Sci Mater Med 2:222–226

Shahidi F, Abuzaytoun R, Steve LT (2005) Chitin, chitosan, and co-products: chemistry, production, applications, and health effects. Adv Food Nutr Res, 49, pp 93–135

Shahidi F, Synowiecki J (1991) Isolation and characterization of nutrients and value-added products from snow crab (Chionoecetes opilio) and shrimp (Pandalus borealis) processing discards. J Agric Food Chem 39:1527–1532

Shao CX, Hu J, Song WJ, Hu WM (2005) Effects of seed priming with chitosan solutions of different acidity on seed germination and physiological characteristics of maize seedling. J Zhejiang Univ Agric Life Sci 1:705–708

Shepherd R, Reader S, Falshaw A (1997) Glycoconj J 14:535–542

Shimahara K, Ohkouchi K, Ikeda M (1982) In: Hirano S, Tokura S (eds) Chitin and chitosan. The Japanese Society of Chitin and Chitosan, Tottori, p 10

Shridhar Pandya (2007) "An Attractive Biocompatible Polymer for pharmaceutical application in various dosage forms – Chitosan", Pharmaceutical Reviews, vol 5, Issue 3

Shuangyun L, Gao W, Hai Ying G (2008) Construction, application and biosafety of silver nanocrystalline chitosan wound dressing. Burns 34:623–628

Sini TK, Santhosh S, Mathew PT (2007) Study on the production of chitin and chitosan from shrimp shell by using Bacillus subtilis fermentation. Carbohydr Res 342(16):2423–2429

Soto-Perlata NV, Muller H, Knorr D (1989) Effect of chitosan treatments on the clarity and color of apple juice. J Food Sci 54:495–496

Spagna G, Pifferi PG, Rangoni C, Mattivi F, Nicolini G, Palmonari R (1996) The stabilization of white wines by adsorption of phenolic compounds on chitin and chitosan. Food Res Intern 29:241–248

Stenberg E, Wachter R (1996) Adv Chitin Sci 1:166

Stevens WF, Win NN, Ng CH, Pichyangukura S, Chandrkrachang S (1998) Adv Chitin Sci 2:40–47

Stössel P, Leuba JL (1984) Effect of chitosan, chitin and some aminosugars on growth of various soilborne phytopathogenic fungi. Phytopathol Z 111:82–90

Struszczyk H (1987) J Appl Polym Sci 33:177–189

Struszczyk H (1994) In: Karnicki ZS, Bzeeski MM, Bykowski PJ, Wojtasz-Pajak A (eds) Chitin world. Verlag für Neue Wissenschsft, Bremerhaven, p 542

Struszczyk H (1998a) In: Muzzarelli RAA, Peter MG (eds) Chitin handbook. Atec Edizioni, Italy, pp 437–440

Struszczyk H (1998b) In: Muzzarelli RAA, Peter MG (eds) Chitin handbook. Atec Edizioni, Italy, p 441

Struszczyk H, Ciechańska D, Wawro D, Stęplewski W, Krucińska I, Szosland L, Van de Velde K, Kiekens P (2004) Some properties of dibutyrylchitin fibres. In: Proceedings of 6th international conference of the European chitin society, EUCHIS'04, Poznan-Poland

Struszczyk H, Kivekas O (1990) Br Polym J 23:261–265

Struszczyk H, Pospieszny H, Kivekas O (1997a) In: Muzzarelli RAA (ed) Chitin enzymology, vol 2. Atec Edizioni, Italy, pp 497–502

Struszczyk MH (2002a) Polimery 47(5):316–325

Struszczyk MH (2002b) Chitin and chitosan III: some aspects of biodegradation and bioactivity of polyaminosaccharides. Polimery 47(9):29–40

Struszczyk MH (2006) Global requirements for medical applications of chitin and its derivatives, Monograph XI, Polish Chitin Society, pp 95–102

Struszczyk MH (2002c) Chitin and chitosan II: applications of chitosan. Polimery 47(6):396–403

Struszczyk MH, Brzoza-Malczewska K (2007) Fibres Text East Eur 5–6:163–166

Struszczyk MH, Halweg R, Peter MG (1997b) Advances in chitin science. In: Muzzarelli RAA, Peter MG (eds) Chitin handbook. Atec Edizioni, Grottammare, pp 40–49

Struszczyk MH, Loth F, Peter MG (2000) In: Peter MG, Domard A, Muzzarelli RAA (eds) Advance in chitin chemistry, vol 4. University of Potsdam, Potsdam, Germany, pp 128–135

Struszczyk MH, Loth F, Pospieszny H, Peter MG (2001a) In: Struszczyk H (ed) Progress on chemistry and application of chitin and its derivatives, vol 7. Polish Chitin Society, Lodz, pp 87–100

Struszczyk MH, Peter MG (1997) Chemistry of Chitosan. In: 4th international workshop on carbohydrates as organic raw materials, Wien, 20.-21.3.(1997), Zuckerindustrie, 131.

Struszczyk MH, Peter MG, DBU in Osnabrück, Materials, Osnabrück, 25/26.11. 1998.

Struszczyk MH, Pospieszny H, Schanzenbach D, Peter MG (1999) In: Struszczyk H (ed) Polish-Russian chitin monograph. Lodz, Polish Chitin Society

Struszczyk MH, Pospieszny H, Schanzenbach D, Peter MG (2001b) In: Uragami T, Kurita K, Fukamoto T (eds) Chitin and chitosan in life sciences. Kodansha, Tokyo, pp 426–427

Struszczyk MH, Ratajska M, Brzoza-Malczewska K (2007) Fibres Text East Eur 2:105–109

Struszczyk MH, Struszczyk KJ (2007) Medical applications of chitin and its derivatives. In: Jaworska M (ed) Progress on chemistry and application of chitin and its derivatives, vol 12. ISSN 1896-5644, pp 139–148

Sudarshan NR, Hoover DG, Knorr D (1992) Antibacterial action of chitosan. Food Biotechnol 6:257–272

Sudheesh Kumar PT, Abilash S, Manzoor K, Nair SV, Tamura H, Jayakumar R (2010) Preparation and characterization of novel α-chitin/nano silver composite scaffolds for wound dressing applications. Carbohydr Polym 80:761–767

Sun W, Payne GF (1996) Tyrosinase-containing chitosan gels: a combined catalyst and sorbent for selective phenol removal. Biotechnol Bioeng 51:79–86

Synowiecki J, Al-Khateeb N (2000) The recovery of protein hydrolysate during enzymatic isolation of chitin from shrimp Crangon crangon processing discards. Food Chem 68(2):147–152

Synowiecki J, Al-Khateeb NA (2003) Production, properties, and some new applications of chitin and its derivatives. Crit Rev Food Sci Nutr 43(2):145–171

Szosland L (1998) In: Muzzarelli RAA, Peter MG (eds) Chitin handbook. Atec Edizioni, Italy, pp 53–60

Szosland L, Struszczyk H (1997) In: Muzzarelli RAA (ed) Chitin enzymology, vol 2. Atec Edizioni, Italy, pp 491–496

Taguchi T, Kishida A, Akashi M (1998) Chem Lett 8:711–722

Tajik H et al (2008) Preparation of chitosan from brine shrimp (Artemia urmiana) cyst shells and effects of different chemical processing sequences on the physicochemical and functional properties of the product. Molecules 13(6):1263–1274

Taked M, Aiba E (1962) Norisho Suisan Koshusho Hokoku 11:339

Teixeira M, Paterson WP, Dunn EJ, Li Q, Hunter BK, Goosen MFA (1990) Ind Eng Chem Res 29:1205–1209

Teng WL, Khor E, Koon T, Lim LY, Ta S. Ch. (2001), Carbohydrate Research 332, pp 305–316

Thanou M, Junginger HE (2005) Pharmaceutical applications of chitosan and derivatives. In: Dumitriu S (ed) Polysaccharides. Structural diversity and functional versatility, 2nd edn. Dekker, New York, pp 661–77

Tokuyasu K, Mitsutomi M, Yamaguchi I, Hayashi K, Mori Y (2000) Biochemistry 39:8837–8843

Tolaimate A et al (2003) Contribution to the preparation of chitins and chitosans with controlled physicochemical properties. Polymer 44(26):7939–7952

Tozaki H, Odoriba T, Okada N, Fujita T, Terabe A, Suzuki T, Okabe S, Muranishi S, Yamamoto A (2002) Chitosan capsules for colon-specific drug delivery: enhanced localization of 5-aminosalicylic acid in the large intestine accelerates healing of TNBS-induced colitis in rats. J Control Release 82(1):51–61

Tsai GJ, Su WH (1999) Antibacterial activity of shrimp chitosan against Escherichia coli. J Food Prot 62:239–243

Tsigos I et al (2000) Chitin deacetylases: new, versatile tools in biotechnology. Trends Biotechnol 18 (7):305–312

Twu Y-K, Chang I-T, Ping C-C (2005) Carbohydr Polym 62:113–119

US patent No. 94-349661 (941205)

US patent No. 2712507

Uppal AK, El Hadrami A, Adam LR, Tenuta M, Daayf F (2008) Biological control of potato Verticillium wilt under controlled and field conditions using selected bacterial antagonists and plant extracts. Biol Control 44:90–100

Uragami T (1989) In: Skjak-Braek G, Anthonsen T, Standford P (eds) Chitin and chitosan. Elsevier Applied Science, New York, pp 783–792

Urbanczyk G, Lipp-Symonowicz B, Jeziorny A, Dorau K, Wrzosek H, Urbaniak-Domagala W, Kowalska S (1997) Progress on chemistry and application of chitin and its derivatives, vol 3, pp 186–187

Van Bennekum AM, Nguyen1 DV, Schulthess G, Hauser H, Phillips MC (2007) Mechanisms of cholesterol-lowering effects of dietary insoluble fibres:relationships with intestinal and hepatic cholesterol parameters. Arie, Shridhar Pandya: An attractive biocompatible polymer for pharmaceutical application in various dosage forms – Chitosan. Latest Rev 5(3)

Vasyukova NI, Chalenko GI, Gerasimova NG, Perekhod EA, Ozeretskovskaya OL, Irina AV, Varlamov VP, Albulov AI (2005) Chitin and chitosan derivatives as elicitors of potato resistance to late blight. Appl Biochem Microbiol 36:372–376, translated from Prik. Biokhim. Mikrobiol. 2000, 36, 433–438

Veroni G, Veroni F, Contos S, Tripodi S, De Bernardi M, Guarino C, Marletta M (1996) In: Muzzarelli RAA (ed) Chitin enzymology, vol 2. Atec Edizioni, Italy, pp 63–68

Vruggink H (1970) The effect of chitin amendment on actinomycetes in soil and on the infection of potato tubers by Streptomyces scabies. Neth J Plant Pathol 76:293–295

Wang B, Tian C, Wang L, Wang R, Fu Y (2010) Chitosan: a green carbon source for the synthesis of graphitic nanocarbon, tungsten carbide and graphitic nanocarbon/tungsten carbide composites. Nanotechnology 21:025606

Wang GH (1992) Inhibition and inactivation of five species of foodborne pathogens by chitosan. J Food Prot 55:916–919

Wang SL, Chio SH (1998) Deproteinization of shrimp and crab shell with the protease of Pseudomonas aeruginosa K-187. Enzyme Microb Technol 22(7):629–633

Wieczorek A, Mucha M (1997) In: Domard A, Roberts GAF, Varum KM (eds) Advance in chitin science, vol 2. Jacques Andre Publisher, Lyon, pp 890–896

Wiśniewska-Wrona M, Kucharska M, Niekraszewicz A, Kardas I, Ciechańska D, Bodek K (2010) Chitosan-alginate biocomposites in the form of films used in bedsores treatment. Polim Med 40(2):57–64

Wojdyła AT (2001) Chitosan in the control of rose diseases – 6 year – trials. Bull Pol Ac Sci Biol Sci 49 (3):243–252

Wojdyła AT, Orlikowski LB, Niekraszewicz A, Struszczyk H (1996) Effectiveness of chitosan in the control of Sphaerotheca pannosa var. rosae and Peronospora sparsa on roses and Myrothecium roridum on dieffenbachia. Med Fac Landbouww Univ Gent 61/ 2a:461–464

Wojdyła AT, Orlikowski LB, Struszczyk H (2001) Chitosan for the control of leaf pathogens. In: Muzzarelli RAA (ed) Chitin enzymology. Atec, Italy, pp 191–196

Wojtasz-Pajak A (1997), "The influence of the parameters of the reaction of deacetylation on the physical and chemical properties of chitosan" in: Progress on Chemistry and Application of Chitin and its Derivatives, Monograph of Polish Chitin Society, vol 3, pp 4–10

WojtaszPajak A (1998) In: Struszczyk H (ed) Progress on chemistry and application of chitin and its derivatives, vol 5. Polish Chitin Society, Lodz, pp 87–93

Wojtasz-Pajak A, Brzeski MM (1998) In: Domard A, Roberts GAF, Varum KM (eds) Advances in chitin science, vol 2. Jacques Andre, Lyon, pp 64–70

World Patent No., WO 9723390 A1 970703

Wu ACM, Bough WA (1978) In: R.A.A. Muzzarelli, E.R. Pariser (eds) Proceedings of 1st international conference on chitin and chitosan. MIT Sea Grant Program, Cambridge, MA, pp 88–102.

Xing R, Liu S, Yu H, Zhang Q, Li Z, Li P (2004) Carbohydr Res 339:2515

Xu Y, Gallert C, Winter J (2008) Chitin purification from shrimp wastes by microbial deproteination and decalcification. Appl Microb Biotechnol 79(4):687–697

Yaghobi N, Hormozi F (2010) Carbohydr Polym 81:892–896

Yamaguchi I, Itoh S, Suzuki M, Osakae A, Tanaka J (2003a) Biomaterials 24:3285–3292

Yamaguchi I, Itoh S, Suzuki M, Sakane M, Osaka A, Tanaka J (2003b) The chitosan prepared from crab tendon: the characterization and the mechanical properties. Biomaterials 24:2031–2036

Yen M-T, Mau J-L (2007) LWT 40:558–563

Yihua YU, Binglin HE (1997) Artif Cells Blood Substit Immobil Biotechnol 25:445–450

Yoshida H, Okamoto A, Kataoka T (1993) Chem Eng Sci 48:2267–2272

Youn DK, No HK, Prinyawiwatkul W (2009) Physicochemical and functional properties of chitosans prepared from shells of crabs harvested in three different years. Carbohydr Polym 78(1):41–45

Yu G, Xu G, Zou H (1991) In: Feng H (ed) C-MRS international symposia proceedings, vol 3. North-Holland, Amsterdam, pp 305–315.

Zhang C, Ping Q, Zhang H, Shen J (2003) Carbohydr Polym 54:137

Zhang M et al (2000) Structure of insect chitin isolated from beetle larva cuticle and silkworm (Bombyx mori) pupa exuvia. Int J Biol Macromol 27(1):99–105

Zhou YG, Yang YD, Qi YG, Zhang ZM, Wang XJ, Hu XJ (2002) Effects of chitosan on some physiological activity in germinating seed of peanut. J Peanut Sci 31:22–25

Polysaccharide-Acting Enzymes and Their Applications

Anu Koivula, Sanni Voutilainen, Jaakko Pere, Kristiina Kruus, Anna Suurnäkki, Lambertus A.M. van den Broek, Robert Bakker, and Steef Lips

Contents

12.1	Introduction	376
12.2	Enzymes Acting on Lignocellulose Polysaccharides	377
12.2.1	Hydrolysis	377
12.2.2	Oxidation	382
12.2.3	Transfer Reactions	383
12.3	Applications of Enzymes in Modification of Lignocellulosic Polysaccharides	383
12.3.1	Applications of Cellulases	383
12.3.2	Applications of Hemicellulases	384
12.3.3	Analytics	386
References		388

A. Suurnäkki (✉)
VTT Technical research centre of Finland, P.O. BOX 1000, 02044 VTT, Finland

VTT Technical research Centre of Finland, VTT Biotechnology, Tietotie 2, Espoo, P.O. Box 1500, Finland
e-mail: anna.suurnakki@vtt.fi

Abstract

Biobased economy is expected to grow substantially in Europe within the coming 20 years. An important part of the bioeconomy is biorefineries in which biomass is processed in a sustainable manner to various exploitable products and energy. Bioeconomy can be seen as an expansion of the biorefinery concept as it also includes the exploitation of biotechnology in processing of non-biological raw materials or production of non-bio products exploiting certain biological principles.

Enzymes offer a selective and efficient means to convert biomass and its components including polysaccharides into chemicals, materials, energy, food and feed in a sustainable manner. Due to their specificity enzymes are powerful tools especially in the targeted modification of biomass components. Furthermore, enzymes can be used to overcome some of the challenges related to the utilisation of biomass. Compared to traditional manufacturing systems, biomass can be processed by enzymes in mild conditions with significantly less energy, water and without the need of aggressive chemicals. A wide variety of potential enzymes suitable for processing and upgrading of lignocellulosic polysaccharides and polysaccharide -based materials is currently commercially available and novel enzymes are actively searched for. The role of these enzymes in future lignocellulosic polysaccharide processing and upgrading is dependent on the value addition and economical feasibility reached as well as the wide industrial acceptance of the bioprocessing technologies.

12.1 Introduction

Polysaccharides, a major structural component found in biomass, are applied in various industrial applications like in food, chemicals and paper as well as for energy purposes. For industrial applications, native polysaccharides are generally chemically modified to provide desired properties for the final product. Chemical modifications mainly include polysaccharide derivatisations and hydrolysis; derivatisations aim at adding functional groups to hydroxyl groups within molecules and acid hydrolysis is used for depolymerisation of the polysaccharide to oligosaccharides or monosaccharides. The conditions used in the chemical reactions are, however, often harsh and cause unwanted, uncontrollable degradation of the polymeric substrate. In addition, selectivity of the chemical reactions is a big challenge especially for the food or pharmaceutical applications in which chemical residues of polysaccharide derivatives are of major concern. Furthermore, processes based on harsh chemicals require high capital and also operational costs. A potential alternative for chemical modification is enzymatic processing especially when aiming at highly specific, structure-retaining and sustainable tailoring of polysaccharides.

Enzymes are proteins with defined molecular structures that catalyse specific chemical reactions. For polysaccharide modification, enzymes are mainly used for catalysis of hydrolysis, oxidation and group transfer reactions. Enzymes useful in these reactions belong to hydrolytic, oxidative and transferase enzyme groups, respectively. The action of an enzyme is substrate specific. The monocomponent enzyme preparations containing only one enzyme activity can thus be used for highly defined, site-specific reactions whereas different enzyme activities are needed for the complete depolymerisation of a polysaccharide. Enzymatic hydrolysis can be used for targeted debranching of side groups from the polysaccharide main chain as in the case of demethylation of pectin by pectin methyl esterase, decreasing the molecular size of the polysaccharide as in the case of amylase treatment of starch used in papermaking or for total hydrolysis of polysaccharide to monosaccharides prior to their further fermentation to ethanol. Oxidation of polysaccharides has been reported to be useful for further modification of biopolymeric carbohydrates (Leppänen et al. 2010). This can be achieved in general by the formation of carbonyl/carboxylic groups during the oxidation process (Kruus et al. 2001; Parikka and Tenkanen 2009). Transfer reactions can be applied in the functionalisation of polysaccharides that can further be used in the modification of e.g., cellulose or fibre materials as in the case of xyloglucan modification of cellulose (Brumer et al. 2004).

In Table 12.1, enzyme groups presently available for modification of the lignocellulosic polysaccharides, cellulose, hemicellulose and pectin are listed.

The selection of the right monocomponent enzyme or enzyme combination is a prerequisite for the successful enzymatic modification of polysaccharides and polysaccharide matrices. Enzyme performance in polysaccharide modification is also affected by the polysaccharide substrate accessibility and reactivity and the various process-related factors such as treatment pH, temperature, consistency and time used, shear forces in the treatment and chemicals present in the process. Thus, a good knowledge of the enzymes and substrate properties, enzyme function in the substrate and the process they are applied for is needed when aiming at optimal result in the enzyme-catalysed modification of polysaccharide materials (Suurnäkki et al. 1997; Bechtold and Schimper 2010).

In the following chapters the main enzyme groups acting on lignocellulose polysaccharides are first presented. Thereafter, some examples of applications of enzymes in lignocellulose polysaccharide-based materials are presented.

Table 12.1 Enzymes useful for lignocellulosic polysaccharide modification

Component	Enzyme group	Tools	Exploitation
Cellulose	Cellulases	Endoglucanases, EG	Depolymerisation Structure modification
		Cellobiohydrolases CBH	Structure modification
		Mix	Total hydrolysis
Hemicellulose	Hemicellulases	Xylanases	Depolymerisation
		Mannanases	Depolymerisation
		Accessory enzymes	Side-group cleaving Transfer reactions for depolymerisation, structure modification and functionalisation
		Mix	Total hydrolysis
Pectin	Pectinases	Pectinases	Cleavage, demethylation
		Hydrophobins Swollenins	Surface hydrophobication Loosening of cellulose fibre structure

12.2 Enzymes Acting on Lignocellulose Polysaccharides

12.2.1 Hydrolysis

12.2.1.1 Cellulolytic Enzymes

Enzymatic Degradation of Cellulose

The efficient degradation of crystalline cellulose requires the action of a battery of enzymes produced by microorganisms habiting the dead or dying plant matter. The cellulolytic microorganisms include protozoa, fungi and bacteria and are ubiquitous in nature. Cellulases are O-glycoside hydrolases (GH) that can cleave the β-1→4 glycosidic bonds in cellulose. Cellulases are classified on the basis of their mode of action into endo- and exoglucanases. Endoglucanases (1,4-β-D-glucan glucanohydrolase; EC 3.2.1.4) can hydrolyse cellulose chains in the middle by making random cuts and are able to attack only the non-crystalline, amorphous regions (Teeri 1997). Endoglucanases efficiently reduce the cellulose chain length and create new chain ends for the action of exoglucanases. Exoglucanases or cellobiohydrolases (1,4-β-D-glucan cellobiohydrolase; EC 3.2.1.- and EC 3.2.1.91) hydrolyse cellulose chains sequentially, by removing cellobiose units from the reducing or non-reducing end of the cellulose chain in a processive manner (Teeri et al. 1998; Wood and Bhat 1988). They are usually active on both crystalline and amorphous parts of the cellulose and decrease the chain length of cellulose very slowly. β-Glucosidases (EC 3.2.1.21) complete the cellulose hydrolysis by cleaving the soluble oligosaccharides and cellobiose into glucose.

The enzyme mixture needed for a complete hydrolysis of crystalline cellulose comprises of different types of cellulases acting in a synergistic manner. Synergism refers to the ability of a mixture of cellulases to have higher activity than the sum of the activities of the individual enzymes. Synergism occurs when the different cellulases attack the substrate at different sites, thus creating or revealing new sites for other enzymes to work upon. In the well-reported endo-exo synergy, endoglucanases provide free chain ends on the cellulose surface for the cellobiohydrolases to act upon.

Most cellulases are modular enzymes consisting of at least a catalytic module and one or more modules involved in substrate binding (CBMs, carbohydrate-binding modules) (Davies and Henrissat 1995). Furthermore, cellulases can have additional modules, required e.g., for attachment to the cell surface. The CBM is responsible for concentrating the enzyme on the substrate surface and improving the action on crystalline substrates while not affecting the activity on amorphous or soluble substrates.

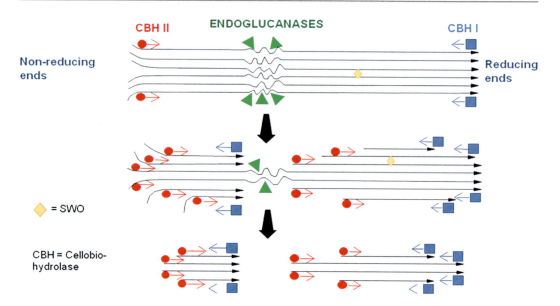

Fig. 12.1 A schematic model of the synergistic enzymatic degradation of cellulose

Classification of Cellulase Modules

Both the catalytic modules (belonging to GH families) and the substrate-binding modules (CBM families) can be classified into different families based on sequence similarities and protein folds (Henrissat et al. 1995; Henrissat and Bairoch 1996). The CAZy (carbohydrate-active enzymes) database classifies the catalytic modules of different glycoside hydrolases into over 120 GH families and the carbohydrate-binding modules (CBMs) into over 60 families. To date, all the GH families to which cellulolytic enzymes have been allocated (GH families 5–10, 12, 44, 45, 48, 51, 61, 74) also contain enzymes for which 3D structural information is available. Cellulose-binding affinity has been described within over ten CBM families.

One GH family can contain catalytic modules derived from both endo- and exoglucanases. Despite of the similar overall fold, however, the active-site architecture of the two different types of enzymes differs. The endoglucanases are commonly characterised by a groove or cleft, into which any part of a linear cellulose chain can fit. On the other hand, the exoglucanases (cellobiohydrolases) bear tunnel-like active sites, which can only accept a substrate chain via its terminus (Fig. 12.1).

12.2.1.2 Hemicellulose-Degrading Enzymes

Enzymatic Degradation of Hemicelluloses

The most common wood hemicelluloses are xylans and glucomannans. The major hemicellulose in hardwoods is xylan, which is composed of b-D-xylopyranosyl units substituted with 4-O-methyl-a-D-glucuronic acid and acetyl groups. Glucomannan is the major hemicellulose in softwoods. Softwoods also contain considerable amounts of arabino-4O-methylglucuronoxylan, in which the xylan backbone is substituted with 4-O-methyl-a-D-glucuronic acid and a-L-arabinofuranosyl residues. Galactoglucomannan backbone consists b-1,4-linked b-D-glucopyranosyl and b D-mannopyranosyl units. Due to the structural complexity of hemicelluloses, the enzymatic degradation of hemicelluloses requires synergistic action of various types of enzymes, generally regarded as hemicellulases.

Hemicellulases are produced by many species of bacteria and fungi and they include endo-acting glycoside hydrolases such as xylanases and mannanases, exo-acting β-1,4-xylosidases, α-L-arabinofuranosidases, α-glucuronidases, β-1,4-mannosidases, α-galactosidases and β-glucosidases as well as esterases which deacetylate xylans and mannans and release fenolic acids from e.g., arabinoxylans (Shallom and Shoham 2003). The endo-acting hemicellulases hydrolyse the internal glycosidic bonds in the hemicellulose backbone by a random mechanism, while the various exo-acting enzymes are responsible for debranching of the hemicellulose polymers as well as digestion of the generated oligosaccharides to monomeric sugars. The action of backbone-degrading enzymes is often affected by solubility and the substitution pattern of the polymer. Their catalytic action is thus interlinked to the activity of debranching enzymes (Berrin and Juge 2008; Moreira and Filho 2008). On the other hand, e.g., acetylation or ferulic acid substitutions may block action of glycosidic hydrolases and different esterases may be required for complete degradation of the polymer (Fazary and Ju 2007; Agger et al. 2010). As such, synergy between different hemicellulolytic enzymes is usually observed.

Xylan and Mannan Backbone-Degrading Enzymes

Endoxylanases (EC 3.2.1.8) randomly cleave the internal 1,4-β-D-xylosidic linkages in xylan polymers, which form the backbone in glucurono-, arabinoglucurono- and arabinoxylans, as well as in complex heteroxylans. Majority of xylanases belong to two structurally different glycoside hydrolase families 10 (e.g., enzymes from *Cellulomonas fimi* and *Thermoascus aurantiacus*) and 11 (e.g., enzymes from *Trichoderma reesei* and *Aspergillus niger*) (Carbohydrate Active Enzymes database, http://www.cazy.org/, Cantarel et al. 2009). A few xylan hydrolysing enzymes have been identified to belong into GH families 5, 7, 8 and 43 (Berrin and Juge 2008). Some xylanases haven reported to have a two-domain structure containing either a cellulose- (Irwin et al. 1994) or a xylan-binding domain (Gilbert and Hazlewood 1993). The effect of binding domains to improve the hydrolysis of fibre-bound xylan is not clear. Family 10 xylanases have a wide substrate specificity and they are able to hydrolyse both highly substituted and linear xylan as well as small molecular weight oligomers, whereas the family 11 xylanases prefer unsubstituted xylan and have low activity on short xylo-oligo-

saccharides (reviewed by Pollet et al. 2010a, b). Most of the GH5 xylanases are bacterial enzymes with specificity to 4-O-methyl-D-glucuronoxylan (e.g., *Erwinia chrysanthemi*, Vrsanska et al. 2007), although recently, GH5 xylanase with activity on non-substituted xylan has been identified from *Bacillus* sp. (Gallardo et al. 2010). All known xylanases in GH8 are also bacterial enzymes, with low affinity on short xylo-oligomers and variable substrate specificity towards substituted xylans (Collins et al. 2005; Pollet et al. 2010b). The only xylan-hydrolysing enzyme characterised in GH7 is a non-specific endo-1,4-glucanase from *Trichoderma reesei* (Cel7B, Kleywegt et al. 1997). The family 43 xylanase from *Paenibacillus polymyxa* acts both as xylanase and α-1,-5, arabinofuranosidase (Gosalbes et al. 1991). Further degradation of xylo-oligosaccharides generated by xylanases is catalysed by β-xylosidases (EC 3.2.1.37). Xylosidases are classified into GH families 3, 30, 39, 43, 52 and 54, and they release xylose from the non-reducing end of the xylo-oligosaccharides (Jordan and Wagschal 2010).

Endomannanases (EC 3.2.1.78) catalyse the hydrolysis of β-D-1,4 mannopyranosyl linkages within the main chain of mannans, glucomannans, galactomannans and galactoglucomannans (Moreira and Filho 2008). Microbial mannanases are classified into GH families 5 and 26. The activity of mannanases is affected by the degree and type of substitutions and in glucomannan and galactoglucomannans by the glucose to mannose ratio in the polymer (Dhawan and Kaur 2007). Structural study of GH5 and GH26 family mannanases from *Bacillus* sp. revealed that the active site of GH5 family mannanase is able to recognise both glucoside and mannoside units, while the GH26 mannanase specifically binds mannoside units (Tailford et al. 2009). The manno- and manno-gluco-oligos released by mannanases are further hydrolysed by β-mannosidases (EC 3.1.1.25) and β-glucosidases (EC 3.2.1.21). Most microbial β-mannosidases belong to the GH families 1 and 2. The characterised GH1 enzymes are from Archea and while both bacterial and fungal GH2 enzymes are characterised. Almost all microbial beta-glucosidases are classified into GH1 and GH3 (Bhatia et al. 2002) and they may

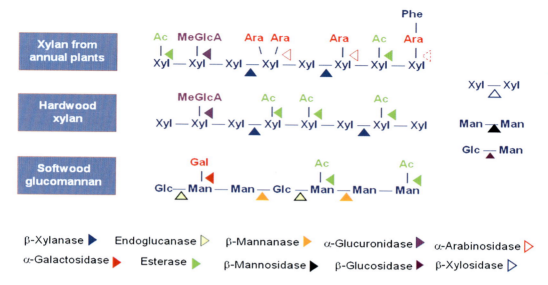

Fig. 12.2 Enzymes acting on hemicelluloses in lignocellulose

be specific for glycans or have a broad substrate specificity (Cairns and Esen 2010).

Debranching/Accessory Enzymes

The carbohydrate side groups connected to xylan and glucomannan main chains are removed by various debranching enzymes. α-Glucuronidases (EC 3.2.1.139) catalyse α-1,2-glycosidic bond cleavage between 4-O-methyl-D-glucuronic acid present in glucurono- and arabinoglucuronoxylans. All xylan-debranching α-glucuronidases are classified into the GH family 67. Most of the glucuronidases release glucuronic acid only from the terminal non-reducing end of the xylo-oligos while some of the enzymes are able to hydrolyse also internal glucuronic acid side groups in the polymers (Siika-aho et al. 1994; Tenkanen and Siika-aho 2000; Biely et al. 2000; de Wet et al. 2006). α-Arabinofuranosidases (EC 3.2.1.55) catalyse the cleavage of α-linked L-arabinofuranosyl residues at the non-reducing ends of the polysaccharides. They are classified into GH families 3, 43, 51, 54 and 62. All of these enzymes catalyses hydrolysis of α-L-arabinofuranoside residues at the non-reducing ends of arabinose contain oligo- or polysaccharides but are differentially affected by the polymer length, substitution pattern and linkage type (reviewed by Numan and Bhosle 2006).

In addition to carbohydrate substitutions, hemicelluloses are often substituted with esterified acetyl or cinnamoyl-groups which can be hydrolysed by various carbohydrate esterases (CE). Acetyl esterase (EC 3.1.1.6) are classified into CE family 16 and they catalyse deacetylation of xylo- and (gluco)manno-oligosaccharides (Tenkanen 1998) whereas acetyl xylan esterases (AXE, EC 3.1.1.72, CE families 1, 4 and 5) are more specific for xylan acetylation (Tenkanen et al. 2003; Biely et al. 2011). Feruloyl esterases (FAE, EC 3.1.1.73) in turn remove ferulic acid from the polysaccharides. Feruloyl esterases are found in CE family 1. Further classifications of FAEs based on sequence, substrate specificity and structural properties have been described (Crepin et al. 2004; Udatha et al. 2011). Functionally, the FAEs differ e.g., in their substrate specificity and ability to act on diferulic acid moieties (Fazary and Ju 2007). Lignin and hemicellulose connecting covalent linkages in plant cell wall also include ester linkages between hydroxyl groups in lignin and carboxyl groups of uronic acids in glucuronoxylan. A novel activity of glucuronyl esterase active on these linkages has been recently reported (Spanikova and Biely 2006). Figure 12.2 gives an overview of the enzymes acting on hemicellulose in lignocellulosics.

12.2.1.3 Pectin-Degrading Enzymes

Pectins are a group of closely related polysaccharides that play an important role in the cell wall binding layer of land plants. There are three major forms of pectin: homogalacturonan, rhamnogalacturonan I (RG-I) and II (RG-II). Homogalacturonan is a linear chain of α-1,4-linked galacturonic acid residues which can be esterified by methanol and/or acetic acid. RG-I consists of a backbone of an alternating disaccharide [α-1, 2 rhamnose → α-1, 4 galacturonic acid]$_n$. The rhamnose units are usually decorated with galactose and/or arabinose side chains (Mohnen 2008; Wong 2008). RG-II is composed of 13 different sugars and contains over 20 different glycosidic linkages (O'Neill et al. 2004). Only enzymes active towards homogalacturonan and RG-I will be discussed here.

Pectinases

Pectinases are defined and classified on the basis of their action towards the homogalacturonan part of pectin. Two main groups are distinguished, namely pectin esterases and pectin depolymerases. The pectin depolymerases can be further divided into hydrolases and lyases depending on the glycosidic bond cleavage mechanism such as hydrolysis and trans-elimination, respectively.

Pectin Esterases

Pectin (methyl) esterase (PE; EC 3.1.1.11) removes methoxyl groups from methylated pectin. Most of these enzymes have a preference for cleaving the methoxyl groups from the non-reducing end of the galacturonan backbone or next to a non-esterified galacturonic acid residue. The plant pectin esterases proceed along the galacturonan backbone like a zipper (processive mechanism) resulting in blocks of non-esterified pectin. The pectin esterases from microbial origin act at random which results in the same effect as obtained by chemical deesterification. The pH optima for plant pectinesterase are in general above pH 7, whereas the pH optima for most microbial pectin esterases are about 4.0–4.5.

Pectin acetyl esterase (PAE; EC 3.1.1.6) is able to cleave acetyl groups from acetylated galacturonic acid residues. The enzyme is produced by plants and microorganisms.

Pectin Depolymerases

Polygalacturonase (PG) exists in two forms, endo-PG (EC 3.2.1.15) and exo-PG (EC 3.2.1.67 and EC 3.2.1.82), depending on their mode of action. The enzymes cleave the α-1,4 linkage between non-esterified galacturonic acid units. Endo-PG cleaves the substrate at random whereas exo-PG is only able to remove mono- or digalacturonic acid units from the non-reducing end. Both types of enzymes are found in plants and microorganisms.

Pectin lyase (PL; EC 4.2.2.10) cleaves the α-1,4 linkage between esterified galacturonic acid units. This results in the formation of unsaturated products. The enzyme is only found in microorganisms. The affinity for the substrate and the enzyme activity is increasing for pectins with a higher degree of methylation. On the other hand, the affinity increases for lower methylated pectins as the pH is decreasing.

Pectate lyase (PAL; EC 4.2.2.2) cleaves the α-1,4 linkage between non-esterified galacturonic acid units. In this case also the formation of unsaturated products is taking place. The enzyme is only produced by microorganisms. The pH optimum is about 8.5–9.5 and the enzyme needs Ca^{2+} for its activity.

Rhamnogalacturonan-Degrading Enzymes

Two endo- and three exo-acting enzymes have been described active towards the RG-I backbone. The enzymes have only been found in microorganisms (Wong 2008). Rhamnogalacturonan hydrolase (EC 3.2.1.-) cleaves the α-1,2 linkage between a rhamnose and galacturonic acid residue and prefers deacetylated RG-I. Rhamnogalacturonan lyase (EC 4.2.2.-) cleaves by β-elimination the α-1,4 linkage between a galacturonic acid and a rhamnose residue, and it also prefers deacetylated RG-I. An unsaturated rhamnogalacturonyl hydrolase (EC 3.2.1.-) has been described able to release the terminal unsaturated galacturonic acid unit of the products obtained after treatment of RG-I with rhamnogalacturonan lyase. Rhamnogalacturonan rhamnohydrolase (EC 3.2.1.40) removes terminal rhamnose units from the non-reducing end of RG-I and RG-I oligosaccharides, whereas rhamnogalacturonan galacturonanhydrolase removes galacturonic acid residues from RG-I and RG-I oligosaccharides. Apart from depolymerases, a rhamnogalacturonan acetyl esterase (EC 3.1.1.-) has been identified. The enzyme shows preference for acetylated RG-I and releases acetic acid although it can also have some activity towards acetylated homogalacturonan (Bonnin et al. 2008).

Many of the RG-I-degrading enzymes have been isolated and characterised at Wageningen UR. For example, rhamnogalacturonan hydrolase was the first enzyme described able to degrade RG-I (Schols et al. 1990) followed by the discovery of rhamnogalacturonan acetyl esterase (Searle van leeuwen et al. 1992). Mutter and Beldman (1994) described in more detail the reaction mechanism of rhamnogalacturonan lyase, rhamnogalacturonan rhamnohydrolase and rhamnogalacturonan galacturonanhydrolase (Mutter and Beldman 1994; Mutter et al. 1996; 1998). All these enzymes were isolated from a commercial enzyme preparation Pectinex Ultra SP-L (Novozymes). Due to unravelling of the DNA sequence of these enzymes, it was possible to produce recombinant enzymes to study their working mechanism in more detail. In this way, for the first time a rhamnogalacturonan hydrolase containing a cellulose-binding domain has been reported (McKie et al. 2001). Another approach was to introduce the genes, coding for pectin-degrading enzymes, into plants and to study their effect on cell walls. For example, the rhamnogalacturonan lyase gene from *Aspergillus aculeatus* was introduced in potato plants. The transgenic plants produced tubers that showed morphological alterations. It was noticed that the cell walls contained much lower amounts of galactose and arabinose residues as in the wild type. From this study it became clear that RG-I was important in anchoring galactans and arabinans (Oomen et al. 2002). Due to the sequenced genomes of *Aspergillus* sp., more information about potential pectinases and RG-I-degrading enzymes became available. For instance, *A. niger* harboured 12 new putative genes encoding for exo-acting glycoside hydrolases that are involved in pectin degradation (Martens-Uzunova and Zandleven 2006).

12.2.2 Oxidation

Currently only a few enzymes are known feasible for oxidisation of polymeric carbohydrates. The success of enzymatic oxidation of polysaccharides depends largely on the efficacy and specificity of the relevant enzymes. Furthermore, quantitative analytical methods are scarce for detecting and separating specific oxidation products. Analysis of the profiles of oxidised groups (carbonyl and carboxyl) relative to the molecular weight distribution of the polymers can, however, be done by group-selective fluorescence labelling techniques applied as precolumn derivatisation in combination with SEC-MALLS (Potthast et al. 2003; Bohrn et al. 2006). Other challenges include the inefficiency in oxygen transportation of the co-substrate in the reaction solution, and therefore, the low yield of polymer oxidation in concentrated solution in general (Parikka et al. 2010).

Due to the challenges in the direct enzymatic oxidation, the combination of chemical and enzymatic methods for polysaccharide oxidation has been proposed. The combined use of TEMPO (2,2,6,6-tetramethylpiperidine-1-oxyl) and laccase/oxygen systems for oxidation of starch and cellulose suspensions has been studied intensively (Bragd et al. 2004). Cellulose can be oxidised in a similar way with the combined action of laccase and TEMPO (Kruus et al. 2001; Patel et al. 2011). On the other hand, galactose oxidase has been known as a potential candidate for enzymatic oxidation on the same substrates (Parikka and Tenkanen 2009).

12.2.2.1 Galactose Oxidase

Galactose oxidase (GAO, D-galactose: oxygen 6-oxidoreductase; EC 1.1.3.9) is an extensively studied carbohydrate oxidase (Parikka et al. 2010; van Hellemond et al. 2006). GAO catalyses the oxidation of galactose-based carbohydrates and primary alcohols into the corresponding aldehydes with the reduction of O_2 to H_2O_2. The regioselectivity of GAO is high as it oxidises the hydroxyl group solely at the C6 position of hexoses. For polymers, GAO oxidises selectively the C-6 hydroxyls of terminal galactose to carbonyl groups in polysaccharides, such as GGM (Parikka et al. 2010).

12.2.3 Transfer Reactions

Specific xyloglucanases or xyloglucan-specific endoglucanases represent a new class of polysaccharide-degrading enzymes which can attack xyloglucans. Xyloglucan endotransglycosylase, XET (EC 2.4.1.207), is able to transfer fragments from donor xyloglucans to suitable acceptors, such as a xyloglucan-derived nonasaccharides. Because of the inherent ability of XET to catalyse transglycosylation rather than hydrolysis, this enzyme has potential for lignocellulosic fibre modification.

12.3 Applications of Enzymes in Modification of Lignocellulosic Polysaccharides

12.3.1 Applications of Cellulases

The microbial cellulolytic enzymes play a crucial role in the carbon cycle on earth as they reduce the most common organic polymer, cellulose, to an exploitable form, glucose. Cellulases, particularly those from filamentous fungi such as *Trichoderma reesei* and *Humicola insolens*, can also be exploited in various industrial applications. In the textile industry, cellulases are used to substitute conventional stonewash methods. They also have applications in the detergent industry for reducing fuzz and pilling of fabrics, and as general cleaning agents for cotton garments (Bhat 2000; Olson and Stanley 1990). In addition, cellulases are used in animal feeds for improving their nutritional quality and digestibility and in the processing of fruit juices (Bhat 2000). Furthermore, cellulases have potential in the paper and pulp industries, where they have e.g., been used to lower the energy consumption in mechanical pulp production (Pere et al. 2000). Reviews of the use of cellulases for pulp and paper (Viikari et al. 2010) and textile applications (Bechtold and Schimper 2010) are available.

Among potential alternative bioenergy resources, lignocellulosics have been identified as the prime source of biofuels and other value-added products. Production of ethanol from annual crops (e.g., starch) is already a well-established technology. Starch- and sugar cane-based ethanol is often referred to as first-generation biofuel (Yuan et al. 2008). However, there is a need to have an alternative, non-food raw material to avoid the conflict between land use for food and for energy production. Cellulose-based ethanol, also referred as the second-generation biofuel, is a promising alternative to reduce oil dependence as well as greenhouse gas emissions, because the plant biomass raw material is renewable, inexpensive and abundantly available. Biomass sources such as agricultural waste (straw), forest wood, side products of the forest industry and energy crops are the main potential sources.

Currently, the processes using lignocellulosic feedstock for bioethanol production are still in the development phase, due to the challenges in hydrolysis of the recalcitrant lignocellulosic substrates. The major obstacles to cost-effective production of ethanol from cellulose are the high enzyme costs due to the high amount of enzymes needed in the process (Viikari et al. 2011).

Bioethanol process concepts can generally be described as separate hydrolysis and fermentation (SHF) or as simultaneous saccharification and fermentation (SSF) (Viikari et al. 2007). The need to reduce the enzyme cost has led to further developments of these process concepts, aiming to ethanol production from biomass in a single process step called consolidated bioprocessing (CBP). CBP combines SSF process and enzyme production utilising ethanologenic yeast or bacteria, which is also able to produce the cellulose enzymes, as reviewed recently by (Xu et al. 2009).

In addition to bioethanol, the pretreatment and enzymatic hydrolysis of lignocellulosic biomass for hydrogen fermentation purposes has been

Table 12.2 Main results for alkaline pretreatment and enzymatic hydrolysis of barley straw at both low and high L/S ratio

Reactor	Stirred tank (lower dm concentration)	Conical screw (higher dm concentration)
Alkaline treatment		
Bone dry straw used for experiment (kg)	0.895	3.686
NaOH dosage (wt % on bone dry straw basis)	10.1	9.9
Liquid/solid ratio at start of alkaline pretreatment	17.4	8.0
Pretreatment Time (h)	6	6
Pretreatment Temperature (°C)	75	75
Yield of washed straw after pretreatment and removing black liquor and other soluble components (%; dry matter straw basis)	56.8	50.4
Enzymatic treatment		
Concentration of pretreated straw at start of enzymatic hydrolysis (%; total wt basis)	12.1	17.3
Liquid/solid ratio	7.3	4.8
Dosage of enzyme on bone dry straw (wt %)	32	31
Total incubation time (hours)	24	24
Incubation temperature (°C)	50	50
Total sugar concentration in hydrolysate (g/l)	41	124
Fermentable sugar polymers in hydrolysate (%; on basis of carbohydrate polymers contained in barley straw)	74.7	52.2

studied. Production of fermentable substrates from alkaline pretreated and enzymatically hydrolysed barley straw previously identified as promising raw material (Panagiotopoulos and Bakker 2008; Panagiotopoulos et al. 2009) was studied in Wageningen UR, Food & Biobased Research group. The main factor of interest in this work was the effect of treatment consistency to the final sugar concentration in lignocellulose-derived hydrolysates. High sugar concentration is desired to efficiently and economically convert carbohydrates into biofuels such as bioethanol, bio-butanol and hydrogen. Alkaline pretreatment of barley straw with 10% sodium hydroxide (NaOH) was first carried out for 6 h at 75 °C. After washing, enzymatic hydrolysis of pretreated material was carried out with the commercial cellulase/hemicellulase cocktail GC220, obtained from Genencor for 24 h at 50 °C. Relevant original data of the two bench-scale experiments are summarised in Table 12.2.

As shown in Table 12.2, conducting pretreatment and enzymatic hydrolysis of barley straw at low liquid/solid ratio c.q. high dry matter concentrations, which is required for an economically feasible pretreatment process, yields high total sugar concentrations in the final hydrolysate. However, the fermentable sugar yield is much lower after the enzymatic hydrolysis in high than low dry matter concentration. This is likely an indication of end-product (glucose, xylose) inhibition of the enzyme. Therefore, enzyme screening and optimisation should be geared towards efficient conversion of pretreatment lignocellulosic biomass at high dry matter concentration. In addition, the simultaneous saccharification and fermentation (SSF) should be pursued, in order to avoid sugar inhibition of the enzyme.

12.3.2 Applications of Hemicellulases

Enzymatic degradation of hemicelluloses in various biomasses has been extensively studied due to the many important applications. These include, for instance, efficient conversion of hemicellulosic biomass to fuels and chemicals (Gírio et al. 2010), delignification of paper pulp (Suurnäkki et al. 1997), improving quality of feed products (reviewed by Bedford 2000) and in baking applications to increase bread volume (Courtin et al. 1999), to improve stability and oven spring

(Spréssler 1997) and dough handling properties (Rouau et al. 1994).

Up to 35% of the biomass in lignocellulosic biomasses may comprise of hemicelluloses. Utilisation of hemicellulose fractions is important when developing sustainable and cost-effective second-generation biofuel production. Utilisation of pentoses in *Saccharomyces* fermentation requires engineering of the yeast strain (Hahn-Hägerdal et al. 2007). Most efficient biomass pretreatment technologies solubilise majority of the hemicellulosic carbohydrates, which then may be hydrolysed to fermentable pentoses by hemicellulases (Gírio et al. 2010). Addition of hemicellulases into the hydrolysis of the pretreated solid biomass also enhances cellulose hydrolysis suggesting that the residual hemicelluloses, even though present in minor amounts, restrict the action of cellulases (Garcıa-Aparicio et al. 2007; Kumar and Wyman 2009a, b).

Xylanase-aided bleaching of chemical pulps is the most widely used and best established enzymatic application in the pulp and paper industry (reviewed by Viikari et al. 2010). Xylanase is used as a bleaching aid prior to the bleaching stage to improve the extractability of lignin. The chemical structure of galactoglucomannan present in TMP water can be further modified with acetyl glucomannan esterase being able to cleave acetyl groups from polymeric glucomannan (Tenkanen 1998). The deacetylation of soluble glucomannan has been found to result in decreased solubility and subsequent adsorption of glucomannan onto the fibres (Thornton et al. 1994).

The specificity of purified enzymes and the mild hydrolysis conditions can be exploited in the characterisation of lignocellulosic polysaccharide materials. Valuable information about the structure and composition of polysaccharides in the wood fibres has been obtained by combining enzymatic treatments to analysis methods such as HPLC, ESCA (Electron Spectroscopy for Chemical Analysis) or NMR. For example, the non-destructive action of enzymes was exploited in identification of the earlier undetected, acid labile hexenuronic acid (Teleman et al. 1995). The organisation of cellulose, hemicelluloses and lignin on the outermost surface of kraft fibres has been studied by combining enzymatic peeling and ESCA analysis (Buchert et al. 1996). Furthermore, the role of hemicellulose components in pulp properties has been studied using specific enzymes (Buchert et al. 1996; Oksanen et al. 1997a, 1997b, 2000). By using gradual enzymatic peeling and HPLC analysis, the location of xylan and glucomannan in the wood pulps has been elucidated (Suurnäkki et al. 1996a). Furthermore, it has been shown that the composition of xylan and glucomannan on the accessible surfaces of fibres differs from the composition of these polysaccharides in the overall pulp (Suurnäkki et al. 1996b).

12.3.2.1 Pectin-Degrading Enzyme Applications

Pectinolytic enzymes are mainly used in the fruit industry to improve juice pressing and clarification and in the pulp and paper industry for enzymatic debarking. In addition these enzymes can play a role in the degradation of biomass in the biorefinery process. The application of alkaline pectinases is, for example, found in degumming of plant bast fibres and textile processing and bioscouring of cotton fibres (Hoondal et al. 2002). Below some examples are given of the use and function elucidation of pectinases and rhamnogalacturonan-degrading enzymes at Wageningen UR.

Pectinases are often used to degrade the pectin in the middle lamella of cell walls resulting in separation of cells (maceration). Pectin lyase is able to macerate potato tuber (van den Broek et al. 1997), and a reactor system has been developed to investigate the enzymatic maceration kinetics of potatoes by pectinases (Biekman et al. 1993). The endogenous pectinases in fruit and vegetables play an important role in maturation, ripening and softening. For instance, these enzymes have been identified in pod development of green beans (Ebbelaar et al. 1996). Blanching is often applied to preserve fruit and vegetables. This treatment has an influence on the endogenous pectinases resulting in their activation and/or inactivation. A mathematical model was designed to describe the enzyme activity of pectin esterases in potatoes, carrots and peaches during blanching (Tijskens et al. 1997; 1999). During storage, the endogenous pectinase also have an effect on the firmness of plant cell walls. Models were developed to describe the activity of the endogenous pectinases (Tijskens et al. 1998; Van Dijk et al. 2006). The model for endo-PG (Van Dijk et al. 2006) was integrated with another model describing the firmness loss of tomatoes during storage (Van Dijk et al. 2006).

Due to the well-defined working mechanism of pectinases and rhamnogalacturonan-degrading enzymes, these enzymes have been used for further elucidation of pectin structures. For example, the degree of blockiness of pectin can be determined using an endo- and exo-PG. This results in information about the blockwise distribution of methylesters in the pectin molecule. It is even possible to determine the size of these blocks and to gain information about how the location of the degradable blocks are distributed over the pectin molecule (Daas et al. 1998; Daas et al. 1999; Daas et al. 2000). This technique is extended to explain the different physical properties of pectins which are chemically similar (Guillotin et al. 2005). Using such an enzymatic strategy (e.g., endo-PG) together with chromatographic techniques, it is also possible to obtain information about the distribution of amide groups in low-esterified pectins (Guillotin et al. 2006).

For further elucidation of the rhamnogalacturonan structure, pectinases can be used. In pectic populations of bilberries and black currants, the present of dimers of RG-II were identified after enzymatic treatment with a cocktail of pectinases (Hilz et al. 2006). A connecting linkage between homogalacturonan/xylogalacturonan and RG-I was identified by mass spectrometry after an acid treatment followed by an enzymatic incubation of endo-PG and rhamnogalacturonan galacturonanhydrolase (Coenen et al. 2007). The structure of okra RG-I was elucidated by using enzymes including rhamnogalacturonan hydrolase (Sengkhamparn et al. 2010). In addition to structure elucidation of cell wall polysaccharides, the enzymes can also be used as an analytical tool to produce oligosaccharides from mutant and wild-type plants. In this way it is possible to discriminate between the different plant types using mass spectrometry and/or capillary electrophoresis by fingerprinting plant cell wall polysaccharide digests (Westphal et al. 2010).

12.3.2.2 Applications of Oxidative Enzymes

Potential applications for oxidised polysaccharides, such as cellulose and galactoglucomannan (GGM), can be found in the food industry as emulsifiers, thickeners, stabilisers and gelling agents (Willfor et al. 2008). Oxidised polysaccharides could also be potential as dry or wet strength additives in various hygienic materials such as diapers (Parikka and Tenkanen 2009).

Oxidation of cellulosic pulps by chemoenzymatic laccase-TEMPO oxidation generates aldehyde groups into cellulose (Patel et al. 2011). This modification could be exploited in improving technical properties such as strength of cellulose pulps or in activation of polysaccharides for further chemical derivatisation. Recently, laccase-TEMPO treatment has been reported to improve the wet strength of the sisal pulp (Aracri et al. 2011).

Aldehyde groups formed to polysaccharide by the enzymatic oxidation can also be used for further chemical modification of the polymer. Leppänen et al. (2010) have shown the feasibility of the simplified chemical allylation reaction for D-α-galactopyranoside pre-oxidised by galactose oxidase. The enzymatic oxidation step can be exploited to enable chemical modification of carbohydrates without tedious drying or protection-deprotection steps in the synthesis.

12.3.3 Analytics

12.3.3.1 Detection of Polysaccharides in Plant Cell Wall

In the study of plant cell wall ultrastructure, specific cytochemical markers are needed. There exist dyes for detecting lignin and pectin, but the situation is complicated for the localisation of cell wall polysaccharides, i.e., cellulose, xylan and glucomannan, within multicomponent plant material. Additionally, chemical composition of wood cell walls varies between plant

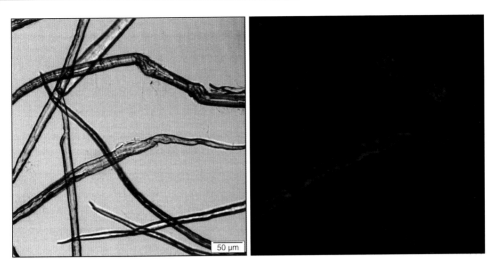

Fig. 12.3 Immunochemical labelling of xylan on Eucalyptus globulus fibres using a polyclonal antibody specific to MeGlsA-Xyl2-3 sub-structure in xylan. Photos courtesy of VTT, Finland

species and depending on the stage of cell wall development. The enzyme-colloidal gold method based on the use of substrate-specific enzymes as probes for localisation of the corresponding substrate was established already in 1980s for ultrastructural visualisation of hemicelluloses in plant cell walls by electron microscopy (Vian et al. 1983; Ruel and Joseleau 1984; Bendayan 1989). The method relies on the purity of the enzyme as well as on the affinity and stability of the enzyme-gold complex. The reliability of the method can, however, be questioned in many cases since non-specific binding cannot be ruled out. More recently, several non-catalytic carbohydrate-binding modules (CBMs) and antibodies, both monoclonal and polyclonal, have been reported to show improved accuracy as specific and versatile markers for carbohydrates and lignin in plant materials (e.g., Awano et al. 1998; Lappalainen et al. 2004; Joseleau et al. 2004; Filonova et al. 2007). The number of natural CBMs is limited, but their exploitation can be widened by molecularly engineered CBMs, which are selected from a molecular library (Filonova et al. 2007). A good antibody is specific and has a high affinity on the antigen. Specificity ascertains binding only to the target antigen, and strength of the binding relies on affinity. In production of antibodies, isolation and purification of the antigen is critical; in that work hydrolytic enzymes are useful tools in preparing defined oligomer substructures of woodderived polymers (Lappalainen et al. 2004). Specificity and cross-reactivity of the antibody always need to be determined prior to use as a probe for ultrastructural studies (Fig. 12.3).

Conclusions

A constantly growing interest in bio-based chemicals, polymers, materials and energy puts a pressure to find new sustainable sources for biomass and efficient methods to modify the biomass components. Lignocellulosics and their polysaccharides, cellulose, hemicelluloses and pectin, are a potential non-food starting material for various applications. To be applicable in various applications, lignocellulose polysaccharides need to be separated and fractionated from the matrix and often thereafter further chemically modified. Biotechnology provides an interesting option for lignocellulose processing and polysaccharide modification, especially when combined with mechanical and chemical technologies.

Sustainable enzymatical methods are available and currently in use in the lignocellulosics and polysaccharide modification in e.g., food and feed, pulp and paper industry and in the ethanol for transfer fuel production. Enzymatic processes or process steps have proven to be efficient when high specificity is needed. Mild reaction conditions used in enzymatic processing enable the retaining of the polymeric structure of the substrate when needed. Challenges for the efficient exploitation of the enzymatic tools in polysaccharide matrix and polymer modification lie in their optimum implementation into the current and future industrial processes.

Plant cell wall is a complex matrix containing many polymeric components strongly interlinked with each other. Better understanding of the chemistry and physical structure of cell wall matrix of various potential lignocellulosics and the exploitation of this information to development of optimal enzyme-aided separation and fractionation processes for various lignocellulose substrates are needed. Thus, efficient disassembly of the plant cell wall structure and exposing the polysaccharides for further enzymatic but also chemical, mechanical or combinatory modifications are of special importance.

Novel, more efficient enzymatic tools for biomass and its polysaccharide modification are also needed. This can be achieved via protein engineering and via enzyme screening utilising the biodiversity. Knowledge of enzyme structure and function is the basis of the selection and further design of the most efficient proteins. To be able to apply and control the enzymes in various applications in an optimal way, it is utterly important also to understand the mechanisms of enzymatic reactions in various polymeric substrates in micro- and macrostructure levels and in process conditions. Thorough understanding of the more powerful enzymes and enzyme process concepts available in the future can open new, sustainable and efficient ways to refine new materials, chemicals and products from the various lignocellulosic biomass sources.

References

Agger J, Viksø-Nielsen A, Meyer AS (2010) Factors affecting xylanase functionality. In: Viikari L, Grönqvist S, Kruus K, Pere J, Siika-aho M, Suurnäkki A. Industrial biotechnology in the paper and pulp sector, In: Industrial Biotechnology (Eds. Soetaert W. and Vandamme, E.), Wiley-VCH Published Online: 29 Apr 2010

Aracri E, Vidal T, Ragauskas A (2011) Wet strength development in sisal cellulose fibers by effect of a laccase–TEMPO treatment. Carbohydr Polym 84:1384–1390

Awano T, Takabe K, Fujita M (1998) Localization of glucuronoxylans in Japanese beech visualized by immunogold labelling. Protoplasma 202:213–222

Bechtold T, Schimper CB (2010) Hydrolysis of regenerated textile fibres and other materials. In: Niedsztrasz WA, Cavaco-Paulo A (eds) Advances in textile biotechnology. Woodhead Publishing Ltd., pp 312–327

Bedford MR (2000) Exogenous enzymes in monogastric nutrition—their current value and future benefits. Anim Feed Sci Technol 86:1–13

Bendayan M (1989) The enzyme-gold cytochemical approach: a review. In: Hayat MA (ed) Colloidal gold. Principles, methods and applications. Academic, San Diego

Berrin J-G, Juge N (2008) Factors affecting xylanase functionality in the degradation of arabinoxylans. Biotechnol Lett 30:1139–1150

Bhat MK (2000) Cellulases and related enzymes in biotechnology. Biotechnol Adv 18:355–383

Bhatia Y, Mishra S, Bisaria VS (2002) Microbial beta-glucosidases: cloning, properties and applications. Crit Rev Biotechnol 22:375–407

Biekman ESA, Kroesehoedeman HI, van Dijk C (1993) Study of the enzymatic maceration kinetics of potatoes in a rotating perforated drum reactor. Food Biotechnol 7(2):127–141

Biely P, Mastihubova M, Tenkanen M, Eyzaguirrec J, Li X-L, Vr'sanska M (2011) Action of xylan deacetylating enzymes on monoacetyl derivatives of 4-nitrophenyl glycosides of β-D-xylopyranose and α-L-arabinofuranose. J Biotechnol 151:137–142

Biely P, de Vries RP, Vrsanska M, Visser J (2000) Inverting character of α-glucuronidase A from $Aspergillus\ tubingensis$. Biochim Biophys Acta 1474:360–364

Bohrn R, Potthast A, Schiehser S, Rosenau T, Sixta H, Kosma P (2006) The FDAM method: determination of carboxyl profiles in cellulosic materials by combining group-selective fluorescence labelling with GPC. Biomacromolecules 7(6):1743–1750

Bonnin E, Clavurier K, Daniel S, Kauppinen S, Mikkelsen JDM, Thibault J-F (2008) Pectin acetylesterases from $Aspergillus$ are able to deacetylate homogalacturonan as well as rhamnogalacturonan. Carbohydr Polym 74(3):411–418

Bragd PL, Van Bekkum H, Besemer AC (2004) TEMPO mediated oxidation of polysaccharides: survey of methods and applications. Top Catal 27(February):49–66

Brumer H, Zhou Q, Baumann MJ, Carlsson K, Teeri TT (2004) Activation of crystalline cellulose surfaces through chemoenzymatic modification of xyloglucan. J Am Chem Soc 126:5715–5721

Buchert J, Carlsson G, Viikari L, Ström G (1996) Surface characterization of unbleached kraft pulps by enzymatic peeling and ESCA. Holzforschung 50:69–74

Cairns JRK, Esen A (2010) β-Glucosidases. Cell Mol Life Sci 67:3389–3405

Cantarel BL, Coutinho PM, Rancurel C, Bernard T, Lombard V, Henrissat B (2009) The Carbohydrate-Active EnZymes database (CAZy): an expert resource for Glycogenomics. Nucleic Acids Res 37:D233–238

Carbohydrate Active Enzyme Database, http://www.cazy.com

Coenen GJ, Bakx EJ, Verhoef RP, Schols H, Voragen AGJ (2007) Identification of the connecting linkage between homo- or xylogalacturonan and rhamnogalacturonan type I. Carbohydr Polym 70(2):224–235

Collins T, Gerday C, Feller G (2005) Xylanases, xylanase families and extremophilic xylanases. FEMS Microbiol Rev 29(1):3–23

Courtin CM, Roelants A, Delcour JA (1999) Fractionation reconstruction experiments provide insight into the role of endoxylanases in bread making. J Agric Food Chem 47:1870–1877

Crepin VF, Faulds CB, Connerton IF (2004) Functional classification of the microbial feruloyl esterases. Appl Microbiol Biotechnol 63:647–652

Daas PJH, Arisz PW, Schols HA, de Ruiter GA, Voragen AGJ (1998) Analysis of partially methyl-esterified galacturonic acid oligomers by high-performance anion-exchange chromatography and matrix-assisted laser desorption/ionization time-of-flight mass spectrometry. Anal Biochem 257(2):195–202

Daas PJH, Meyer-Hansen K, Schols HA, de Ruiter GA, voragen AGJ (1999) Investigation of the non-esterified galacturonic acid distribution in pectin with endopolygalacturonase. Carbohydr Res 318(1–4):135–145

Daas PJH, Voragen AGJ, Shcols H (2000) Investigation of the galacturonic acid distribution of pectin with enzymes part 2 - Characterization of non-esterified galacturonic acid sequences in pectin with endopolygalacturonase. Carbohydr Res 326(2):120–129

Davies G, Henrissat B (1995) Structures and mechanisms of glycosyl hydrolases. Structure 3:853–859

de Wet BJM, van Zy WH, Prior BA (2006) Characterization of the Aureobasidium pullulans α-glucuronidase expressed in *Saccharomyces cerevisiae*. Enzyme Microb Technol 38:649–656

Dhawan S, Kaur J (2007) Microbial Mannanases: an overview of production and applications. Crit Rev Biotechnol 27:197–216

Ebbelaar MEM, Tucker GA, Laats MM, Dijk C, Stolle-Smits T, Recourt K (1996) Characterization of pectinases and pectin methylesterase cDNAs in pods of green beans (*Phaseolus vulgaris* L). Plant Mol Biol 31(6):1141–1151

Fazary AE, Ju Y-H (2007) Feruloyl esterases as biotechnological tools: current and future perspectives. Acta Biochim Biophys Sin 39:811–828

Filonova L, Gunnarsson LC, Daniel G, Ohlin M (2007) Synthetic xylan-binding modules for mapping of pulp fibres and wood sections. BMC Plant Biol 7:54–63

Gallardo O, Fernandez-Fernandez M, Valls C, Valenzuela SV, Roncero MB, Vidal T, Díaz P, Pastor FIJ (2010) Characterization of a family GH5 xylanase with activity on neutral oligosaccharides and evaluation as a pulp bleaching aid. Appl Environ Microbiol 76:6290–6294

Garcıa-Aparicio M, Ballesteros M, Manzanares P, Ballesteros I, Gonzalez A, Negro JM (2007) Xylanase contribution to the efficiency of cellulose enzymatic hydrolysis of barley straw. Appl Biochem Biotechnol 137–140:353–65

Gilbert H, Hazlewood G (1993) Bacterial cellulases and xylanases. J Gen Microbiol 39:187–194

Gírio FM, Fonseca C, Carvalheiro F, Duarte LC, Marques S, Bogel-Łukasik R (2010) Hemicelluloses for fuel ethanol: A review. Bioresour Technol 101:4775–4800

Gosalbes MJ, Pérez-González JA, González R, Navarro A (1991) Two beta-glycanase genes are clustered in Bacillus polymyxa: molecular cloning, expression, and sequence analysis of genes encoding a xylanase and an endo-beta- (1,3)-(1,4)-glucanase. J Bacteriol 173:7705–7710

Guillotin SE, Bakx EJ, Boulenguer J, Mazoyer J, Schols HA, Voragen AGJ (2005) Populations having different GalA blocks characteristics are present in commercial pectins which are chemically similar but have different functionalities. Carbohydr Polym 60(3):391–398

Guillotin SE, Mey N, Ananta E, Boulenguer P, Schols HA, Voragen AGJ (2006) Chromatographic and enzymatic strategies to reveal differences between amidated pectins on a molecular level. Biomacromolecules 7(6):2032–2037

Hahn-Hägerdal B, Karhumaa K, Fonseca C, Spencer-Martins I, Gorwa-Grauslund MF (2007) Towards industrial pentose-fermenting yeast strains. Appl Microbiol Biotechnol 74:937–953

Henrissat B et al (1995) Conserved catalytic machinery and the prediction of a common fold for several families of glycosyl hydrolases. Proc Natl Acad Sci USA 92:7090–7094

Henrissat B, Bairoch A (1996) Updating the sequence-based classification of glycosyl hydrolases. Biochem J 316:695–696

Hilz H, Williams P, Williams P, Doco T, Schols HA, Voragen AGJ (2006) The pectic polysaccharide rhamnogalacturonan II is present as a dimer in pectic populations of bilberries and black currants in muro and in juice. Carbohydr Polym 65(4):521–528

Hoondal GS, Tiwari RP, Tewari R, Dahiya N, Bek QK (2002) Microbial alkaline pectinases and their industrial applications: a review. Appl Microbiol Biotechnol 59(4–5):409–418

Hyvolution http://www.hyvolution.nl

Irwin D, Jung E, Wilson D (1994) Characterization and sequence of a Thermomonaspora fusca xylanase. Appl Environ Microbiol 60(3):763–770

Jordan DB, Wagschal K (2010) Properties and applications of microbial β-D-xylosidases featuring the catalytically efficient enzyme from *Selenomonas ruminantium* Appl. Microbiol Biotechnol 86:1647–1658

Joseleau J-P, Imai T, Kuroda K, Katia Ruel K (2004) Detection in situ and characterization of lignin in the G-layer of tension wood fibres of *Populus deltoides*. Planta 219(2):338–345

Kleywegt GJ, Zou JY, Divne C, Davies GJ, Sinning I, Ståhlberg J, Reinikainen T, Srisodsuk M, Teeri TT, Jones TA (1997) The crystal structure of the catalytic core domain of endoglucanase I from *Trichoderma reesei* at 3.6 A resolution, and a comparison with related enzymes. J Mol Biol 272:383–397

Kruus K, Niku-Paavola M-L, Viikari L (2001) Laccase—a useful enzyme for modification of biopolymers. In: Chiellini E, Gil H, Braunegg G, Buchert J, Gatenholm P, van der Zee M (Eds) Biorelated polymers: sustainable polymer science and technology. . Kluwer Academic/Plenum. New York , pp 255–261

Kumar R, Wyman CE (2009a) Effect of xylanase supplementation of cellulase on digestion of corn stover solids prepared by leading pretreatment technologies. Bioresour Technol 100:4203–4213

Kumar R, Wyman CE (2009b) Effects of cellulase and xylanase enzymes on the deconstruction of solids from pretreatment of poplar by leading technologies. Biotechnol Prog 25:302–314

Lappalainen A, Tenkanen M, Pere J (2004) Specific antibodies for immunochemical detection of wood-derived hemicelluloses. In: Gatenholm P, Tenkanen M (eds) Hemicelluloses: science and technology, ACS Symposium Series 864. American Chemical Society, Washington, DC, pp 140–156

Leppänen AS, Niittymäki O, Parikka K, Tenkanen M, Eklund P, Sjöholm R, Willför S (2010) Metal mediated allylation of enzymatically oxidized methyl α-D-galactopyranoside. Carbohydr Res 345: 2610–2615

Martens-Uzunova ES, Zandleven JS (2006) A new group of exo-acting family 28 glycoside hydrolases of Aspergillus niger that are involved in pectin degradation. Biochem J 400:43–52

McKie VA, Vincken JP, Voragen AGJ, Broek LAM, van den Stimson E, Gilbert HJ (2001) A new family of rhamnogalacturonan lyases contains an enzyme that binds to cellulose. Biochem J 355:167–177

Mohnen D (2008) Pectin structure and biosynthesis. Curr Opin Plant Biol 11(3):266–277

Moreira LR, Filho EX (2008) An overview of mannan structure and mannan-degrading enzyme systems. Appl Microbiol Biotechnol 79:165–178

Mutter M, Beldman G (1994) Rhamnogalacturonan alpha-L-rhamnopyranohydrolase. A novel enzyme specific for the terminal nonreducing rhamnosyl unit in rhamnogalacturonan regions of pectin. Plant Physiol 106(1):241–250

Mutter M, Beldman G, Pitson SM, schools HA, Voragen AGJ (1998) Rhamnogalacturonan alpha-D-galactopyranosyluronohydrolase - An enzyme that specifically removes the terminal nonreducing galacturonosyl residue in rhamnogalacturonan regions of pectin. Plant Physiol 117(1):153–16

Mutter M, Colquhoun IJ, Schols HA, Beldman G, Voragen AGJ (1996) Rhamnogalacturonase B from *aspergillus aculeatus* is a rhamnogalacturonan alpha-L-rhamnopyranosyl-(1->4)-alpha-D-galactopyranosyluronide lyase. Plant Physiol 110(1):73–77

Numan MT, Bhosle NB (2006) Alpha-L-arabinofuranosidases: the potential applications in biotechnology. J Ind Microbiol Biotechnol 33:247–260

Oksanen T, Pere J, Buchert J, viikari L (1997a) The effect of *Trichoderma reesei* cellulases and hemicellulases on the paper technical properties of never-dried bleached kraft pulp. Cellulose 4:329–339

Oksanen T, Buchert J, Viikari L (1997b) The role of hemicelluloses in the hornification of bleached kraft pulps. Holzforschung 51(4):355–360

Oksanen T, Pere J, Paavilainen L, Buchert J, Viikari L (2000) Treatment of recycled kraft pulps with Trichoderma reesei hemicellulases and cellulases. J Biotechnol 78:39–48

Olson L, Stanley P (1990) Compositions and methods to vary color density Patent WO 90/02790

O'Neill MA, Ishii T, Albersheim P, Darvill AG (2004) Rhamnogalacturonan II: Structure and function of a borate cross-linked cell wall pectic polysaccharide. Annu Rev Plant Biol 55:109–139

Oomen R, Doeswijk-Voragen CHL, Bush MS, Vincken J-P, Borkhardt B, van den Broek LAM, Corsar J, Ulvskov P, Voragen AGJ, McCann M, Visser RGF (2002) In muro fragmentation of the rhamnogalacturonan I backbone in potato (*Solanum tuberosum* L.) results in a reduction and altered location of the galactan and arabinan sidechains and abnormal periderm development. Plant J 30 (4):403–413

Panagiotopoulos IA, Bakker RR (2008) Comparative study of different lignocellulosic feedstocks enzymatic hydrolysis for fermentable substrates production. Proceedings of the 16th European Biomass Conference & Exhibition. Valencia, Spain, pp 1749–1752

Panagiotopoulos IA, Bakker RR, Budde MA, de Vrije T, Claassen PAM, Koukios EG (2009) Fermentative hydrogen production from pretreated biomass: a comparative study. Bioresource Technol 100:6331–6338

Parikka K, Tenkanen M (2009) Oxidation of methyl alpha-D-galactopyranoside by galactose oxidase: products formed and optimization of reaction conditions for production of aldehyde. Carbohydr Res 344:14–20

Parikka K, Leppanen A-S, Pitkanen L, Reunanen M, Willfor S, Tenkanen M (2010) Oxidation of polysaccharides by galactose oxidase. J Agric Food Chem 58:262–271

Patel I, Ludwig R, Haltrich D, Rosenau T, Potthast A (2011) Studies of the chemoenzymatic modification of cellulosic pulps by the laccase-TEMPO system. Holzforschung 65(4):475–481

Pere J, Siika-aho M, Viikari L (2000) Biomechanical pulping with enzymes: response of coarse mechanical pulp to enzymatic modification and secondary refining. TAPPI J 83:85–185

Pollet A, Delcour JA, Courtin CM (2010a) Structural determinants of the substrate specificities of xylanases from different glycoside hydrolase families. Crit Rev Biotechnol 30:176–191

Pollet A, Schoepe J, Dornez E, Strelkov SV, Delcour JA, Courtin CM (2010b) Functional analysis of glycoside hydrolase family 8 xylanases shows narrow but distinct substrate specificities and biotechnological potential. Appl Microbiol Biotechnol 87:2125–2135

Potthast A, Röhrling J, Rosenau T, Borgards A, Sixta H, Kosma P (2003) A novel method for the determination of carbonyl groups in cellulosics by fluorescence labelling. 3. Monitoring oxidative processes. Biomacromolecules 4(3):743–749

Rouau X, El-Hayek ML, Moreau D (1994) Effect of an enzyme preparation containing pentosanes on the bread making quality of flours in relation to changes in pentosan properties. J Cereal Sci 19:259–272

Ruel K, Joseleau J-P (1984) Use of enzyme-gold complex for ultrastructural localization of hemicelluloses in the plant cell wall. Histochemistry 81:573–580

Schols HA, Geraeds C, Searle-van Leeuwen MF, Kormelink FJM, Voragen AGJ (1990) Hairy (ramified) regions of pectins. 2. Rhamnogalacturonase—a novel enzyme that degrades the hairy regions of pectins. Carbohydrate Res 206(1):105–115

Searle van leeuwen MJF, Broek AM, Schols HA, Beldman G, Voragen AGJ (1992) Rhamnogalacturonan acetylesterase—A novel enzyme from *Aspergillus aculeatus*, specific for the deacetylation of hairy (ramified) regions of pectins. Appl Microbiol Biotechnol 38(3):347–349

Sengkhamparn N, Sagis LMC, de Vries R, Schols HA, Sajjaanantakul T, Voragen AGJ (2010) Physicochemical properties of pectins from okra (*Abelmoschus esculentus* (L.) Moench). Food Hydrocolloids 24(1); 35–41

Shallom D, Shoham Y (2003) Microbial hemicellulases. Curr Opin Microbiol 6:219–228

Siika-aho M, Tenkanen M, Buchert J, Puls J, Viikari L (1994) An α-glucuronidase from *Trichoderma reesei* RUT C-30. Enzyme Microb Technol 16:813–819

Spanikova S, Biely P (2006) Glucuronoyl esterase—Novel carbohydrate esterase produced by *Schizophyllum commune*. FEBS Lett 560:4597–4601

Sprössler BG (1997) Xylanases in baking. In: Angelino SAGF, Hamer RJ, van Hartingsveldt W, Heidekamp F, van der Lungt JP (eds) 1st European symposium of enzymes in grain processing. TNO Food and Nutrition, Zeist NL, pp 177–187

Suurnäkki A, Heijnesson A, Buchert J, Tenkanen M, Viikari L, Westermark U (1996a) Location of xylanase and mannanase action in kraft fibres. J Pulp Paper Sci 22(3):J78–J83

Suurnäkki A, Heijnesson A, Buchert J, Tenkanen M, Viikari L, Westermark U (1996b) Chemical characterization of the surface layers of unbleached pine and birch kraft pulp fibres. J Pulp Paper Sci 22(2): J43–J47

Suurnäkki A, Tenkanen M, Buchert J, Viikari L (1997) Hemicellulases in bleaching of chemical pulps. Adv Biochem Eng Biotechnol 57:261–287

Tailford LE, Ducros VM-A, Flint JE, Roberts SM, Morland C, Zechel DL, Smith N, Bjoernvad ME, Borchert TV, Wilson KS, Davies GJ, Gilbert HJ (2009) Understanding how diverse β-mannanases recognize heterogeneous substrates. Biochemistry 48:7009–7018

Teeri TT (1997) Crystalline cellulose degradation: new insight into the function of cellobiohydrolases. Tibtech 15:160–167

Teeri TT, Koivula A, Linder M, Wohlfahrt G, Divne C, Jones TA (1998) *Trichoderma reesei* cellobiohydrolases: why so efficient on crystalline cellulose? Biochem Soc Trans 26:173–178

Teleman A, Harjunpää V, Tenkanen M, Buchert J, Hausalo T, Drakenberg T, Vuorinen T (1995) Characterization of 4-deoxy-ß-L-threo-hex-4-enopyranosyluronic acid attached to xylan in pine kraft pulp and pulping liquor by ^{1}H and ^{13}C NMR spectroscopy. Carbohydr Res 272:55–71

Tenkanen M (1998) Action of *Trichoderma reesei* and *Aspergillus oryzae* esterases in the deacetylation of hemicelluloses. Biotechnol Appl Biochem 27: 19–24

Tenkanen M, Siika-aho M (2000) An alpha-glucuronidase of *Schizophyllum* commune acting on polymeric xylan. J Biotechnol 78:149–161

Tenkanen M, Eyzaguirre J, Isoniemi R, Faulds CB, Biely P (2003) Comparison of catalytic properties of acetyl xylan esterases from three carbohydrate esterase families. In: Mansfield S, Saddler JN (eds) Application of enzymes to lignocellulosics, ACS Symposium Series vol. 855, pp. 211–229

Thornton J, Tenkanen M, Ekman R, Holmbom B, Viikari L (1994) Effects of alkaline treatment on dissolved carbohydrates in suspensions of Norway spruce thermomechanical pulp. J Wood Chem Technol 14 (2):176–194

Tijskens LMM, Rodis PS, Hertog MLAT, Kalantzi U, van Dijk C (1998) Kinetics of polygalacturonase activity and firmness of peaches during storage. J Food Engineer 35(1):111–126

Tijskens LMM, Rodis PS, Hertog MLAT, Proxenia N, van Dijk C (1999) Activity of pectin methyl esterase during blanching of peaches. J Food Engineer 39 (2):167–177

Tijskens LMM, Waldron K, Ng IA, van Dijk C (1997) The kinetics of pectin methyl esterase in potatoes and carrots during blanching. J Food Engineer 34 (4):371–385

Udatha DBRKG, Kouskoumvekaki I, Olsson L, Panagiotou G (2011) The interplay of descriptor-based computational analysis with pharmacophore modeling builds the basis for a novel classification scheme for feruloyl esterases. Biotechnol Adv 29:94–110

van den Broek LAM, den Aantrekker ED, Voragen AGJ, Beldman G, Vincken J-P (1997) Pectin lyase is a key enzyme in the maceration of potato tuber. J Sci Food Agric 75(2):167–172

Van Dijk C, Boeriu C, Peter F, Stolle-Smits T, Tijskens LMM (2006) The firmness of stored tomatoes (cv. Tradiro). 1. Kinetic and near infrared models to describe firmness and moisture loss. J Food Engineer 77(3):575–584

Van Hellemond E, Leferink NGH, Heuts DPHM, Fraaije MW, van Berkel WJH (2006) Occurrence and biocatalytic potential of carbohydrate oxidases. Adv Appl Microbiol 60(60):17–54

Vian B, Brillouet JM, Satiat-Jeunemaitre B (1983) Ultrastructural visualization of xylans in cell walls of hardwood by means of xylanase-gold complex. Biol Cell 49:179–182

Viikari L, Alapuranen M, Puranen T, Vehmaanpera J, Siika-Aho M (2007) Thermostable enzymes in lignocellulose hydrolysis. Adv Biochem Eng Biotechnol 108:121–145

Viikari L, Grönqvist S, Kruus K, Pere J, Siika-aho M, Suurnäkki A (2010) Industrial biotechnology in the paper and pulp sector. In: Soetaert W, Vandamme E (eds) Industrial biotechnology. Wiley-VCH Published Online: 29 APR 2010

Viikari L, Vehmaanperä J, Koivula A (2012) Lignocellulosic ethanol: from science to industry. Biomass Bioenerg. Available on line. http://dx.doi.org/10.1016/j.biombioe.2012.05.008

Vrsanska M, Kolenova K, Puchart V, Biely P (2007) Mode of action of glycoside hydrolase family 5 glucuronoxylan xylanohydrolase from *Erwinia chrysanthemi*. FEBS J 274:1666–1677

Westphal Y, Schols HA, Voragen AGJ, Gruppen H (2010) MALDI-TOF MS and CE-LIF fingerprinting of plant cell wall polysaccharide digests as a screening tool for Arabidopsis cell wall mutants. J Agric Food Chem 58(8):4644–4652

Willfor S, Sundberg K, Tenkanen M, Holmbom B (2008) Spruce-derived mannans—a potential raw material for hydrocolloids and novel advanced natural materials. Carbohydr Polym 72(2):197–210

Wong D (2008) Enzymatic deconstruction of backbone structures of the ramified regions in pectins. Protein J 27(1):30–42

Wood TM, Bhat KM (1988) Methods for measuring cellulase activities. Methods Enzymol 160:87–112

Xu Q, Singh A, Himmel ME (2009) Perspectives and new directions for the production of bioethanol using consolidated bioprocessing of lignocellulose. Curr Opin Biotechnol 20:364–371

Yuan JS, Tiller KH, Al-Ahmad H, Stewart NR, Stewart CN Jr (2008) Plants to power: bioenergy to fuel the future. Trends Plant Sci 13:421–429

Index

A

Acid functions, polysaccharides
　agar-agar, 56
　alginate, 55–56
　carrageenan, 56–57
　fucoidan/fucogalactan, 56
　pectins
　　characteristics, 52
　　commercial sources, 52
　　food industry, 52
　　homogalacturonan (HG), 53
　　plant cell wall, 52
　　rhamnogalacturonan type I (RG-I), 53–54
　　rhamnogalacturonan type II (RG-II), 54
　　structural variations, 52
　　xylogalacturonan, 54–55
Agar-agar, 56
Alginate, 55–56
Amino functions, polysaccharides
　chitin and chitosan, 49–50
　glycosaminoglycans (GAG), 50–51
　hyaluronan/hyaluronic acid, 50
　murein, 51
Amylopectin
　alternative models, 36
　amaranth, 35
　basic labelling, 34
　DP, 32
　molecular mass, 31
　molecular structure, 32
　unit chain distribution, 34
Analytics, polysaccharide-acting enzymes, 386–387
Antimicrobial functionalisation
　cationic functional polysaccharides (CAT-PS), 199–200
　dissociable groups, viscose fibres, 203
　pK value, 201
　point of zero charge (PZC), 201
　viscose nontreated and chitosan-treated (impregnated) fabrics, 201, 204
　XPS, 202
Arabinoxylans (AXs), 45
(Arabino)glucuronoxylans (AGX), 44–45

B

Bemberg silk, 156
(1→3)-β-Glucans
　Callose, 31
　chemical structure, 29
　Curdlan, 29, 30
　Laminaran, 31
　Paramylon, 31
　schizophyllan, 29
　Scleroglucan, 29
Biomass production, polysaccharide, 2
Bleaching
　chemical pulps
　　alkali extraction, 267
　　chlorine dioxide, 266
　　elementary chlorine free (ECF), 265–266
　　hardwood and softwood, 268
　　hydrogen peroxide, 266–267
　　oxidant-reinforced alkaline extraction, 267
　　oxygen delignification, 266
　　ozone, 267
　　peracetic acid, 267–268
　　sequences, 265
　　totally chlorine free (TCF), 265–266
　mechanical pulps
　　brightness gain, 265
　　hydrogen peroxide, 264–265
　　objective, 264
　　oxidative/reductive reactions, 264
　　sodium dithionite, 264
BOKU work
　capillary electrophoresis
　　borate complexation scheme, 78
　　detection method, 80
　　electropherogram, 79–80
　　reductive amination, 79
　CCOA and FDAM method
　　carbonyl structures, 71, 72
　　carboxyl structures, 71, 72
　　conventional methods, 73
　　fluorescence labeling and carbonyl/carboxyl profiling, 74
　　gel permeation chromatography (GPC), 74

BOKU work (*cont.*)
    molecular weight distribution (MWD), 73
    obstacles, 73
  cellulose morphology, 80–82
  trace chromophores
    cellulosic sources, 76
    classical extraction, 76
    CRI method, 77–78
    isolation, 75
    off-white discoloration, 75
    structure and stabilization, 77

C
Callose, 31
Capillary electrophoresis, BOKU work
  borate complexation scheme, 78
  detection method, 80
  electropherogram, 79–80
  reductive amination, 79
CarbaCell process, 156, 159
Carbazole-carbonyl-oxyamine (CCOA), BOKU work
  carbonyl structures, 71, 72
  carboxyl structures, 71, 72
  conventional methods, 73
  fluorescence labeling and carbonyl/carboxyl profiling, 74
  gel permeation chromatography (GPC), 74
  molecular weight distribution (MWD), 73
  obstacles, 73
Carbohydrate-binding modules (CBMs), 387
Carrageenan
  etymology, 18–19
  acid functions, polysaccharides, 56–57
Cationic functional polysaccharides (CAT-PS), 199–200
Cellnet, 3
Cellulases, 383–384
Cellulolytic enzymes, 377–378
Cellulose
  amorphous phase, 92
  anhydroglucose units (AGU), 25
  characteristics, 91–92
  crystal structures, 27
  degree of polymerisation (DP), 25
  derivatising pathways, 93–94
  etymology, 16
  hydrogen bonding system, 27, 28
  intra-and intermolecular hydrogen bond, 25, 27
  ionic liquid solutions
    cellulose–EMIMAc viscosity, 123–125
    diffusion, $[EMIM]^+$ and $[Ac]^-$ Ions, 126–127
    materials and methods, 121–122
    phase diagram and solubility limit, 127–128
    steady-state and intrinsic viscosity, 122–123
  lignin mixtures
    aqueous 8 % NaOH, 140–143
    blending, 139–140
    cellulose acetate butyrate (CAB), 140
    coagulation, ligno-aerocellulose morphology, 143–144
    papermaking, 139
    properties, 139
  macroscopic mechanisms
    gradient of solubility, 99–100
    overall dissolution mechanism, 100–101
  mechanism, 92
  model representation, 28
  molecular structure, 27
  non-derivatising compounds
    LiCl-based solvents, 96
    N-methylmorpholine-N-oxide/water, 96–97
    phosphoric acid-based solvents, 95–96
    rheology of solutions, 97–98
  occurrence, 25
  polysaccharide blends
    advantages, 133
    AFM analysis, 137, 138
    contact angle measurements, 137–138
    dissolution screenings, 134–135
    expected effects, 134, 135
    mechanical properties and sugar composition, 136, 137
    microscopic images, 135, 136
    rheological investigations, 136
    solubility and miscibility, 135
    solvent routes, 134
    spinning, 136
    x-ray flat film, 137–138
    zeta potential (ZP) measurements, 138
  sodium hydroxide–water solutions
    dissolution in NaOH–Water, 102–120
    mercerisation, 101
  stabilisation
    solution quality and solution state, 128–130
    thermostability, 130–133
  swelling and dissolution mechanism, 92
Cellulose morphology, BOKU work, 80–82
Cellulose nanocrystals (CNC), 219
CELSOL®/BIOCELSOL® process
  alkalization, 170
  enzymatic modification, 171–172
  feasibility studies, 174–175
  film forming and fibre spinning, 172–173
  hydrothermal treatment, 171
  preparation, alkaline solutions, 172
  properties, 173–174
  steam explosion, 171
Chemical characterization, polysaccharides
  alien groups and end groups, 70
  BOKU work
    capillary electrophoresis, 78–80
    CCOA and FDAM method, 71–75
    cellulose morphology, 80–82
    trace chromophores, 75–78
  chemical analysis, 65–67
  crystallinity, 71

DP and distribution, 70–71
extraction and purification, 67
monomers and building blocks, 67–68
side chains, linkage pattern and anomeric configuration, 69–70
substituents, 68–69
Chemical pulps
  bleaching, 265–268
    alkali extraction, 267
    chlorine dioxide, 266
    elementary chlorine free (ECF), 265–266
    hardwood and softwood, 268
    hydrogen peroxide, 266–267
    oxidant-reinforced alkaline extraction, 267
    oxygen delignification, 266
    ozone, 267
    peracetic acid, 267–268
    sequences, 265
    totally chlorine free (TCF), 265–266
  separation method
    kraft pulping, 262–263
    sulfite pulping, 261–262
Chemimechanical pulping, 259–261
Chitin and chitosan
  alternative methods, 334–336
  amino functions, polysaccharides, 49–50
  applications
    bacteria, 359
    bionanotechnology, 358
    cholesterol-lowering effects, 345
    films, 346–347
    foliar treatment agents, 360–361
    food additives, 345–346
    fungi and oomycetes, 359
    hair care, 362
    hemostatic topical agents, 354
    implantable application, 354–358
    medical devices regulation, 347–348
    oral care, 362
    plant disease control, 359–360
    preservatives, 347
    primary and secondary wound dressings, 349–354
    seed coating agents, 360
    skin care, 362
    soil amendment, 361
    viruses, 358–359
    wound dressings, 349
  characteristics, 329–330
  crustacean waste, 337–338
  etymology, 18
  forms
    applications, 340–343
    derivatives, 339–340
  future research, 338
  future trends, 363–364
  isolation procedure, 330–334
  microorganisms and application, 336–337
  research, EPNOE, 364
Chromophores, BOKU work

cellulosic sources, 76
classical extraction, 76
CRI method, 77–78
isolation, 75
off-white discoloration, 75
structure and stabilization, 77
Collaborative research activities and phd/postdoc mobility, 8–9
Complex heteroxylans (CHX), 45–46
Coordination and Support Action (CSA), 10–11
Curdlan, 29, 30

D

Degree of polymerisation (DP)
  amylopectin, 32
  cellulose, 25
  chemical characterization, polysaccharides, 70–71
  starch, 34
  viscose rayon process, 157
Dendronization
  ATR-FTIR spectroscopy, 316
  dendron, 315
  propargyl-polyamidoamine (PAMAM), 319
Derivatization, cellulose
  carbonyl and carboxyl groups, oxidation, 306
  chemical modification, 284
  dendronization
    ATR-FTIR spectroscopy, 316
    dendron, 315
    propargyl-polyamidoamine (PAMAM), 319
  1,3-dipolar cycloaddition reaction, azide-and alkyne-containing cellulosics, 315
  esterification
    aliphatic acid, 289–291
    dicarboxylic acids and unsaturated acids, 291–293
    Dimethyl Sulfoxide (DMSO)/Tetrabutylammonium Fluoride (TBAF), 286–288
    homogeneous, solvents, 284
    ionic liquids, 288
    new tools, 293–297
    N,N-Dimethyl Acetamide (DMAc)/LiCl, 284–286
    sulfation, 297–300
  oxidation, 307–312
  regioselective functionalization
    application, orthogonal protecting groups, 305–306
    3-O-ethers, 304–305
    trialkylsilyl ethers, 302–304
    triphenylmethyl ethers, 301–302
  structure–property relationships, 312–315
Dextran
  anhydrous dextran structure, 39, 40
  methylation analysis, 39
  molecular structure, 37, 38
1,3-dipolar cycloaddition reaction, azide-and alkyne-containing cellulosics, 315
Dissemination, EPNOE, 9

E

Education road map, EPNOE, 9
Ellipsometry, 225
EPNOE network, research and education
   achievements, 7
   applied R&D research, 8
   building
      Cellnet, 3
      location, academic/research partners, 3, 4
      members, 3, 5
      objectives, 5
      scientific expertise, 3, 4
   challenges and opportunities, 10–11
   collaborative research activities and phd/postdoc mobility, 8–9
   Coordination and Support Action (CSA), 10–11
   dissemination, 9
   education road map, 9
   Joint Communication and Involvement, Scientific, Policy and Industrial Communities, 9–10
   organisation
      associate members, 5
      dedicated meetings, 6–7
      partner databases access, 6
      regular members, 5
      research information, 6
      strategic and technological watch data, 6
      structure, 6
   research road map, 7–8
   tool box, 9
Esterification
   aliphatic acid, 289–291
   dicarboxylic acids and unsaturated acids, 291–293
   dimethyl sulfoxide (DMSO)/tetrabutylammonium fluoride (TBAF), 286–288
   homogeneous, solvents, 284
   ionic liquids, 288
   new tools, 293–297
   N,N-dimethyl acetamide (DMAc)/LiCl, 284–286
   sulfation, 297–300
Etymology
   carrageenan, 18–19
   cellulose, 16
   chitin, 18
   heparin, 20
   hyaluronan, 20
   Indo-European representation, mildness/sweetness
      metathesis, 16
      mild, pleasant, 16–17
      from millstone to mildness, 17
   inulin, 19
   levan, 20
   murein, 20
   pectin, 19–20
   pullulan, 20
   saccharide and sugar
      apiose, 15
      arabinose, 15–16
      derivation, 14
      mannose, 15
      mildness and sweetness, 15
      polymerized sugars, 14
      Rhamnus frangula, 15–16
      ribose, 16
      Sorbus domestica, 15
      suffix-*ose*, 14–15
   starch
      Germanic languages and in Finnish, 17–18
      Greek, Latin and Roman Languages, 17
      Slavic languages, 18
European Union (EU), 2–3

F

Fluorenyl-diazomethane (FDAM), BOKU work
   carbonyl structures, 71, 72
   carboxyl structures, 71, 72
   conventional methods, 73
   fluorescence labeling and carbonyl/carboxyl profiling, 74
   gel permeation chromatography (GPC), 74
   molecular weight distribution (MWD), 73
   obstacles, 73
Fucoidan/fucogalactan, 56
Functionalization methods, pulp fibers
   chemical pulps, bleaching
      alkali extraction, 267
      chlorine dioxide, 266
      elementary chlorine free (ECF), 265–266
      hardwood and softwood, 268
      hydrogen peroxide, 266–267
      oxidant-reinforced alkaline extraction, 267
      oxygen delignification, 266
      ozone, 267
      peracetic acid, 267–268
      sequences, 265
      totally chlorine free (TCF), 265–266
   low-consistency refining, 268–269
   mechanical pulps, bleaching
      brightness gain, 265
      hydrogen peroxide, 264–265
      objective, 264
      oxidative/reductive reactions, 264
      sodium dithionite, 264
   papermaking chemicals, 269–271

G

Galactose oxidase (GAO), 382–383
Gel permeation chromatography (GPC), 74
Glucans
   (1→3)-β-Glucans
      Callose, 31
      chemical structure, 29
      Curdlan, 29, 30
      Laminaran, 31
      Paramylon, 31
      schizophyllan, 29
      Scleroglucan, 29

cellulose
  anhydroglucose units (AGU), 25
  crystal structures, 27
  degree of polymerisation (DP), 25
  hydrogen bonding system, 27, 28
  intra-and intermolecular hydrogen bond, 25, 27
  model representation, 28
  molecular structure, 27
  occurrence, 25
dextran
  anhydrous dextran structure, 39, 40
  methylation analysis, 39
  molecular structure, 37, 38
glycogen
  average chain length (CL), 35
  average exterior chain length (ECL), 35
  average interior chain length (ICL), 35
  molecular structure, 34, 37
  structural parameters, 37, 38
  Whelan's model, 35, 37
pullulan
  chemical structure, 41
  number average molecular weight (M_n), 42
starch
  A-and B-type, 31, 33
  alternative models, 33, 36
  amaranth amylopectin, 33, 35
  amylopectin, 32–34
  degree of polymerisation (DP), 34
  molecular structure, 31, 32
(Glucurono)arabinoxylan (GAX), 44–45
Glucuronoxylans, 42–44
Glycogen
  average chain length (CL), 35
  average exterior chain length (ECL), 35
  average interior chain length (ICL), 35
  molecular structure, 34, 37
  structural parameters, 37, 38
  Whelan's model, 35, 37
Glycosaminoglycans (GAG), 50–51

H
Hemicellulases, 384–386
Hemicellulose-degrading enzymes, 378–380
Heparin, 20
Homogalacturonan (HG), 53
Homoxylans, 42
Hyaluronan, etymology, 20
Hyaluronan/hyaluronic acid, 50
Hydrolysis, lignocellulose
  cellulolytic enzymes, 377–378
  hemicellulose-degrading enzymes, 378–380
  pectin-degrading enzymes, 381–382

I
Indo-European representation, mildness/sweetness
  metathesis, 16
  mild, pleasant, 16–17
  from millstone to mildness, 17

Interaction ability, cellulose
  anionisation, 235–236
  cationisation, 236–238
  cellulose model films, 230
  computations, molecular modelling, 244–246
  conductometric titration, 227
  contact angle and surface energy determination, 226
  modified polymer surfaces, 241–244
  nanocomposites, 231–232
  nanoparticle synthesis, 232
  polyelectrolyte titration, 227
  polysaccharide hybrid materials, 238–241
  potentiometric titration, 226–227
  preparation, cellulose nanocrystals, 231
  quartz crystal microbalance (QCM), 228–229
  textile and pulp cellulose fibres and foils, 232–235
  z-potential (ZP), 227–228
Inulin
  etymology, 19
  molecular and supramolecular structures and terminology, 57, 58
Ionic liquid solutions, cellulose
  diffusion, [EMIM]$^+$ and [Ac]$^-$ ions, 126–127
  EMIMAc viscosity, 123–125
  materials and methods, 121–122
  phase diagram and solubility limit, 127–128
  steady-state and intrinsic viscosity, 122–123

J
Joint Communication and Involvement, Scientific, Policy and Industrial Communities, 9–10

K
Kraft pulping, 262–263
Kunstseiden und Acetat-Werke, 155

L
Laminaran, 31
Levan
  etymology, 20
  molecular and supramolecular structures and terminology, 58
Lignin mixtures, cellulose
  aqueous 8% NaOH, 140–143
  blending, 139–140
  cellulose acetate butyrate (CAB), 140
  coagulation, ligno-aerocellulose morphology, 143–144
  papermaking, 139
  properties, 139
Ligno-aerocellulose morphology, coagulation, 143–144
Lignocellulose
  hydrolysis
    cellulolytic enzymes, 377–378
    hemicellulose-degrading enzymes, 378–380
    pectin-degrading enzymes, 381–382
  oxidation, 382–383
  polysaccharide modification, 376, 377
  transfer reactions, 383

Low-consistency refining, 268–269
Lyocell process, 156
    dissolution, NMMO, 159–160
    innovation, 162–163
    ionic liquids, 165–168
    modification, 163–165
    new technologies, 168–170
    shaping, 160–162

M

Macroscopic mechanisms, cellulose
    dissolution mechanism, 100–101
    gradient of solubility, 99–100
Magnetic nanoparticles, functionalisation
    preparation, magnetic nanocoatings, 204–205
    synthesis, magnetite particles, 204
Mannans
    galactomannans, 47
    glucomannans, 48
Mechanical pulps
    bleaching
        brightness gain, 265
        hydrogen peroxide, 264–265
        objective, 264
        oxidative/reductive reactions, 264
        sodium dithionite, 264
    separation method, 259–261
Mixed-linkage β-glucans, 49
Molecular and supramolecular structures and terminology, polysaccharides
    with acid functions
        agar-agar, 56
        alginate, 55–56
        carrageenan, 56–57
        fucoidan/fucogalactan, 56
        pectins, 52–55
    with amino functions
        chitin and chitosan, 49–50
        glycosaminoglycans, 50–51
        hyaluronan/hyaluronic acid, 50
        murein, 51–52
    glucans
        (1→3)-β-Glucans, 28–31
        cellulose, 25–28
        dextran, 37–39
        glycogen, 34–37
        pullulan, 39–42
        starch, 31–34
    inulin, 57, 58
    levan, 58
    mannans
        galactomannans, 47
        glucomannans, 48
    mixed-linkage β-glucans, 49
    structural features
        glycosidic linkage, 24
        heteroglycans, 25
        structure, 25, 26
        sugar building blocks, 24

    xanthan gum, 58
    xylans
        (glucurono)arabinoxylan (GAX), 44–45
        arabinoxylans (AXs), 45
        complex heteroxylans (CHX), 45–46
        glucuronoxylans, 42–44
        (arabino)glucuronoxylans (AGX), 44–45
        homoxylans, 42
        xyloglucans (XG), 48–49
Molecular modelling, cellulose
    ab initio methods, 229–230
    molecular mechanics force field, 229
    semiempirical methods, 229
Murein
    amino functions, polysaccharides, 51–52
    etymology, 20

N

Non-derivatising compounds, cellulose
    LiCl-based solvents, 96
    N-methylmorpholine-N-oxide/water, 96–97
    phosphoric acid-based solvents, 95–96
    rheology of solutions, 97–98

O

Oxidation
    cellulose derivatization, 307–312
    lignocellulose, 382–383

P

Papermaking chemicals, 269–271
Paramylon, 31
Pectin-degrading enzymes, 381–382
Pectins
    characteristics, 52
    commercial sources, 52
    etymology, 19–20
    food industry, 52
    homogalacturonan (HG), 53
    plant cell wall, 52
    rhamnogalacturonan type I (RG-I), 53–54
    rhamnogalacturonan type II (RG-II), 54
    structural variations, 52
    xylogalacturonan, 54–55
Plasma activated regenerated cellulose fibres, 196–199
    application, 199
    dyeing and printing properties, 198
    modification, 196, 197
    Owens–Wendt–Rabel–Kaeble (O–W) calculation method, 197
    surface energies (SFE), 197
    water contact angles, 198
Point of zero charge (PZC), 201
Polysaccharide-acting enzymes
    applications
        analytics, 386–387
        cellulases, 383–384
        hemicellulases, 384–386
        chemical modifications, 376

definition, enzymes, 376
hydrolysis, lignocellulose
    cellulolytic enzymes, 377–378
    hemicellulose-degrading enzymes, 378–380
    pectin-degrading enzymes, 381–382
lignocellulose polysaccharide modification, 376, 377
oxidation, lignocellulose, 382–383
transfer reactions, lignocellulose, 383
Polysaccharide blends, cellulose
    advantages, 133
    AFM analysis, 137, 138
    contact angle measurements, 137–138
    dissolution screenings, 134–135
    expected effects, 134, 135
    mechanical properties and sugar composition, 136, 137
    microscopic images, 135, 136
    rheological investigations, 136
    solubility and miscibility, 135
    solvent routes, 134
    spinning, 136
    x-ray flat film, 137–138
    zeta potential (ZP) measurements, 138
Polysaccharide fibres, textiles
    accessibility and reactivity, 194–195
    advantages, 191
    antimicrobial functionalisation
        cationic functional polysaccharides (CAT-PS), 199–200
        dissociable groups, viscose fibres, 203
        pK value, 201
        point of zero charge (PZC), 201
        viscose nontreated and chitosan-treated (impregnated) fabrics, 201, 204
        XPS, 200
    bio-based material, 191
    characterisation, sorption behaviour, 192
    characteristics, 188, 189
    comparison, 189, 190
    effects, alkali type, 192–194
    estimated total production volume, 188, 189
    functionalisation, magnetic nanoparticles
        preparation, magnetic nanocoatings, 204–205
        synthesis, magnetite particles, 204
    future importance, 188
    modification, 191
    plasma activated regenerated cellulose fibres
        application, 199
        dyeing and printing properties, 198
        modification, 196, 197
        Owens–Wendt–Rabel–Kaeble (O–W) calculation method, 197
        surface energies (SFE), 197
        water contact angles, 198
    research and development, 210
    shaping process, 190
    spun-dyed cellulosics
        dispersed colourants, 207
        dissolved colourants, 206–207
        mass colouration, Lyocell, 207
        proposed techniques, 207
        vat dyes, 205–206
    structural assembly, 195–196
Profilometry, cellulose, 225
Propargyl-polyamidoamine (PAMAM), 319
Pullulan
    chemical structure, 41
    etymology, 20
    number average molecular weight (M_n), 42
Pulp fibers, papermaking and cellulose dissolution
    challenges and opportunities
        2011–2012, 275
        2030, 278–279
        2060, 280–281
        chemicals and functional materials, 276–280
        education, 278, 280
        fuels and energy, 276, 279
        value chain, 275–276, 279
    chemical pulps, 254
    functionalization methods
        chemical pulps, bleaching, 265–268
        low-consistency refining, 268–269
        mechanical pulps, bleaching, 264–265
        papermaking chemicals, 269–271
    properties and uses
        dissolution, 272–273
        paper, tissue, and packaging materials, 272
    raw materials
        wall structure, 257–258
        wood structure and chemical composition, 255–257
    separation methods
        chemical pulping, 261–263
        mechanical and chemimechanical pulping, 259–261
        recycled fibers, 263

Q
Quartz crystal microbalance (QCM), 228–229

R
Recycled fibers, 263
Regeneration, cellulose
    Bemberg silk, 156
    CarbaCell process, 156, 159
    CELSOL®/BIOCELSOL® process
        alkalization, 170
        enzymatic modification, 171–172
        feasibility studies, 174–175
        film forming and fibre spinning, 172–173
        hydrothermal treatment, 171
        preparation, alkaline solutions, 172
        properties, 173–1744
        steam explosion, 171
    history, man-made cellulosic fibres, 154–155
    Kunstseiden und Acetat-Werke, 155
    Lyocell process, 156

Regeneration (cont.)
    dissolution, NMMO, 159–160
    innovation, 162–163
    ionic liquids, 165–168
    modification, 163–165
    new technologies, 168–170
    shaping, 160–162
  molecule chain, 154–155
  shaping, technological approaches, 154–155
  ultra-lightweight and highly porous cellulose aerogels
    cellulose I aerogels, 176–178
    cellulose II aerogels, 178–181
  viscose rayon process, 157–159
  viscose spinning syndicate (VSS), 154
Regioselective functionalization
  application, orthogonal protecting groups, 305–306
  3-O-ethers, 304–305
  trialkylsilyl ethers, 302–304
  triphenylmethyl ethers, 301–302
Research and education, polysaccharide
  biomass production, 2
  carbohydrates, 1
  EPNOE network
    achievements, 7
    applied R&D research, 8
    building, 3–5
    challenges and opportunities, 10–11
    collaborative research activities and phd/postdoc mobility, 8–9
    Coordination and Support Action (CSA), 10–11
    dissemination, 9
    education road map, 9
    Joint Communication and Involvement, Scientific, Policy and Industrial Communities, 9–10
    organisation, 5–7
    research road map, 7–8
    tool box, 9
  European Union (EU), 2–3
  factors, renewable biomass, 1
  uses, 2
Research road map, EPNOE, 7–8
Rhamnogalacturonan type I (RG-I), 53–54
Rhamnogalacturonan type II (RG II), 54

S

Saccharide and sugar, etymology
  apiose, 15
  arabinose, 15–16
  derivation, 14
  mannose, 15
  mildness and sweetness, 15
  polymerized sugars, 14
  Rhamnus frangula, 15–16
  ribose, 16
  Sorbus domestica, 15
  suffix-*ose*, 14–15

Sarfus, 225–226
Scanning electron microscopy (SEM), 224
Scleroglucan, 29
Separation methods, pulp fibers
  chemical pulping
    kraft pulping, 262–263
    sulfite pulping, 261–262
  mechanical and chemimechanical pulping, 259–261
  recycled fibers, 263
Sodium hydroxide–water solutions, cellulose
  dissolution in NaOH–water
    additives, 105
    coagulation kinetics, 116–118
    enzymatic activation, 118–120
    phase diagram, 102, 103
    solubility, 102, 104
    structure, 107–113
    viscosity, 110–111
    ZnO influence, 113–116
  mercerisation, 101
Spun-dyed cellulosics
  dispersed colourants, 207
  dissolved colourants, 206–207
  mass colouration, Lyocell, 207
  proposed techniques, 207
  vat dyes, 205–206
Stabilisation
  solution quality and solution state, 128–130
  thermostability
    actions and degradation reactions, 130, 131
    iminodiacetic acid sodium salt (ISDB), 132–133
    onset temperatures, comparison, 132, 133
    reaction pathway, 130, 131
    ring degradation products, 132–134
Starch
  A-and B-type, 31, 33
  alternative models, 33, 36
  amaranth amylopectin, 33, 35
  amylopectin
    alternative models, 36
    amaranth, 35
    basic labelling, 34
    DP, 32
    molecular mass, 31
    molecular structure, 32
    unit chain distribution, 34
  degree of polymerisation (DP), 34
  molecular structure, 31, 32
Starch, etymology
  Germanic languages and in Finnish, 17–18
  Greek, Latin and Roman Languages, 17
  Slavic languages, 18
Sulfite pulping, 261–262
Surface analytical methods using time-of-flight mass spectrometry, 224
Surface properties and characteristics, cellulose
  atomic force microscopy (AFM), 224–225

cellulose nanocrystals (CNC), 219
ellipsometry, 225
interaction ability
    anionisation, 235–236
    cationisation, 236–238
    cellulose model films, 230
    computations, molecular modelling, 244–246
    conductometric titration, 227
    contact angle and surface energy determination, 226
    modified polymer surfaces, 241–244
    nanocomposites, 231–232
    nanoparticle synthesis, 232
    polyelectrolyte titration, 227
    polysaccharide hybrid materials, 238–241
    potentiometric titration, 226–227
    preparation, cellulose nanocrystals, 231
    quartz crystal microbalance (QCM), 228–229
    textile and pulp cellulose fibres and foils, 232–235
    z-potential (ZP), 227–228
model polysaccharide surfaces, 216–219
molecular modelling
    ab initio methods, 229–230
    molecular mechanics force field, 229
    semiempirical methods, 229
profilometry, 225
pulp fibres, 220–222
Sarfus, 225–226
scanning electron microscopy (SEM), 224
surface analytical methods using time-of-flight mass spectrometry, 224
textile cellulose fibres, 219–220
x-ray photoelectron spectroscopy (XPS), 223–224

T
Textile cellulose fibres, 219–220
Tool box, EPNOE, 9
Transfer reactions, lignocellulose, 383

U
Ultra-lightweight and highly porous cellulose aerogels
    cellulose I aerogels, 176–178
    cellulose II aerogels, 178–181

V
Viscose rayon process
    degree of polymerization (DP), 157
    derivatisation, 157, 158
    developments, 158
Viscose spinning syndicate (VSS), 154

W
Whelan's model, glycogen, 35, 37

X
Xanthan gum, 58
X-ray photoelectron spectroscopy (XPS)
    antimicrobial functionalisation, 200
    surface properties and characteristics, 223–224
Xylans
    (glucurono)arabinoxylan (GAX), 44–45
    arabinoxylans (AXs), 45
    complex heteroxylans (CHX), 45–46
    glucuronoxylans, 42–44
    (arabino)glucuronoxylans (AGX), 44–45
    homoxylans, 42
Xylogalacturonan, 54–55
Xyloglucans (XG), 48–49

Printed by Publishers' Graphics LLC